Annual Review of Chinese Architectural Design Works

2010–2011
中 国 建 筑 设 计 作 品 年 鉴

（上）

《中国建筑设计作品年鉴》编委会 编

江苏人民出版社

图书在版编目（CIP）数据

2010～2011中国建筑设计作品年鉴 / 《中国建筑设计作品年鉴》编委会编. -- 南京 : 江苏人民出版社, 2011.12
ISBN 978-7-214-07746-2
Ⅰ. ①2··· Ⅱ. ①中··· Ⅲ. ①建筑设计—作品集—中国—2010～2011 Ⅳ. ①TU206

中国版本图书馆CIP数据核字(2011)第258497号

2010—2011中国建筑设计作品年鉴（上、下册） 《中国建筑设计作品年鉴》编委会 编

责任编辑：刘 焱 艾 璐
责任监印：彭李君
美术设计：郝泽星 彭 丹 贾 妍
出　　版：江苏人民出版社（南京湖南路1号A楼 邮编：210009）
发　　行：天津凤凰空间文化传媒有限公司
销售电话：022-87893668
网　　址：http://www.ifengspace.cn
集团地址：凤凰出版传媒集团（南京湖南路1号A楼 邮编：210009）
经　　销：全国新华书店
印　　刷：深圳市彩美印刷有限公司
开　　本：889毫米*1194毫米 1/12
印　　张：80（上：41，下：39）
字　　数：1000千字
版　　次：2012年1月第1版
印　　次：2012年1月第1次印刷
书　　号：ISBN 978-7-214-07746-2
定　　价：1058.00元（上、下册）
(USD 200.00)

（本书若有印装质量问题，请向发行公司调换）

主　　管：中华人民共和国住房和城乡建设部
主　　办：中国建筑文化中心
编　　辑：《中国建筑设计作品年鉴》编委会
北京主语空间文化发展有限公司
主　　编：陈建为
执行主编：肖 峰
编辑部主任：王 伟
编　　辑：代 君 郭 晨 金 峰 林 朋 冷娜英 李 悦
李 妍 刘笑艳 欧 阳 齐 阳 张 辉 张 澜
编辑部地址：北京市海淀区三里河路13号
中国建筑文化中心712室
邮　　编：100037
电　　话：+86-10-88151966/+86-10-88153600/13910120811
传　　真：+86-10-88151958
网　　址：www.archrd.com
E-mail：zgjsmail@126.com

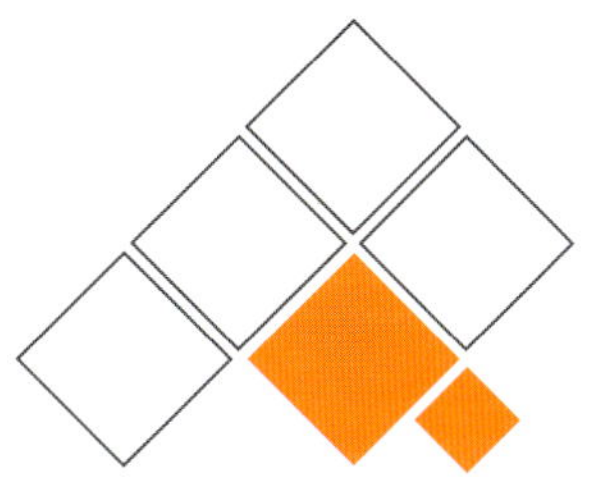

2010—2011
中国建筑设计作品年鉴

谨以此年鉴献给为中国建筑设计行业发展而辛勤付出的设计师们!

《2010—2011中国建筑设计作品年鉴》（以下简称《年鉴》）
收录了国内外180余家设计机构的近2000件优秀建筑设计作品，
这些不同类型、不同设计理念的作品基本反映了当前建筑设计行业多元化的发展状况和实践成果，
从中可以直观地领略到中国建筑设计行业融汇中西、贯通古今的设计思想。
对于各地建设主管部门和从事城市规划、建筑设计、景观设计、工程建设、房地产开发人员以及相关科研教育机构掌握建筑设计行业发展趋向、了解新的设计理念，《年鉴》会有很好的参考、借鉴作用。
《年鉴》在编辑过程中，得到了各地建设主管部门和众多设计院所、设计师的大力支持和帮助，
在此，谨向为《年鉴》提供帮助和支持的单位和个人表示衷心的感谢!
《年鉴》在国内公开发行，并委托中国图书进出口（集团）总公司在国外和国内港、澳、台地区发行。
在《年鉴》的编辑过程中，不免会有疏漏或错误之处，敬请读者指正，并提出宝贵意见，
以便我们不断提高《年鉴》的编辑水平，满足读者的需求。

《中国建筑设计作品年鉴》编辑部
2011年11月

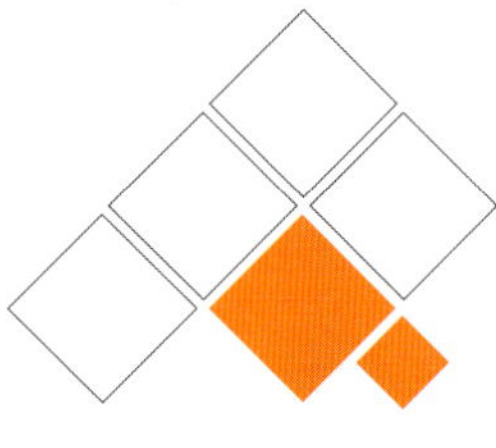

编委题词

Editorial Committee Inscription

Saving our environment is the most vital issue that humankind must address today leading into our fears that this millennium may be our last. An ecological approach to our businesses and design is ultimately about environmental integration. Ecodesign is designing for bio-integration.

建筑师在本世纪所担负最大的课题就是用生态设计的方法去解决我们人类社会所面临的环境问题。我们的终极目标就是实现建筑与环境的和谐共生。

杨经文

希望"年鉴"成为全面总结和推动中国建筑创作发展的优秀典籍。

何镜堂

愿《中国建筑设计作品年鉴》成为中国设计力量在世界设计之林崛起的推动力！

推陈出新

把祖国建设得更美好

办年鉴，年年明鉴，推精品，精益求精，树大师，不遗余力。

齐康

希望"年鉴"成为高品质的专业年鉴。

孟建民

贺"中国建筑设计年鉴"出版

作为中国建筑设计的记录 终将成为世界建筑史的辉煌篇章

程泰宁

宗白华先生言，"一切艺术综合于建筑，而礼乐诗歌舞剧之表演，亦与建筑背景协调，成为一完美的生活，所以每一文化的强盛时代，莫不有伟大的建筑计划，以容纳和表现这一丰富之生命。"先生此言发自半个多世纪以前，今日我们更应致力于这一和谐的美的实现早日来临

吴良镛

历史纪录 时代鉴证

刘景樑

祝贺《中国建筑设计作品年鉴》发行

良师益友

追求卓越 宁静致远

贺《中国建筑设计作品年鉴》出版发行

2010—2011

中国建筑设计作品年鉴

愿《中国建筑设计作品年鉴》在中国设计界更有权威性、代表性。

潜心研究市场需求
专注挖掘项目价值

认真解决建筑功能
努力创造优秀建筑

林怀文

为了中国建筑设计的崛起，为了本土建筑师的声望与力量，请有志同仁加入到"新东方主义"建筑设计运动中，也请《中国建筑设计作品年鉴》作好此方面的引导。

设计，随其法，都是一种选择：在喧杂中选择宁静，在盲目中选择理性，在平庸中选择创意，在浮华中选择真诚。

希望本年鉴不是建筑师或设计单位的"目录"，而是大家彼此观摩学习的园地。

建筑文化是一个最能感染其他的文化，而建筑设计是一个最没有国界的产品。

中国建筑设计的发展，从过去20多年来，由抄袭西方，模仿西方，到最近两年来的探索中国建筑文化，可以说是逐渐的找到了"自我"。但我很希望中国设计师们，不要重复義和团的"自大"。

的确，如何"古为今用"，"洋为中学"，进而"中为西传"，是我们设计人员未来的重要工作。共勉之。

立足澳门，背靠祖国，面向世界，共创繁荣。

希望中国建筑师的职业化水准迅速提升；希望中国建筑师有更多更好的相互交流的平台；希望中国建筑师所面对的设计市场环境更开放、更公平。

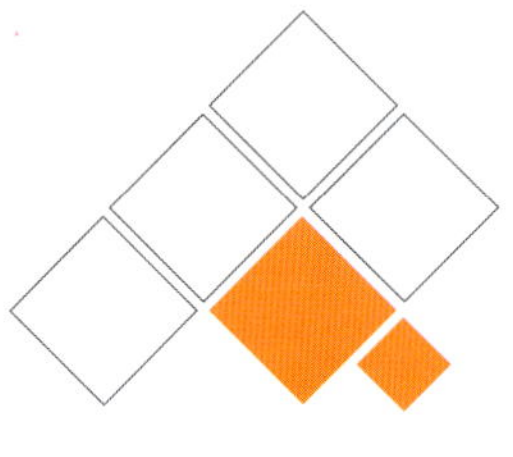

编委题词

Editorial Committee Inscription

科学设计
服务社会

[illegible]

回顾新中国60年的建筑设计历程，努力进取，迎接中国建筑创作更加美好的未来。

[illegible]

佳作 拙作 重写作
前鉴 后鉴 重写鉴

AIM 陈晓宇

追求品质、倡导原创、宏扬个性、力争创新

[illegible]

建筑是一种艺术
唯有时间+空间+人类创为
方证明其优秀

[illegible]

建筑是思想的容器

[illegible]

纵贯三江
笔涂四海

顺德建筑设计院
梁[illegible]书

浓缩精华，展示精品

建筑文化[illegible]

[illegible]

愿《中国建筑设计作品年鉴》在中国建筑界树起中国建筑文化的大旗。

James C. Wang

愿《中国建筑设计作品年鉴》成为东西方建筑文化交流的平台。

[illegible]建

特邀编委
(按姓氏首字母排序)

目录 上 卷

目录 上卷

目录 下卷

设计作品分类索引 卷

设计作品分类索引 上卷

公共建筑

宾馆、酒店、餐厅、招待所

医院、医疗建筑

交通建筑

展览、展示建筑

设计作品分类索引 上卷

旅游、休闲建筑

院校、培训建筑

文化、艺术建筑

体育建筑

纪念馆、古建筑

工业建筑

设计作品分类索引 上卷

居住建筑

设计作品分类索引 卷

城市规划设计、景观规划设计

设计作品分类索引 上卷

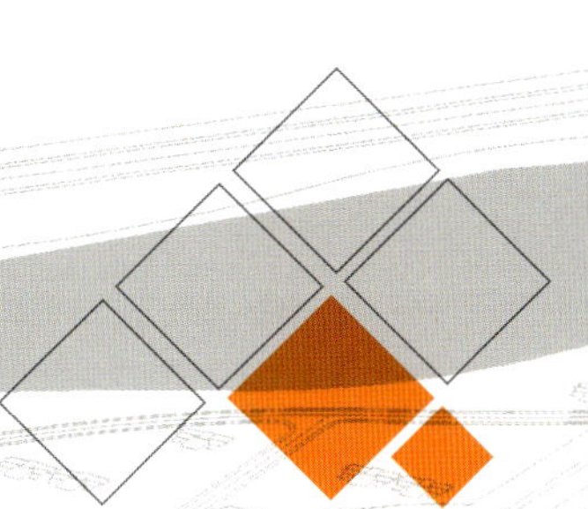

2010—2011
主要建筑设计作品
Main Architectural Design Works
Annual Review of Chinese Architectural Design Works

NITA

荷兰NITA设计集团

荷兰NITA设计集团是由荷兰6家顶级设计团队联合而成的国际顶级设计机构，总部设在荷兰阿姆斯特丹。在大型综合空间、城市环境及主题公园、居住社区、旅游度假与娱乐项目的规划设计方面所展现出的无与伦比的创造能力得到了广泛的认可。NITA创新性地采用LEGO（乐高）协作设计模式，汇集了近千名专业设计人员，包括资深景观师、管理顾问、规划师、建筑师、工程师及在交通、能源、生态等领域的众多国际专家。

1999年荷兰NITA设计集团进入中国，在昆明园艺博览会的荷兰园设计获得最高奖项——金奖后，NITA开始在中国境内开展设计业务；2002年在上海成立中国总部，开始为高速发展的中国全面提供国际领先的景观、规划、建筑、旅游和生态策略的设计服务。

2006年，荷兰NITA设计集团"回归设计与绿色技术"融合共生的绿色城市理念，获得中国领导及专家的肯定，参加了上海2010世博会总体景观规划国际竞赛，并被上海世博事务协调局聘请为2010上海世博园区（浦东、浦西）景观设计及绿化工程总体管理单位。

伴随着在中国最为精彩的上海世博系列作品的建成，NITA正以其国际级的绿色创意设计和解决多学科复杂设计的能力传播并实践"绿色城市"的理想。为了让中国更多城市分享"绿色城市"的经验与成果，NITA相继在上海、杭州、南京、宁波等城市设立了多个分支机构以及工程技术管理、生态科研研发、植物繁殖培育等专项事业部，汇聚了各个专业领域400多名技术人才，NITA努力为中国不同区域的人们带来国际领先的绿色规划设计服务。

荷兰NITA设计集团一直关注自然、城市与人的关系，尤其在全球变暖、自然环境多样性遭到严重破坏的今天，NITA致力研究并建立一种经济与自然环境和谐共生的绿色发展模式。基于此，NITA把全球领先的"回归设计"理念与欧洲最著名生态国家荷兰的绿色技术相融合，我们从荷兰启程到阿联酋再到中国，NITA在全球不同国家实践着"绿色城市"的理想。

2010年，由荷兰NITA设计集团担当总体设计与管理控制的5.28平方千米的上海绿色世博园区，向全球崭露绿色英姿。上海世博园区是"绿色城市"在中国的最佳示范实践，获得了全球的广泛认可与赞誉。NITA正以中国世博作为"绿色城市"实践的新起点，努力为中国不同区域的人们改善生活环境，积极传播"绿色城市"理念。

NITA的绿色理想是让中国更多的城市分享并实践绿色城市——"经济发展与自然环境和谐共生"成功经验，让蔚蓝地球重新焕发勃勃的绿色生机。

NITA坚信："回归设计与绿色技术"融合共生的"绿色城市"理念，是使地球恢复并保持健康发展的解决之道。

NITA Design Group, an international institute composed of 6 top design teams from Europe, is recognized widely for its creative idea and ideas on urban planning, residential complex, theme park, resort and recreation project, etc. Founded its headquarter in Amsterdam, NITA Design Group which borrows LEGO method, is an elite team of nearly a thousand international professional landscape architects, administrative consultants, urban planners, architects, engineers and other specialists on transportation, energy and ecology.

NITA Design Group started its business in China by participating the Kunming International Horticultural Exposition in 1999 and won the top prize. In 2002, NITA founded the China Headquarter in Shanghai and have provided professional services on landscape, architecture, urban planning, resort planning and ecological projects.

In 2006, NITA entered the international competition on the landscape master plan for the Shanghai EXPO 2010 and gained high praises from the authority and expertise for its Green City idea. NITA was appointed as the General Consultant of landscape design and Principal Consultant of environmental construction for EXPO 2010 by the Shanghai World EXPO Bureau.

NITA practices his 'Green City' idea and apply it on solving difficult, complicated problems for the projects in China and all over the world. After the completion of the remarkable EXPO projects, NITA shares its idea and experiences with more Chinese cities. To provide professional landscape architecture service for more Chinese clients, NITA has founded branch offices, engineering management firms, ecology research facilities and plant cultivation firms in Shanghai, Hangzhou, Nanjing and Ningbo, and has recruited over 400 talents in various fields.

NITA deeply concerned about the relationship among the nature, the cities and the humans, especially the global warming and the environmental destructions. NITA devotes itself in finding a solution which will be compatible both economies and the nature. Based on the world leading Regression Design concept, NITA creates and promotes the Green City idea, and integrating with the Green Technology from Netherlands, a country which is famous for its green ecology. NITA has sailed out from Netherlands to UAE, and all the way to China, bringing its Green City idea to the cities all over the world.

In 2010, NITA actively promotes the Green City idea in constructing the EXPO site of over 5.28 square kilometers, which receives high recognition from Chinese and foreign leaders. The EXPO Projects are the most successful cases of the Green City idea. While the EXPO projects set a new milestone, now NITA set out for a more difficult task, spreading the Green City idea, restoring the environment and improving the habitat condition for all Chinese.

NITA aims to promote and spread its Green City idea to more Chinese cities, shares the successful experience on economy and environment development, and bring life and green back to earth.

NITA firmly believes that, the very solution to restore and maintain the health condition of our earth, is to practicing the Green City idea, which is integrating with the Regression Design and the Green Technology.

地址：上海市徐汇区田林路142号G座4楼
邮编：200233
电话：+86–21–31278900
传真：+86–21–31278901
邮箱：info@nitagroup.com
网址：www.nitagroup.com/

Add: Fourth Floor, G Tower, No.142 Tianlin Road, Xuhui District, Shanghai
P.C.: 200233
Tel: +86–21–31278900
Fax: +86–21–31278901
E-mail: info@nitagroup.com
http://www.nitagroup.com/

2010上海世博会

世博公园及园区总体景观设计

地点：上海
客户：上海世博土地控股有限公司
服务：总体设计

中国2010年上海世博会会址位于上海市中心，跨越黄浦江两岸，它是上海黄浦江两岸开发、旧区改造和产业布局调整的重点地区，也是上海新一轮城市空间拓展、城市综合服务功能提升的重要地段。

上海2010世博公园是为2010年上海世博会召开而启动的项目，其规模及功能因其所处位置的特殊性及开发的原因，担负着特殊的功能及形象特征，需体现世博的形象特征且满足展会召开期间的功能使用。项目规划用地面积约29公顷，另外相关设计用地包括公共活动中心用地和演艺中心用地，总面积约42公顷。

江南广场公园景观设计

地点：上海
客户：上海世博土地控股有限公司
服务：景观设计

江南广场公园及浦西滨江景观绿地规划设计用地面积约为25.39公顷。规划设计范围由江南广场公园和滨江景观绿地两部分组成。江南广场作为临时性项目，主要为世博会期间的集会、庆典、观演等活动提供大型场地。场地内的3个原江南造船厂的保留船坞是江南公园最大的特色。黄浦江沿岸防汛岸线长约2 900米。在基本保留原有驳岸和防汛墙的基础上，对其进行改造和整理，保证滨水景观通廊的连续性。设计提出从“中国制造”到“中国创造”的设计概念。设计节选了基地中两个极具代表性的历史活动片断，用以概括基地的本质属性和特征。

2010上海世博会 白莲泾公园景观设计

地点：上海
客户：上海世博土地控股有限公司
服务：景观设计

白莲泾公园位于世博园区浦东段的北侧，北接黄浦江，南至雪野路桥，西起世博园区浦东中心绿地"世博公园"，东接世博园区世博村及配套设施，规划设计用地面积约20.18公顷。

该区域曾经是莲藕荡漾生态自然的美丽河道，随着中国工业的快速发展，见证着上海港口城市的辉煌。

规划的最核心出发点就是把"空间公共化"；把上海最宝贵的黄浦江的滨江景观资源公共化。规划强调在地块的使用中应满足不同的人群，创造不同的空间，在最大程度上让使用者享受到黄浦江最美丽的风景。

2010上海世博会 世博村景观设计

地点：上海
客户：上海世博土地控股有限公司
服务：景观设计

世博村规划用地位于浦明路以南、白莲泾以东、浦东南路以西地段，总用地面积约373 000平方米。项目由10个部分组成，性质有五星级酒店、公寓式酒店、物流仓库、公寓式酒店、商业和办公等。

世博村将在世博会筹备及举办期间，为各参展国家和国际组织人员提供住宿、办公、餐饮、购物、娱乐、物流、后勤等服务。世博会后转变为领馆区或涉外区，为区内及周边居民提供良好的居住、商务、办公、接待、购物、休闲等功能，将成为上海又一个特色鲜明的国际化街区。

芜湖市滨江景观公园

芜湖市滨江景观公园位于芜湖市西部的长江沿岸，全长9.5千米，宽度100～200米不等。青弋江自东向西在此汇入长江，将滨江公园分为南北两段。已经建成的北段为一期工程，位于芜湖市中心区，全长2.35千米，占地面积约36万平方米，主要包含生态居住区、主题公园景观区、文化艺术展示区、休闲商业区、商贸办公区及江畔公园等。

规划设计深入挖掘滨水空间的文化特色，体现芜湖的文脉延续。“港口将世界给了城市，城市把烙印留给码头”。只有留下城市的历史足迹，才会让心灵有归宿；强调文化性，充分利用城市历史的个性、文化遗产；在设计中尊重领域感，在寻求地方文脉与精神时发现作为滨江回忆的码头，利用码头文化作为滨江文脉的主线，赋予滨江丰富的想象空间。

沣河项目

沣河位于西安市西部，发源于秦岭北麓的沣峪沟，属渭河的一级支流，全场70.5千米，流域面积1 460平方千米，常年水量充沛，是长安八水之一，沣河综合治理项目是沣渭新区重点建设项目之一，在沣渭新区区域空间内主要分布着西周的沣京遗址和镐京遗址，秦代大阿房宫遗址及汉代的汉城遗址和建章宫遗址。

规划场地位于沣京遗址和镐京遗址之间，属于沣镐文化圈，因此，把规划区域的主题文化定位为西周文化。场地设计定位以水为魂，以绿为魄，打造生态城市滨水景观。

济宁北湖湿地项目

项目位于济宁市北湖省级旅游度假区。北湖位于济宁市城南6千米处，地处南四湖之北而得名。北湖区域拥有丰富的动植物资源与水资源，是得天独厚的生态自然区。济宁市是历史文化名城，有深厚的历史文化底蕴，孔孟文化、运河文化、佛教文化、水浒文化等灿若星辰。

规划设计以"天人合一"为思想理念，以"道法自然"为处事原则，从"道法自然、天人合一"过渡到自然形成有机论，完成从哲学到科学的演绎。

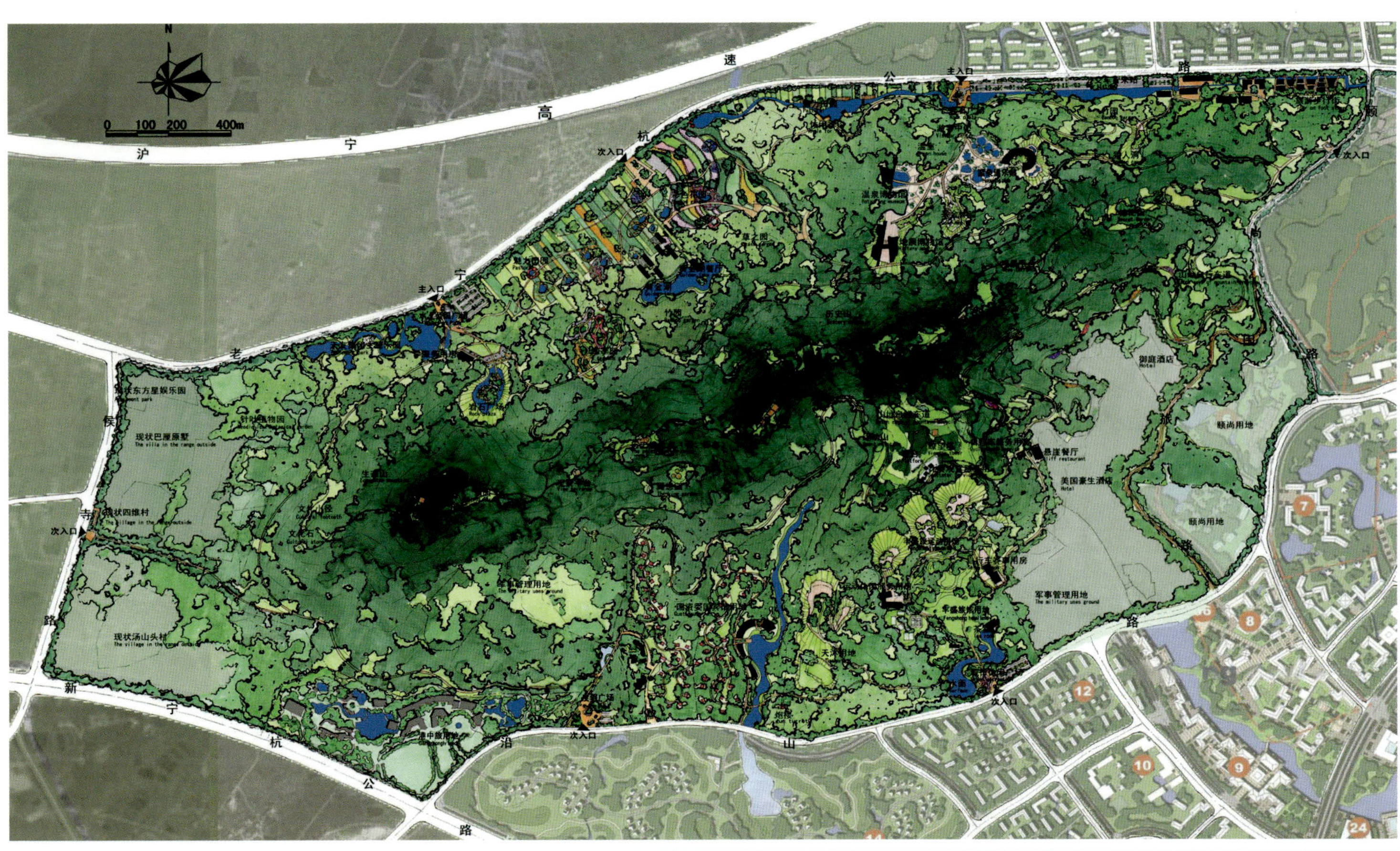

南京汤山地质公园

南京汤山地质公园风景区规划用地总面积约580公顷，包括大部分汤山山体。景观设计将汤山地质公园定位为长三角地区典型地质地貌、山林水系等典型自然风貌展现的体验平台，与南京历史文化名城相呼应的历史自然交融的休闲旅游度假基地，集休闲、度假、文化、展示、运动、体验为一体的综合配套服务中心。

设计形成“一自然生态核”、“四休闲旅游区”（城市休闲区、水上公园区、自然区、地质区）的总体结构。

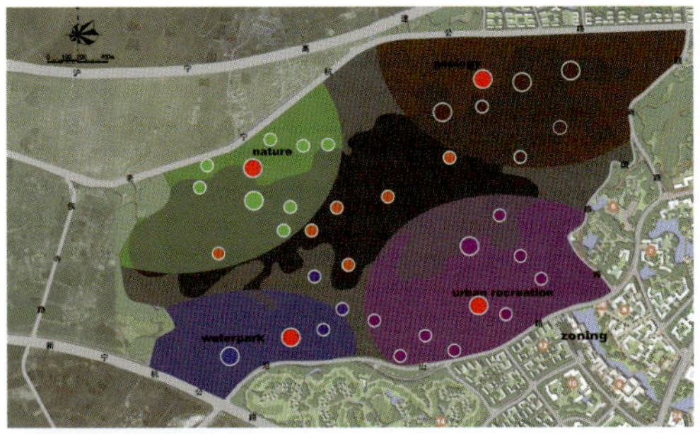

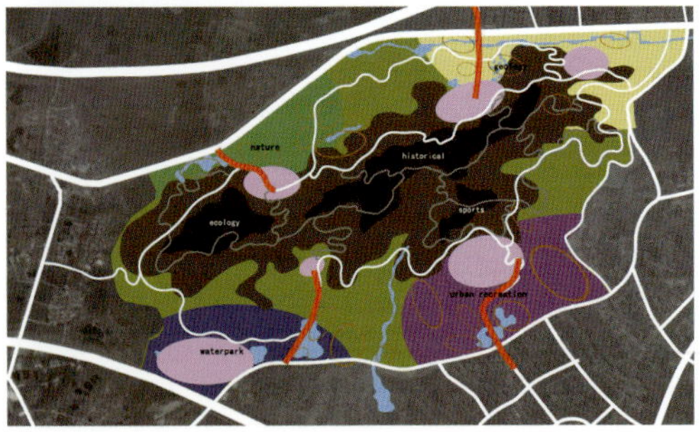

长春高新技术开发区南区景观设计

长春高新技术开发区南区，一个“现代科技”的概念被引入硅谷大街，设计原则依循科技性、创新性、文化性、生态性，同时兼顾其经济性、可操作性，务必希望能够塑造出与当地产业与文化紧密结合的并可持续发展的现代式城市街景系统。

一系列的规划设计目标推动了硅谷大道景观设计的形式和功能，在街道空间中设置足够的公共文化交流空间，将在方案中实现体验现代、科技、产业的文化交流空间及景观系统。这些设计将成为一个整体，服务于游客和城市居民。

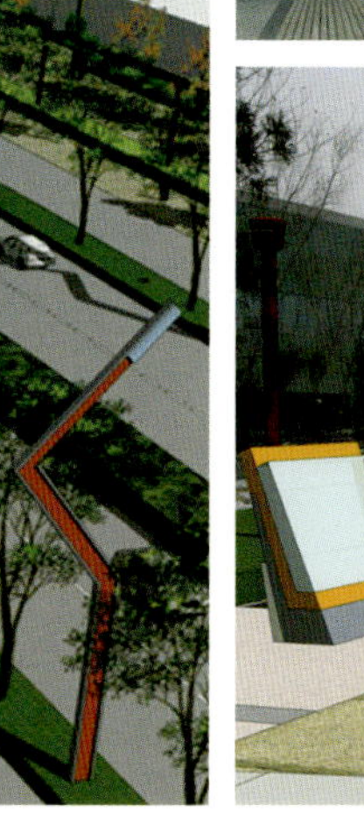

珠海横琴岛总体规划

珠海横琴岛总体规划的理念根基是来源于岛屿自然发展的特征，而不仅仅来自灵感。规划遵循岛上的自然发展状态，维护原有生态平衡，从而使横琴岛维持可持续的发展方式。

我们将横琴岛喻为南中国海上的一支圣洁的马蹄莲，是基于NITA一贯倡导的“GREEN CITY”（绿色城市）理念，结合“呼吸”这个生命存在的必然方式而产生的。我们希望将来的横琴岛是个绿色的、美丽的、怡然的、蓬勃的生命有机体，像一支纯洁、高雅、精致的马蹄莲。通过花的功能作比喻，我们将这些语汇融于设计，让在花中“呼吸”的人们体会它的生命。

哈尔滨何家沟项目

何家沟由东河沟、西河沟和干沟三段组成，河道形态呈“人”字形，贯穿哈尔滨西部，东、西河沟汇合于干沟经顾乡大坝西侧汇入松花江。

规划用地河道全长26.5千米，规划范围为以河道中心线向两侧各返45米的狭长范围，总用地238.5公顷。

何家沟被定位为哈尔滨市最重要的水系绿道。设计通过水环境和两岸公园绿地建设，使何家沟形成有机的、可持续发展的活力水系。其滨河空间不但能提供多种生物栖息和城市居民的游憩地，同时也成为松花江清洁的支流和水源。

规划设计力争把何家沟打造成为城市绿色标杆——倡导城市生态文明；都市触媒——聚焦城市独特个性；活力界面——引领城市品质生活。

海棠湾GLOIA度假酒店

海棠湾GLOIA度假酒店位于海南省三亚市海棠湾南区，为“国家级海岸疗养康复基地”。总用地面积约610 000平方米。设计将地域性人文脉络和场所精神有机组织，对基地休闲、商务度假及休养的基本功能进行重点设计，将现代人健康高尚的生活模式与东方哲学融合，寻求宾客身心与自然的对话和平衡。

技术与感官的结合，造就独特的地域景观体验。在特有的环境中，将文化中不可割舍的情感元素与独特的文化精神，体现于各级景观之间，通过高差以及植被等处理手法等进行的借景或分割，展现出热带特色的同时创造出具有丰富层次的优质景观。

美国HOOP建筑设计有限公司

HOOP ARCHITECTURAL DESIGN CONSULTANTS. INC.

美国HOOP建筑设计有限公司于2000年在深圳、上海设立分支机构，负责开拓中国大陆市场。在中国，拥有一支经验丰富、精益求精的高水平、国际化专业设计团队，致力于建筑设计、规划设计、景观设计、策划咨询等诸多领域的发展，在高端住宅和综合商业设计领域，表现尤为突出。凭借雄厚的设计实力和国际化的设计理念，HOOP建筑在中国境内已完成大量优秀设计作品并付诸实施，在业内取得了较高的知名度和良好的声誉。

以"专注创新、专业高效、责任至上"为专业宗旨，以"人文精神、团队合作、工作激情"为发展动力，始终致力于为客户提供"独创性、高质量、高效益"的设计服务。

我们与国内众多知名开发商及政府机构都建立了长期战略合作伙伴关系。把客户的信赖视为企业的生命，以出色的专业技能在项目的各个操作阶段为客户提供及时、周到的服务。同时也非常感谢客户为我们提供宝贵的创作机会，使得我们的设计理想得以实现。

HOOP ARCHITECTURAL DESIGN CONSULTANTS,INC, founded in California of USA in 1995, is an influential architectural design consultant in California. It has series of sophisticated works in USA, Canada and Southeast Asia.

HOOP ARCHITECTURAL DESIGN CONSULTANTS,INC, setting up its representative offices in Shenzhen, and Shanghai in 2000 and 2003 respectively, open up China Mainland market. Thanks to marvelous ability, advanced idea, and sincere cooperation of triple parties, we made remarkable achievements.

We emphasize creation of design idea, use best solution to solve problems of aesthetic, technology and economy. We consider staffs' teamwork and natural responsibilities as most important, and regard customers' dependence as our lives. Brand-new design idea, high-level service and ambition to surpass works before constantly are guarantee to finish high quality projects.

USA
Add: 70 Pacific St.Room 615
Cambridge, MA 02139
U.S.A 94086
Tel: 001-617-4525166 Fax: 001-617-4525166
Http://www.hoop-china.com

中国 上海
地址：浦东新区世纪大道1589号
长泰国际金融大厦9楼08–11单元.8楼05单元
Shanghai China
P.C.: 200122
Tel: +86-21-58783137 Fax: +86-21-58782763
Http://www.hoop-china.com

中国 深圳
南山区
高新南四道与科技南十路交汇处
高新技术产业园迈科龙大厦3楼
Shenzhen China
P.C.: 518000
Tel: +86-755-83458622 Fax: +86-755-83458633
Http://www.hoop-china.com

西塘保利项目
开发单位：保利房地产开发有限公司
建设地点：浙江 嘉善
建筑规模：282 000平方米
设计时间：2010年
项目状况：实施阶段

Xitang Poly Residence
Client: Poly Real Estate Group Co.,Ltd.
Project Location: Jiashan, Zhejiang
Plot Area: 282,000 m^2
Design Time: 2010
Project State: Implementation stage

南京保利紫晶山
开发单位：保利房地产开发有限公司
建设地点：江苏 南京
建筑规模：340 000平方米
设计时间：2009年
项目状况：竣工完成

Nanjing Poly Xianlin Residence
Client: Poly Real Estate Group Co.,Ltd.
Project Location: Nanjing, Jiangsu
Plot Area: 340,000 m^2
Design Time: 2009
Project State: Completed

常州绿地世纪城三期

开发单位：绿地集团
建设地点：江苏 常州
建筑规模：258 926平方米
设计时间：2010年
项目状况：实施阶段

Changzhou Greenland Residence-Century City Phase III
Client: Greenland Group
Project Location: Changzhou, Jiangsu
Plot Area: 258,926 m^2
Design Time: 2010
Project State: Implementation stage

常州龙湖香缇漫步

开发单位：龙湖地产
建设地点：江苏 常州
建筑规模：580 000平方米
设计时间：2010年
项目状况：竣工完成

Changzhou Longhu Project
Client: Longfor Properties Co., Ltd.
Project Location: Changzhou, Jiangsu
Plot Area: 580,000 m^2
Design Time: 2010
Project State: Completed

宜兴天氿御城

开发单位：苏宁环球
建设地点：江苏 宜兴
建筑规模：800 000平方米
设计时间：2010年
项目状况：实施阶段

Yixing Tianqiuyucheng City
Client:
Suning-universal
Project Location:
Yixing, Jiangsu
Plot Area:
800,000 m^2
Design Time:
2010
Project State :
Implementation stage

重庆华润置地二十四城

开发单位：华润（重庆）有限公司
建设地点：重庆
建筑规模：600 000平方米
设计时间：2010年
项目状况：实施阶段

ChongQing CR Land Twenty Four City
Client: China Resources Land
Project Location: Chongqing
Plot Area: 600,000 m^2
Design Time: 2010
Project State: Implementation stage

江阴爱家上院

开发单位：上海爱家房地产开发有限公司
建设地点：江苏 江阴
建筑规模：165 000平方米
设计时间：2010年
项目状况：实施阶段

Jiangyin Aijia Residence
Client: Aijia Real Estate Group Co.,Ltd.
Project Location: Jiangyin, Jiangsu
Plot Area: 165,000 m^2
Design: Time 2010
Project State: Implementation stage

苏州太阳城三期

开发单位：苏州雅戈尔置业有限公司
建设地点：江苏 苏州
建筑规模：373 000平方米
设计时间：2009年
项目状况：实施阶段

Suzhou Sun City Phase III
Client: Youngor Suzhou
Project Location: Suzhou, Jiangsu
Plot Area: 373,000 m^2
Design Time: 2009
Project State: Implementation stage

南通文峰广场

开发单位：南通新景置业有限公司
建设地点：江苏 南通
建筑规模：243 000平方米
设计时间：2007年
项目状况：实施阶段

Nantong Wenfeng Plaza
Client: Nantong Xinjing Real Estate Co.,Ltd.
Project Location: Nantong, Jiangsu
Plot Area: 243,000 m^2
Design Time: 2007
Project State: Implementation stage

苏州相城中央商贸区

开发单位：苏州市相城城市建设有限责任公司
建设地点：江苏 苏州
建筑规模：198 000平方米
设计时间：2009年
项目状况：竣工完成

Central Business District, Suzhou City
Client: Chengtou Suzhou
Project Location: Suzhou, Jiangsu
Plot Area: 198,000 m^2
Design Time: 2009
Project State: Completed

宁波城投滨江商业区

开发单位：宁波城投置业
建设地点：浙江 宁波
建筑规模：250 000平方米
设计时间：2010年
项目状况：建筑方案

Ningbo Riverside Commercial Center
Client: Corporation Investment Chengtou Ningbo
Project Location: Ningbo, Zhejiang
Plot Area: 250,000 m^2
Design Time: 2010
Project State: Building design

湖州市便民中心

开发单位：湖州市人民政府
建设地点：浙江 湖州
建筑规模：105 000平方米
设计时间：2011年
项目状况：建筑方案

Huzhou City Convenience Center
Client: Huzhou Government
Project Location: Huzhou, Zhejiang
Plot Area: 105,000 m^2
Design Time: 2011
Project State: Building design

南京苏宁天润城商业地块

开发单位：苏宁环球股份有限公司
建设地点：江苏 南京
建筑规模：900 000平方米
设计时间：2010年
项目状况：实施阶段

Nanjing Tianrun Cheng Commercial Plot
Client: Suning Group Co.,Ltd.
Project Location: Nanjing, Jiangsu
Plot Area: 900,000 m^2
Design Time: 2010
Project State: Implementation stage

武汉东湖万达商业广场

开发单位：万达集团
建设地点：湖北 武汉
建筑规模：102 000平方米
设计时间：2010年
项目状况：实施阶段

Wuhan Wanda Commercial Center
Client: Wanda Group Corporation Ltd.
Project Location: Wuhan, Hubei
Plot Area: 102,000 m^2
Design Time: 2010
Project State: Implementation stage

2010—2011 主要建筑设计作品

Main Architectural Design Works

Annual Review of Chinese Architectural Design Works

澳大利亚柏涛（墨尔本）建筑设计有限公司
PEDDLE THORP ARCHITECTS

澳大利亚柏涛（墨尔本）建筑设计有限公司是澳大利亚最大的建筑设计公司之一。柏涛的历史可以追溯到1889年，一个多世纪以来，柏涛一直走在世界建筑设计行业的前列。

杰出、实用和经济是柏涛公司设计的原则；设计上的创新和技术上的更新是公司的宗旨；技术上的可靠和设计上的独特更是公司长期的声誉之所系。庞大的技术资源，广泛、丰富的经验，使柏涛公司能够承担各种规模、各种类型的区域规划设计和各类建筑的设计。

柏涛建筑设计集团除在澳大利亚几个大城市外，还在东南亚、欧洲和美洲设有分支机构，设计业务遍布世界各地。柏涛墨尔本公司集中了一大批优秀的建筑师和相关专业人员，除开展通常的建筑设计业务之外，还对体育、医疗、住宅建筑设有专门的研究机构。主要作品有澳大利亚国家网球中心、澳大利亚墨尔本奥林匹克公园及自行车赛馆、ESSO澳洲总部、墨尔本水底世界水族馆、马来西亚运动中心，以及众多建于澳洲本地与国外的酒店、商业大厦、政府大厦、写字楼工程、居住区建筑、医疗设施等。

柏涛墨尔本公司在中国的设计机构拥有众多高素质的中外建筑师，国际化的先进设计理念、本地化的优秀团队服务，使公司业务发展迅速。到目前为止，业务范围已覆盖了中国境内26个省、市、自治区，并在上海常设实力雄厚的设计机构。在澳洲本部的支持下，公司有能力在大规模的城市区域规划设计、大型公共建筑设计（包括办公楼、商业中心、酒店、教育行政文化设施、运动娱乐设施、医疗设施等）以及住宅规划设计、建筑设计及园林景观设计等方面，以独特的设计手法、先进的技术和丰富的经验，活跃在国际建筑设计舞台上，并始终如一地为客户提供一流的服务。

Founded in 1889, Peddle Thorp Pty Ltd. has been one of the largest architectural design firms in Australia. For over one hundred years, Peddle Thorp has always been at the forefront of the architectural world.

Pre-eminent, practical and efficient design works are the principles; design innovation and technical renovation are treated tenets; established reputation has been based on technical reliability and unique design. With superior technical resources, advanced computer skills and comprehensive experiences, Peddle Thorp has been specializing in regional planning and architectural design in various scales and categories.

Peddle Thorp Group has practices worldwide with offices in Australia, South East Asia, Europe and America. Supported by a number of outstanding architects and experts, Peddle Thorp provides professional design service for sports, health and various residential projects. Major works are Australian National Tennis Centre, Melbourne Olympic Park and Velodrome, ESSO Headquarters, Melbourne Underwater World Aquarium, Malaysia Sports Centre, and a series of hotels, retail projects, offices, residences, and health facilities both in Australia and internationally.

Peddle Thorp Asia has established business in 26 provinces, autonomous regions and cities throughout China and a branch office in Shanghai, to accommodate a quality team of both Foreign and Chinese architects, introducing an internationally advanced design philosophy. In collaboration with the Melbourne Office, Peddle Thorp Shenzhen is able to provide a wide range of professional planning, architectural design, residential design, and landscape design by utilizing a unique approach. Advanced skills and comprehensive experience commit Peddle Thorp to provide a specialized service for our clients worldwide.

澳大利亚柏涛（墨尔本）建筑设计有限公司亚洲分公司
PEDDLE THORP ARCHITECTS ASIA

地址：广东省深圳市南山区华侨城生态广场A栋302
邮编：518053
电话：+86-755-2692 8866　传真：+86-755-2690 5186
邮箱：main@szpta.sina.net　网址：www.ptma.com.cn

Add: Room 302, Building A, Ecological Square,OCT, Shenzhen, Guangdong
P.C.: 518053
Tel: +86-755-2692 8866　Fax: +86-755-2690 5186
Email: main@szpta.sina.net　Http://www.ptma.com.cn

1 >> 滕王阁项目【厦门】　2 >> 华润银湖 · 蓝山【深圳】　3 >> 中信城项目　【长春】

4 >> 中建 · 开元壹号【西安】

5 >> 大理山水间【大理】

6 >> 畅园【昆明】

7 >> 丰城市 C1–3 地块规划【江西】

8 >> 曦城上苑【合肥】

9 >> 远洋地产长春净月项目【长春】

10–11 >> 长沙天麓项目【长沙】

12 >> 中粮天津六纬路项目体验中心【天津】

13 >> 金地宝山项目【上海】

14—15 >> 中铁逸都国际【贵阳】

12

14

13

15

16

17

16 >> 华润苏州项目【苏州】

17 >> 中星无锡项目【无锡】

18 >> 河北易水城概念规划【河北】

18

19 >> 国贸长江项目【芜湖】

20 >> 中建 · 国际港【北京】

21 >> 北京中粮后沙峪 A–10 地块【北京】

22 >> 北京 · 丰台商业【北京】

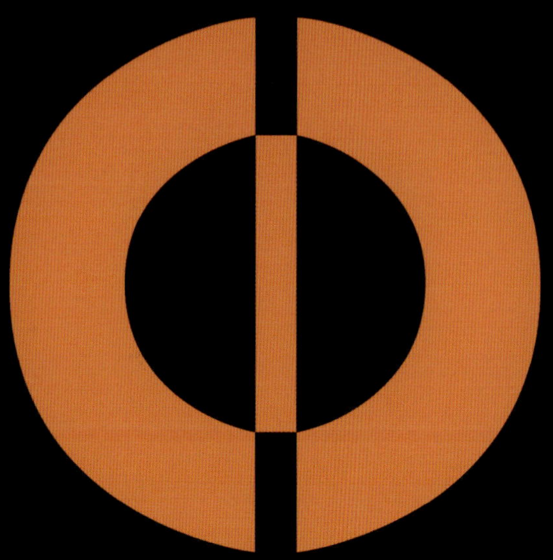

美国开朴建筑设计顾问(深圳)有限公司

CAPA Architecture Designing Consultant (SHENZHEN) Ltd. USA

美国开朴建筑设计顾问（深圳）有限公司注册于美国马里兰州华盛顿行政区，国内总部设于深圳。公司致力于办公楼、高档居住区、城市综合体及旅游度假村、科技园区等的规划与方案设计，以市场为导向，通过产品创新来提升核心竞争力和与环境和谐共生、传承文化、持续发展的生命力。公司作品具有强烈的时代个性和艺术前卫感，服务于国内众多品牌上市公司，打造专业精品是公司长期坚持不变的经营理念。

CAPA Architecture Designing Consultant (SHENZHEN) Ltd. USA. Registered in Maryland, Washington Administrative Region, China headquarters in Shenzhen. Company is committed to planning and designing of office, high-end residential, urban complexes and tourist resorts, science and technology park, Formed close to the market as guide, product innovation and core competitiveness of harmony with the environment, cultural heritage, sustainable development vitality. The company whose works with a strong personality and artistic avant-garde sense of the times serves many domestic brands listed companies. Professional quality remains unchanged as a long-term business philosophy.

地址：广东省深圳市福田区金田路3038号现代国际大厦2001室
电话：+86–755–82557831　传真：+86–755–82557636
网址：www.capadesign.com
邮箱：capa2000@vip.sina.com

Add: Room 2001, Modern International, Jintian Road, Shenzhen, Guangdong
Tel: +86–755–82557831　Fax: +86–755–82557636
http: //www.capadesign.com
E-mail: capa2000@vip.sina.com

1732 CASTLE ROCK RD FRNDERICK MARYLAND AMERICA
Tel: 410–767–1350　888–246–5941
http: //www.capadesign.com
E-mail: capa2000@vip.sina.com

1–2 成都世纪都会项目

建设地点：四川 成都

占地面积：39 260平方米

建筑面积：196 294平方米

1-2 Chengdu Century Metropolis Project

Construction Location: Chengdu, Sichuan

Site Area: 39,260 m^2

Building Area: 196,294 m^2

3–4 龙光深圳综合体概念设计

建设地点：广东 深圳

占地面积：28 424平方米

建筑面积：104 700平方米

3-4 Logan Shenzhen Complex Conceptual Design

Construction Location: Shenzhen, Guangdong

Site Area: 28,424 m^2

Building Area: 104,700 m^2

成都世纪都会项目定位为引领时尚、领先国际潮流。项目功能为高品质住宅、甲级写字楼、星级艺术酒店、大型购物中心。集大型百货、国际品牌旗舰店、国际影院、高档餐饮、特色文化酒吧为一体的大型国际都市综合体，主要面对的客户为高档商务人士、城市高级白领。

“龙凤呈祥”在中国传统观念里，龙和凤象征着吉祥如意，代表人们对美好生活的向往。项目用地位于成都古城东南方，用地呈“L”形，形似凤展翅，又似龙腾飞，面向成都古城飞去。故方案规划以超高层建筑为头，带领商业、住宅向城市区腾飞，奔腾而去。综合楼的表面设计——韵律、时尚、梦幻般的高山流水境界，其在办公建筑面朝主市区的方向，建筑表皮顺墙面有韵律地延续下来，秉承了高山流水的文化概念，表达了建筑的动态感觉，并吸取高雄大立广场、伦敦30 St Mary Axe的立面意境，运用高科技的现代建筑技术，实现综合楼的时代价值。

1

2

4

5

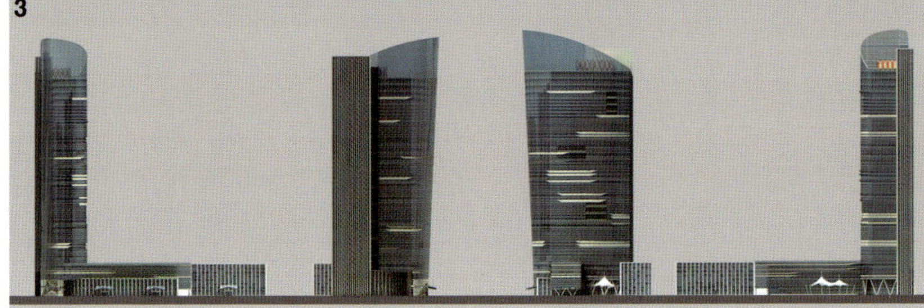
3

1–3 张家港金茂创业大厦建筑设计

建设地点：江苏 张家港

占地面积：15 379平方米

建筑面积：79 166平方米

1-3 Zhangjiagang Jinmao Chuangye Tower Building Design

Construction Location: Zhangjiagang, Jiangsu

Site Area: 15,379 m^2

Building Area: 79,166 m^2

4–5 深圳天佶国际广场概念设计

建设地点：广东 深圳

占地面积：25 114平方米

建筑面积：432 600平方米

4-5 Shenzhen Tianji International Plaza Conceptual Design

Construction Location: Shenzhen, Guangdong

Site Area: 25,114 m^2

Building Area: 432,600 m^2

1–2 张家港港城大厦

建设地点：江苏 张家港

占地面积：22 273平方米

建筑面积：69 727平方米

1-2 Zhangjiagang City Building

Construction Location: Zhangjiagang, Jiangsu

Site Area: 22,273 m^2

Building Area: 69,727 m^2

1

2

3–4 徐州行政中心

建设地点：江苏 徐州

占地面积：127 034平方米

建筑面积：116 469平方米

3-4 Xuzhou Administration Center

Construction Location: Xuzhou, Jiangsu

Site Area: 127,034 m^2

Building Area: 116,469 m^2

3

4

1–3 成都锦江琉璃场项目

建设地点：四川 成都

占地面积：47 190平方米

建筑面积：235 833平方米

1-3 Chengdu Jinjiang Glass Field Project

Construction Location: Chendu, Sichuan

Site Area: 47,190 m^2

Building Area: 235,833 m^2

4–5 西安紫薇意境

建设地点：陕西 西安

占地面积：92 598平方米

建筑面积：398 243平方米

4-5 Xi'an Ziwei Garden

Construction Location: Xi'an, Shaanxi

Site Area: 92,598 m^2

Building Area: 398,243 m^2

1–5 怀来鸿威瑞云文化艺苑

建设地点：河北 怀来

占地面积：1 115 470平方米

建筑面积：721 587平方米

1-5 Huailai Homeway Ruiyun Culture Garden

Construction Location: Huailai, Hebei

Site Area: 1,115,470 m^2

Building Area: 721,587 m^2

规划方案立足于本项目山地地形，自由而富有韵律，形成丰富的景观，使得空间不断地变化，富有趣味，有移步换景的效果。峡谷地形更加丰富了景观的层次；借景于峡谷，突出空间的趣味性和吸引力，也增添地块的独特性。独特的地形和地势将造就该项目的独特性和与众不同的品质。

5

4

1

2

3

1–2 深圳中粮一品澜山

建设地点：广东 深圳

占地面积：53 113平方米

建筑面积：126 610平方米

1-2 COFCO Shenzhen Yipin Lanshan

Construction Location: Shenzhen, Guangdong

Site Area: 53,113 m^2

Building Area: 126,610 m^2

3–4 南京建邺金鼎湾状元府

建设地点：江苏 南京

占地面积：58 000平方米

建筑面积：170 627平方米

3-4 Global Golden Tripod Bay Champion's Mansion, Nanjing

Construction Location: Nanjing, Jiangsu

Site Area: 58,000 m^2

Building Area: 170,627 m^2

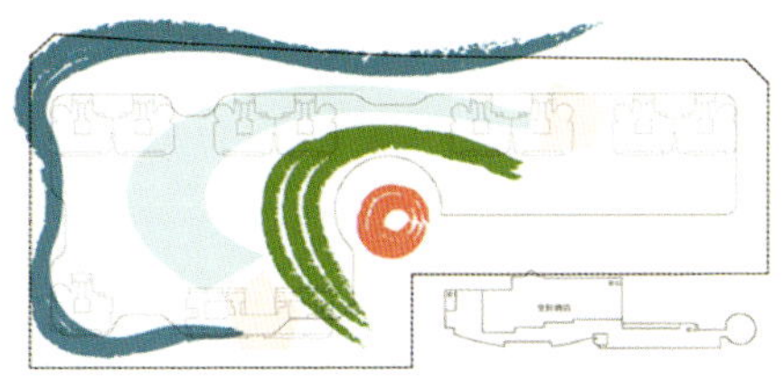

柔和的建筑形式

7.3的容积率，3.386万平方米的用地

本方案以曲线的形式呈现在市民面前，以柔和的建筑形式消除高容积率下建筑群带来的压迫感。

游龙

蜿蜒曲折的退台、层层叠加的立体绿化，就像一条灵动的游龙盘踞在基地内。方案呈对称布局，基地中心是项目的焦点——圆形退台下沉庭院，代表了龙珠。

立体公园

本方案容纳了皇岗村尊重历史、尊重自然环境的生活习惯，休闲娱乐的功能，开放的空间特色。同时，所有功能块均面对立体绿化，这必将成为人与自然相互融合的典范。

1

2

1–2 皇岗庄式花园

建设地点：广东 深圳

占地面积：33 860平方米

建筑面积：326 505平方米

1-2 Huanggang Village Garden

Construction Location: Shenzhen, Guangdong

Site Area: 33,860 m^2

Building Area: 326,505 m^2

3–5 西安紫薇尚层

建设地点：陕西 西安

占地面积：37 467平方米

建筑面积：179 606平方米

3-5 Ziwei Future Residential Community, Xi'an

Construction Location: Xi'an, Shaanxi

Site Area: 37,467 m^2

Building Area: 179,606 m^2

3

4

5

CAPA

美国开朴建筑设计顾问（深圳）有限公司

1 龙光东部曦城

建设地点：广东 惠州

占地面积：117 000平方米

建筑面积：258 000平方米

2—3 休宁·新安江国际旅游养生 度假村概念规划设计

建设地点：安徽 黄山

占地面积：699 109平方米

建筑面积：420 412平方米

1 Eastern Xi City of Dragon Optical

Construction Location: Huizhou, Guangdong

Site Area: 117,000 m^2

Building Area: 258,000 m^2

2-3 International Travel Health Resort Xiuning Xin'anjiang Conceptual Planning and Design

Construction Location: Huangshan, Anhui

Site Area: 699,109 m^2

Building Area: 420,412 m^2

1—5 聚豪会高尔夫庄园概念规划

建设地点：广东 深圳

占地面积：148 591平方米

建筑面积：60 000平方米

1-5 Conceptual Design Of Juhaohui Golf Manor

Construction Location: Shenzhen, Guangdong

Site Area: 148,591 m^2

Building Area: 60,000 m^2

6 中粮·北纬28°

建设地点：湖南 长沙

占地面积：396 517平方米

建筑面积：654 700平方米

6 28 ° North Latitude COFCO

Construction Location: Changsha, Hunan

Site Area: 396,517 m^2

Building Area: 654,700 m^2

AIM International
[加拿大]
亚瑞建筑景观
[中国甲级]

2009–2011 中国最具影响力境外设计机构
2008–2011 年度联华国际集团最佳战略同盟伙伴

AIM国际设计集团是加拿大、中国香港注册的跨国品牌，是中国、瑞士、加拿大合资设立的甲级建筑设计单位。JET是AIM集团在加拿大的分支机构。广州总部聚集了近300名国际国内专业人才，形成涵盖地产策划、城市规划、建筑设计、室内设计、景观设计、广告设计、项目管理、工程施工八大领域的全程服务产业链。项目覆盖美国、澳洲、加拿大、泰国、中国等国家，项目达数百个。

AIM亚瑞建筑景观设计引进先进的国际项目管理模式，开放、包容、激情的公司文化吸引了众多才华横溢的明星建筑师，在"优秀的设计、优质的服务"理念指导下，已经为近百个项目提供了专业水准的服务。AIM亚瑞伴随客户业主们的成功而迅速成长，同时，多元化的集团模式使AIM亚瑞在策划设计理念和技术手段上也引领着市场方向，逐步赢得世界口碑。

部分合作伙伴

中信地产、恒大地产、万达地产、龙光地产、保利地产、联华国际、广晟地产、佛奥集团、光大地产、中南恒展集团、中力集团、吉林大学、鼎峰地产、龙华地产、集安投资、建南集团。

荣誉

2009–2011年中国最具影响力境外设计机构
2008–2011年度联华国际集团最佳战略同盟伙伴

AIM Group International is a multi-disciplinary design firm originally formed in Canada. JET is a branch of AIM Group in Canada. We have offices in Canada, Hong Kong China and Guangzhou China. The department in Guangzhou has over 300 staffs including employees from Canada, Britain and Hong Kong, China. Our service projects are all over the world and mainland China. We cross the boundaries of architecture, interior, landscapes and urban design, modeling, graphic design and project management. AIM consists of three departments which are ARUR, Idea and M-team.

AIM ARUR Architectural Design keeps our management method open and flexible to encourage staff to be creative and competitive in a growing firm. With involvement in hundreds of projects across the field of commercial, residential and public buildings around the world, we have gained trust form clients and continued to win a world reputation.

Some of its partners

CITIC Real Estate, Evergrande Real Estate, Wanda Real Estate, LOGAN Real Estate, Poly Real Estate, Lanwa International, Rising Real Estate, Foao Group, Everbright Real Estate, CENTENIO Group, Zhongli Group, Jilin University, D.F Real Estate, Longhua Real Estate, Ji An Investment, Jiannan Group.

Honors

2009-2011 Annual, China's most influential design agencies outside
2008-2011 Annual, Best strategic alliance partners of Lanwa International Group

中国建筑景观部：
T +86–20–3881 9168
+86–20–3881 1627
+86–20–3884 8126
F +86–20–3881 2825
E A@AIMgi.com（建筑）
La@AIMgi.com（景观）
Q 1124955421

中国室内部：
T +86–20–3881 3815
+86–20–3881 3785
+86–20–3886 4882
F +86–20–3881 1267
+86–20–3438 7365（广美）
E M@AIMgi.com
Q 491240981

中国商业地产策划（广告部）：
T +86–20–3884 8831
+86–20–3884 8832
+86–20–3884 8823
F +86–20–3880 9480
E I@AIMgi.com
Q 49448826

国际部（加拿大）
T +416–4919 988
F +416–5677 803

国际部（香港）
T +852–2411 6631
F +852–8949 7386

联系我们
[中 国] 广东省广州市体育西路173号天河大厦综合楼二楼、四楼、五楼
[中国香港] Rm.306 Golden Gate Commercial Building, 136 Austin Road, Tsimshatsui, Kowloon, Hong Kong, China
[加拿大] 91 Glen worth Road, Toronto, Canada

2F

4F
5F

1 联华 | 星河传说荷塘月色二期

广东 东莞
联华集团
20 000平方米
规划设计 | 建筑设计

GALAXY LEGEND LOTUS POND II
Dongguan, Guangdong
LIANHUA Group
20,000 m^2
Planning & Architecture Design

2 中信 | 德方斯公寓

广东 东莞
中信集团
90 000平方米
规划设计 | 建筑设计
已竣工

CITIC DEFANGSI MANSION
Dongguan, Guangdong
CITIC Group
90,000 m^2
Planning & Architecture Design
Completed

3 中力 | 大一山庄

广东 广州
中力集团
150 000平方米
建筑设计

RISING FRESHMAN HILL
Guangzhou, Guangdong
ZHONGLI Group
150,000 m^2
Architecture Design

4 SUNSHINE GARDEN
佛奥 | 阳光花园

广东 中山
佛奥集团
500 000平方米
规划设计 | 建筑设计
已竣工

Zhongshan, Guangdong
FOAO Group
500,000 m^2
Planning & Architecture Design
Completed

1

2

3

4

5 中信 | 凯旋国际

广东 东莞
中信集团
300 000平方米
规划设计 | 建筑设计
已竣工

CITIC TRIUMPH INTERNATIONAL
Dongguan, Guangdong
CITIC Group
300,000 m^2
Planning & Architecture Design
Completed

5

1 中南恒展 | 碧江帝景

广东 清远
中南恒展
112 976平方米
规划设计 | 建筑设计

BIJIANG DIJING
Qingyuan, Guangdong
CENTERIO GROUP
112,976 m^2
Planning & Architecture Design

2 中信 | 商业广场

广东 东莞
中信集团
89 000平方米
规划设计 | 建筑设计
已竣工

CITIC BUSINESS PLAZA
Dongguan, Guangdong
CITIC Group
89,000 m^2
Planning & Architecture Design
Completed

3 龙华 | 尚墅名门

广东 佛山
龙华集团
47 000平方米
规划设计 | 建筑设计
已竣工

LONGHUA SUNSU GARDEN
Foshan, Guangdong
LONGHUA Group
47,000 m^2
Planning & Architecture Design
Completed

4 联华 | 花园城

广东 东莞
联华集团
500 000平方米
规划设计 | 建筑设计
已竣工

LIANHUA GARDEN CITY
Dongguan, Guangdong
LIANHUA Group
500,000 m^2
Planning &Architecture Design
Completed

5 广晟 | 海韵兰庭

广东 广州
广晟投资集团
140 000平方米
规划设计 | 建筑设计
已竣工

WAVES & ORCHIDS GARDEN
Guangzhou, Guangdong
GUANGSHENG Group
140,000 m^2
Planning & Architecture Design
Completed

1–2
沙面迎亚运全岛综合整治工程

广东省外事办[左]
鹅潭宾馆[上右]
广东 广州
沙面岛全岛（112栋建筑）
建筑设计 | 景观设计
已竣工

THE FACADE RECONSTRUCT IN SHAMIAN FOR THE ASIAN GAMES
GUANGDONG FOREIGN AFFAIRS OFFICE [LEFT]
ETAN HOTEL[RIGHT]
Guangzhou, Guangdong
All Island(112)
Architecture & Landscape Design
Completed

3 东莞道滘客运站

广东 东莞
30 000平方米
规划设计 | 建筑设计
已竣工

Dongguan Daojiao Central Station
Dongguan, Guangdong
30,000 m^2
Planning & Architecture Design
Completed

4 东莞铂尔曼酒店

广东 东莞
25 000平方米
规划设计 | 建筑设计
已竣工

Dongguan Pullman Hotel
Dongguan, Guangdong
25,000 m^2
Planning & Architecture Design
Completed

5-6
流花公园展馆

广东 广州
1 175平方米
规划设计 | 建筑设计 | 景观设计
已竣工

LIUHUA EXHIBITION CENTER
Guangzhou, Guangdong
1,175 m^2
Planning & Architecture
& Landscape Design
Completed

1

2

1–2
顺德新粤丰商业广场

广东 顺德
3 000平方米
规划设计｜建筑设计

SHUNDE XINYUEFENG SHOPPING MALL
Shunde, Guangdong
3,000 m^2
Planning & Architecture Design

3

3 沙河商业中心

广东 广州
8 000平方米
建筑设计
已竣工

SHAHE SHOPPING MALL
Guangzhou, Guangdong
8,000 m^2
Architecture Design
Completed

4 广州国际皮具中心

广东 广州
43 781平方米
规划设计｜建筑设计

INTERNATIONAL LEATHER OUTLET
Guangzhou, Guangdong
43,781 m^2
Planning & Architecture Design

5 番禺汇珑新天地

广东 广州
137 158平方米
规划设计｜建筑设计
施工中

PANYU HUILONG SUNNYDAY MALL
Guangzhou, Guangdong
137,158 m^2
Planning & Architecture Design
Under Construct

4

5

2010—2011
主要建筑设计作品
Main Architectural Design Works
Annual Review of Chinese Architectural Design Works

烟台万科海云台 Vanke Chefoo Island, Yantai

CONCORD DESIGN GROUP CDG国际设计机构 城 市 区 域 规 划 建 筑 设 计 环 境 景 观 设 计

地址：北京市海淀区长春桥路11号万柳亿城中心A座10层/13层
电话：+86-10-58816545
58816546
58816547
传真：+86-10-58816156
邮箱：cdg@cdgcanada.com
网址：www.cdgcanada.com

长春远洋戛纳小镇 SINO-OCEAN Cannes Town, Changchun

青岛万科东郡 Vanke East County, Qingdao

公司简介

CDG国际设计机构是由多家境外设计事务所组成的专门针对中国市场的设计机构，2001年在加拿大不列颠哥伦比亚省首府维多利亚市注册成立。

CDG国际设计机构利用其主要合伙人对中国市场的了解及与长期合作客户的良好互信的伙伴关系，整合境外及本土的优秀设计资源，为不同需求的客户提供最优化的设计组合。

CDG国际设计机构目前在北京设有办事处，设计人员100余人。境外组合机构包括建筑设计及环境景观设计机构，服务范围涉及城市区域规划、建筑设计、环境景观设计等。

CDG INTERNATIONAL DESIGN LTD. consists of several overseas design firms, all focusing on the Chinese market. It was registered in 2001 in Victoria B.C., Canada.

The main partners of CDG are very familiar with the Chinese market, and they all have long established mutual trusting relations with a wide variety of clients. The Group integrates excellent overseas and native design resources, providing the best design combination for each client with differing demands. With a mastery of European, American and Chinese architectural theory and practice, CDG successfully collaborates with a wide variety of designers, introduces new ideas and rigor into Chinese projects.

At present there is a main branch of CDG in Beijing, including about 100 designers. All of the branches provide services in the fields of urban planning, architectural design, landscape design.

廊坊华夏铂宫　Noble Asset, Langfang

CONCORD DESIGN GROUP　CDG国际设计机构　城市区域规划　建筑设计　环境景观设计

公司近年主要项目一览表

综合区域规划/建筑设计

1、沈阳万科惠斯勒小镇
2、中铁余杭项目
3、富力北京旧宫项目
4、富力上海昆山项目
5、中建北京燕京桥项目
6、中铁长春西湖项目
7、吉林广泽紫晶城
8、美克天津嘉美湾
9、济宁北湖项目
10、北京永定河孔雀英国宫
11、青岛万科东郡

临淄太公湖高尔夫球会别墅 Taigong Lake Golf Villa, Linzi

哈尔滨半岛世家 The Peninsula Mansion, Harbin

12、烟台万科假日风景
13、北京潮白河孔雀英国宫
14、廊坊公务员小区
15、沈北亚泰新城
16、中铁北京马坡项目
17、烟台万科海云台
18、沈阳信达理想新城
19、长春万科惠斯勒小镇
20、天津万科张家窝片区规划
21、天津万科士林苑
22、泰辰鲅鱼圈海湾馨城居住区
23、廊坊华夏铂宫
24、长春中东净月湾
25、松原中东班芙小镇
26、通化中东帕萨迪娜
27、青岛万科城市花园
28、天津万科金色雅筑
29、鞍山万科惠斯勒小镇
30、廊坊锦瑞尚城
31、天津嘉铭中新生态城红树湾
32、中铁置业沈阳人杰水岸
33、万科长春228厂地块概念设计
34、沈阳万科金域兰湾
35、青岛万科金色城品
36、廊坊华元和庭
37、沧州大合庄京旧城改造
38、东莞同沙项目
39、北京大学肖家河居住区
40、北京阿凯笛亚别苑
41、鞍山万科城
42、北京密云中加会馆
43、万科长春净月2#地
44、长春吉林农大棚户区改造
45、东莞香树丽舍居住区
46、金隅万科城概念设计
47、东莞锦盛年华居住区
48、贵阳花溪大地之舞大剧院地块
49、廊坊嘉轩凤凰城
50、北京金隅小营西路项目
51、东莞凯逸豪庭
52、长春伊通河地王居住区
53、北京海棠公社
54、廊坊华夏水晶城
55、东莞凯旋门江滨居住区
56、东莞天骄峰景居住区
57、北京首创京棉一厂改造
58、北京首创吉普厂改造概念设计
59、北京归谷园二期概念设计
60、北京通州漷县镇中心概念规划
61、东莞松山湖商住中心区规划
62、东莞理想0769家园
63、东莞海怡花园三期
64、东莞金鳌湾江岸公寓
65、北京中海高尔夫花园
66、东莞盛世华南居住区
67、北京鹿港嘉苑
68、北京博雅西园

长春万科惠斯勒小镇　Vanke Whistler Town, Changchun

CONCORD DESIGN GROUP　CDG国际设计机构　城 市 区 域 规 划　建 筑 设 计　环 境 景 观 设 计

商业、公共建筑设计

1、长春远洋戛纳小镇
2、白山广泽城市综合体
3、张家口融侨龙泉广场
4、张家口融侨水泉沟项目
5、哈尔滨天邑蓝湾
6、长春中东南部新城项目
7、长春中东铁北项目
8、青岛天泰城美立方
9、吉林中东凯悦公馆
10、通化滨江戴思酒店
11、松原中东新天地商场&帕萨迪娜
12、北京国际青年创意中心概念设计
13、廊坊东115项目
14、通化规划大厦

海口香格里拉国宾馆　Shangri-La Resort, Haikou

沈北亚泰新城项目　Shenbei Yatai New Town, Shenyang

15、通化中东新天地商场
16、大连正源华南餐饮城
17、成都海峡两岸科技开发园中心商务区
18、长春万科兰乔圣菲商业公寓
19、张家口五一广场
20、广饶商业中心区
21、廊坊一号
22、东莞横岗水库度假村
23、北京悦港城
24、东莞豪凯五星大酒店
25、东莞松山湖锦绣山河城市中心
26、东莞华南商场生活城E&F区
27、廊坊第六大街
28、胜芳中心广场
29、北京龙腾文化广场
30、东莞樟木头沃尔玛广场
31、宜昌长江帝景6号综合楼
32、东莞亿兆厚街商业广场
33、丹东边境经济合作A区-E区
34、北京彼岸青滩高尔夫会所
35、贵阳半岛国际大酒店
36、廊坊第八大街西区
37、北京望京六佰本随便消费区
38、北京后沙峪政府办公楼
39、北京酒仙桥风情商业街
40、东莞广华商贸中心区
41、东莞万江区行政中心
42、北京卷石天地
43、华南MALL

别墅

1、临淄太公湖国际高尔夫别墅
2、海口香格里拉国宾馆
3、吉林松花湖别墅
4、中体潍坊高尔夫社区别墅
5、长沙中铁水映加州
6、唐山万科南湖别墅
7、河东蓟县别墅
8、哈尔滨半岛世家
9、沈阳半山湖别墅
10、长春昂展净月别墅
11、长春观澜湖别墅
12、沈阳岭墅
13、北京阿凯笛亚庄园二期别墅
14、长春万科潭溪别墅
15、长春国信·美邑
16、北戴河总部基地
17、沈阳万科兰乔圣菲三期\四期\五期
18、东莞松山湖锦绣山河二期别墅
19、东莞紫园别墅
20、长春天安第一城三期
21、北京中海安德鲁斯庄园
22、北京阿凯笛亚庄园
23、东莞松山湖锦绣山河一期别墅
24、东莞江畔花园游艇别墅
25、北京银湖别墅

造型设计项目

1、天津中天首府
2、北京翰庭
3、北京金隅七零九零
4、潍坊白浪金沙商业广场
5、北京福泰华城
6、北京总部公寓
7、北京半岛国际公寓
8、北京金隅丽港城
9、北京博雅园

潮白河 · 孔雀英国宫　UK Palace, Chaobai River

永定河 · 孔雀英国宫　UK Palace, Yongding River

CONCORD DESIGN GROUP　CDG国际设计机构　城市区域规划　建筑设计　环境景观设计

景观设计

1. 临淄太公湖国际高尔夫别墅
2. 哈尔滨半岛世家
3. 河东蓟县别墅
4. 哈尔滨天邑蓝湾
5. 沈北亚泰新城
6. 长春昂展净月别墅
7. 唐山万科南湖别墅
8. 廊坊华夏铂宫
9. 长春观澜湖别墅
10. 天津中天首府
11. 北京阿凯笛亚别苑
12. 长春万科潭溪别墅
13. 北戴河总部基地
14. 鞍山万科城
15. 长春国信 · 美邑
16. 沈阳万科兰乔圣菲别墅三期/四期
17. 北京酒仙桥风情商业街
18. 宜昌长江帝景
19. 北京阳光大厦广场

中铁置业沈阳人杰水岸 Zhongtie Elite Bayshore, Shenyang

中铁余杭项目 Zhongtie Yuhang Project

荣誉

CDG国际设计机构荣获 2010年住房和城乡建设部中国建筑文化中心“中国最具影响力商业地产设计机构”奖
CDG国际设计机构荣获 2010年中国房地产年度总评榜“中国最具风格规划设计机构”奖
中铁人杰水岸荣获 2010年中国房地产年度总评榜年度“城市规划设计”大奖
CDG国际设计机构荣获 2009年住房和城乡建设部中国建筑文化中心“中国最具影响力境外设计机构”奖
“沈阳万科金域兰湾”荣获 2008年中国人居典范建筑规划“设计方案规划金奖”
“国信·美邑”荣获 2008年中国住交会“中国名盘·优秀别墅”大奖
CDG国际设计机构荣获 2007年亚洲房地产峰会“亚洲建筑规划设计区域十大品牌机构”奖
中海安德鲁斯庄园"荣获 “2007年詹天佑大奖住宅区金奖”
林世彤荣获 2007年亚洲房地产峰会“亚洲建筑规划设计区域十大创新人物”称号
“华南MALL”荣获 “2004年中国建筑艺术奖”
2006年被美国新闻周刊评为“人类新七大奇迹之一”

美国EBU建筑设计咨询（中国）有限公司

EBU ARCHITECTURE DESIGN CONSULTATION CO., INC.

EBU建筑设计咨询有限公司是美国最富前瞻性与创新力的设计公司之一，总部位于美国南部工业城市达拉斯。

多年来，EBU在城市及区域规划、旅游地产规划、办公建筑、酒店建筑、居住建筑、社区建筑、高层建筑、绿色建筑、商住综合体建筑以及景观和室内设计等方面成功设计了许多令人瞩目的优秀工程项目，积累了丰富的设计及工程经验。

主要的合伙人和设计骨干，除了依托其丰富的设计经验和专业鉴别力亲历亲为，负责公司的各类工程项目外，还积极参与建筑业的学术团体、政府机构和社区服务活动，在美国具有广泛的影响。

EBU于2000年进入中国市场，在经历了多年与中国同行的合作设计后，于2004年3月在中国上海注册成立了美国EBU建筑设计咨询（中国）有限公司(中文名：东筑建筑设计)。EBU在中国上海拥有近40人的团队规模，通过国际化的项目管理和技术培训锤炼出一批对中国建设市场和文化背景有着深刻认识的专业团队，并依托卓越的设计实力，与国内多家国家级设计机构建立起紧密的合作关系。

EBU（中国）多年来致力于提供高素质和全套完整的设计服务，并以国际化的设计观念和实务运作享誉业界。公司以“全球的参与……无限的构思……”为指导思想，充分利用公司强大的设计团队、丰富的管理经验和广泛的技术资源，针对每个项目的个性化问题与不同挑战，量身配置项目团队，纵深剖析客户要求，睿智思考，缜密构想，在设计中充分考虑建筑造价、地域性文化的延续、人文与生态环境等因素对项目的影响，完美提供创造性的解决方案，为城市、建筑的发展提出自己的见解。我们将全球领先的设计理念与中国境内市场条件相结合，寻求服务的最高性价比，精心呵护每一个项目，并以我们的专业知识和实践经验为基础全力推进设计质量，以实现“设计创造价值”的服务目标。

作为一个不断发展的设计公司，EBU凭借强大的设计实力、先进的设计理念以及团队成员的精诚合作，已经在中国境内的18个省市重点区域完成了众多项目的设计工作，取得了骄人的成绩。现在，EBU正在稳步构建一个以达拉斯和上海为中心的综合类设计集团，以开发项目全流程设计服务为核心，全力以赴并始终如一地为全球客户提供一流的服务。

EBU Architecture Design Consultation Co., Inc. was registered in Delaware and the head office was in Dallas, one of the famous industrial centers in US. EBU is trying to build one of the most creative and foresighted design companies.

EBU had a lot of practical and successful experience in urban/regional planning, resort, office, hotel, residential, community, high-rise, green building, residential/commercial mixture building, landscape design and interior design in the past years. The partners and chief designers, relying on their rich experience, not only operate and responsible for various projects but also are active participants in architecture, planning and landscape academic groups, public and community services by which built a respectful reputation in USA.

EBU first explored Chinese market in 2000. After several years' cooperation with Chinese partner, EBU Architecture Design Consultant (China) Co., Inc. (Dongzhu Architecture Design Co. Inc. in Chinese) was registered in Shanghai, China in March 2004. EBU had a great understanding of local designing market and culture background in China. By managing international project and inputting in professional training, the team has been grown up to a mature team with approximately 40 practioners and kept strategically cooperation with several national level local design institutes.

For several years, EBU (China) is well-known for committed to providing high-quality and full construction services through global leading design concepts and practical operations. EBU takes “Global Participant and Infinite Idea” as company's ideology, fully utilizes strong design team, rich management experiences and wide range of technical sources, especially aiming at customizing each project with different challenges, team organization, in-depth analysis of customer requirements, smart ideas, sophisticated thinking for effects of construction cost, regional culture continuity, humanistic and ecological factors to provide ideal creative solutions by bringing our professional construction and development ideas. In order to achieve our service target as"design to create value", we will combine the global design concept, Chinese wisdom and market to maximize cost-effective service, well-care of our every project, and enhance our design quality through professional knowledge and practical experiences.

As a growing design company, EBU had been involved as a high-reputation FSP (Full-service-process) Design Company and finished successfully in 18 different provinces in China. EBU (China) is an active advocator in building a Dallas-Shanghai based international professional design group to dedicate ourselves to provide high quality service for all the clients.

中国
地址： 上海市卢湾区局门路550号8号桥
创意园三期7号楼7202-7203室
电话： +86-21-33315277
传真： +86-21-33315399
邮箱： ebu@ebuarchitects.com
网址： www.ebuarchitects.com

CHINA
Add: Rm. 7202-7203 Block 7, Phase III Bridge 8,
No. 550 Jumen Rd. Shanghai 200023 P.R.C.
Tel: +86-21-33315277
Fax: +86-21-33315399
E-mail: ebu@ebuarchitects.com
http://www.ebuarchitects.com

US
Add: 7457 Glass House Walk.Frisco
Texas.U.S.A. 75035
Tel: 001-214-387-0704
E-mail: ebu@ebuarchitects.com

办公商务类
Commercial Office

苏陕国际金融中心

业　　主：中登集团
设计内容：规划、建筑方案
建设地点：陕西 西安
用途：超高层甲级写字楼、五星级酒店
用地面积：42 005平方米
总建筑面积：225 022平方米
设计日期：2009年
项目情况：方案设计

Su Shan International Finance Centre

Owners: Zhongdeng Group
Design Elements: Planning, Building Programs
Building Location: Xi'an, Shaanxi
Uses: High-rise, Five Star Hotels
Land Area: 42,005 m^2
Total Floor Area: 225,022 m^2
Design Time: 2009
Project Status: Design

旅游地产
Tourism Real Estate

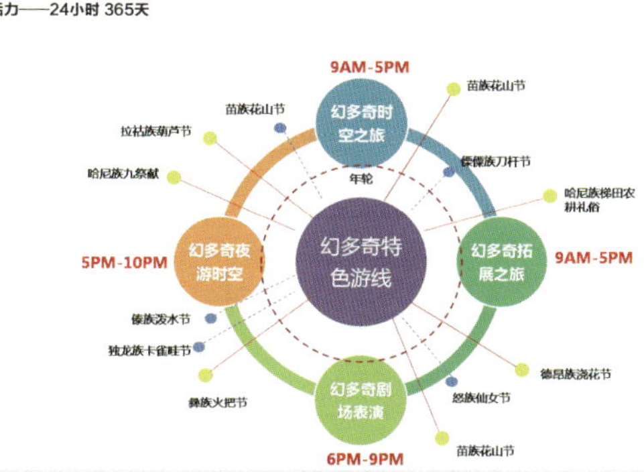

石林幻多奇主题公园概念规划与设计

用地面积：88.6公顷
建筑面积：169 600平方米
建设地点：云南 石林
设计时间：2010年
设计功能：旅游地产／特色性主题公园

Concept Master Plan and design of Shilin Huanduoqi Theme park

Land Area: 88.6 ha
Building Area: 169,600 m^2
Building Location: Shilin, Yunnan
Design Time: 2010
Design Features: Tourism, Real Estate / Theme Park

石林旅游服务区综合服务中心修建性详细规划

用地面积：50公顷
建筑面积：257 000平方米
建设地点：云南 石林
设计时间：2010年
设计功能：综合型旅游服务中心

Detailed Plan for Complex Service Center Stone Forest

Land Area: 50 ha
Building Area: 257,000 m^2
Building Location: Shilin, Yunnan
Design Time: 2010
Design Features: Integrated Tourism Service Center

高端物业类
High-end Property

云南石林高尔夫别墅项目

设计内容：规划与建筑设计
建设地点：云南 石林
用　　途：高尔夫度假别墅
建设地点：91 037平方米
容 积 率：0.24
设计日期：2009年

Yunnan Shilin Golf Villa

Design Content: Planning and Design
Building Location: Shilin, Yunnan
Uses: Golf Villa
Land Area: 91,037 m^2
Plot Ratio: 0.24
Design Time: 2009

洛阳国龙龙城项目规划与建筑设计

业　　主：洛阳国龙置业有限公司
设计内容：规划与建筑设计
建设地点：河南 洛阳
用　　途：高层住宅
用地面积：17.45公顷
容 积 率：5.0
设计日期：2010年

Architecture Design for Apartment of Guolong Project, Luoyang

Owners: Luoyang Guolong Properties Company Limited
Design Content: Planning and Design
Building Location: Luoyang, Henan
Uses: High-rise Residential
Land Area: 17.45 ha
Plot Ratio: 5.0
Design Time: 2010

南方东银重庆建设厂项目

业　　主：南方东银集团
设计内容：规划、建筑方案、
　　　　　初设、施工图配合
建设地点：重庆
用　　途：滨江豪宅
用地面积：8.01公顷
总建筑面积：369 920平方米
容 积 率：3.90
设计日期：2009年

Project of Architecture Design for Factory Building

Owner: Nanfang Dongyin Property Company, Chongqing
Design Elements: Planning, Building Programs,
Preliminary Design, Working Drawings
Building Location: Chongqing
Uses: Riverside Luxury
Site Area: 8.01 ha
Total Floor Area: 369,920 m^2
Plot Ratio: 3.90
Design Time: 2009

城市商业综合体 Urban Commercial Complex

石林生态酒店式公寓修建性详规

业主信息：云南圆通投资有限公司
用地面积：28.67公顷
建筑面积：193 191平方米
建设地点：云南　石林
设计时间：2011年
设计功能：高尔夫度假公寓及石林景
　　　　　区商业配套

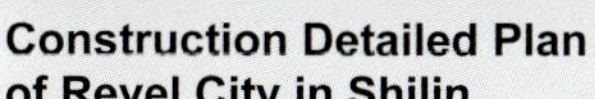

Construction Detailed Plan of Revel City in Shilin

Owners Information: Yunnan Yuantong
Investment Co., Ltd.
Land area: 28.67 ha
Building Area: 193,191 m^2
Building Location: Shilin, Yunnan
Design Time: 2011
Design Features: Stone Forest Scenic Spot Golf
Resort Apartments
and Commercial Facilities

大理风城星座

业主信息：大理金洲房地产开发有限公司
设计内容：规划、建筑方案
建设地点：云南　大理
用　　途：住宅、酒店公寓、商业、酒吧
用地面积：29 198.2平方米
地上建筑面积：146 023平方米
容 积 率：5.0
设计日期：2009年

Constellation Dali Windy City

Owner: Dali Jinzhou Real Estate Development Co., Ltd.
Design Elements: Planning, Building Programs
Building Location: Dali, Yunnan
Uses: Residential, Hotel Apartments, Commercial, Bar
Site Area: 29,198.2 m^2
Ground Floor Area: 146,023 m^2
Plot Ratio: 5.0
Design Time: 2009

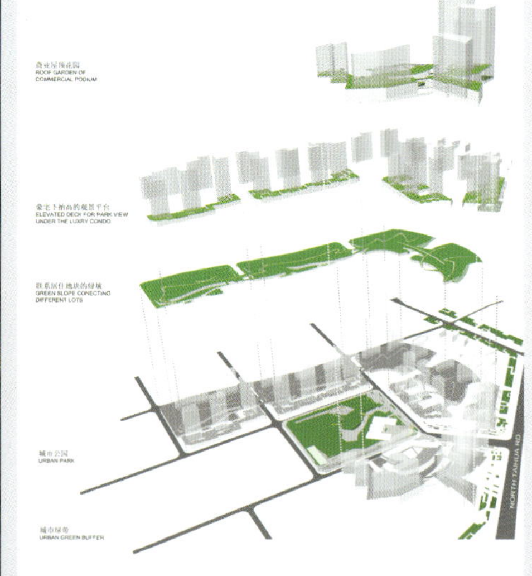

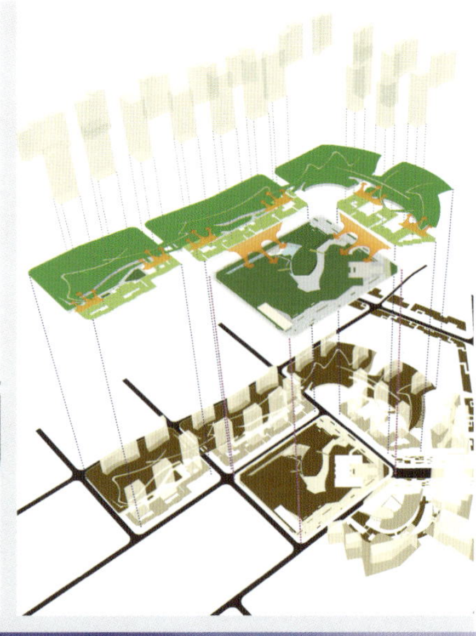

陕西华岭地产西安“中央公园”项目概念性规划设计

业　　主：陕西华岭地产
建设地点：陕西 西安
设计内容：规划设计，建筑设计
用　　途：大型综合性商业、办公、度假酒店、
　　　　　酒店公寓、豪宅、会所
总用地面积：22.4公顷
地上筑面积：830 000平方米
容 积 率：3.7
设计日期：2011年

Conceptual General Planning Design for Central Park Project

Owner: Hualing, Shanxi Real Estate Co., Ltd.
Building Location: Xi'an, Shaanxi
Design Elements: Planning and Design, Architectural Design
Uses: Large-scale Comprehensive Commercial, Office, Resort Hotels,
Hotel Apartments, Luxury, Club
Total Land Area: 22.4 ha
Ground Floor Area: 830,000 m^2
Plot Ratio: 3.7
Design Time: 2011

绿色建筑 Green Buildings

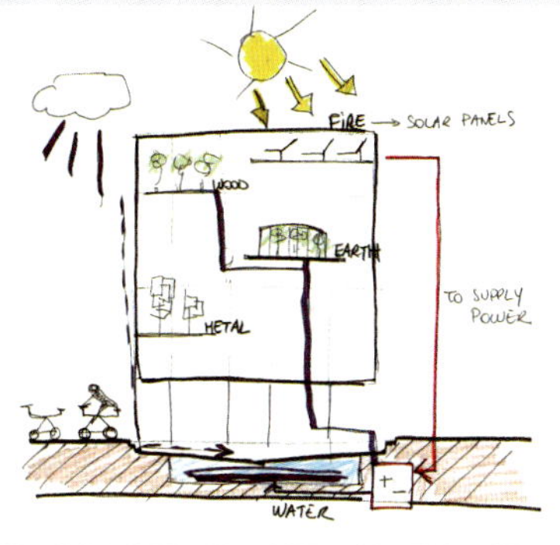

昆山煤气厂棕地改造项目竞标

业主信息：昆山城市建设投资发展有限公司
用地面积：14 738平方米
建筑面积：4 198平方米
建设地点：上海
设计时间：2010年
设计功能：棕地项目改造、生态公园、商业配套

基地原址为煤气工厂，由于它的迁走给这个城市和城市人口带来新的景象。

本项目包括了保留过去的印象和促使人们对于自然更多的关注。

项目既是商业也是公园，我们想在这两者之间建立一个“健康”的合作关系。

我们沿用了圆形的工厂基地来创造成为一个广场。目标是将人们集合到这样一个温暖的地方。人们可以在这里感受到浓浓绿意，可持续发展的生态科技，并且可以充分享受自由购物的快乐。

工厂基地的广场中心表现了自然的力量。三个入口实现了通行，北边主要服务于住宅区，南边面对主路成为主要的商业入口，西北边则是一个较为私密的入口。

为了创造出一个自然公园，圆形的基址将被商店，有机的绿色覆盖层包围着。这个公园也包括了各种各样的植物与咖啡厅。

煤气罐本身是竖向的自然，运用了中国金、木、水、火、土“五行”的概念。

意义上是运用自然物质自身动态的性能，隐喻着一些自然现象。

第一层代表了“水”。作为底层，可以收集雨水、电池和植物净化。

第二层代表了“土”，由温室组成，包含了一些农业植物，比如空心菜和泰国罗勒。

第三层的高度为13.80米，是一个太阳能板的平台，共放置了6块太阳能板。这一层代表了“火”。

Kunshan Brown Land Project Competition

Owners Information: Kunshan City Construction Investment Development Co., Ltd.
Land Area: 14,738 m^2
Building Area: 4,198 m^2
Building Location: Shanghai
Design Time: 2010
Design Features: Brown Land Reform Projects, Ecological Parks, Commercial Facilities

KUNSHAN PROJECT-OPTION 1: “REBIRTH”

The project consists in keeping the image of the past and motivate the people to be aware of the nature.

The program is both commercial and park so we thought about making “pleasant” collaboration between them.

The site was a gas factory, and we respected the original round shape in order to create a plaza. The goal is to gather people in a warm place to enjoy the green feeling, sustainable techniques and the shopping in an open-minded environment.

The park contains a great diversity of plants and coffee places.

The center of the plaza in place of the factory is a cylinder of vertical nature, implementing the Chinese concept "wuxing" about the 5 senses: wood, fire, earth, metal, water.

The meaning was to use the natural substance for their dynamic property, in this way the cylinder allows to met aphorize some natural phenomenon.

The ground floor represents “WATER”. The basement receives the rainwater, the battery and purifying plants.

The first floor is the image of the “METAL” for exhibition of sculptures made with recycle materials found on the site.

The second floor is a green house containing some farming as water spinach, Thai basil. It is represents the sense “EARTH”.

The third floor at 13.80m is the platform of solar energy. Indeed 6 solar panels are established. This floor is the “FIRE” representative.

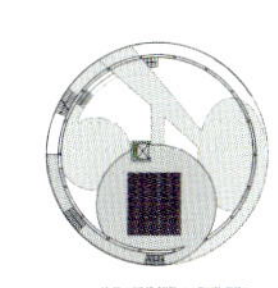

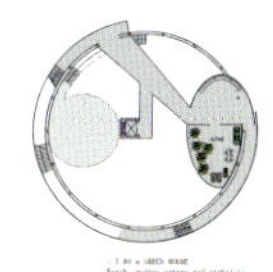

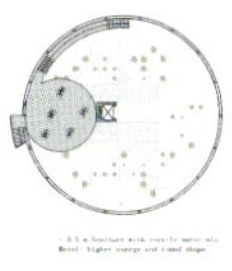

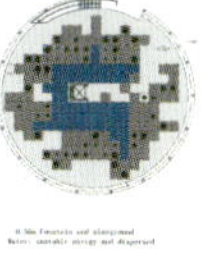

景观园林 Landscape

石林生态民族运动场撒尼生态运动员公寓项目景观设计

业　　主：云南圆通房地产开发有限公司
设计内容：景观设计
建设地点：云南 石林
用　　途：运动员公寓
用地面积：91 037平方米
地上建筑面积：21 849平方米
容 积 率：0.24
设计日期：2010年

Landscape Design for Competitor Apartment in Ecological and Ethnological Stadium, Shilin

Owner: Yunnan Yuantong Real Estate Development Co., Ltd.
Design Content: Landscape Design
Building Location: Shilin, Yunnan
Uses: Athletes Apartments
Land Area: 91,037 m^2
Ground Floor Area: 21,849 m^2
Plot Ratio: 0.24
Design Time: 2010

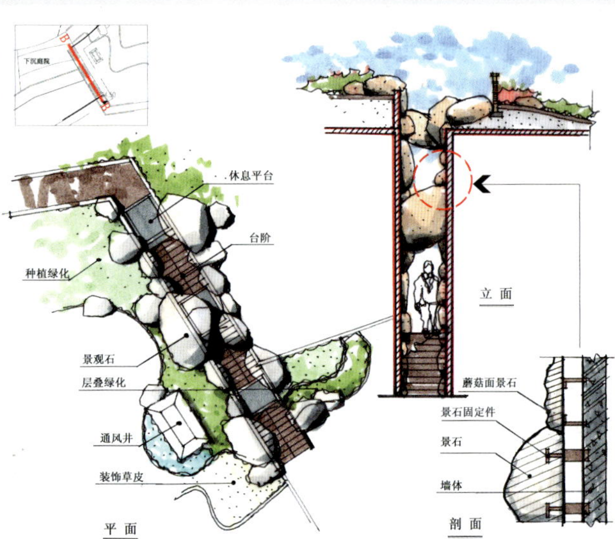

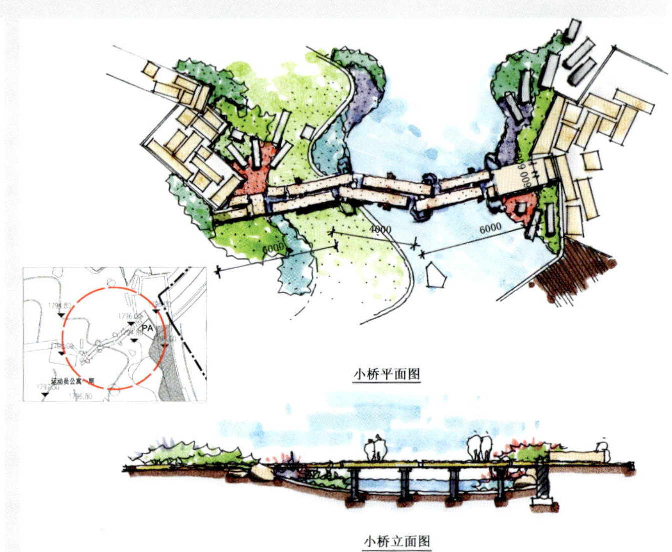

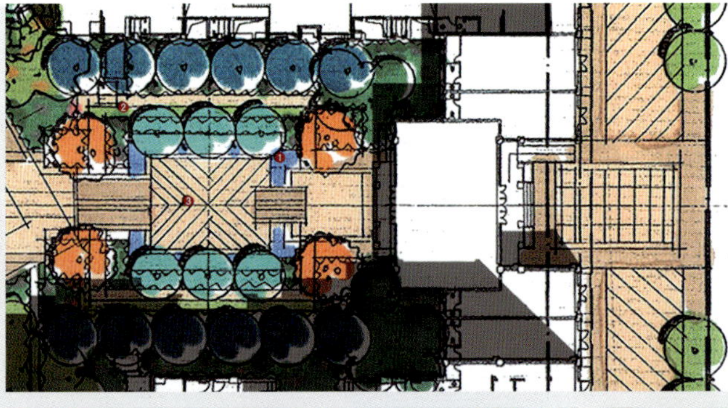

龙城项目TDJYX–2009–21号地块景观设计

业　　主：洛阳国龙置业有限公司
设计内容：景观设计
建设地点：河南 洛阳
用　　途：高层住宅
用地面积：60 473平方米
地上建筑面积：3 023平方米
容 积 率：5.00
设计日期：2010年

Landscape Design for TDJYX-2009-21 District of Longcheng Project

Owners: Luoyang Guolong Properties Company Limited
Design Content: Landscape Design
Project Location: Luoyang, Henan
Uses: High-rise Residential
Land Area: 60,473 m^2
Ground Floor Area: 3,023 m^2
Plot Ratio: 5.00
Design Time: 2010

2010—2011
主要建筑设计作品

Main Architectural Design Works

Annual Review of Chinese Architectural Design Works

CPP International Architects Ltd.

瑞士坎培建筑设计顾问公司

瑞士坎培建筑设计顾问公司由Mario Campi教授1962年创立于瑞士卢加诺，并设苏黎世、瑞典及中国分部。公司作为在现代建筑史上具有重要地位的"Ticino"学派的典型代表为世人展示了大量融传统建筑文化、地域环境特征与现代建筑风格于一体的建筑作品，对现代建筑运动产生了深远的影响。近半个世纪以来，公司在瑞士及欧洲境内完成了大量的公共及居住建筑，是瑞士一流的规划建筑设计公司之一。

CPP坎培建筑设计顾问公司从2004年开始进入中国市场，已完成多项优秀设计工程，业务范围涉及城市规划、办公研发、高档住宅建筑以及景观设计。作品多次获得奖项，其中包括联合国环境署颁发的第二届亚洲人居环境规划设计奖。凭借优秀的设计团队，我们愿与客户分享我们的专业经验和良好的敬业精神，为广大的中国客户提供最优质的服务。

设计理念

1. 简洁、理性和有效的原则——注重建筑语汇的清晰和设计逻辑的严谨。
2. 城市设计是建筑设计的初始点——强调整体环境的价值提升。
3. 主动式绿色建筑体系——绿色建筑不应该仅仅被理解成为技术手段，更应该是一个设计概念通过设计自身创造出可持续发展的建筑和环境。

电话：+86-25-83699511
传真：+86-25-83699512
邮箱：cpp_nanjing@126.com

Tel: +86-25-83699511
Fax: +86-25-83699512
E-mail: cpp_nanjing@126.com

CPP International Architects Ltd. was founded by Prof. Mario Campi in Lugano, Switzerland in 1962, with branches in Zürich (Switzerland), Sweden and China. As typical of "Ticino" school which plays an important role in modern history of architecture, CPP exhibits a huge number of building works that integrate architectural culture, geological environmental features and modern building styles. Since nearly half a century ago, CPP has been a first-class architectural planning & design company in Switzerland, having completed a lot of public and residential buildings within Switzerland and Europe.

CPP enters into Chinese market since 2004, having completed a number of excellent design projects, covering municipal planning, office R&D, hi-end residential building and landscaping. Its works have won many honors, including the 2nd Asia Habitat Environmental Planning & Design Award granted by the UNEP. With CPP excellent design team, we are ready to share our expertise and professionalism with customers, and offer the best quality service to Chinese customers.

CPP Design Ideas:

1. "Concise, Rational and Effective" principle – clear terminology of architecture and prudent logic of design.
2. Urban design is the inception point of architectural design – highlighting the value promotion of entire environment.
3. Active green building system – green building is not merely a technology, but also a conception, which produces sustainable buildings and environment by the design itself.

南京汉西110kV变电站综合体
建筑面积：7 000平方米　　项目状态：建成　　设计时间：2009年

Hanxi 110kV Transformer Substation, Nanjing
Floor Area: 7,000 m^2　Status: Completed　Design Time: 2009

南京鼓楼国际外包服务大楼立面改造及景观设计
建筑面积：6万平方米　　项目状态：建成　　设计时间：2010年

Façade Reconstruction & Landscaping of International Outsourcing Service Building, Gulou District, Nanjing
Floor Area: 60,000 m^2　Status: Completed　Design Time: 2010

南京工大科技园侯村园区
建筑面积：10万平方米　　项目状态：委托设计　　建成时间：2011年

Houcun Area of NJUT Technical Park, Nanjing
Floor Area: 100,000 m^2　Status: Design　Completion Time: 2011

南大苏福特软件公司研发楼
建筑面积：8万平方米　　项目状态：建成　　设计时间：2008年

NJUSOFT R&D Building, Nanjing
Floor Area: 80,000 m^2　Status: Completed　Design Time: 2008

江苏新城地产南京浦口商业项目
建筑面积：4万平方米　　项目状态：委托设计　　设计时间：2011年

Jiangsu Future Land Pukou Commercial Project, Nanjing
Floor Area: 40,000 m^2　Status: Design　Design Time: 2011

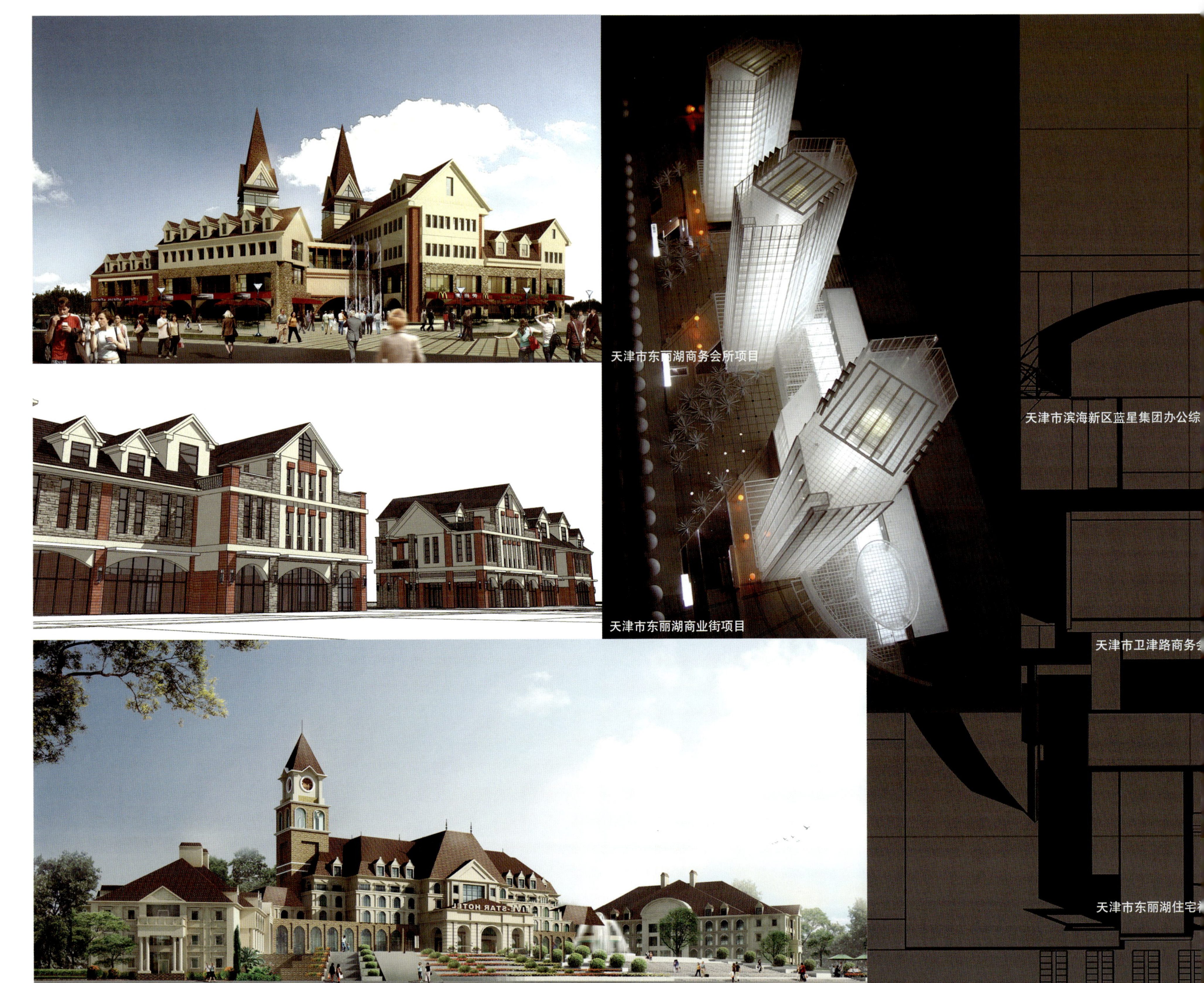

天津市东丽湖商务会所项目
天津市滨海新区蓝星集团办公综
天津市东丽湖商业街项目
天津市卫津路商务会
天津市东丽湖住宅
天津市某五星级酒店
OARCH
欢 迎 设 计 精 英 加 盟
天津市河西区卫津路 电话: +86-22-23358796
新金龙大厦南楼10层 传真: +86-22-23358796
邮编: 300060 邮箱: oarch@126.com 网址: www.oarch.cn

张家港软件园白领公寓

张家港软件园综合体

美国JY建筑规划设计事务所
锦杰建筑规划设计咨询(上海)有限公司

美国JY建筑规划设计事务所，锦杰建筑规划设计咨询（上海）有限公司于1983年创立于美国洛杉矶，提供建筑规划设计及相关行业的专业咨询服务，创立人俞锦杰(Yu Jinjie)先生。在1983年至1993年间，JY与多名美国著名的建筑师如Jackson K.Walters、Margaret Courtney、Phil Stivers等合作，作品遍布南加州，包括规划、公建和高级住宅区设计。

美国JY建筑规划设计事务所1992年正式进入中国市场，并成立了上海代表处，致力于规划、建筑、景观等多领域，凭借着丰富的国际经验和高水准的专业素养，在大型规划设计、高标准办公楼和高端住宅设计方面表现突出，已建成大量代表作品，在行业内积累了较高的知名度和美誉度。

美国JY建筑规划设计事务，锦杰建筑规划设计咨询（上海）有限公司拥有专业建筑师、规划师、设计师40余人。专业团队包括规划、建筑设计、景观设计的专家，形成了国际、国内多专家合作共同完成项目的设计风格。这种风格保证了项目整体设计方案的完美性。JY事务所以专业化的操作模式对设计流程进行高效率的科学管理，严格控制每一设计阶段的设计质量，充分发挥团队合作的力量和优势，务必使每个作品能体现事务所的整体设计水平。

建筑是百年大计，每一项建筑对城市环境与社会人文将产生长远的影响力。建筑设计是一种专业的服务，JY公司把"精益求精，精心出精品"作为公司一贯的专业宗旨，这也是职业设计师的一种最基本的职业精神。"认真、敬业、诚信、创新"这八个字就是JY事务所坚守的专业信念和设计理念。

JY事务所总裁俞锦杰先生有着多年在美国同美国房产开发商及政府合作的经验，参与了大量土地开发早期的策划工作，因此JY事务所在项目操作上一贯保持强烈的市场意识和策划型设计的优势，善于站在发展商的角度，和市场营销策划部门一起以多元的思维和互通的语言进行相互沟通协调，不但在设计理念方面，而且在开发顺序、产品定位等方面全方位地为开发商提供富有价值和建设性的顾问服务。

2004–2005年度JY事务所被设计界颇为权威的《设计新潮》杂志评为民用建筑设计市场全国排名第51名，民用建筑设计企业分类业务全国排名第30名，景观设计全国排名第23名，民用建筑设计企业效率全国排名第9名。

JY Design Planning, Inc. was established in 1983 in Los Angeles, California. JY has engaged in architecture, planning and design consultancy services. From 1983 to 1993, JY focused on planning, public works and high end residential building types in Southern California, in cooperation with well known architects such as Jackson K. Walters, Margaret Courtney, Phile Stivers.

In 1992 JY established a Shanghai Representative Office. The firm focused on planning, architectural and landscape design. Due to its experience, JY has completed grade A and high end residential design projects. The firm has developed an extensive list of work and is well regarded in the professional field in the region.

JY employs over 40 professional architects, planners and designers. The team is composed of foreign and domestic experts, ensuring a well balanced design, taking advantage of the experience abroad as well as a deep knowledge of the Chinese market. Every project is unique and it's regarded as high quality delivery for the firm.

We strive to provide great architecture and produce good solutions for our clients' investments. As a result our products influence positively the urban environment and social conditions of the end users. The company's motto is: Striving for perfection, creation of high quality products and timeless design.

President Mr. Jinjie Yu of JY has a vast experience and knowledge with the American real estate development companies, procurement and land development. Due to this experience, he has been able to apply his knowledge to the Chinese market and dynamics.

2004-2005, JY ranked 51st in the Civil Construction Design Market, 30th in the Civil Construction Design Enterprise Classifying Business, 23rd in the Landscape Design and 9th in Civil Construction Design Enterprise Efficiency in China by the authoritative *Design Trends* Magazine.

地址：上海市打浦路88号海丽大厦25楼
电话：+86-21-53011100
传真：+86-21-53010105
邮箱：info@jy-design.com.cn
网址：www.jy-design.com.cn

Add: F25, Haili Building, No.88 Dapu Road, Shanghai
Tel: +86-21-53011100
Fax: +86-21-53010105
E-mail: info@jy-design.com.cn
http:// www.jy-design.com.cn

上海交响乐团

建设地点：上海　　基地总面积：16 300平方米
楼　　层：3层　　建筑面积：16 550平方米

Shanghai Symphony Orchestral

Project Location: Shanghai　　Total Area of Base: 16,300 m^2
Floors: 3　　Floor Area: 16,550 m^2

青岛某高端项目

建设地点：山东 青岛　　占地面积：28 800平方米
楼　　层：3—7层　　建筑面积：133 897平方米

A High-end Project, Qingdao

Project Location: Qingdao, Shandong　　Land Area: 28,800 m^2
Floors: 3-7　　Floor Area: 133,897 m^2

东郊半岛花园（原机场镇项目）

建设地点：上海
占地面积：78 816平方米
建筑面积：45 200平方米
楼　　层：2–4层
竣工时间：2011年

East Suburb Peninsula Garden (Project of Airport Town)

Project Location: Shanghai
Land Area: 78,816 m^2
Floor Area: 45,200 m^2
Floors: 2-4
Completion Time: 2011

长沙东湖壹号

建设地点：湖南 长沙
占地面积：224 617平方米
建筑面积：283 536平方米
楼　　层：2–33层
设计时间：2008年
竣工时间：2011年

East Lake No.1, Changsha

Project Location: Changsha, Hunan
Land Area: 224,617 m^2
Floor Area: 283,536 m^2
Floors: 2-33
Design Time: 2008
Completion Time: 2011

东海御庭

建设地点：上海
占地面积：212 150平方米
建筑面积：98 101平方米
楼　　层：2-3层
设计时间：2008年
竣工时间：2011年

East Sea Yuting

Project Location: Shanghai
Land Area: 212,150 m^2
Floor Area: 98,101 m^2
Floors: 2-3
Design Time: 2008
Completion Time: 2011

长沙巨隆超高层项目

建设地点：湖南 长沙
用地面积：39 150平方米
建筑面积：271 770平方米
楼　　层：68层
设计时间：2011年

Julong Ultra High-rise Project, Changsha

Project Location: Changsha, Hunan
Land Area: 39,150 m^2
Floor Area: 271,770 m^2
Floors: 68
Design Time: 2011

方案一

方案二

大连运达科技城

建设地点：辽宁 大连
用地面积：23 399平方米
建筑面积：203 500平方米
楼　　层：30–39层

Yunda Sci-Tech Town, Dalian

Project Location: Dalian, Liaoning
Land Area: 23,399 m^2
Floor Area: 203,500 m^2
Floors: 30-39

春华街老城区改造 Reconstruction of Chunhua Street Old Town Areas

项目位于大连市中山区，总基地面积为83 514平方米，其中回搬区37 710平方米，商住区45 804平方米，商住区北临春华街，西南侧靠炮台山，东接城市规划道路，两侧环山；回搬区西临规划道路，南侧紧邻炮台山，与商住区仅隔一条规划路，整体地势南高北低，离东港湾近2.5千米，整体空间环境非常优越，空气清新，安静优雅，不良干扰因素较少，非常适宜人居。商住区地上规划建筑面积104 300平方米，含6 000平方米商业配套设施等；回搬区103 400平方米，包括一所幼儿园。

基地北面能俯瞰美丽的大连湾，商住区根据地形大致分成3个不同标高的层次，布置有12幢7–31层的高层住宅，北部春华街南侧布置了2–3层楼的商用，前面留有开阔的购物广场，方便了周边及小区内部的居民，也是居民休息活动的理想场所；回搬区基地被港院街分为东西两个部分，西边布置了8幢24–33层的高层住宅，东侧依地形布置了4幢30–33层的高层住宅楼，住宅楼中间是幼儿园。

上海松江洋江花园 Yangjiang Garden, Songjiang, Shanghai

本项目位于松江区九亭镇九新公路五号地块。地块西临九新公路，道路红线24米，东临友谊河，北临松江医院九亭分院，南邻村民别墅。用地呈梯形，北边界长159米，南边界长145米，南北向长178米。基地平整，友谊河水质较好，河边绿化较好，未来会有很大的改造空间。九新公路对面为云润家园，多层住宅，临街有两层商用，建筑为新古典风格。

基地内地势平坦，绝对标高在3.85米至4.2米之间。

规划总用地面积26 092平方米，总建筑面积59 400平方米，其中地上建筑面积46 966平方米，容积率1.8。拟建4栋21－22层高层住宅，配以精致、优美宜人的景观，为本地高档豪华住宅小区。

成都国光项目 Guoguang Project, Chengdu

建设地点：四川 成都
占地面积：29 317平方米
建筑面积：12 323平方米
楼　　层：2—3层
设计时间：2011年

Project Location: Chengdu, Sichuan
Land Area: 29,317 m^2
Floor Area: 12,323 m^2
Floors: 2-3
Design Time: 2011

LOOK architects

Architecture & Urban Design

LOOK Architects Pte Ltd
LOOK 建筑设计公司
18 Boon Lay Way
#09-135 TradeHub 21
Singapore 609966
Tel : + 65 6316 8676
Fax : + 65 6316 8690
Email : office@lookarchitects.com.sg

LOOK建筑设计公司成立于1993年，公司重视设计实践，以严格的分析和研究完成新颖的标志性建筑和城市设计。

公司创始人骆文义和黄素香以加强公司的合作环境为出发点，不断地萌发全新的设计理念，在东南亚地区形成一个能全面、敏锐地开拓新领域的设计公司。作为2009年新加坡年度总统设计奖的获得者，骆文义以他对设计的热情，引领着公司迈向更加令人振奋的明天。

公司的多元化极大地体现在完成的多种不同类型的项目上，包括私人住宅、公寓大楼、大学机构、社区图书馆、商业总部、濒水步道和桥梁，以综合的设计理念和创新意识来强化建筑细部、利用生态材料以及空间和环境的关系进行合理的规划设计。

客户个人和集体的经验，创造性地被阐释为设计的表现形式和观点，以丰富每个项目的设计理念，并连同对地方特色的可持续性，开拓出一个人与环境和谐共生、生活气息鲜明的生态设计领域。

Founded in 1993, LOOK Architects is a design-intensive practice committed to rigorous analysis and research to produce innovative, iconic buildings and urban design.

Founders Look Boon Gee and Ng Sor Hiang foster a collaborative studio environment to germinate and develop design ideas on a broad spectrum of works in the Southeast Asian region, defining LOOK Architects as a practice keen on breaking new ground in various fields of design. Recipient of The President's Design Award – Designer of the Year 2009, Look Boon Gee's passion for his craft continues to lead the practice towards exciting directions.

Versatility in handling projects of vastly different natures is evident in completed works, an array that includes private residences, apartment tower, university institution, community library, commercial headquarters, waterfront promenade and bridges. Embracing a design philosophy of the integrated whole, a high degree of inventiveness goes into construction detailing, ecological use of building material and planning of spatial relationships in relation to the environment.

Clients' personal and collective experiences are creatively interpreted to form perspectives that enrich the design approach of each project, and together with sustainable responses to the distinctive character of place, design enters the realm of eco-poetry, a lively dialogue that reverberates between nature and aspirations of man.

PRESIDENT*S DESIGN AWARD SINGAPORE
DESIGNER OF THE YEAR 2009

www.lookarchitects.com

新加坡科技设计大学

LOOK建筑设计公司与盛邦新业集团携手合作，赢得了新加坡科技设计大学(SUTD)B区——学生宿舍与体育场的建筑设计方案。该设计方案从中国国画中的"留白"艺术中获得灵感，开发了一系列穿梭在工作、学习与生活环境之间的重重空间，以创造出独一无二的学习环境。为了体现出"简约中的深奥"的设计理念，没有使用过多的修饰，而只采取设计精髓。该方案具有很强的灵活性，单元的排列模式可以使建筑适应未来的远景规划。方案致力于激发青少年的活力与热情，为每一位大学生创造了实现梦想的平台。

占地面积：52 662 平方米
建筑面积：46 524 平方米
地　　点：新加坡

SINGAPORE UNIVERSITY OF TECHNOLOGY AND DESIGN PLOT B—SPORTS & HOUSING

LOOK Architects as Principal Designer, in collaboration with Surbana International Consultants has won the design competition for the Plot B—Student Housing and Sports Complex, SUTD (Singapore University of Technology and Design) Campus. To foster a truly unique learning experience, we have developed a design that embraces creativity, possibilities and fraternity among the students and faculty members by weaving the collaborative spaces within the living environment. Ascribing to the design philosophy of sophistication in simplicity, we have pared down excesses to distill the essential building blocks for a system based design, which has the flexibility to adopt various permutations and adapt to future changes. Ultimately, the design aims to celebrate the youthful spirit & enthusiasm of students and provide a platform for each and every aspiring mind to dream and excel at SUTD.

Site Area: 52,662 m^2
Gross Floor Area: 46,524 m^2
Location: Singapore

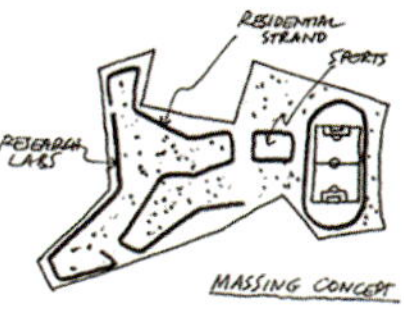

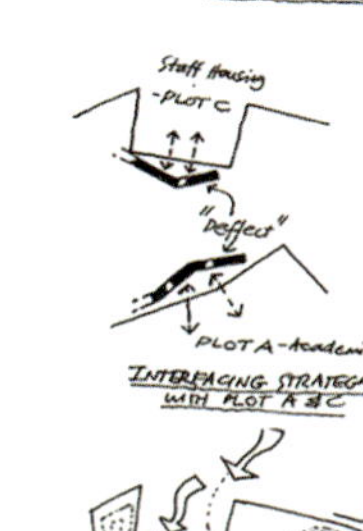

西蒙港公寓

在设计初期，我们萌发了通过设计较小的楼面和高度通风的理念，使建筑朝向主要的季风方向，创造了季风走廊。同时，南北朝向也有效地避免了太阳直射，从而降低了辐射和空调的使用频率。被动设计是长期可持续发展的重要手段。在建筑高度的限制下，我们非常严谨地考虑并设计了独特的屋顶格架，可以有效地减低屋顶接受的热能，减少电负荷，创造更舒适的生活环境。

占地面积：9 110平方米
建筑面积：12 754平方米
地　　点：新加坡

SIMON LANE CONDOMINIUM

The idea of creating a highly porous slice of urban fabric populated by small scale blocks allows for "ventilation breezeways" between the individual blocks. All of the housing blocks are also oriented in the north-south direction to minimise solar gain and reduce air-conditioning cooling loads. Emphasis is placed on passive modes of climatic control as a long-term sustainability strategy. For instance, height control imposed by planning parameters on the site has been carefully considered and interpreted a unique high-level shading element is introduced to act as an additional thermal buffer to the roof.

Site Area: 9,110 m^2
Gross Floor Area: 12,754 m^2
Location: Singapore

3 BALMORAL 项目

该项目是一座位于乌节路购物区附近的12层高的公寓住宅楼，配有半地下室的停车场。针对不断增长的城市人口，设计师的主要目标是探索以建筑材料来“解材”建筑体的方式——在对建筑立面悬臂式阳台护栏的处理上，使用了半透明的穿孔铝板。从顶楼复式单位的天台花园可以鸟瞰良木山及附近乌节路一带夜晚繁华的景象。为了美化街景，设计了绿化斜坡，以取代传统的实心围墙，最终使建筑与周围环境和谐共生。

占地面积：2 283 平方米
建筑面积：3 652 平方米
地　　点：新加坡

3 BALMORAL

A 12-storey apartment block with semi-basement carpark, in a residential enclave on the outskirts of the Orchard shopping belt. In response to the condition of ever-increasing density in the urban environment, the building mass is de-materialized by translucent perforated aluminium screen panels swathing cantilevered balconies on the facade. Rooftop gardens of duplex penthouse units offer panoramic views of nearby Goodwood Hill and the vibrant night lighting along Orchard Road. To enhance the public streetscape, sloping greenery instead of solid walls define the site boundary, creating an amiable, permeable interface between the private development and its surroundings.

Site Area: 2,283 m^2
Gross Floor Area: 3,652 m^2
Location: Singapore

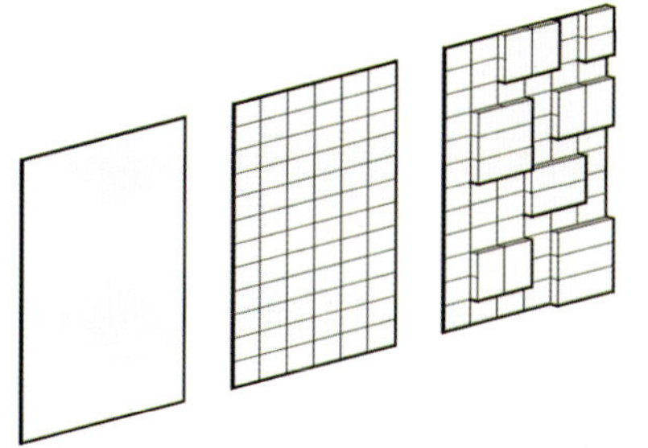

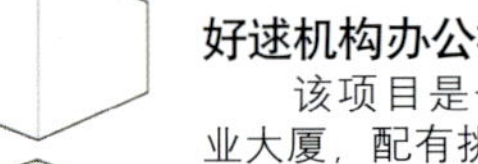

好逑机构办公楼

该项目是一座七层楼的多用户工业大厦，配有挑高两层的停车场。建筑形式的表达，例如一个双层的开放式阳台，创造出宜人的绿色空间。精心的外观设计呼应了内部空间的使用要求，传达一种独特的整体设计形象。

占地面积：2 092 平方米
建筑面积：5 230 平方米
地　　点：新加坡

HOR KEW CORPORATE OFFICE

7-storey multiple-user building with elevated carpark spanning 2 storeys. Articulation of the building mass creates welcome green spaces for tenants, such as that of a large double - volume open terrace. The facade is carefully composed to complement programmatic demands and convey a distinctive overall image of the development.

Site Area: 2,092 m^2
Gross Floor Area: 5,230 m^2
Location: Singapore

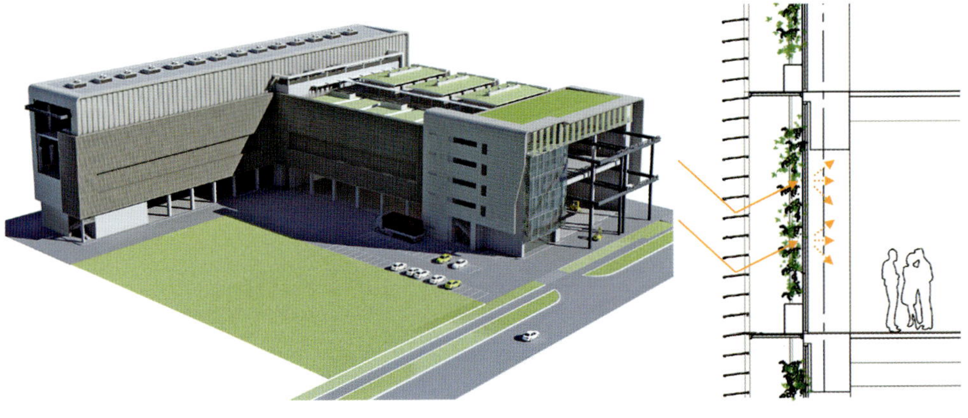

长成预制科技中心

该项目为位于大士南部的多层预制科技中心，提供预制混凝土组件。中心还设有办事处和员工宿舍等设施。为了体现国家最先进的自动化预制技术，此工业建筑采用了绿色环保立面，显得更加与众不同。垂直绿化带、穿孔铝板和预制混凝土的使用，最大限度地减少了建筑的热吸收。

占地面积：19 606 平方米
建筑面积：13 970 平方米
地　　点：新加坡

TIONGSENG PREFAB HUB

Multi-storey concrete precasting factory with offices and workers' dormitory facilities at Tuas South. Echoing state-of-the-art automated precasting technology, the industrial building distinguishes itself with an impressive green facade more than 100 metres in length—a composition of vertical green panels, perforated aluminium screens and precast concrete modules that work collectively to minimise heat gain in the building.

Site Area: 19,606 m^2
Gross Floor Area: 13,970 m^2
Location: Singapore

Boston International Design Group

美国波士顿国际设计集团

波士顿国际设计集团成立于2004年，延续了美国史塔宾建筑事务所国际最高水准的设计声誉，总部位于美国马萨诸塞州剑桥市，毗邻著名的哈佛大学和麻省理工学院。公司的主要创始人及主要的设计人员均来自两所名校，因此波士顿国际设计集团有着浓郁的学院气氛。

波士顿国际设计集团一直致力于把项目的规划设计与业主的要求最大限度地保持一致。我们的主要目的是和业主、合作者分享高标准的专业设计和学术成果。我们并不孤立地看重为自己的作品创建特殊的美学风格，我们的设计旨在以平等的合作关系、积极向上的工作来为业主提供杰出的设计，并创造最终的经济效益。我们在世界各地的项目设计已经超越了地理和文化的界限，以价值和兴趣为基础，其核心包括：很大程度上尊重当地的历史和文化。通过规划和设计，对增进当地城市生态环境作出实质的贡献；通过合作，业主将获得建立在他们的需求和期望上的卓越的设计。这使得我们能够与业主建立长期的合作，我们甚至为有的业主服务了30多年。每当接受一项设计委托后，我们就会和业主一起反复研究、磨合，以确保每一个新建筑都能够成为一座独特的、难忘的成功作品。

我们的创造力和充分理解业主的全部意图是密不可分的。场地的条件和业主及社会的利益是两个同等重要的因素。新的场地条件和不同的业主要求使得每个项目都是非常特殊的。我们从来不局限于某一种特定的模式，而是结合分析场地、文脉、经济、社会机遇等一系列因素后，根据项目的实际情况来找寻最合适的解决方案。因此，我们的项目总是在超越、在创新。

Boston International Design Group was set up in 2004. It continued the highest international standard design reputation of the Stubbins Associates of American, and headquartered in Massachusetts Avenue Cambridge, MA, USA, where adjacent to the famous Harvard University and Massachusetts Institute of Technology. The main founders and designers of the company are all from these two schools, that's why BIDG is full of collegial atmosphere.

BIDG has been committed to matching project planning and designing with requirements of customers as much as it could. Our main purpose is to share high standards of professional design and academic achievements with customers and partners. We do not focus separately on creating unique aesthetic style for our works, but aim to provide outstanding designs with equal partnership, positive work to our customers and make the ultimate economic benefits. Our design projects around the world have gone beyond the geographic and cultural boundaries, and are based on values and interest, which mainly include respecting local history and culture to a large extent. We make substantial contribution to local urban environment through planning and designing; customers will get excellent design based on their needs and expectations through our cooperation. This allows us to establish long-term cooperation with the customers and even over 30 years for some customers. Every time we accept a design commission, we discuss and adjust with customers over and over again to ensure that every new building can be a unique and unforgettable one.

Our creativity cannot be separated with fully understanding of customers' whole intention. Site condition and interests of customers and the society are two important factors, and new site condition and different requirements from customers make every project very special. We are never limited to a specific model of certain type, but find the most appropriate solution with combined analysis of a series of factors like context, economy, social opportunities, etc. Therefore, we are always exceeding ourselves and innovating in our projects.

地址：上海市浦东新区福山路33号建工大厦15楼
电话：+86-21-51327266
传真：+86-21-51327269
网址：www.bidg.com.cn

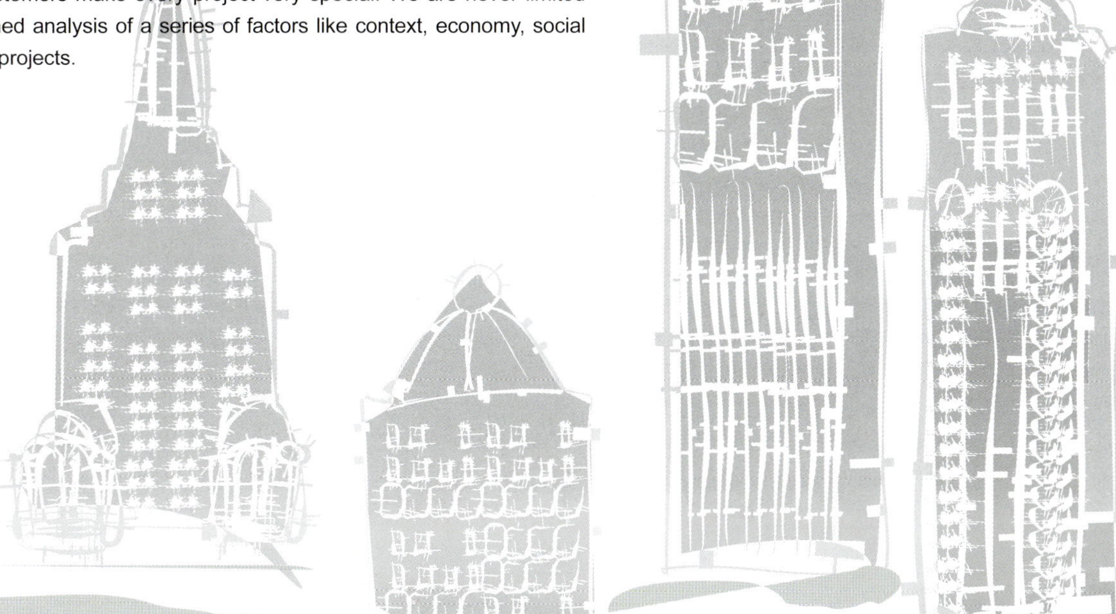

北京海淀西山文化大道

建设地点：北京
建筑面积：298.2万平方米
占地面积：734.5公顷
设计时间：2010年
竣工时间：待定

基地位于海淀区四季青地区，自紫竹院路西段延伸至杏石口路沿线一带，北京西四环与五环之间。规划用地面积约734公顷，建筑面积约298.2万平方米。为了将这个项目建成为一个高起点、高水平的文化产业聚集区，要考虑文化的内涵和品质，以及社会效益和经济效益。

杭州银泰海·威国际喜来登酒店

建设地点：浙江 杭州
建筑面积：166 060.6平方米
占地面积：23 143平方米
设计时间：2010年
竣工时间：待定

基地位于杭州市钱塘江南岸，三桥与四桥之间，是一个集酒店、办公、住宅于一体的综合体。规划布局充分利用临江的优势，使江景房占据四分之三的比例，呈现高品质办公空间、全江景视觉盛宴。办公区独享尊贵空中接待大厅，160米的高度临江俯瞰杭城全景。酒店的立面造型简洁挺拔，过目难忘，双层呼吸式玻璃幕墙树立了杭州节能建筑的新标杆。

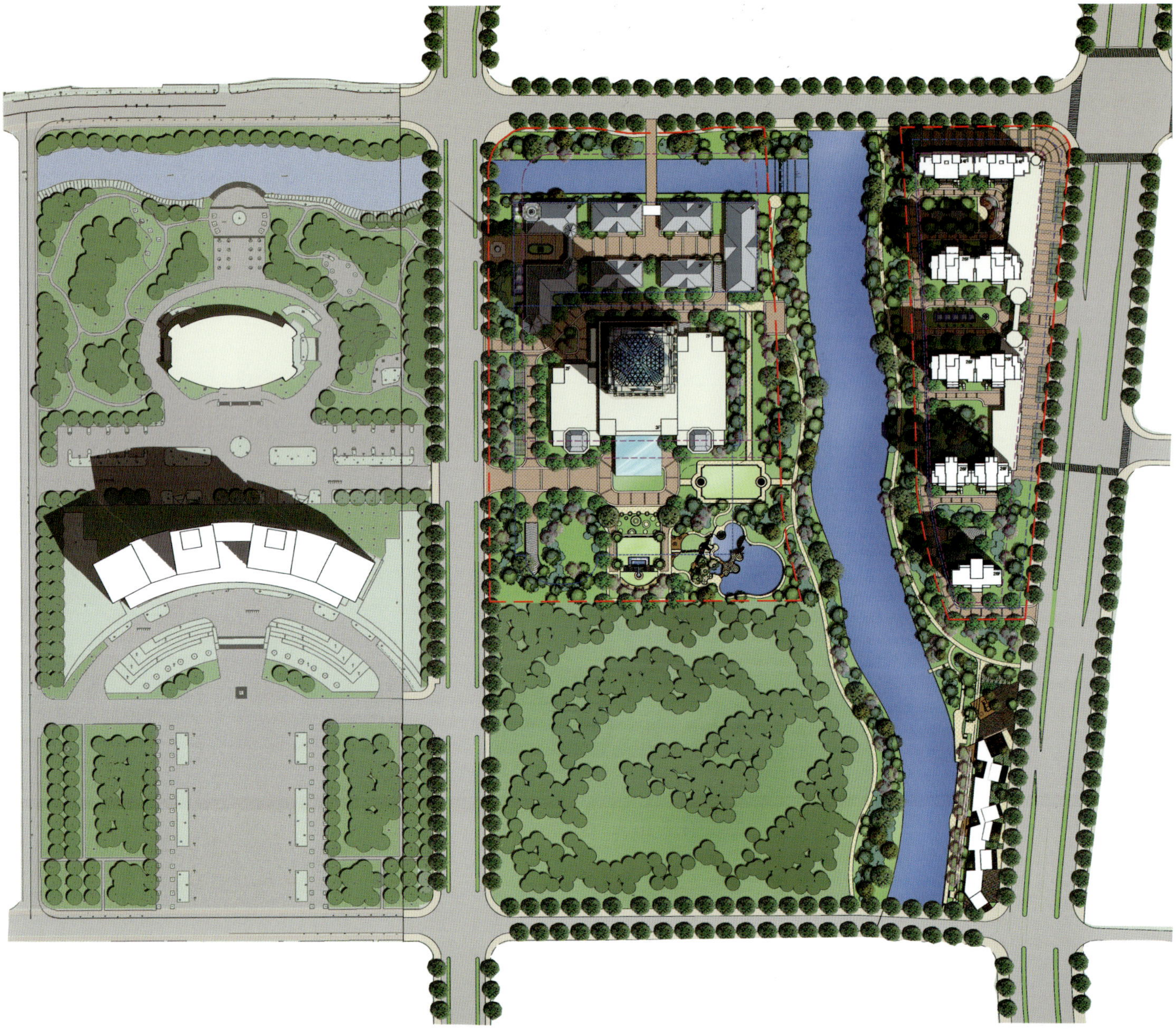

湖州长兴万豪酒店

建设地点：浙江 湖州
建筑面积：185 807.8平方米
占地面积：16 935.9平方米
设计时间：2010年
竣工时间：待定

湖州长兴万豪酒店项目位于长兴县龙山新区，长兴县行政中心东侧，南邻长兴大剧院，是长兴县重点发展的核心区域。交通便利，地理位置优越。以湖州飞英塔“七层八面”为原型，根据建筑功能需求和严谨的立面比例关系，建筑立面层层拔高，逐渐收分，整体形象高峻挺拔，极具特色。精心设计的酒店屋顶细部，来自钻石切割工艺的灵感，使酒店具有强烈的可识别性，无论白天还是黑夜，都将成为长兴人民瞩目的焦点。

三亚鹿回头滨海商业小镇项目设计

建设地点：海南 三亚
建筑面积：80 958.18平方米
占地面积：42 994.20平方米
设计时间：2011年
竣工时间：待定

该项目位于海南省三亚市半山半岛。三亚半山半岛，位于三亚小东海鹿回头半岛，整个半岛由鹿回头公园、鹿回头岭两山和小东海、鹿回头湾两湾组成，是三亚市西至海坡、东至亚龙湾的50余千米海岸线中唯一一个待开发的半岛，有着优质的沙滩和背山临海的绝顶自然资源。

重庆观音桥商圈规划

建设地点：重庆
占地面积：规划总用地面积42.42公顷
设计时间：2010年
竣工时间：待定

观音桥地区位于重庆市地理核心，是重庆最有活力的商圈，将规划成长江上游地区"购物之都"的购物时尚中心，成为西部最具影响力的消费乐园。

在本次概念规划中，我们充分研究了现状条件，对比国际著名的商业街区，整合区域功能，以街为轴强调连续的商业氛围，利用不同的街道属性形成不同的商业体验，并结合重庆山城特色进行立体式开发，形成地上、地下立体的商业网络。尤其突出中央绿地，打造西部独一无二的"大公园生态商圈"。勾画魅力城市天际线，全力提升中心城区城市形象。

重庆湖广会馆历史街区保护规划

建设地点：重庆
建筑面积：165 309平方米
占地面积：65 539平方米
设计时间：2010年
竣工时间：待定

湖广会馆历史街区位于渝中区中心，东临长滨路，西临解放东路东段陕西路，长江索道位于基地南侧，北侧为东正街，路网密集，道路交通发达。能通过城市快速路快捷地通向江北、南岸、渝北等其他行政区，并方便地通往外围城市区。

ÉTÉ 翌德国际设计机构

été lee et associés architectes urbanistes

■ 法国翌德国际设计机构 ■ 上海翌德建筑规划设计有限公司

来自法国的翌德国际设计机构（ÉTÉ Lee et Associés architectes, urbanistes）是城市规划、建筑与景观设计领域的实践先锋与思想先行者，该机构的项目遍布法国、西班牙、美国、韩国等国家。近十年来，在中国的50余座大中城市主持完成了300余项设计实践。其中，上海2010年世博水门、上海市南外滩地区、上海市张家浜楔形绿地、上海芦潮港海滨国际花城、上海市外高桥商务别墅区，以及应用绿色建筑技术的北京鼎嘉恒苑、上海市金地·格林郡、成都市华新锦绣尚郡、宁波盛世天城等项目的建成和投入使用，充分体现了翌德国际设计机构（ÉTÉ）"适用·和谐·动人"的创作理念和对自然、文化、经济环境的高度责任感与洞察力，以及在新技术、新材料的运用方面的领先地位。

迄今为止，机构荣膺国家、省部级奖励30余项，并先后获得了由联合国人居署、中国建筑学会颁发的集体与个人国际贡献嘉奖。翌德国际设计机构（ÉTÉ）拥有完善的项目管理体系，健康、活跃的企业文化和创作氛围，多年来，致力于将欧洲先进的设计理念与中国的实际情况充分结合，为中国的城市建设提供高质量的专业服务和可持续发展的技术方案。其国际化的专业协作团队将对社会、经济、环境与时效的综合考虑融入创作过程，为每个项目带来超凡的品质和长远的效益。全方位的效益评估和缜密的设计思路，使翌德国际设计机构（ÉTÉ）成为众多地方政府、投资集团的理想合作伙伴。

ÉTÉ Lee et Associés Architectes Urbanistes, comes from France, is the practice pioneer and thought leadership of urban planning, architecture and landscape design in planning, architecture and landscape designing field, its institution and projects throughout France, Spain, the United States, South Korea and other countries. Over the past 10 years, it has hosted and completed more than 300 design practices in more than 50 medium-sized cities of China. Including Shanghai World Expo 2010, Watergate, Shanghai the South Bund Area, Shanghai Zhangjiabang green wedge, Shanghai Luchaogang Coastal International Flower City, Shanghai Waigaoqiao Business Villa and the application of green building technology, Beijing Ding Jia Heng Yuan, Shanghai Gold land Green County, Chengdu Huaxin Fairview County, Ningbo Flourishing Days City, the completion and put into use of these projects, fully embodies "apply • harmony • moving." the creative concept of ÉTÉ Lee et Associés Architectes Urbanistes and the high sense of responsibility and insight of natural, cultural and economic environment, and the leadership of the use of new technologies, new materials.

So far, ÉTÉ has won more than 30 national, provincial and ministerial level awards, and gained the international contribution awards for group and individual by the UN Habitat and Architecture Society of China. ÉTÉ has a perfect project management system, healthy, vibrant corporate culture and creative atmosphere, over the years, it is committed to integrate advanced design concepts and the actual situation of China, provide high-quality professional services and technology solutions for sustainable development for Chinese urban construction.

The international team of professional collaboration will consider taking the society, economic, environment and time into the creative process, bring exceptional quality and long-term benefits for each project. Comprehensive assessment of the benefits and careful design ideas to make ÉTÉ Lee et Associés Architectes Urbanistes to be a ideal partner of many local governments and the investment groups.

巴黎
地址：98,rue Quincampoix, Paris, France
邮编：75003
电话：0033(0)142760104
传真：0033(0)142067867

上海
地址：上海市静安区乌鲁木齐北路480号万泰国际21楼
邮编：200040
电话：+86-21-53082775
传真：+86-21-53082776
邮箱：etelee@163.com

成都
地址：四川省成都市高新区府城大道西段399号天府新谷5号楼1206
邮编：610041
电话：+86-28-85358788
传真：+86-28-55358788
邮箱：etechengdu@163.com

Paris
98, rue Quincampoix, 75003 Paris, France
Tel: 0033(0)142760104
Fax 0033(0)142067867

Shanghai
21th Floor WanTai Mansion No.480, North Urumqi Road, Jing'an District, Shanghai
P.R.China 200040
Tel: +86-21-53082775
Fax: +86-21-53082776
E-mail: etelee@163.com

Chengdu
Rm.1206, Bldg.5, No.399, Fucheng Road, Chengdu, Sichuan
P.R.China 610041
Tel: +86-28-85358788
Fax: +86-28-55358788
E-mail: etechengdu@163.com

www.etelee.com

乌鲁木齐市高铁片区城市设计

建设地点：新疆 乌鲁木齐
设计时间：2011年
项目规模：总用地面积833.4公顷
　　　　　总建筑面积5 980 000平方米

该设计以"产城融合"和"首位城区"为两大核心理念，把产业、功能、空间、交通、环境紧密相连，突出对乌昌地区乃至新疆产业发展、城市建设的引领和示范作用，打造面向国际的产业高地和商务舞台。

Urban Design of High Speed Railway Area, Urumqi

Construction Location: Urumqi, Xinjiang
Design Time: 2011
Project Scale: Land Area 833.4 ha　　Building Area 5,980,000 m^2
"The city and industry integration" and the "first city" for the two core concepts, the industry, function, space, traffic, environment are closely linked. The Leading and exemplary role is highlighted in the industry development and construction of Urumqi-Changji Region and Xinjiang, building a stage for international industry and commerce.

上海市民生路码头改造

建设地点：上海
设计时间：2010年
建筑面积：140 000平方米

建于一百多年前的民生路码头（时为英商蓝烟囱码头）是当年亚洲最大的码头之一，码头内拟保留的典型工业建筑代表了上海近代城市工业的历史。

以海上繁华作为设计导向，展现"星光璀璨•时尚文化信息发布平台"的主题构思，改造后的民生路码头作为上海这座国际大都市的时尚活动地标，能够吸引高中端消费者，同时开放滨江休闲绿地，促进社会和谐。

Transformation of Minsheng Road Wharf, Shanghai

Construction Location: Shanghai
Design Time: 2010
Building Area: 140,000 m^2

One hundred years ago, Minsheng Road Terminal (for British Smoke Chimney Pier) is one of the Asia's largest terminals. The retaining typical Industrial building represents the modern industrial history of Shanghai.

With the prosperity as a design guide, "sparkling, fashion culture and information publishing platform" as the main theme, Minsheng Road Terminal after the transformation is taken as the landmark of this international metropolis, attracting high end consumers, at the same time opening Riverside leisure space, to promote social harmony.

合肥华邦·光明世家建筑设计

建设地点：安徽 合肥
设计时间：2008年
建筑面积：430 947平方米

住宅立面设计风格秉承装饰艺术设计理念，建筑设计风格清新典雅、端庄大方。单体讲究线脚的丰富细腻，讲究形体变化的韵律，讲究窗、门、阳台等建筑构件组合的比例关系。

Architectural Design of Hua Bang·Bright Family, Heifei

Construction Location: Hefei, Anhui
Design Time: 2008
Building Area: 430,947 m^2

Residential facade design style uphold, Art Deco design concept as the main theme, fresh and elegant, dignified and generous. Single building stresses architrave rich and delicate, exquisite shape changes and the proportion between building components such as windows, doors, and balconies.

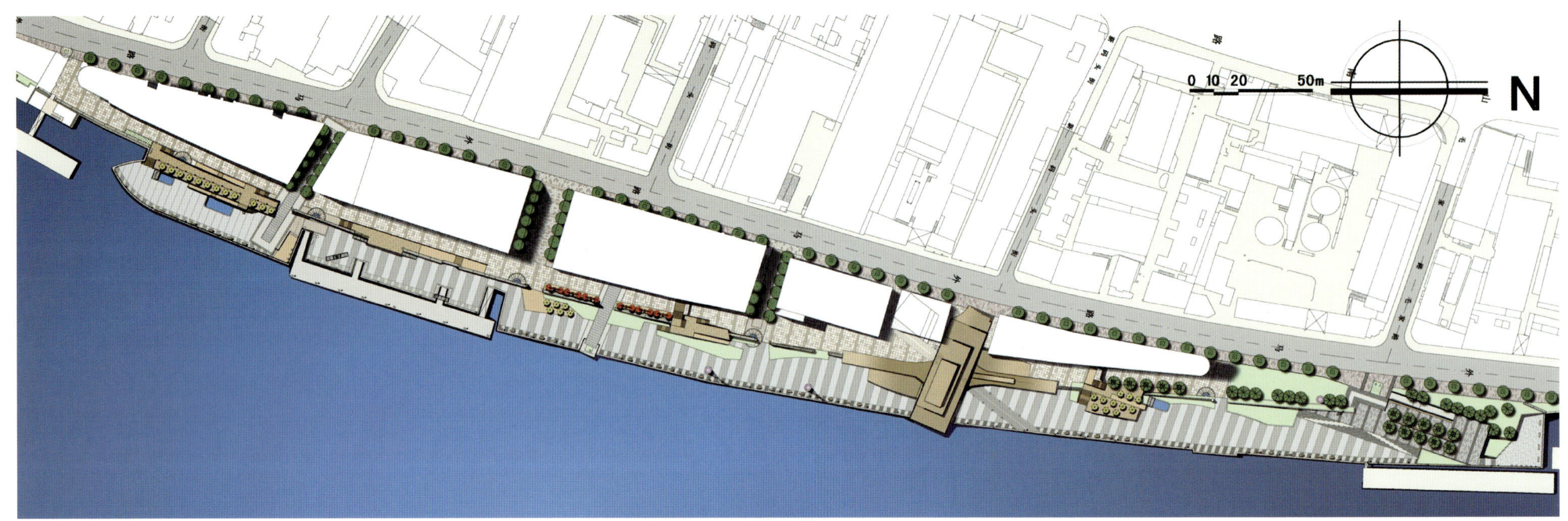

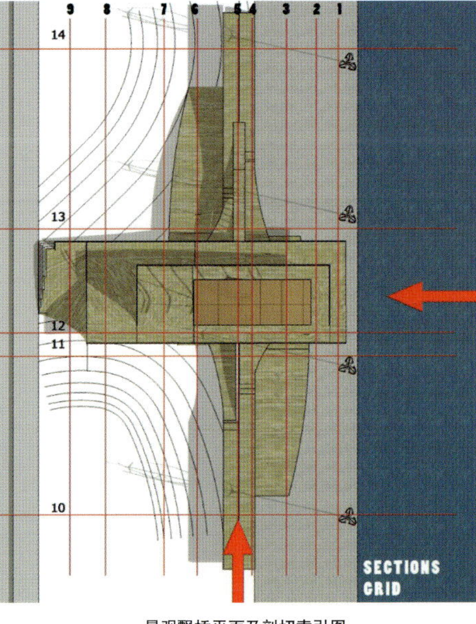

景观飘桥平面及剖切索引图

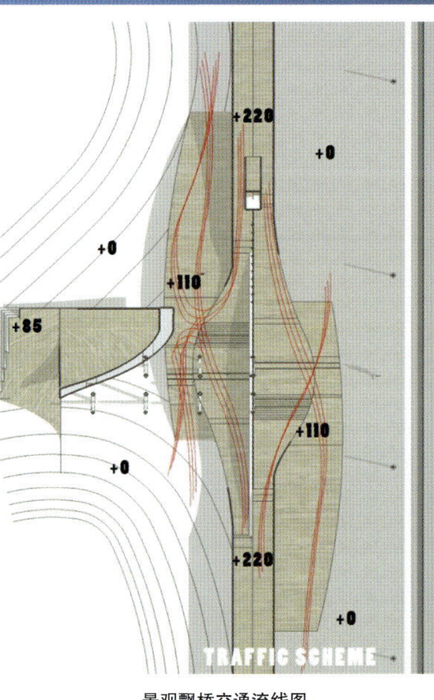

景观飘桥交通流线图

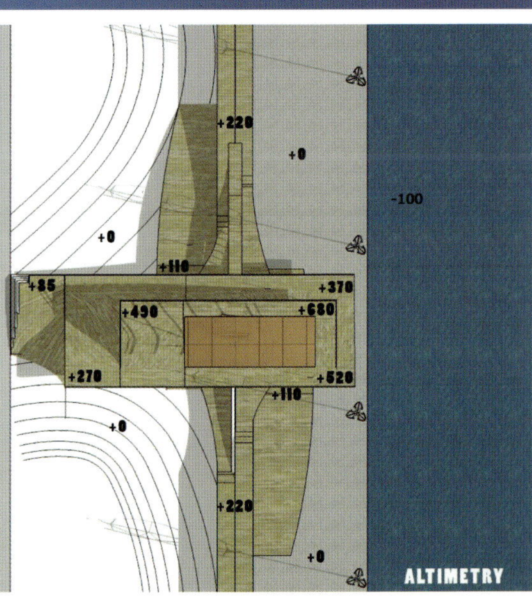

景观飘桥竖向平面图

上海市复兴码头景观设计

建设地点：上海
设计时间：2010年
景观设计面积：32 000平方米

项目定位为城市亲水公共活动及休闲空间。

其承载的社会功能有：时尚秀、主题发布会、室外咖啡吧、亲水休闲活动、眺望江景等。

Landscape Design of Fuxing Wharf, Shanghai

Construction Location: Shanghai
Design Time: 2010
Landscape Design Area: 32,000 m^2

The project is positioned as a city hydrophilic public activities and leisure space. The main social functions: fashion show, theme conference, outdoor cafe bar, hydrophilic leisure activities, overlooking the river scene.

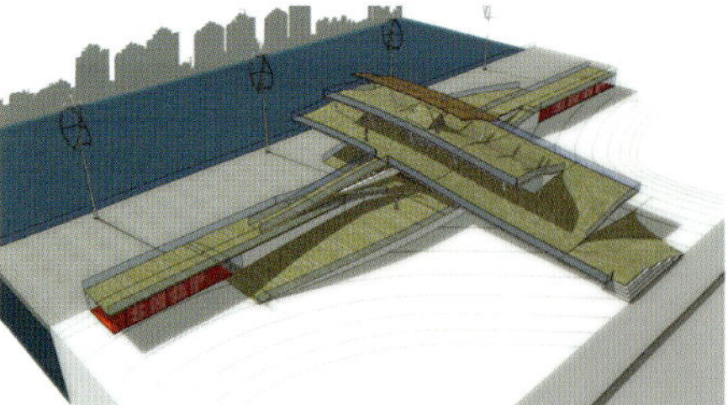
景观飘桥鸟瞰图一

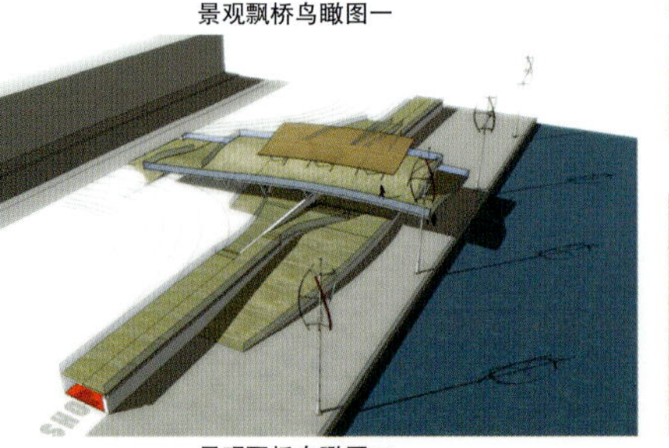
景观飘桥鸟瞰图二

景观飘桥透视图一

景观飘桥透视图二

景观飘桥透视图三

景观飘桥透视图四

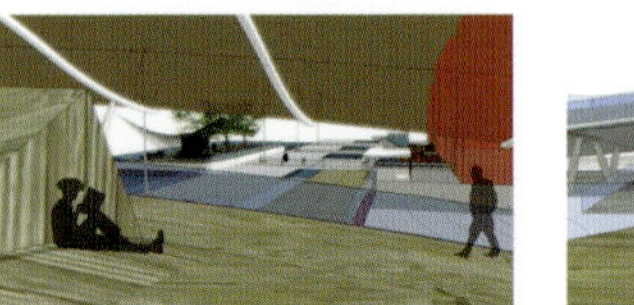
景观飘桥透视图五

景观飘桥透视图六

水系
+
绿地
+
建筑
=
水绿编织的低碳城市

上海西虹桥商务区规划设计

建设地点：上海
设计时间：2010年
用地面积：18.9平方千米

结合本地区的优势基础，规划确定了该地区的主导功能：

· 国家会展中心；
· 以会展产业为集聚核心，向外拓展贸易金融服务业、总部办公、创意产业等相关产业，既能为主功能区提供专业服务，又形成主功能区的产业延伸；
· 同时为主功能区提供文化展示、居住配套、生态支撑等服务；
· 与主功能区共同构成国际贸易中心的重要载体。

Planning Design of West Hongqiao Business District, Shanghai

Construction Location: Shanghai
Design Time: 2010
Land Area: 18.9 km^2

Combining the advantages of the region, the planning determined the dominant features of the region:

• National Convention Centre;
• Taking the exhibition industry as the cluster core, outward expand the trade and financial services, headquarters office, creative industries and other related industries, it can both provide professional services for the main functional areas and also form an extension of the main functional areas of the industry;
• Simultaneously provide cultural exhibition, residential facilities, ecological support and other services for the main functional areas;
• Together with the main functional areas constitutes an important carrier of the International Trade Center.

澳大利亚 HYN 建筑设计顾问有限公司（境外）
HYN ARCHITECTURE DESIGN & CONSULTING PTY LTD.AUSTRALIA
深圳市汉方源建筑设计顾问有限公司（境内）

澳大利亚 HYN 建筑设计顾问（深圳）有限公司

澳大利亚 HYN 建筑设计顾问（深圳）有限公司总部设在澳大利亚新南威尔士，近年来在深圳设立了亚洲办事处。其业务范围包括项目前期策划研究和建筑设计。项目类型涉及住宅区，别墅，办公，学校，商业、酒店、城市综合体建筑及旅游地产等。

HYN 坚持充分理解业主和市场的需求，在方案前期同步融入项目策划研究工作，协助业主明确项目定位及经营开发理念。坚持产品研发创新，最大化地挖掘提升项目价值，力争为业主创造出高素质、高附加值、艺术化的个性产品。

HYN 以“敬业、诚信、创新”为企业理念，将国外设计工程全程服务机制引入中国，在国内与知名结构水电设计、景观设计等专业公司长期紧密配合，创建多行业互动合作的工作模式。HYN 在设计过程中与客户保持快速有效的沟通和反馈，注重项目后期服务，特别是材料选型及施工制作工艺，增强项目实际运作的可操作性。

HYN ARCHITECTURE DESIGN & CONSULTING LTD.

HYN is a progressive, multi-disciplinary, design-led architecture and structural engineering practice. The company takes a holistic approach with an integrated focus to create inspirational, environmentally sustainable solutions.

HYN recently set up an Asian branch in Shenzhen, China., to cater to the local socio-economical conditions. With our local team, we are able to fully understand and accurately reinterpret client's requirements, transforming ideas into reality, which encompasses maximized return for the client, comfort for the users, and positive contribution to the local community.

In HYN, we value “professionalism, credibility, creativity”. We put a lot of emphasis on materiality and building technique, extending our services and expertise throughout the project's implementation process. From our well-established collaboration with local E&M firms landscaping and interior design firms, we have created an efficient communication channel which enables us to give timely feedback to our clients, and deliver desirable results on time.

单位名称：澳大利亚 HYN 建筑设计顾问（深圳）有限公司
单位地址：广东省深圳市南山区华侨城东方花园 F28 栋
联系电话：13316855618
+86–755–26004743
传　　真：+86–755–26004743–613
邮　　箱：hynsz@sina.cn
网　　址：www.hyndesign.com

旅游地产

文博宫二期国际养生创意园

建设地点：广东 深圳
用地面积：2 380 064.6 平方米
建筑面积：380 051 平方米
容 积 率：0.16
主要设计：刘 臻、覃伙计、蓝聪尧、沈振杰、康文华

酒店及办公

海南文昌市天成国际酒店

用地面积：9 500 平方米
建筑面积：35 561 平方米
容 积 率：2.99
主要设计：李 帅

珠海水湾头酒店

用地面积：7 232.52 平方米
建筑面积：51 000 平方米
容 积 率：7.05
主要设计：张 伟

深圳侨香路南方工作站

用地面积：4 902.28 平方米
建筑面积：54 151.34 平方米
容 积 率：8.5
主要设计：蓝聪尧、沈振杰

城市综合体

深圳南头关口片区旧城改造专项规划设计

建筑用地面积：114 503 平方米
建筑面积：750 829 平方米
容 积 率：6.56
主要设计：蓝聪尧

赣州中心城项目

用地面积：52 967.5 平方米
建筑面积：287 777.47 平方米
容 积 率：3.5
主要设计：刘 臻、唐 波、朱 埔

合肥万向城项目

用地面积：14 950 平方米
建筑面积：118 000 平方米
容 积 率：7.8
主要设计：沈振杰

肇庆华生中心三期项目

用地面积：49 200 平方米
建筑面积：179 699 平方米
容 积 率：4.42
主要设计：唐 波、陈栩淮

商　业

中国西南城食品博览综合体

用地面积：3 340 000.5 平方米
建筑面积：4 000 000 平方米
容 积 率：0.95
主要设计：骆 清、明 芬

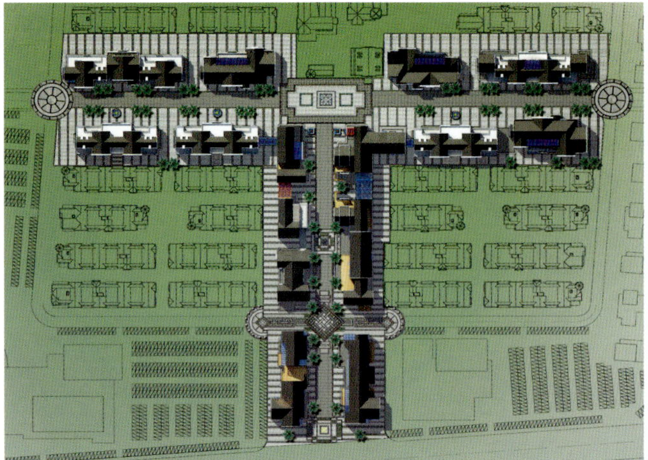

四川绵阳仙海湖别墅

建筑单位：四川绵阳鼎浩实业集团
景观面积：127 386 平方米
主要设计：骆 清

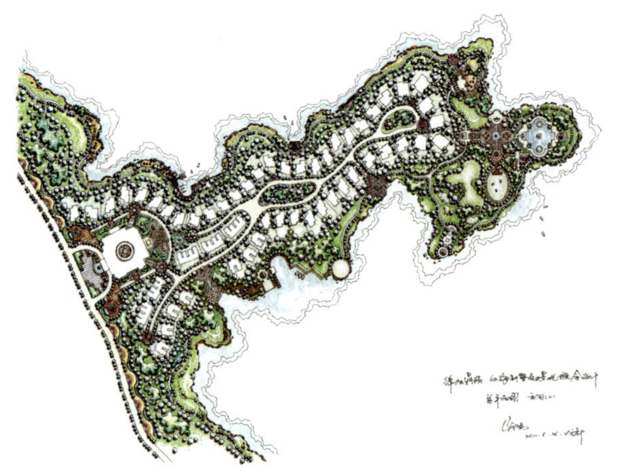

住　宅

中山坦洲优越花园项目

用地面积：198 173 平方米
建筑面积：594 519 平方米
容 积 率：3.0
主要设计：刘 臻、覃伙计

中山新都汇二期项目

用地面积：33 920 平方米
建筑面积：174 691 平方米
容 积 率：5.1
主要设计：唐 波、陈栩淮

深圳坂田金和成广场

用地面积：19 433 平方米
建筑面积：116 500 平方米
容 积 率：4.5
主要设计：刘 臻、马国勇

住 宅

中山大信海岸家园项目

用地面积：84 364.40 平方米
建筑面积：350 652.38 平方米
容 积 率：3.43
主要设计：刘 臻、蓝聪尧

河南信阳河一号

用地面积：
36 380 平方米
建筑面积：
130 000 平方米
容 积 率：3.5
主要设计：
刘 臻、蓝聪尧

江门市"泰山泊天下"项目

用地面积：61 566 平方米
建筑面积：186 136 平方米
容 积 率：2.5
主要设计：刘 臻、朱 埔

肇庆端州一路项目

用地面积：46 046 平方米
建筑面积：167 200 平方米
容 积 率：3.8
主要设计：刘 臻、明 芬

规　划

成都长秋山国际会议中心项目

用地面积：306 663 平方米
建筑面积：840 000 平方米
容 积 率：0.36
主要设计：骆 清

CPG Corporation Pte Ltd.

新加坡CPG集团

新加坡CPG集团是亚太地区主要的建筑管理与咨询服务专业公司，是世界银行和亚洲开发银行旗下的国际注册咨询单位。前身是新加坡公共工程局（PWD），成立于1833年，于1999年企业化。在过去的近两个世纪，CPG作为新加坡公共建筑及基础设施建设的主要管理承担者，为新加坡的国家建设作出了卓越的贡献，以不凡的设计理念塑造了新加坡“花园城市”的形象。CPG参与了新加坡多项标志性建筑的设计与管理，其中最具代表性的项目有：新加坡滨海艺术中心、新加坡滨海湾公园、樟宜机场、总统府、国会大厦、国立大学、南洋理工大学、陈笃生医院、邱德拔医院以及新加坡赛马场等。

新加坡CPG集团自2003年成为澳大利亚上市公司道纳（Downer EDI）旗下子公司。在中国、澳大利亚、新西兰、印度、越南、菲律宾及中东等国家和地区设有多个分支机构，承接项目遍及全球20多个国家和地区。

新加坡CPG集团完成的项目涵盖了各领域，并取得了令人瞩目的成就，例如：投资达80亿人民币的新加坡樟宜国际机场，已连续十几年被评为世界最佳机场；新加坡Solaris纬壹科技城先后荣获绿色标志白金奖和空中绿意一等奖；投资25亿人民币的新加坡邱德拔医院建成三年内荣获绿色标志白金奖、空中绿意一等奖和绿色建筑领袖奖等7项大奖。

新加坡CPG集团自1999年进入中国市场，已在国内40多个城市和地区承接了500多个项目，为各领域客户提供了全方位规划和建筑咨询服务，已与各地政府机构、上市地产公司、知名国企及民营地产开发商合作并成为长期战略合作伙伴。

新加坡CPG集团提供的服务内容：总体规划与城市设计、建筑设计与咨询、环境规划、土木与结构工程、机械与电气工程、环保设计、交通运输工程、工程造价与合约管理、项目管理、产业管理；提供规划和综合性设计与咨询的服务领域：总体规划、城市设计、专项规划、机场设计、住宅建筑设计、教育机构建筑设计、商务办公建筑设计、医疗机构建筑设计、政府机构建筑设计、绿色环保建筑设计、公共设施建筑设计。

CPG Corporation Pte Ltd.(新加坡CPG集团)
238B Thomson Road #18-00 Tower B Novena Square Singapore 307685
Tel: +65-6357-4888
Fax: +65-6357-4188
http://www.cpgcorp.com.sg

中国区总部——新工工程咨询（上海）有限公司
地址：上海黄浦区西藏中路585号新金桥广场九楼
电话：+86-21-63517888
传真：+86-21-63519888
网址：www.cpgcorp.com.sg

新加坡南洋理工大学艺术、设计与媒体学院
School of Art, Design and Media, Nanyang Technological University
2011年 新加坡建设局绿色建筑标志奖（白金奖）

媒体城@新加坡纬壹科技城
Mediapolis@One-North
2011年 新加坡建设局绿色建筑标志（优金奖）

新加坡邱德拔医院 Khoo Teck Puat Hospital

2009年 新加坡建设局绿色建筑标志奖（白金奖）
2010年 新加坡建筑师协会及国家公园局空中绿意一等奖
2010年 艾默生杯特别感谢奖
2011年 FuturArc绿色领袖奖（公共建筑类）
2011年 新加坡建设局通用设计金奖
2011年 新加坡建筑师协会设计奖（医疗建筑类）
2011年 新加坡建筑师协会年度最佳建筑奖
2011年 美国波士顿第七届“设计与健康”全球大会国际学院奖

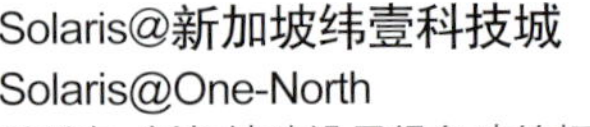

Solaris@新加坡纬壹科技城
Solaris@One-North
2010年 新加坡建设局绿色建筑标志奖（白金奖）

广东顺德喜来登大酒店

建设地点：广东 顺德
客户名称：佛山市顺德区华财企业投资有限公司
基本功能：酒店、高级公寓
竣工日期：2009年
用地面积：50 082平方米
建筑面积：145 500平方米

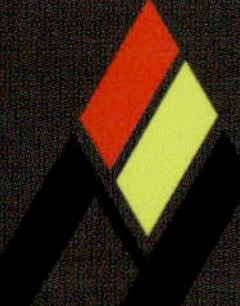

AU-SINO澳华（悉尼）建筑景观设计机构

AU-SINO(Sydney) ARCHITECTURAL & LANDSCAPE DESIGN PTY. LTD.

AU–SINO澳华（悉尼）建筑景观设计机构是从事城市规划、建筑设计、房地产咨询及建设项目管理服务的专业设计机构。公司成立于澳大利亚悉尼，在中国杭州设有分支机构，并在南昌、济南设立了办事处。公司由专业设计师组成，是具备较高建筑艺术修养与强烈工作责任心的创作团体。

AU–SINO澳华设计机构自进入中国设计市场以来，凭借工程设计与项目服务相结合的业务优势，积极参与城市建设与商业房地产开发，将先进及独创性的设计理念融入所承担的项目，创造出众多具有独特人文地域特点，且能有效运作建造的项目，无论在政府业主还是私人投资商业主方面，AU–SINO澳华（悉尼）建筑设计机构都赢得了良好的口碑。多年来，AU–SINO澳华设计机构一直与各相关部门建立并保持了良好的合作关系。

AU–SINO（悉尼）建筑设计机构采用境外公司的设计管理模式，并习惯于将策划的概念引入规划设计中，以为投资商如何通过规划的方式取得最大的经济效益和社会效益回报作为宗旨。

随着国内工程建设项目日益国际化和规范化，为确保每个项目拥有最专业的设计水准，力求建筑品质的完美无缺，AU–SINO澳华（悉尼）建筑设计机构在国内先后选择了设立分支机构和办事处，以保证全国每一个项目都能有一流的设计师进行全程创作和跟踪，使每一个方案都能在设计品质和完成质量上保持高度一致性，并通过直接、详尽、科学的管理体系，让项目在获得独创设计与合理建造成本的同时，创造更多的品牌附加价值。使业主得益于与众多有创造性而又注重实效的建筑师合作，从每个项目中共同寻找独特机会，为市场创立新的建设理念，获得新的成功。

As a professional design company, AU-SINO (Sydney) dedicates itself to urban planning, building design, real estate consulting and project management. Founded in Sydney, we have established a regional headquarters in Hangzhou, and offices in Nanchang and Jinan. AU-SINO architects pride themselves in their architectural aesthetics and committed working attitude.

Ever since AU-SINO ventured into the Chinese market, we demonstrated prominent strength in our active participation in local urban planning and commercial development by integrating both the design and design-related services and applying state of the art design concept in our projects. We have a well-established reputation among the government sectors, clients and private investors through our unique vernacular design and its proven feasibility. For years, our connections in China grew soundly.

AU-SINO has inherited and applied our international design management standard locally and has competently combined both strategic planning and design planning for our ultimate goal of maximizing the client's profit and positive social gain.

We have set up regional headquarters and offices in order to maintain our standard of services worldwide and provide a better platform for interactions between our professionals and the clients. We strive to ensure that every single project is designed and followed up by our first class architects through a more straightforward, detailed and scientific management system. Our design is thus made across-the-board unique, feasible and beyond. Our innovative and practical professionals are all-time ready to offer their expertise to our clients in search of unique social and economical opportunities for every project.

▲【宜春天沐温泉谷】
项目位置：江西 宜春　项目规模：200万平方米
项目简介：世界华人高端温泉养生住区

▲【海南平海东郊椰林项目】
项目位置：海南 文昌　项目规模：40万平方米
项目简介：超五星级酒店群

Tel: +86-571-28880052　Fax: +86-571-28886630　Mail: au-sino@sohu.com　www.au-sino.com

RESORT·HILLSIDE VILLA OFFICE·COMMERCIAL COMPLEX HIGH-END RESIDENTIAL

【旅游度假·山地别墅】

▲【宿迁运河湾世知酒店】
项目位置：江苏 宿迁 项目规模：20万平方米
项目简介：运河湾五星级度假酒店

▲【宜春天沐温泉度假酒店】
项目位置：江西 宜春 项目规模：1万平方米
项目简介：以现代艺术来阐释中国江南风格的温泉度假酒店

▼▲【杭州普达海动漫产业园】
项目位置：浙江 杭州 项目规模：200万平方米
项目简介：包括产业基地、产业配套和地产开发的综合性项目

▲【海南平海逸龙湾】
项目位置：海南 文昌 项目规模：39.5万平方米
项目简介：五星级酒店式海景公寓

▲【天台浅水湾御景山庄】
项目位置：浙江 天台 项目规模：27万平方米
项目简介：大型景观别墅区

▲【海南星河湾度假酒店】
项目位置：海南 文昌 项目规模：4万平方米
项目简介：滨海度假式五星级酒店

◀【大同德和山庄】
项目位置：山西 大同
项目规模：2万平方米
项目简介：晋宅大院、餐饮主题度假山庄

▼【金华欧源原塑】
项目位置：浙江 金华 项目规模：30万平方米
项目简介：大型现代人文理念亲水居住社区

AU-SINO ARCHITECTURAL DESIGN

RESORT·HILLSIDE VILLA | OFFICE·COMMERCIAL COMPLEX | HIGH-END RESIDENTIAL

【办公·商业综合体】

AU-SINO ARCHITECTURAL DESIGN

▲【绍兴越隆集团大厦】
项目位置：浙江 绍兴　项目规模：4万平方米
项目简介：绍兴柯桥中央商务核心区高档办公楼

▲【杭州华立西溪综合体】
项目位置：浙江 杭州　项目规模：5万平方米
项目简介：杭州城西商业综合体——旋转的盒子

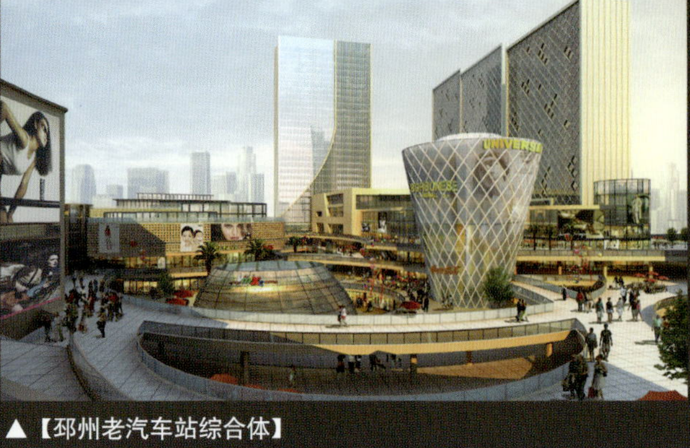

▲【邳州老汽车站综合体】
项目位置：江苏 邳州　项目规模：22万平方米
项目简介：大型城市综合商业体

▲【杭州复地复城国际】
项目位置：浙江 杭州　项目规模：30万平方米
项目简介：市中心高端物业综合体

▲【杭州中控集团大厦】
项目位置：浙江 杭州　项目规模：7万平方米
项目简介：高标准纯写字楼

【景德镇帝王大厦】▲
项目位置：江西 景德镇　项目规模：7万平方米
项目简介：瓷都中心坐标的城市综合体

▲【天津安利欧铂城】
项目位置：天津　项目规模：5万平方米
项目简介：高端商业综合体

◀【杭州现代集团大厦】
项目位置：浙江 杭州　项目规模：6万平方米
项目简介：杭州黄龙商业区纯写字楼

▼【杭州天辰国际广场】
项目位置：浙江 杭州　项目规模：12万平方米
项目简介：综合物业大厦

▼【千岛湖新城商务中心】
项目位置：浙江 杭州　项目规模：3万平方米
项目简介：新中式风格综合体

RESORT·HILLSIDE VILLA | OFFICE·COMMERCIAL COMPLEX | HIGH-END RESIDENTIAL

【高端住宅区】

▲【上饶博能久仰山河】
项目位置：江西 上饶 项目规模：25万平方米
项目简介：信江河畔的高端住宅小区

▲【柯桥赞成香林花园】
项目位置：浙江 绍兴 项目规模：32万平方米
项目简介：新装饰艺术风格的纯水岸住宅区

▲【南昌平海九里象湖城】
项目位置：江西 南昌 项目规模：63万平方米
项目简介：高端品质的大型综合性社区

▲【哈尔滨新洲碧水庄园】
项目位置：黑龙江 哈尔滨 项目规模：70万平方米
项目简介：松花江边园林生态楼盘

▲【杭州复地复城国际】
项目位置：浙江 杭州 项目规模：30万平方米
项目简介：市中心高端物业综合体

▲【宁波浅水湾蔚蓝海岸】
项目位置：浙江 宁波 项目规模：11万平方米
项目简介：纯高层人文居住小区

▲【潍坊亚特尔凤凰太阳城】
项目位置：山东 潍坊 项目规模：89万平方米
项目简介：超大规模，综合类高品质

【苏州冠城观湖湾】▶
项目位置：江苏 苏州
项目规模：25万平方米
项目简介：高档临湖生态小区

【柯桥万达广场住宅】◀
项目位置：浙江 绍兴
项目规模：37.5万平方米
项目简介：万达广场的高端大户型住区

ARCHITECTURAL DESIGN AU-SINO

Gensler

Gensler是一家全球建筑、设计、规划和战略咨询公司，专业从事各企业、机构和公共组织所拥有或使用的各种建筑及设施的设计咨询。我们提供全方位的建筑服务，从初始策划到设计、实施和管理。我们坚持客户至上的理念，深入了解客户目标和策略，力求通过我们的工作和服务显著增加客户的企业价值。

Gensler于1965年成立于美国旧金山。为了确保与客户密切互动，公司从一间办公室成长为一家拥有38个办公室和一个专业智囊团的大公司，人员超过2900人。

Gensler获得著名的美国《商业周刊》设计大奖评选出的多项荣誉，该奖项旨在评选出以战略性商业目的为导向的、创新设计解决方案。2000年，美国建筑师学会授予Gensler "年度最佳事务所" 荣誉，是其授予合作事务所的最高奖项，并将Gensler列为 "21世纪设计事务所典范" 。《工程新闻记录》(Engineering News–Record)杂志及《世界建筑》(World Architecture)杂志将Gensler列为全球顶级建筑事务所。2006年，美国绿色建筑商会(US Green Building Council)授予Gensler "领袖奖" 。

Gensler is a global architecture, design, planning, and strategic consulting firm that specializes in a wide range of building and facilities owned or used by businesses, institutions, and public agencies. Our services engage the full building cycle from initial planning through design, implementation, and management. We focus on our clients, understand their goals and strategies, and seek to add substantial value to their enterprise through our work and services.

Gensler was founded in San Francisco in 1965. To ensure close interaction with its clients, the firm has grown from one office to a broad-based organization with 38 locations and a professional resource in excess of 2,900 people.

Gensler received the Year 2000 Architecture Firm Award, the AIA's highest honor to a firm that has consistently produced distinguished architecture. Michael J. Stanton, FAIA, then President of the AIA, said in his announcement, "Gensler is America's foremost collaborative practice. The firm exemplifies how the creative mix of disciplines, all with 'place' as their focus, adds richness and value to buildings and their settings. Year 2000 is an appropriate time to honor a firm that has consistently pushed the boundaries of architecture."

地　址：上海市卢湾区湖滨路222号企业天地1号楼908室
邮　编：200021
联系人：沈 杰
电　话：+86–21–61351922 Direct
+86–21–61351900 Main
传　真：+86–21–61351999
网　址：www.gensler.com/shanghai

Add: Room 908, Qiye Tiandi No.1 Building, Hubin Road 222, Luwan District, Shanghai
P.C.: 200021
Contact: Jay Shen
Tel: +86–21–61351922 Direct
+86–21–61351900 Main
Fax: +86–21–61351999
Http://www.gensler.com/shanghai

21世纪中心大厦

总 面 积：110 750平方米
占地面积：12 000平方米
建成时间：2010年

The 21st Century Center Building

The Total Area: 110,750 m^2
Land Area: 12,000 m^2
Completion Time: 2010

21世纪中心大厦位于上海市最大的商务及金融区陆家嘴，占地面积近12 000平方米，大厦规划共48层。届时，四季酒店及服务式公寓将进驻21世纪中心大厦。

根据现有规划，21世纪中心大厦将分为四部分，包括2层裙楼、26层以下为50 000平方米的甲级写字楼、中段10层为四季酒店的190间客房、顶部约为60套服务式公寓。

The 21st Century Center Building locates in Shanghai's biggest business and financial district-Lujiazui covering approximately 12,000 square meters. We plan to build this building with 48 stories. Then, the Four Seasons Hotel and service apartment will start business here.

According to the current master plan, this building will be divided into 4 sections, including 2-story podium, the lower section under 26th floor that is for the Grade A office building part with approximately 50,000 square meters, the 10 stories in the middle section that is for the Four Seasons Hotel with about 190 guest rooms and about 60 suites of service apartment on the top floor.

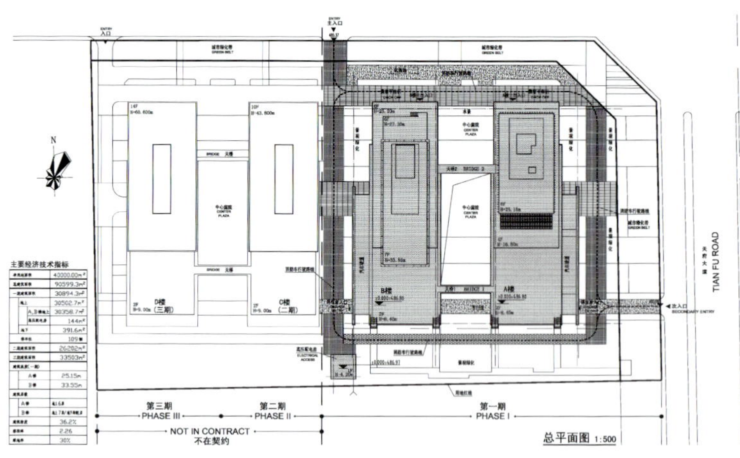

成都赛门铁克总部办公一期

建筑面积：80 000平方米
建成时间：2011年

Symantec Chengdu Campus Phase I

Construction Area: 80,000 m^2
Completion Time: 2011

赛门铁克新的研发园区位于天府路和新星路拐角处，并且以宽大的建筑形式成一系列台阶状排列，形成整个园区。这些好似行进中的士兵将从十字路口排列升起，也使得道路得到最大的视野，并且提供了不同的比例尺度。

每个建筑由侧面所连接景观庭院。青葱优美的绿化环境以屋顶绿化梯台形式连续延伸攀爬至整个方案的每一处。这些花园给人们一个导向：从园区室内的工作环境通往自然环境。这样的设计也给整个园区提供了大、中、小三种户外空间，以便人们用餐、放松或交谈。

建筑顶层最宜人的功能：高大宽敞的结构给人们活力与无与伦比的感受。这些宜人的功能便于建筑间的流通和促进整个园区内部的循环。

秉承可持续发展的理念：整个园区将努力提供一个建筑设计的解决方案来处理对环境、使用者和来访者的影响。

赛门铁克新的研发园区将成为一个著名的科技中心：位于成都高科技区，吸引中国乃至世界的精英汇聚此地，使它成为首屈一指的商务软件区。

The Symantec New Research & Development Park Zone locates at the corner of Tianfu Road and Xinxing Road with large buildings to form a series of step-like arrangement covering the entire park. These step-shaped buildings like the marching soldiers rise from the intersection. Such design makes the road enjoy more broad view and provide different scales at the same time.

Each building is connected by its sides and the landscape yard. The Lush and beautiful environment covers the continuous extension to everywhere of this program with green roof ladder platform. These gardens function as a guide leading people from indoor work environment of the zone to natural environment. Such design also provides the whole park zone with large, medium and small sized outdoor space for people to dine, relax or chat.

The most comforting functions of the top floor of the building and the capacious structure give us a kind of distinctive feeling of vitality. These satisfying functions are convenient to the building's inner communication and promote the inner circulation throughout the whole zone.

Adhering to the concept of sustainable development, we will endeavor to provide the certain architectural design solutions of the whole park zone to cope with the influence of environment, users and visitors.

The Symantec new Research and Development Zone will become a distinguished technology industrial center located in Chengdu high-tech District attracting talents throughout China and all over the world. We aim to make it the premier business software district.

招商银行

建筑面积：202 000平方米
地　　点：上海
设计时间：2006年
建成时间：2011年

Merchants Bank

Construction Area: 202,000 m^2
Location: Shanghai
Design Time: 2006
Completion Time: 2011

国际性的金融机构需要国际性的人才：为了吸引业界精英的加盟，招商银行目前正建造总面积202 000平方米的总部大厦，以提供客户及员工所需要的各种设施。

我们在建筑设计、室内设计及景观设计中力求彰显企业文化，使设计和谐统一，天衣无缝，正如招商银行服务多样的银行卡中心，竭诚提供令客户满意的舒适服务。我们的设计力求做到形象统一，正如一张发给客户的信用卡，而且也能时时处处提醒员工身处招商银行的大家庭之中。

简洁、优雅与宁静是设计意向的三大核心概念。坚实可靠的石材象征着有力的基础，而清爽自然的布局则体现了招商银行合作无间的精神。

The international financial institute demands the international talents. In order to attract the elites of this industry to join in, at present, China Merchants Bank invests a 202,000 square-meter headquarters building to provide different kinds of amenities desired by clients and staff.

In the process of the architectural design, interior design and landscape design, we aim to highlight corporate culture and to make the design harmoniously and seamlessly. That's like the bank's card center which could offer a variety of services dedicating to provide the satisfying and comfortable services to customers. Our design strives to achieve the unifying image like a credit card, which is sent to customers, could remind every staff to be in the big family of the bank all the time.

Simplicity, grace and tranquility are the three core concepts of this design. The solid and reliable stone symbolizes the strong base, and the fresh and natural layout embodies the close cooperation spirits of China Merchants Bank.

J.A.O.Design International Architects & Planners Limited

美國龍安建築規劃設計顧問有限公司

美国龙安建筑规划设计顾问有限公司1984年成立于纽约，是由美籍华人、国际著名建筑规划专家饶及人先生创建的国际性建筑规划设计服务公司，作为在中国已成功经营了10余年的外籍设计公司，其规划与建筑作品遍及31个省、市、自治区的80多个城市，以及中东、加勒比海地区，并赢得117个知名设计奖项。自2004年起，该公司连续多年获得“中国十大影响力设计单位”的殊荣，凭借自身优秀的设计能力和良好的市场口碑，规划设计的作品达300余个，为海内外投资者的开发提供了大量成功典范。美国龙安公司凭借扎实稳健的设计实力及先进的设计理念、丰富的经济学识、突出的工作业绩，在业界树立了良好的专业品质和企业形象，是在中国将“国际设计理念”落地的专业单位。

美国龙安建筑规划设计顾问有限公司作为美国龙安集团旗下最核心的品牌公司，始终将集团“以城市规划设计为切入点进行城市运营，为中国百姓提供世界级的低碳绿色建筑”的发展理念作为己任，积极推动中国城市化发展的进程。

美国龙安公司拥有多种文化背景的专业设计人员，为客户提供以下四种专业服务。（1）规划设计：城市规划（新市镇规划、旧区改建规划、旅游度假区规划、市政规划、CBD设计）及各专项规划。（2）建筑设计：甲级写字楼、购物中心、住宅楼群、博物馆、会展中心、国际学校、综合大楼、五星级酒店等。（3）景观设计：公园、居住区环境景观设计、游乐场、广场等。（4）室内设计：星级酒店、营业厅、各商业场所、别墅、公寓、售楼处、会所等。迄今为止，城市规划总面积2 808平方千米（约46个纽约曼哈顿岛）；住宅区规划设计总建筑面积约为2 180万平方米（约为21万个家庭住宅）；公建设计总建筑面积1 150万平方米（相当于71个国家大剧院）；旧城改建街道总长度11 300米（相当于6条半法国香榭丽舍大道）；景观设计总面积776公顷（约2个半纽约曼哈顿中央公园面积）；室内设计总面积13万平方米（约430个样板间）。

2009年，美国龙安成功收购了住房和城乡建设部甲级设计单位北京众拓建筑工程设计有限责任公司（与住建部科技发展促进中心共同持股，其中北京龙安控股70%，资质号A111000883），成为目前国内为数不多的具有甲级资质的外籍建筑设计企业之一。双方整合优势资源形成核心竞争力，强强联合进一步巩固了美国龙安在中国建筑规划设计领域内的领先地位。

与其他国际设计公司不同，美国龙安公司作为跨国经营的事务所，中西合璧的理念结合得最好，在中国获奖最多，成为中国最具影响力的国际设计事务所；由于充分了解与熟悉中国的体制与管理，并能和中国各城市主管领导进行无障碍沟通，美国龙安公司结合自身的国际资源优势与实际经济情况，同时通过集团总裁饶及人先生的充分整合，集中调动国际基金资源，介绍与带进投资，让项目得到落地执行。

提供专业服务，为客户量身定制配套顾问服务

美国龙安建筑规划设计顾问有限公司以“精品设计”、“量身定制”为座右铭，针对不同种类的项目需求，向客户提供建设开发、投资可行性研究，以及项目投资、融资等配套顾问服务。

中西文化兼容并蓄，客户意愿完美体现

美国龙安建筑规划设计顾问有限公司着重于两栖文化的结合，塑造“中西合璧”的理念，成功地将先进的设计思路与商贸哲学运用于各项规划和建筑设计项目中，使客户的意愿得到完美体现。公司凭借多年实战经验与丰富的社会资源为客户提供高品位、高回报率的专项服务。

中西交流的桥梁

美国龙安建筑规划设计顾问有限公司能将国际先进的规划设计理念引入，并通过资金板块、外商资源，使这些理念成为现实。对经过规划后，一些基础较好、前景较明朗的项目，可以协助或自主对国外招商、引入境外投资机构包括亚洲开发银行(ADB)、世界银行(WD)的资金或引进大型金融机构参与项目，使整个项目从设计构思到实施完成达成完美的统一。美国纽约曼哈顿区驻中国商务代表处选择于美国龙安建筑规划设计顾问有限公司挂牌设立，即为外界对其在中国业界和社会地位的肯定范例之一。

卓越成绩，无上光荣

美国龙安建筑规划设计顾问有限公司和饶及人先生近两年所获部分重要奖项及荣誉包括：

2011年 首开绵阳仙海湖及沈阳沈北新区首开乐活小镇荣获“绿色建筑之星最佳绿色建筑奖”

2011年 饶及人总裁荣获由中国住房和城乡建设部主管、住房和城乡建设部优秀期刊《中国建设信息》杂志颁发的“2010年度房地产杰出人物”奖

2010年 美国龙安被住房和城乡建设部、中国建筑文化中心评为“中国最具业主满意度设计机构”

2010年 石家庄·易水龙脉项目荣获“中国资源节约型、低碳社区影响力示范楼盘”称号

2010年 饶及人先生被搜狐焦点网选为搜狐焦点“2009年度房地产意见领袖”

2010年 饶及人先生于“和谐中国2009年度影响力人物”颁奖盛典上荣获“十大影响力人物”

2009年 在“2009年中国城市建设与管理高峰论坛暨中国城市建设与管理60年表彰活动中”，饶及人先生被中国城市经济学会、《建筑》杂志社授予“中国城市规划60年十大杰出人物”荣誉称号

2009年 饶及人先生于2009中国·青田华侨总部经济发展论坛上被青田县委、县政府授予青田侨界“六十杰”十大创业精英荣誉称号

2009年 美国龙安集团获得由中国国际城市化发展战略研究委员会颁发的“2009年中国城市化进程科学发展特别奖”

2009年 “中国移动通信集团江西有限公司红角洲生产基地”项目获得全国工商联房地产商会第六届精瑞住宅科学技术奖规划建筑设计优秀奖

2009年 云南昆明五华区王家桥片区城市设计，获国际竞赛第一名

2009年 唐山南湖生态城总体概念规划项目获得联合国人居署2009HBA中国范例卓越贡献最佳奖

2009年 唐山市南湖生态城国际定向咨询，国际方案征集获奖实施方案

2009年 在中国商业地产行业年会颁奖盛典上，美国龙安建筑规划设计顾问有限公司获得“2008中国知名商业设计机构”奖项

2008年 昆明城市设计概念性方案征集荣获优胜奖

2008年 “2008中国城市住宅与房地产业发展高峰论坛”中，饶及人先生获得由中国房地产业协会、中国建筑节能减排产业联盟联合颁发的“2008中国责任地产年度人物”奖

2008年 第二届中国城市化国际峰会中，饶及人先生获得由中国国际城市化发展战略研究委员会颁发的“2008年度中国城市化进程十大贡献人物”奖

2008年 全国工商联房地产商会主办的第五届精瑞住宅科学技术奖活动中，“南昌东方·海德堡”项目荣获住区规划设计优秀奖

2008年 中国台湾第16届中华建筑金石奖中，北京通州宋庄南部文化创意产业集聚区城市概念设计、内蒙古呼和浩特大南和大北街蒙藏文化景观街整治项目、遂宁市行政中心建筑方案设计分获国际组设计大奖

2008年 在《2007—2008中国建筑设计作品年鉴》中，美国龙安公司被评为优秀设计机构并成为特邀编委单位。

2008年 美国龙安公司荣获中国房地产创新规划建筑设计杰出成就奖

2008年 中国·内蒙古呼和浩特市回民区伊斯兰建筑特色景观街项目荣获联合国人居署迪拜国际范例推动设计奖

2008年 中国·宋庄文化创意产业集聚区项目荣获联合国人居署迪拜国际范例推动设计奖

2007年 饶及人先生荣获“CIHAF2007年度中国最具影响力设计师”称号，美国龙安公司荣获“CIHAF2007年度中国最具影响力规划建筑设计机构”称号

2007年 美国龙安公司获得全国工商联房地产商会首批认证“全国绿色生态建筑示范项目签约规划设计机构”

2007年 美国龙安公司荣获国际房地产协会及国际建设联盟等机构授予的2007年度“中国房地产十大建筑设计公司”称号

2007年 内蒙古呼和浩特多元文化特色景观街项目在庆祝内蒙古自治区成立60周年的活动中，荣获最佳建筑工程设计奖。饶及人先生获得“市长特别奖”

中国总部(北京)：北京市朝阳区光华路5号世纪财富中心东座20层 邮编：100020

J.A.O.Design International Architects & Planners Limited

美國龍安建築規劃設計顧問有限公司

J.A.O Design International Architects & Planners Limited was established in 1984 in New York, providing internationally celebrated architecture and planning services. It was founded by Chinese American Mr. James Jao, a famous expertise in architecture and planning. As a certified foreign enterprise that has more than a decade years' running experience in China, its planning and construction works has been throughout more than 80 cities in 31 provinces, municipalities and autonomous regions, as well as the Middle East, the Caribbean and other regions, which have won 111 known design awards. Since 2004, it has been honored with "China Top Ten impact Power Design Unit "award for 6 years by virtue of its excellent design capability and good market reputation. The company has planned and designed more than 300 works which could provide a good many of success paragons for both domestic and foreign investment. With solid strength solid design and advanced design concepts, knowledge-rich economy, outstanding job performance, J.A.O establishes a good professional quality and corporate image in the industry, as a professional unit bringing "international design" into China.

As the company's core brand branch, J.A.O Design International Architects & Planners Limited, always holds the development ideas that "to take urban planning and design as a focal point for urban operations, to provide Chinese people with world-class low-carbon green buildings" as its mission, actively promoting the process of urbanization in China.

J.A.O employs professional designers with various cultural and national background. J.A.O Design provides professional services to its clients:

(1) Programming. Urban Planning: including new town planning, Urban reconstruction planning, city planning, tourism planning, Municipal planning, CBD design, and other specific planning.

(2) Architecture Design: including Class A office buildings, shopping centers, residential buildings, museums, exhibition & convention centers, international schools, mixed-use apartments, and five star-rated hotels.

(3) Landscape Architecture: including gardens, parks, plazas, etc.

(4) Interior Design: Star-rated hotels, business hall, commercial establishments, villas, apartments, sales offices, clubs, etc.

So far, its city planning area has covered 2,746 square kilometers (equivalent to 45 Manhattan in New York); approximately 1,785 million square meters total floor area of residential planning and design (equivalent to 17 million households in housing); a total construction area of 9.3 million square meters in public designed building (equivalent to 58 National Grand Theatre); total length of the street of 11,300 meters in Urban reconstruction planning (equivalent to six and a half in France Champs Elysees); landscape design covering a total area of 776 hectares (about two and a half Manhattan's Central Park area in New York); interior design covering a total area of 130,000 square meters (about 430 showroom).

In 2009, J.A.O successively took acquisition of a Class A design unit, the Beijing Zhongtuo Construction Engineering Design Limited Liability Company (010518-sj), become one of the few foreign architectural design firms with Class A qualification. After integration of advantageous resources, the two parties formed a core competitiveness and powerful alliances to further strengthen the leadership status of J.A.O in Chinese architectural planning and design industry.

Other than international design firms, J.A.O as a multinational company sets up the best combination of Chinese and Western philosophy, winning the most awards in China, and becomes the most influential international design firm in China. Fully understanding and being familiar with China's system and management, along with the obstacle-free communication with competent leadership in all Chinese cities respectively, J.A.O combines international resources of it itself and the actual economic situation, relying on the fully integration under Group CEO, Mr. James Jao, and intensively mobilizes international financial resources to introduce and bring investment to enable the project to implement.

Providing customers with professional services with customized supporting consulting services

J.A.O takes "fine design" and "customized" as the motto. Based on projects demand of different types, it provides customers with supporting advisory services involving building development, investment feasibility studies, project investment and financing.

Inclusive Chinese and western culture, perfect embodiment of customers' wills

J.A.O focuses on the combination of amphibious culture, and shapes the concept of "when East meets West". It has successfully applied advanced design ideas and business philosophy to the planning and architectural design projects, so that customers' wills are perfectly embodied. By many years' practical experience and abundant social resources, it provides customers with high-grade and high rate of return special services.

Bridge for Chinese and Western communication

J.A.O can bring in the international advanced planning design concepts and make these ideas a reality through the funding plate and foreign resources. For those projects which have good base and bright perspective after planning, the company can help or independently attract foreign investment, and introduce foreign investment institutions including the funds of the Asian Development Bank (ADB), World Bank (WD). It can introduce large financial institutions involved in the project, so that the whole project achieves the perfect unity of completion from design concept to implementation. Business Representative Office of Manhattan, New York in China chose to be set up in J.A.O., which is one of the affirmation examples for its social status in the Chinese industry from public opinions.

Important awards and honors won by J.A.O. Design International Architects & Planners Limited and Mr. James Jao in recent two years:

2011 Capital Development Xianhai Lake Project (Mianyang) and Capital Development Lohas Town (Shenbei New District, Shenyang) won the prize of "Star of Green Building – The Best Green Building".

2011 CEO Mr. James Jao won the prize "Year 2010 Outstanding Real Estate Practitioner" granted by *China Construction Information Journal*, an excellent periodical under the administration of the Ministry of Housing and Urban-Rural Development.

2010 J.A.O. Design International Architects & Planners Limited won the title of "Owners' Most Satisfied Design Institute of China" awarded by Ministry of Housing and Urban-Rural Development, China Architectural Culture Center.

2010 Yishui Longmai, Shijiazhuang Project won the title of "Resource Conservation, Low-Carbon Community Influence Demonstration Building".

2010 Mr. James Jao was selected as Sohu Focus -- "2009 Leader of Real Estate Reviews" by Sohu Focus Website.

2010 Mr. James Jao was selected as one of the "Top 10 Most Influential Persons" on the awards ceremony of "Harmonized China – 2009 Influential Persons".

2009 In "2009 Urban Construction & Management Summit Forum of China – the 60 Years Awarding Activity of China's Urban Construction & Management", Mr. James Jao was honored as "Top Ten Outstanding Persons of China's Urban Planning" by *Architecture and Construction*.

2009 In the 2009 Economy Development Forum of Headquarter of Overseas Chinese of China – Qingtian, Mr. James Jao was honored as the "Top 60 Elites" of Overseas Chinese of Qingtian by Qingtian County Party Committee and Qingtian County Government.

2009 J.A.O. Design International Architects & Planners Limited won the "Special Award for Urbanization Process Scientific Development of China, 2009" issued by China International Urbanization Development Strategy Research Committee.

2009 "Red Delta Production Base of China Mobile Limited (Jiangxi Branch)" won the Excellent Award for Planning Architectural Design of 6th Elite Award for Housing Technology issued by China Commercial Real Estate Commission.

2009 Urban Design of Wangjiaqiao Area of Wuhua District, Kunming, Yunnan won the First Prize of International Competition.

2009 Overall Conceptual Planning of South Lake Eco-City, Tangshan won the Best Prize for Excellent Contribution of HBA - China's Demonstration Project, 2009 issued by United Nations Human Settlements Programme.

2009 International Orient Consulting and International Scheme Collection of South Lake Eco-City, Tangshan won awards for Construction Scheme.

2009 In Anniversary Awards Ceremony of China Commercial Real Estate Commission, J.A.O. Design International Architects & Planners Limited won the award of "China Famous Commercial Design Institute, 2008".

2008 Conceptual Scheme Collection of Urban Design, Kunming won the award for Excellent Scheme.

2008 In "2008 Urban Housing & Real Estate Development Summit Forum of China", Mr. James Jao won the award of "Person of the Year 2008 of China Responsible Real Estate" issued by China Real Estate Association and China Building Energy-Saving Emission Reduction Industry Alliance.

2008 In the 2nd China Urbanization International Summit, Mr. James Jao won the award of "Top 10 Contributors for China Urbanization Process, 2008" issued by China International Urbanization Development Strategy Research Committee.

2008 In the 5th Elite Award for Housing Technology issued by China Commercial Real Estate Commission, "Nanchang Oriental – Heidelberg" won Excellent Award for Planning and Design of Residential Community.

2008 In 16th Architectural Golden Stone Award of Taiwan, China, Urban Conceptual Design of Culture and Creativity Integrating Area in Southern Area of Songzhuang, Tongzhou, Beijing, Renovation Project of South and North Mongolian-Tibetan Cultural Landscape Street, Hohhot Municipality, Inner Mongolia and Architectural Scheme Design of Administration Center, Suining all won Design Awards of International Team.

2008 In *China Architectural Design Works Annuals, 2007 – 2008*, J.A.O. Design International Architects & Planners Limited was honored as Excellent Design Institute and became the Invited Editorial Unit.

2008 J.A.O. Design International Architects & Planners Limited won the Outstanding Achievement Award for Creative Planning Architectural Design of China Real Estate.

2008 Landscape Street with Islam Building Characteristic in Muslim Region of Hohhot Municipality, Inner Mongolia, China won the Promoting Design Award for Dubai International Example issued by United Nations Human Settlements Programme.

2008 Culture and Creativity Integrating Area of Songzhuang, China won the Promoting Design Award for Dubai International Example issued by United Nations Human Settlements Programme.

2007 Mr. James Jao won the title of "CIHAF Most Influential Designer of China, 2007". J.A.O. Design International Architects & Planners Limited was selected as "CIHAF the Most Influential Planning & Architectural Design Organization of China, 2007".

2007 J.A.O. Design International Architects & Planners Limited was one of the earliest "National Green Eco-Building Demonstration Project Contract Planning & Design Organizations" authenticated by China Commercial Real Estate Commission.

2007 J.A.O. Design International Architects & Planners Limited was selected as "Top 10 Large Real Estate Architectural Design Companies of China, 2007" by International Real Estate Association, International Construction Alliance and other organizations.

2007 In the celebration ceremony of Inner Mongolia Autonomous Region's 60th Birthday, the Project of Multi-Cultural Characteristic Landscape Street, Hohhot Municipality, Inner Mongolia won the "Best Construction Engineering Design Award". Mr. James Jao won the "Mayor Special Award".

Tel: +86–10–8587 5222 Fax: +86–10–8587 5922 E-mail: jaobj@jaodesign.com Http://www.jaodesign.com

阿克苏拜城县总体规划（2010年）

Aksu Baicheng County Master Plan (2010)

建设地点：新疆 拜城
建设规模：20平方千米
服务范围：城市总体规划、城市设计、控制性详细规划

Location: Baicheng, Xinjiang
Construction Scale: 20 Square kilometres
Service Scope: Master Plan, Urban Design, Regulatory Planning

拜城县地处新疆阿克苏地区，具有良好的生态环境与城市发展基础，本规划从总体规划、城市设计以及控制性详细规划3个层面展开，运用适宜的生态技术手法，提出一套将生态理念落地于城市的方法，通过各个层面的逐级控制，保障生态城市的落地建设，更加有效地指导当代我国城市的发展和建设。

本规划在尊重现有生态格局基础上，创建宏观的生态网络，建立生态区块的连接，规划确立"生态核心+生态廊道+生态斑块"的生态网络系统。组团及片区间不同层级的生态中心形成全覆盖的、多功能的绿色空间体系。控制城市街区规模，打造人性化可步行的城市，同时大力发展绿色交通、绿色产业等，引进新兴生态技术，诸如太阳能光电、地源热泵、中水回收等，建立起真正的西部生态城市。

核心区效果图

城北新区鸟瞰

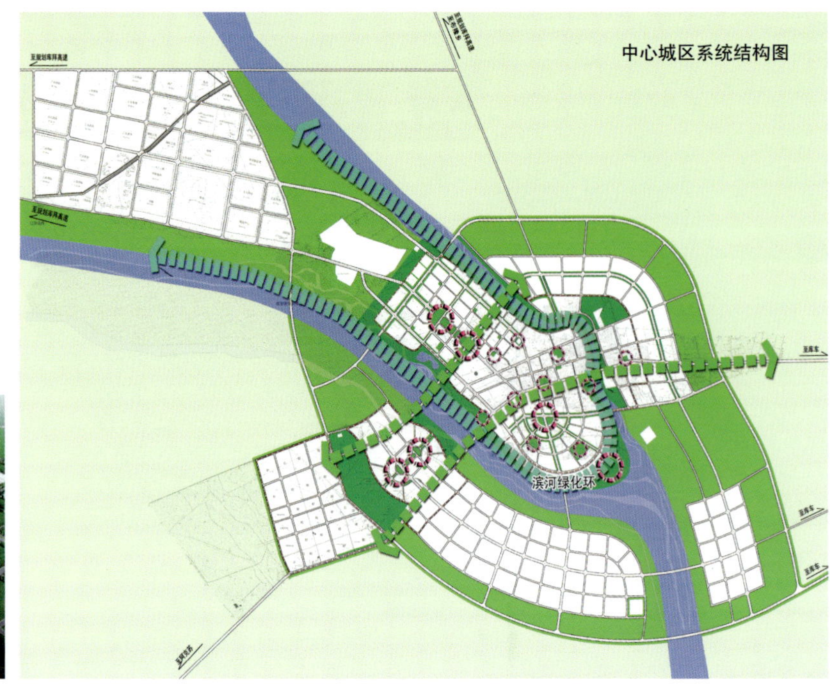

土地利用

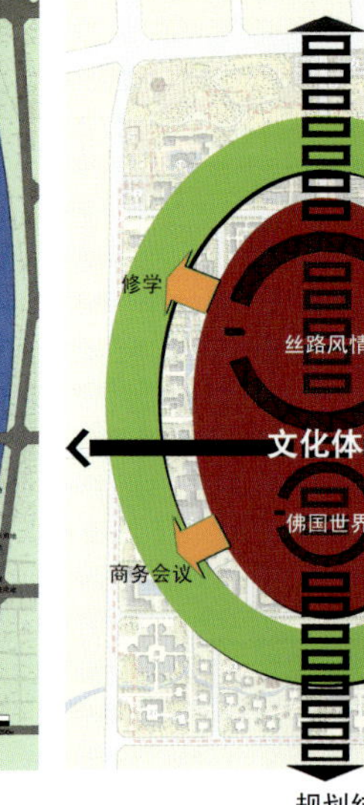

规划结构

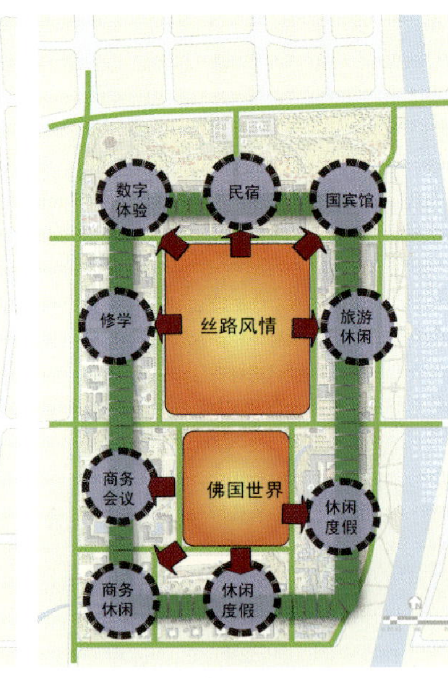

功能组织图

敦煌沙州古城复建规划项目（2009年）

Recovery Planning Design of Shazhou Ancient Ruins, Dunhuang (2009)

建设地点：甘肃 敦煌
建设规模：115公顷
服务范围：修建性详细规划

Location: Dunhuang, Gansu
Construction Scale: 115 ha
Service Scope: Site Plan

随着旅游业的日益兴盛，敦煌凭借其无与伦比的文化魅力将迅速拓展市场，敦煌古城借力于莫高窟的人文影响以及鸣沙山月牙泉独特的自然风光，通过对城市文化内涵的深度发掘，重塑敦煌古城的形象。

本次规划以古城之魂为线索，从古城众多的文化中，淬炼出其最具核心竞争力部分，即丝路文化与宗教文化，明确古城的资源禀赋，再现古城的辉煌。

本规划以华戎交汇为主题，从丝路文化以及宗教文化出发，建立核心文化体验圈层，以汉唐及西域风情深度体验为目的，弘扬中国国学文化以及佛教文化，集中展现古城文化的更替与交汇融合的文明的传承；对传统文化进行挖掘的同时，形成次生旅游休闲圈层，衍生出修学、商务会议、休闲度假等不同功能，完善敦煌市旅游基础设施，提升整体旅游环境。

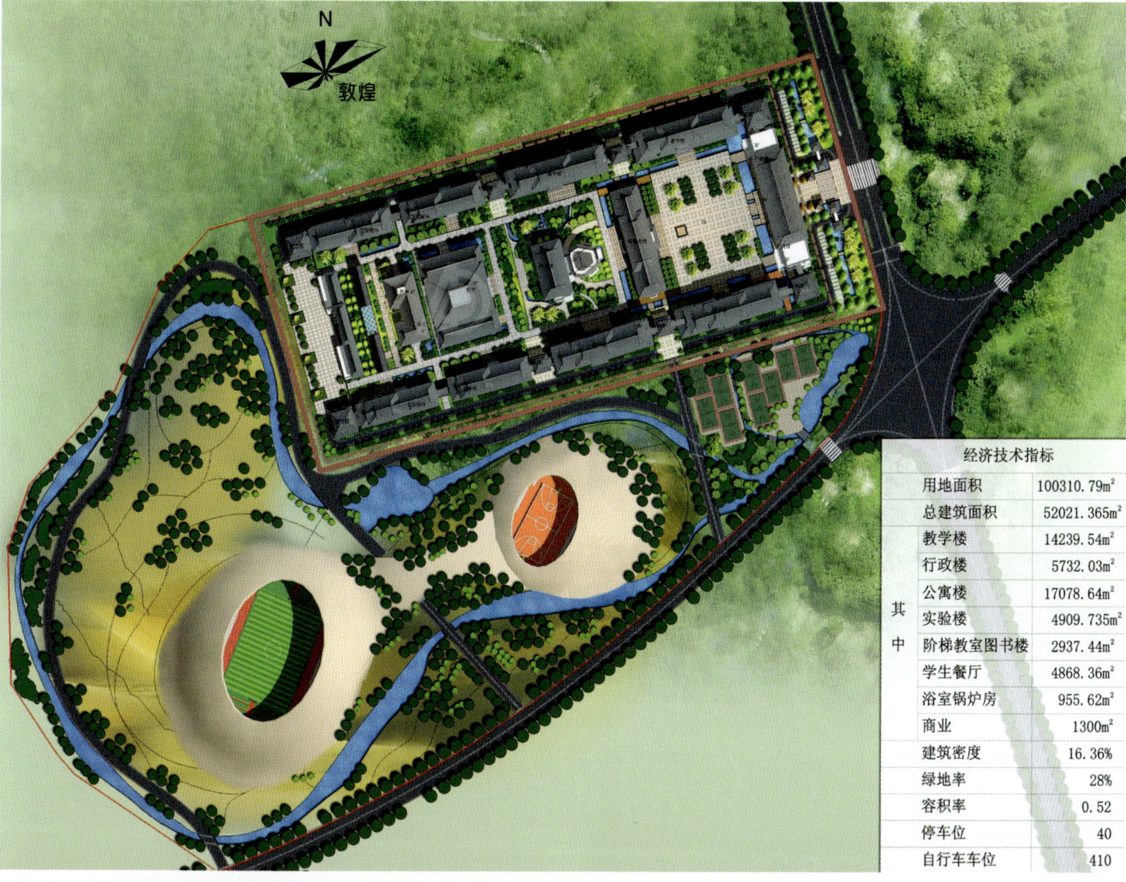

经济技术指标		
	用地面积	100310.79m²
	总建筑面积	52021.365m²
其中	教学楼	14239.54m²
	行政楼	5732.03m²
	公寓楼	17078.64m²
	实验楼	4909.735m²
	阶梯教室图书楼	2937.44m²
	学生餐厅	4868.36m²
	浴室锅炉房	955.62m²
	商业	1300m²
	建筑密度	16.36%
	绿地率	28%
	容积率	0.52
	停车位	40
	自行车车位	410

敦煌四中（2010年）

建设地点：甘肃 敦煌
建设规模：50 647平方米
服务范围：概念方案设计、
实施方案设计

Dunhuang Number Four Middle School (2010)

Location: Dunhuang, Gansu
Construction Scale: 50 647 m²
Service Scope: Conceptual Design,
Architectural Design

本项目位于甘肃省敦煌市，东临鸣沙山路，南侧为规划中的城市体育馆。场地地势平坦开阔。

1. 用历经千古劲风洗礼的美丽而神奇的沙漠之丘，雕塑成我们的主体育场和篮球馆；

2. 用水、花和绿树展示着顽强的生命和残酷的大漠千古斗争的和谐；

3. 把整个教学用地和体育用地，就其地形视做一个巨大的大漠宝石；

4. 把我们的四中——“敦煌书院”用古老而现代的书院格局雕刻在这宝石上，形成一颗巨大的艺术和文明之印，很中国！很经典！很敦煌！

5. 用水、树、花草的巧布妙植构成千古大漠美丽的生态新貌；

6. 曲直相融、通畅相宜的交通组织体现出书院古今相贯的气魄和宏达；

7. 有序的功能分区与组织构成书院生命永恒楼、阁、廊的有机布置，又为莘莘学子提供着万种求知立志空间；

8. 以汉唐建筑之雄，现代建筑之亲和构成独到的古今建筑相和谐的建筑风格；

9. 高低错位、组群井然、高雅而富有青春活力，无不象征新一代学子之风范；

10. 学校体育设施，利用近邻的敦煌市体育文化中心统一设置，以使城市资源共享。

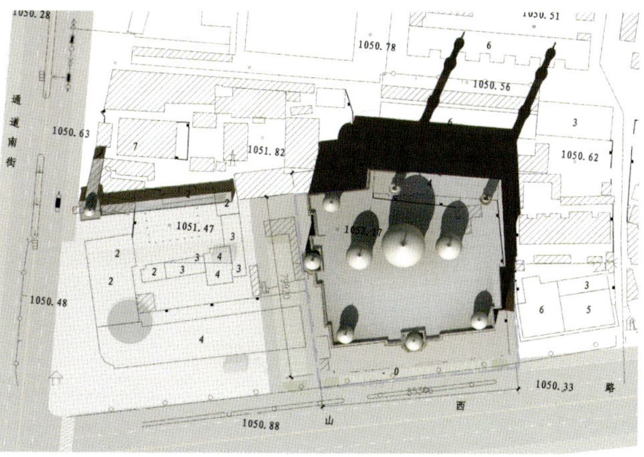

呼和浩特市回民区阿拉伯宫（2008年）

建设地点：内蒙古 呼和浩特
建设规模：30 600平方米
服务范围：方案设计、初步设计、施工图设计

Arabian Palace in Hohhot Inner Mongolia (2008)

Location: Hohhot, Inner Mongolia
Construction Scale: 30,600 m²
Service Scope: Architectural Design, Primary Design,
Construction Drawing Design

项目位于呼和浩特市中山西路与通道南路交汇处东北角，在行政区的划分上位于呼和浩特市回民区，属老城区范围。场地紧临城市主干道中山西路，占据商业中心的有利位置。场地周边为原始的回民部落，拥有独特的民族文化。

项目为集餐饮、表演及购物的综合体，力图打造呼和浩特新的商业中心，为老城区注入新的活力和生机。设计地上五层，地下一层：地下一层为停车场，一至三层为商场，四、五层为餐饮娱乐中心。整体造型选择了阿拉伯风格，穹顶和高塔的组合勾勒出清晰的建筑表情和丰富的天际线，符合呼和浩特市回民区的地域特征。

江西南昌东方海德堡（2007年）

建设地点：江西 南昌
占地面积：8公顷　建筑面积：143 000平方米
服务范围：建筑、景观方案，初步设计

Jiangxi Nanchang Oriental Heidelberg Community (2007)

Location: Nanchang, Jiangxi
Plot Area: 8 ha　Building Area: 143,000 m^2
Service Scope: Architecture and Landscape Design & Preliminary Design

项目规划设计立足于从人文、社会、环境与文化特点出发，根据以人为本及可持续发展的原则，充分考虑现代社会住宅使用者对居住环境的多层次需求，创造一个"地标性高尚居住区"。规划设计强调尊重环境、以人为本，着力体现邻里和谐，倡导交流的现代社区精神，打造"景观、文化、情感、交流"的主题，使最多的住户同时享有"阳光、绿化、水体和场所"，在满足人与人交往的同时，休闲、文化、绿化环境成为社区生活的主旋律，以适应本社区多元化、差异化、不同层次和人群的特殊需求。

本项目的建筑、街道、绿地、人行步道、自然驳岸、景观水体等元素，通过富于想象力的设计手法，有效地整合在一起，充分体现生态理念，绿色广场的设计表现、自由水体的运用和对阳光的重视，以及为邻里和谐营造良好的交流氛围，塑造了建筑形态的标志性、时尚性和可识别性。建筑外立面采用新古典主义的设计手法，玻璃和其他材料的良好结合运用，使之具有时代感和韵律感。在节约能源和资源的循环利用方面，采用极具价值的雨水收集系统、绿色生态系统、植被降尘系统，太阳能节能系统等等，提高住区绿色建筑的品位。

美国龙安集团新址室内设计（2011年）

建设地点：北京
建筑面积：2 402平方米
服务范围：方案设计、施工图设计

Interior Design of New Premises of J.A.O. Design International Group (2011)

Location: Beijing
Building Area: 2,402 m^2
Service Scope: Architectural Design, Construction Drawing Design

该项目将“乐活”的生活理念引入住宅市场，必将带动新的市场方向，成为区域领跑者。从沈阳及沈北新区发展来看，蒲河新城将得到大力发展，其生态化的城市环境必将吸引大量置业及投资人群。乐活小镇项目力争打造新城的首座生态宜居小镇，创造特色生态居住环境，提升地区城市形象，树立虎石台地区居住标杆，打造蒲河新城及沈北新区的居住品牌。

良好的人居环境，是打造生态小镇的前提条件。首开·国风润城（乐活小镇）的设计切入点“乐活”即来源于此。“乐活”是一种生活态度，是乐观和包容，生活的质量不仅取决于经济基础，更取决于全新的生活观念。

“乐活”由LOHAS音译而来，LOHAS是英语Lifestyles of Health and Sustainability的缩写，意为以健康及自给自足的形态生活，强调“健康、可持续的生活方式”，“健康、活力、生态、和谐”是乐活的核心理念。

首开·国风润城（乐活小镇）项目适合打造完整的小镇空间和功能结构，形成完善的小镇人文气息和情调氛围，充分展现小城镇应有的活力和生命力。以“乐活”为开发理念，打造一个生活的天堂、充满活力的微型都市，引领沈阳新城市生活方式的风情宜居小镇。在这里，居住环境将对居住者的社交活动、生活方式等产生一系列的重要影响；在这里，我们可以体验到一种生活的享受，可以感触到家园的认同感与归属感。

沈阳首开·国风润城（乐活小镇）（2010年）
建设地点：辽宁 沈阳
占地面积：1 078 000平方米
建筑面积：493 000平方米
服务范围：方案设计、初步设计

Shenyang Shoukai · National Custom Happy City (Lohas Town) (2010)
Location: Shenyang
Plot Area: 1,078,000 m^2
Building Area: 493,000 m^2
Service Scope: Architectural Design, Preliminary Design

首开通州马驹桥住宅区（2011年）
建设地点：北京
建设规模：296 490.62平方米
服务范围：总体规划方案、景观概念设计、
建筑方案设计、初步设计、施工图设计

BCDH Tongzhou Majuqiao Residential Area (2011)
Location: Beijing
Construction Scale: 296,490.62 m^2
Service Scope: Master Plan, Landscape Design, Architectural Design,
Preliminary Design, Construction Drawing Design

本项目的建筑群落，像一只只蓝天下自由飞翔的白天鹅，它们在蓝天下，或单或对或成行，构成一朵朵、一丝丝美丽动人的祥云。在这祥云下面，是美丽、自然、生态的绿洲。天鹅们要栖息在这里——带着它们深深的眷恋和甜甜的梦想！我们的建筑群落，像一座座振奋人心的现代都市雕塑，标志着首都新型建筑形态的开端，代言片区城市风貌发展的永恒经典，领跑片区文明和经济，示范中国农迁房与都市和谐发展！我们的建筑群落，系统而全面地贯穿绿建（低碳、环保、节约、友好）国际精神，并首次实施整个社区生态公园化，彻底人车分流，把安全、从容的原生态体系归还人类的栖息场所，同时把社区建设成为现代智能化社区，并做到完善的现代生活配套，充分展示一个生态体系上享尽现代文明的时尚社区。这就是我们首开——国企、大企的英雄气概，把党和政府亲民爱民、建设城乡一体化的意志落实到成果，把高尚和品质贡献给社会，把美丽和经典留给历史——这是我们的责任，这是我们的开发宗旨，这是我们让这片土地承载的伟大价值！

首开绵阳熙悦湾规划项目（2010年）

建设地点：四川 绵阳
建设规模：360公顷
服务范围：修建性详细规划

Mianyang Xianhaihu Lake, Capital Development Company (2010)

Location: Mianyang, Sichuan
Construction Scale: 360 ha
Service Scope: Site Plan

此项规划将仙海湖度假旅游与人文生态有机整合，将城市稀缺生活资本——湖域风光、山地人文、国际人居价值共冶一炉，造就既满足现代人旅游度假、休闲娱乐的需求，更实现了一种纯美湖居生活的理想，创造自然、健康、休闲人居新模式，塑造集休闲、商务、度假、居住于一体的旅游度假目的地。

通过对市场的研究，该规划确定：以高尔夫为楔入点，提升知名度与影响力；以旅游度假为主题，带动区域人气的集聚；以高尚居住为主体，塑造高品质旅游度假目的地，树立项目在区域的标杆性形象，塑造为仙海度假区、绵阳市乃至中国西北部的名片，带动地区经济发展。

石家庄易水龙脉（2010年）

建设地点：河北 石家庄
占地面积：256 640平方米
建筑面积：493 431平方米
服务范围：方案设计

Yishui Longmai Residential Community, Shijiazhuang (2010)

Location: Shijiazhuang, Hebei
Plot Area: 256,640 m^2
Building Area: 493,431 m^2
Service Scope: Scheme Design

石家庄易水龙脉项目位于石家庄正定县塔元庄，距石家庄市12千米，距正定县3千米。项目紧邻京石高速和胜利北街，南侧是16千米滨水景观带——紧邻滹沱河1号水面的67公顷生态岛（滹沱河整治工程，政府正在建设中）。住宅建设用地275 000平方米，容积率为1.2。

本项目配合地方建设，立足于资源节约、绿色低碳的设计理念。规划设计的各个细部都围绕这一主题展开。

项目规划设计着眼于营造跨世纪的人居环境，体现以人为本和新经济时代的社会精英阶层对成功和创意生活的双重追求，突出人与大自然的融合。项目的规划设计从居住行为功能出发，由环境和空间的组成入手，体现对人的理解和关怀，强调经济效益、环境效益和社会效益的有机结合，实现人居环境建设的健康性、安全性、环保性和与自然环境的亲和性，从而营造出高标准、高技术、充满情感的智能化生态型人居环境社区。

中策尚德府邸（2010年）

建设地点：浙江 宁波
占地面积：3.33公顷
建筑面积：94 700平方米
服务范围：方案设计、初步设计

Shangde Palace (2010)

Location: Ningbo, Zhejiang
Plot Area: 3.33 ha
Building Area: 94,700 m^2
Service Scope: Architectural Design, Preliminary Design

本项目位于宁波北商务区的总部魔方区，东临金山路，西近望山路，南起规划路，北至长兴路。地块位于4号轻轨线与绕城高速相交的核心商务区，西接规划中的华东商场与慈城新城，北枕狮子山休闲公园，向东紧靠江北工业园区本部，向南毗邻以规划中的奥林匹克体育中心为核心的未来商务区，是一块以新兴中小企业总部为依托，承载宁波北门户综合商务区历史使命的商业旺地。地块总用地面积3.33公顷，东侧与北侧道路已修建完成，配套条件成熟。

项目为商务办公集合的中小企业的总部基地，具体形态主要表现为独栋低密度花园办公区、带有配套性质酒店的甲级写字楼、行政商务楼；通过建筑与景观的规划以及户型的创新设计、服务性酒店及配套商业的设置等多方面的烘托来打造江北门户区标杆性作品；目标客户群主要定位为工业园区内对办公品质有较高要求的中小企业主。

设计以合理布局、品位高尚、注重细节、空间丰富为主要设计思路，将地块的功能用到极致，打造具有浪漫欧式气质的高尚办公区，提供高雅、丰富的功能设施。

青岛虹字河水库周边地区概念规划项目（2010年）

建设地点：山东 青岛
建设规模：12平方千米
服务范围：概念性规划

Conceptual Plan Project of Qingdao Hongzi River Reservoir Areas (2010)

Location: Qingdao, Shandong
Construction Scale: 12 square kilometers
Service Scope: Conceptual Plan

项目具有毗邻机场、城市门户节点的战略位置，具有非常优良的生态资源。

规划以绿色生活为发展基础，以会议会展、商务酒店等为核心产业，延伸产业链条，通过康体休闲、特色度假等配套产业建设，促进片区服务业发展，完善城阳区的产业结构，强化与周边地区的协同互补关系，引导地区服务业的特色化、差异化发展，构建出一个建构于生态文明理念、成长于现代服务经济、引领休闲时代的现代生态型会议休闲城。

北立面展开图1:300

西立面图1:300

天津滨海圣光皇冠假日酒店（2006年）

建设地点：天津
建设规模：150 000平方米
服务范围：建筑方案设计、初步设计

Crowne Plaza, Binhai Tianjin (2006)

Location: Tianjin
Construction Scale: 150,000 m^2
Service Scope: Architectural Design, Preliminary Design

本项目位于空港物流加工园区内的核心位置，毗邻空港物流加工区行政中心和已建的18洞温泉高尔夫球场。北侧紧邻园区中心湖面，隔湖是规划的公建区，东侧为园区中心环岛和规划的居住区，西侧隔路与保税区相望。

项目用地呈比较规则的长方形，长边近200米，短边长为185米。规划用途为五星级酒店和酒店式公寓，占地面积约3.5公顷，容积率为3.18，绿地率35%，建筑密度37.2%，建筑限高43米。

五星级酒店10层、酒店式公寓及写字楼11层，均为框架–剪力墙结构。地上总面积112 692平方米，其中酒店部分建筑面积40 513平方米，客房总钥匙数372，写字楼部分建筑面积28 390平方米，酒店式公寓部分建筑面积41 506平方米，总户数为469。

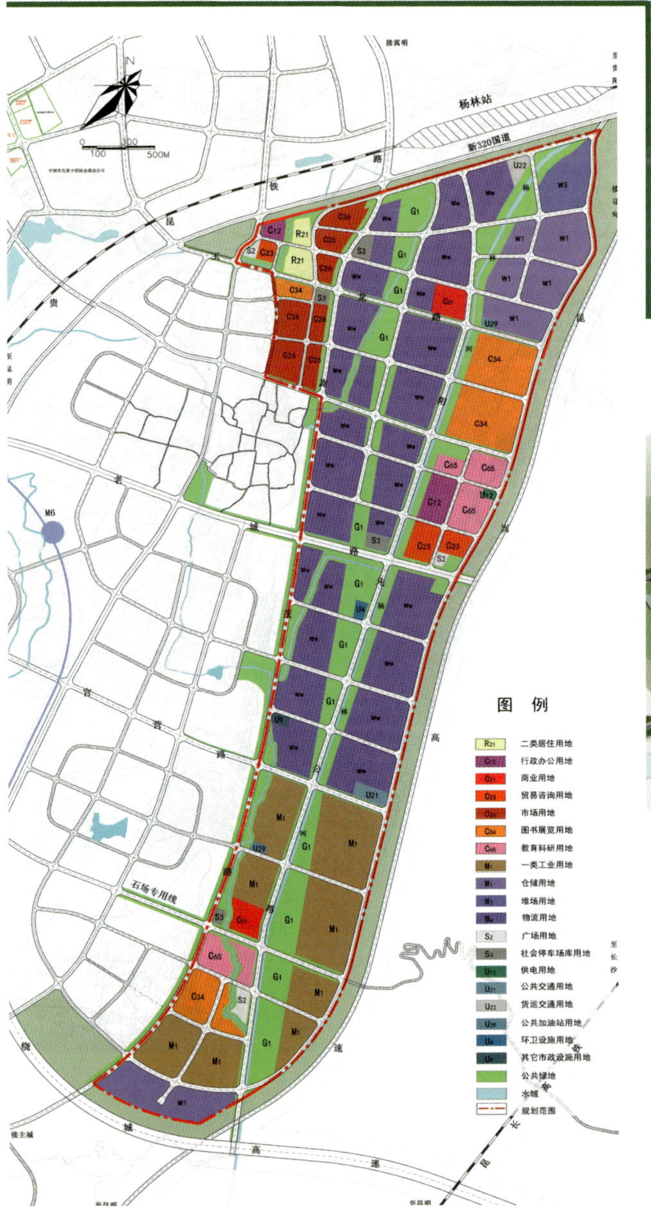

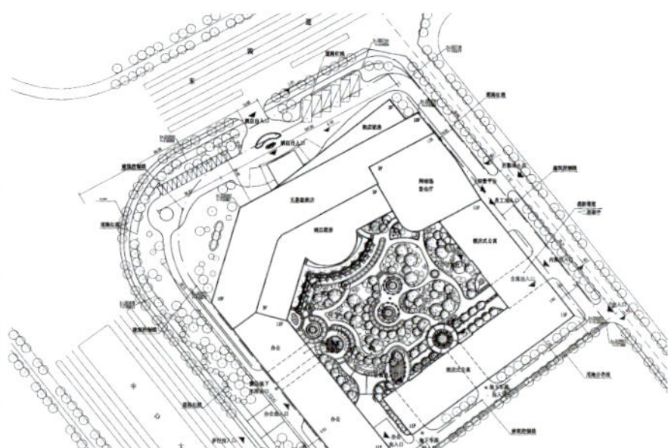

云南省昆明市杨林泛亚商贸物流园与泛亚家具产业园规划（2010年）

建设地点：云南 昆明
建设规模：800公顷
服务范围：控制性详细规划

Regulatory Plan of Fanya Trade Logistics Park & Fanya Furniture Industrial Park, Yanglin, Kunming, Yunnan (2010)

Location: Kunming, Yunnan
Construction Scale: 800 ha
Service Scope: Regulatory Plan

在云南省二强一堡战略、昆明市建设面向西南桥头堡战略的宏观背景下，位于空港新城的杨林泛亚商贸物流园与泛亚家具产业园迎来了跨越式发展机遇。规划把握竞争特色优势，结合当前物流产业与家具产业发展趋势，引入新产业理念，将本次产业园建设为引领物流新风向的国际陆港、创新型示范家具产业园。

其中，泛亚商贸物流园以仓储、配送、铁路公路联运等现代物流为主体功能，涵盖商贸、展示、交易、中介服务等现代商贸职能，形成以专业化市场为基础、仓储配送为特色、商贸信息为核心、第三方物流为支撑的昆明地区主要的商贸物流中心。

泛亚家具产业园以创意、研发、制造、展示为核心组成部分，辅以商贸信息等服务职能，汇聚“学、研、产、商”一体化资源，加速昆明及云南家具产业升级，立足昆明、面向全国、辐射南亚和东南亚家居市场，立体化打造全国一流的“创新型示范园区”。

通道南路伊斯兰特色风情街（2006年）

建设地点：内蒙古 呼和浩特
建设规模：1 150米
服务范围：建筑方案、施工图设计

Tongdao South Road Islamic Feeling Street (2006)

Location: Hohhot, Inner Mongolia
Construction Scale:1,150 m
Service Scope: Architectural Design, Construction Drawing

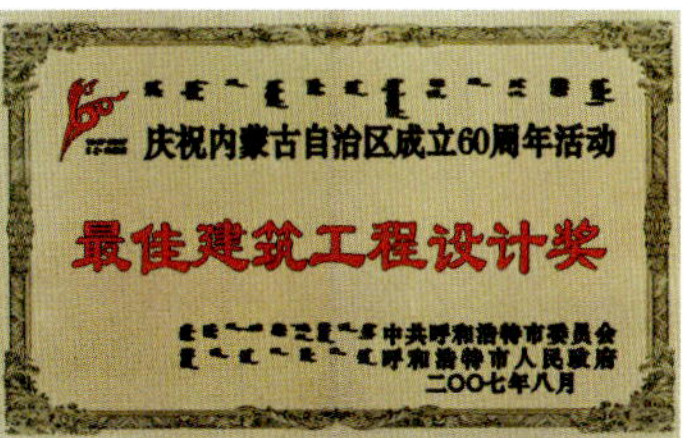

九鹏商厦原貌

九鹏商厦改造后日景

清真北寺原貌

清真北寺改造后日景

赵莉大师私宅（2010年）

建设地点：北京
建筑面积：2 152.7平方米
服务范围：建筑方案、施工图设计

Private House of Master Zhao Li (2010)

Location: Beijing
Building Area: 2,152.7 m^2
Service Scope: Architectural Design, Construction Drawing

本项目位于南北长、东西窄的条形地块上，总占地面积约1 775平方米，整体布局犹如北京四合院的院落，设计为两进院落三重建筑的房子。

由南侧入户，第一重建筑主要为主人的接待区，包括客厅、厨房、餐厅和一间会议室。

第二、三重建筑为画廊，其中第二重建筑主要以贵宾休息厅和画家的工作室为主，第三重建筑主要为画家的画廊，为一开敞挑空的空间，主要用于画家平时绘画和展廊用。三重建筑中间，布置了一大一小两个精致的院子，南侧院子稍大，用于平时接待客人，周边房间布置为餐厅和多功能厅，尽可与三五好友在此享受到远离市区的宁静。

北侧院子是主人为自己留下的私人小院，不大不小的院子，建筑的尺度刚刚好，周边房间是主人的私人收藏，只有主人的私密客人才能到达此处。

二层北侧是首层画廊的挑空，空间开阔，中央位置设计了一座桥，既可俯瞰整个画廊，也可作为空间的分隔。南侧两重建筑均为主人的生活起居空间，南侧建筑主要以公共起居空间为主，中央建筑以起居为主，包括几间卧室。

正如赵莉本人，人淡如菊，这个小小的美术馆，像一朵秋天的雏菊，淡淡地开放。

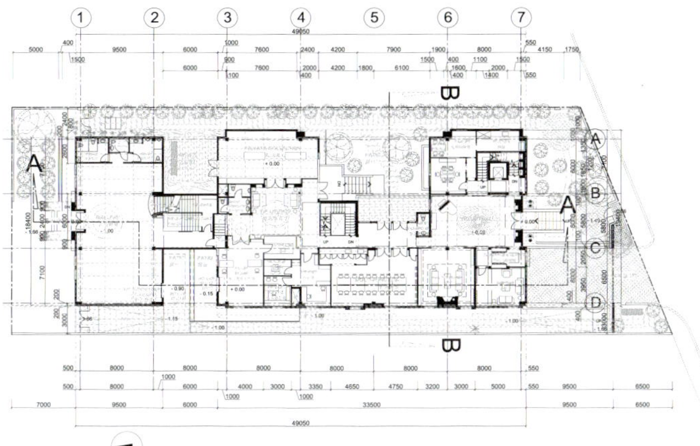
总平面图

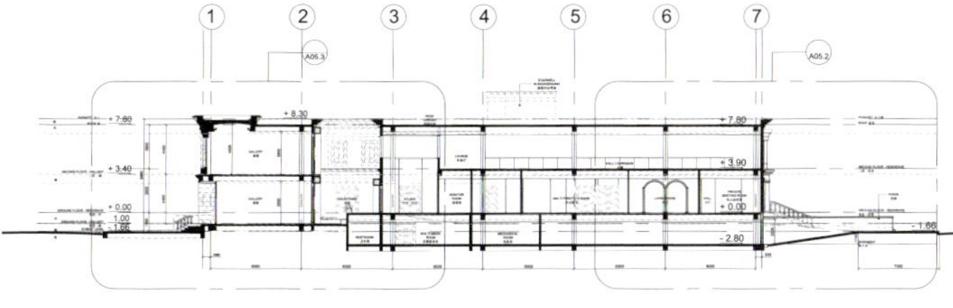
纵向剖面图

青岛索菲亚大酒店地下水城室内设计项目

建设地点：山东 青岛
建设规模：5 000平方米
服务范围：方案设计、施工图设计

Interior Design of Underground SPA Club in Sophia International Hotel, Qingdao

Project Location: Qingdao, Shandong
Construction Scale: 5,000 m^2
Service Scope: Conceptual Design, Construction Drawing & Design

一区四楼地面图
比例 1:200

烟台海阳旭家会所室内设计项目

建设地点：山东 烟台
建设规模：1 500平方米
服务范围：方案设计、施工图设计

Interior Design Project, Tiber Beach Golf Club, Yangtai

Project Location: Yantai, Shandong
Construction Scale: 1,500 m^2
Service Scope: Conception, Construction Drawing & Design

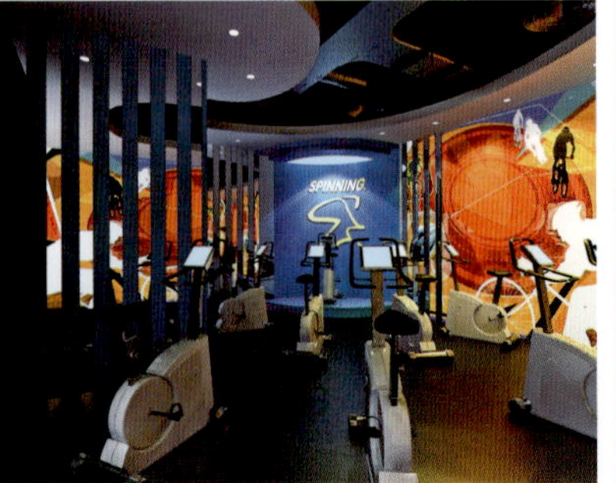

HMD. 汉米敦（中国） HMD CHINA

汉米敦建筑设计有限公司注册于英国伦敦，是一间提供多专业工程咨询服务的设计公司。为了更高效地为中国城市化建设提供专业服务，分别在上海、北京、深圳设立了分公司及办事处。汉米敦（中国）的业务范围涵盖了从城市总体规划、城市设计、建筑设计、景观环境设计到室内设计的专业内容。通过不懈的努力，公司在星级酒店、城市综合体、商业地产、旅游地产以及大型房地产项目上获得了不俗的业绩。汉米敦（中国）公司的董事及其核心团队，曾共同服务于英国阿特金斯中国公司，尤其是董事总经理徐石先生，设计董事保罗莱斯及李剑波先生，都参与过阿特金斯中国公司的初期发展及重大项目的设计工作，如上海松江泰晤士小镇、上海杨浦大学城创智天地等。北京公司带磊涛先生亦有着丰富的专业经验，在他们的领导下，由近百位来自不同国家和地区的专业人士所组成的国际化团队，为客户提供着国际水准的高质量咨询服务。

汉米敦的专家们所追求的是将国际先进的设计理念、高端的建筑科技应用到城市发展中，并使之与中国的市场及传统文化融合，进而创造出具有独特文化价值及建筑语言的设计作品。公司推崇开放及创新、荣誉及责任、品质及卓越的核心价值观，提倡一个多元化的、共同创作的工作氛围，并时刻要求设计师们注重对社会、对环境、对客户的责任和义务。

HMD is an international multidisciplinary design consultancy with offices in Shanghai, Beijing and Shenzhen. HMD China's business scope covers Planning, Urban Design, Architectural Design, Landscape Design and Interior Design. HMD's work is focused on hotel, commercial, retail development, tourism and large-scale residential projects. The three directors Mr. Jeff Xu (Managing Director), Mr. Paul Rice (Design Director) and Mr. Lee Jianbo (Partner and Senior Architect) with the core team of HMD staff have all worked together in China for many years. The HMD China team, now numbers more that one hundred professional staff.

HMD's goal is to form a socially responsible professional business model, integrating international best practice and Chinese cultural values. It is the aim of the company to create design work of long term value, practicality and elegance.

HMD 上海

上海市常德路800号八佰秀9号楼6楼
6/F, Building 9, No.800 Changde Road, Shanghai
Tel: +86–21–32553266
Tax: +86–10–32553266–801

HMD 深圳

广东省深圳市福田区益田路6009号新世界中心2606室
Room 2606, New World Centre, No.6009 Yitian Road, Futian District, Shenzhen
Tel: +86–755–83216996
Fax: +86–755–22211820

HMD 北京

北京市朝阳区建国路乙118号招商局京汇大厦1002室
Room 1002, NO.B-118, The Exchange Beijing, Jianguo Avenue, Chaoyang District, Beijing
Tel: +86–10–65677929
Fax: +86–10–65677929–801

沿海沈阳国际中心（四季酒店）

建设地点：辽宁 沈阳
占地面积：31 757平方米
建筑面积：241 135平方米
设计时间：2010年
客户名称：沈阳沿海荣天置业有限公司

该设计以整体发展社区的模式，通过内部庭院、广场和连廊，将各个建筑在地下层、一层组合串联。同时尽可能多地使用建筑高度，减少覆盖率，将室外空间释放出来形成城市广场和绿地。 设计旨在打造一个完整的城市综合体，融合了住宅、公寓、酒店和商业。在布局上各种业态相对独立、流线互不干扰，以不同的业态促进彼此的发展。

Coastal Shenyang International Center

The design adopts a model of developing the community in all aspects, which connects every building underground and on the first floor via internal yard, square and connected corridors. Meanwhile the design is trying to make use of building heights as much as possible to reduce coverage rate, releasing outdoor spaces to form urban squares and green land. The design aims at building a complete urban complex which integrates residence, apartment, hotel and commerce and each part of the complex is relatively separate in respect of layout and their streamline will not obstruct each other, which will lead to promotion of the development of each part.

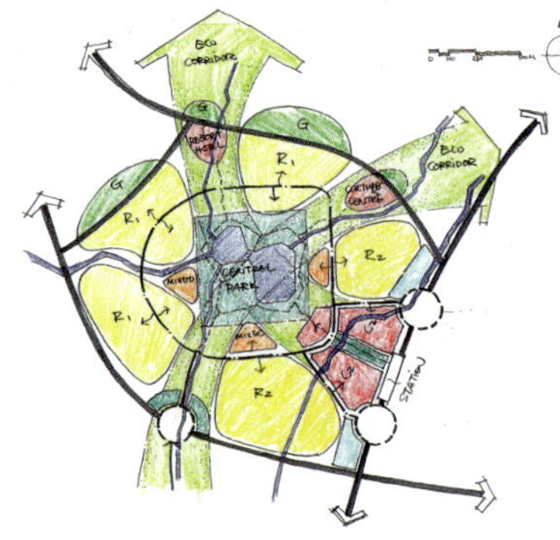

朝阳区孙河组团规划设计

规划面积：2.83平方千米

地理位置的优势加快了孙河项目与中心城区的功能对接，促进了孙河项目与周边区域产业的资源互补、产业协调和区域互动。依托项目便利的交通、优越的生态环境、大量的居住人群，规划区域内将重点发展集商业、文化、休闲相结合的综合性功能区，包括健身、购物、娱乐、休闲、品牌餐饮、商务酒店等六大功能，重点推进音乐厅、体育场、图书馆等文体设施的建设，将孙河项目打造成生态功能、经济功能和谐统一的朝阳区商业文化休闲区，成为温榆河绿色生态走廊发展的重要组成部分，为北京周边新城建设树立"世界城市生态发展的示范区"。

Planning and Design of Sunhe County Group, Chaoyang District

Planned Area: 2,830,000 m^2

Geographical advantage of Sunhe project speeds up its connection with downtown functions, and promotes resource complement, industry coordination and regional interaction with industries of surrounding areas. Relying on convenient transportation, excellent ecological environment and a large number of residents, the planned area will be developed into a multi-functional area integrated with business, culture and leisure, including six major functions: fitness, shopping, entertainment, leisure, chain restaurants and business hotel. The project focuses on promoting constructions of concert hall, stadium, library and other sports and cultural facilities. Sunhe County will be built into a business cultural leisure area of Chaoyang District with unified ecological and economic functions and become an important part of the development of Wenyu River Green Corridor to establish a "Demonstration Area for Urban Ecological Development of the World" for new town construction around Beijing.

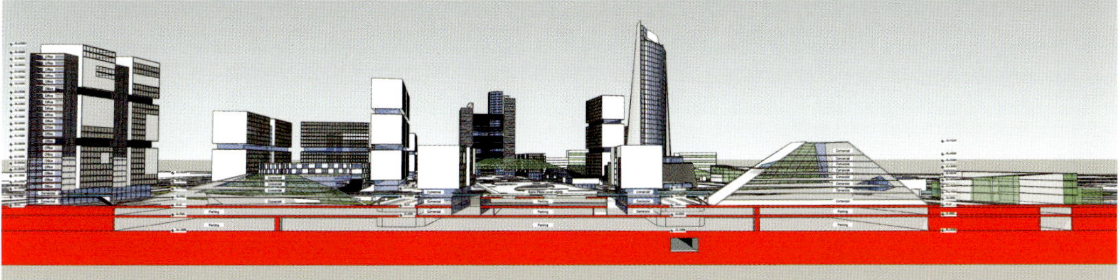

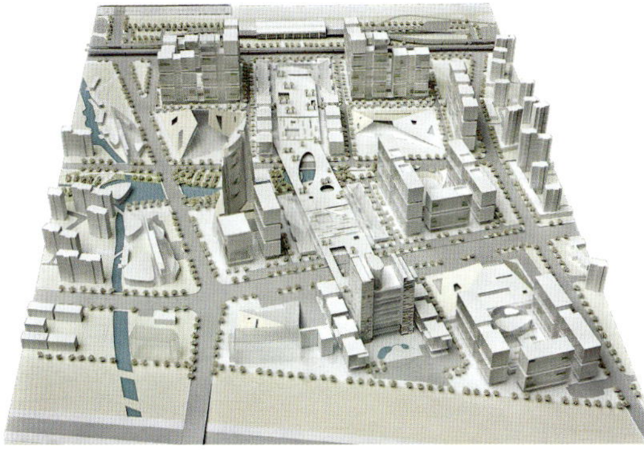

苏州工业园区城铁综合商务区城市设计

建设地点：江苏 苏州
占地面积：1 500 000 平方米
建筑面积：3 000 000平方米
设计时间：2011年
客户名称：苏州工业园区城市重建有限公司

本项目基地地理位置优越，交通便利，是苏州高端商务客流进出的主要门户。项目将依托规划地铁、城市快速路，快捷联系城市中心，打造成苏州北部新城都心以及高铁门户服务高地。规划确立三轴：横向生态绿轴、纵向商业商务轴和滨水文化轴；四片区：站前商业商务区、行政综合区、文化休闲区和生态居住区。站前商业商务区具有活力、充满朝气和拥有宜人尺度的城市轴线结合中心绿廊形成一个多样化空间、步行可达、立体交通以及可持续发展的生态低碳的示范城区。

Urban Design Of General Business District Of Inter-City Railway Station at Suzhou Industrial Park

The project is aimed at building up a new city center and a service center for the high-speed railway in northern Suzhou. There are three axes that will be developed according to the plan: a landscape ecological green belt, a lengthwise retail and commercial area and a river bank cultural belt. There will be four zones: the commercial and retail zone next to the station, an administrative and comprehensive zone, a central cultural and recreational zone and ecological residential zone. The commercial and retail zone in front of the station is traversed by a city axis which creates vitality and a pleasant scale. Combined the central green belt with a multi-dimensional, ecological and low-carbon demonstration city area a sustainable development will be built up. In this district people can get to anywhere by foot and can also enjoy a variety of public transport system.

武汉巴登城高尔夫度假别墅

建设地点：湖北 武汉
占地面积：233 410平方米
建筑面积：116 702平方米
设计时间：2010年
客户名称：武汉巴登城投资有限公司

在设计中充分考虑了建筑与景观的有机融合，古典建筑与现代生活的完美融合，在酒店和水疗中心等的高端配套设施中打造出一个顶级休闲居住空间。 本项目总体规划布局呈现出叶脉般的内在逻辑结构。主脉为居住区道路，串联三大居住组团和各公共服务设施，细胞单元阵列沿绿地景观依次展开，各细胞单元间的“空隙”仿佛生态景观廊道，使生态景观资源能够更有效地向社区方向渗透。

Wuhan Badeng City Golf Resort Villa

The design has fully considered the organic integration of buildings and landscape, the perfect combination of ancient buildings and modern life, with the aim of building up a top-level recreational residential space out of the premium accessory facilities of the hotel and SPA center. A leaf vein-shaped internal logical structure is represented in the general planning and layout of the project. The main vein contains roads of residential areas, which connects three main residential groups and various public service facilities. The branch units are spread respectively along green belt and landscape and the "gaps" between branch units look like ecological visual corridors, which enable ecological landscape resources to infiltrate into the community more effectively.

正立面

背立面

万科武汉红郡

建设地点：湖北 武汉
占地面积：230 000平方米
建筑面积：346 000平方米
设计时间：2009年
客户名称：万科房地产有限公司

新城市主义理念在红郡的设计中得到了充分的展现：井然有序、配套齐全、绿意盎然、尺度宜人、其乐融融。开放的城市核心路、社区中心、街区、院落的有序组合，使传统武汉老街坊邻里的融洽氛围和浓厚的生活气息又重新展现。红郡整体上来说，同时延续了城市花园和万科的气质：平凡、理性、成熟、执著、创造。它也许并不会让人眼前一亮，但静静品味、慢慢融入，则能体会出其特别之处。

Wuhan, China Vanke Red County

The Red County project demonstrates concepts of new urbanism: an orderly, green, properly scaled space is formed. The combination of core roads, community centers, neighbourhoods, and courtyards makes a harmonious atmosphere and clear structure that will enhance the traditional neighbourhood space The Red County continues the style of the city garden and the rationale of Vanke project development.

武汉沿海黄狮海酒店商业综合区（洲际皇冠会展酒店）

建设地点：湖北 武汉
占地面积：7 760 000 平方米
建筑面积：163 000平方米
设计时间：2010年
客户名称：沿海地产投资（中国）有限公司

该项目依托沿海品牌，既包括以四星级湖景假日酒店为核心，集休闲、娱乐、度假、会议于一体的高端商务、居住综合体，也包含黄狮海岸生态、健康、舒适型的生态花园式居住社区。力求将该项目打造成城市新地标、塑造城市突出形象的同时，形成个性居住文化。

Wuhan Coastal Huangshi Hai Hotel Commercial Area (Intercontinental Crown Convention Hotel)

The project is a high-end commercial and residential center, including a four-star Holiday Hotel with the existing lakescape and other types of facilities for leisure, entertainment, vacation and meeting. The project also includes an ecological garden residential community with a new pedestrian retail street environment. By separating the vehicle and pedestrian systems, maximizing the waterfront views and creating a significant new landscape space the project creates a type of people orientated living style.

集士港镇宝龙城市广场概念方案设计

建设地点：浙江 宁波
用地面积：169 134平方米
建筑面积：309 041平方米
设计时间：2011年
客户名称：宝龙实业发展有限公司

Ji Shi Harbor Town Bao Long City Plaza Concept Design

Project Location: Ningbo, Zhejiang
Land Area: 169,134 m^2
Floor Area: 309,041 m^2
Design Time: 2011
Owner: Baolong Industrial Development Co., Ltd.

重庆万科河运校住宅

建设地点：重庆
占地面积：105 463平方米
建筑面积：645 373平方米

本项目的设计宗旨是“以人为本、健康生活、高品质居住环境”。规划从环境入手，通过综合分析小区周边环境因素，确立社区组成：社区以两种完全不同的生活理念出现在人们面前，一种是无遮挡的高层体量尽量往高空发展，以完好的朝向使人们能够充分享受到阳光；而另一种方式则是根植于大地，尽可能贴近地面，同时将环境及绿化通过露台发展到三维的空间中去，使居住于此的人们能够充分地享受到接近大地的益处。

Wanke Heyun School Community, Chongqing

Project Location: Chongqing
Land Area: 105,463 m^2
Floor Area: 645,373 m^2

The purpose of this project is “People-oriented, Healthy-living and High-quality Environment”. Starting from environment, through a comprehensive analysis of environmental factors of surroundings, the planning defines composition of community: community comes in people’s lives in two completely different ways. One is non-blocking high-rise building as high as possible, which ensures good sunlight with good orientations; the other is rooted in the earth, as close to the ground as possible, which meanwhile brings the environment and greening into a three-dimensional space through terrace, so that people who live there can fully enjoy the benefits of living close to earth.

James Wang Design Associates, Inc
美国James · 王建筑师事务所
James Wang Design Associates
北京杰地亚建筑咨询有限公司

地址：北京市朝阳区新源里16号琨莎中心2座1101室
邮编：100027
电话：+86-10-88510711
传真：+86-10-88510892
邮箱：jwda@vip.163.com

Add: Suite 1101, Tower 2, Kunsha Building, 16 Xinyuanli, Chaoyang District, Beijing
P.C.: 100027
Tel: +86-10-88510711
Fax: +86-10-88510892
E-mail: jwda@vip.163.com

美国James · 王建筑师事务所创办人James Wang是美籍华人、中国台湾淡江大学建筑学学士，美国伊利诺大学建筑硕士，是美国建筑师协会会员。1987年在洛杉矶创立本事务所，2000年在北京成立北京杰地亚建筑咨询有限公司，设计人员近百人。

美国James · 王建筑师事务所下设规划建筑设计部和室内装潢设计部。业务从规划到建筑单体乃至室内设计都具有多元化设计经验，每一个设计作品都有各自独特的理念，并领先于本行业。善于根据业主的需求将产品做到温馨、舒适、豪华、低碳、环保。美国James · 王建筑师事务所工作的方针及服务宗旨：与客户充分沟通，了解客户所需，设计出客户最满意的作品，在提供舒适的使用空间的同时达到综合投入回报最佳组合。

下设两个部门的设计业务涉及：住宅地产、商业地产、旅游地产及综合类地产等行业。产品细分为：高层住宅及公寓、私人会所、豪华别墅、度假酒店大型商业区及高层办公楼等。

美国James · 王建筑师事务所个性化的设计特点是预算分析与设计方案相结合，所做的每一个项目，都经过仔细的预算及估价分析。在整个设计过程中，该事务所会提供不同的设计构思，检验可行性，并确定工程造价及持久性同设计目标吻合。事务所通常会同业主及估价师审核所有系统及每一个细部环节。此做法，在政府机构及私人企业项目上，已证明非常成功。事务所努力的方向是在不牺牲设计目标的情况下，保证绝对能在预定时间、指定预算内设计出理想的作品。十多年在中国的经营，已经设计创造作品数千件，覆盖全国多个省市地区，James · 王的工作成就得到了多方面的认可，在中国获得全国和北京的多项优秀设计奖项。特别是在2011年荣膺第八届地产年度风云榜“中国最具影响力设计师大奖”。

James Wang Design Associates, Inc. (JWDA) is created by James Wang, a Chinese American born in Taiwan. He is Bachelor of Architecture from Tamkang University (Taiwan), Master of Architecture from Illinois University (USA), and member of American Institute of Architects. In 1987, he set up JWDA in Los Angeles; in 2000, he set up Beijing JWDA Architectural Consulting Co., Ltd. in Beijing, with nearly 100 designers.

JWDA sets up architectural planning design division and interior decorative design division. We have diverse design experiences from planning to single building or even to interior design. Every design work has its own unique philosophy, leading the industry. We are good at making warm, comfortable, luxury, low-carbon, and environment-friendly products according to owners' demand. Our working guideline and service purpose are: to fully communicate with customers, to understand what customers need, and to design the most satisfactory works, with the best overall ROI in providing comfortable space for use.

The two subordinate divisions operate businesses involving: residential property, commercial property, tourism property and comprehensive property, etc. The products include: high-rise residence & apartment, private club, luxury villa, resort hotel and large commercial community and high-rise office building etc.

Our design is featured by the combination of budget analysis and design scheme. Every project has gone through careful budget and evaluation analysis. In the whole process of design, we will provide different conceptions, discuss the feasibility, and determine the engineering cost and whether the endurance agrees with the design target. We usually review all systems and every detailed process with owners and evaluators. This practice has proven very successful in the projects of governmental agencies and private enterprises. We are committed to designing ideal works within budget and schedule, without prejudicing the design target. With more than 10 years' operation in China, we have designed thousands of design works, covering many provinces and municipalities of China, and we are recognized widely by all social communities. We have won many excellent design awards from Beijing and all over China. In particular, we won the "Chinese Most Influential Designer Award" in the 8th Real Estate Yearly Fame Board in 2011.

James Wang Design Associates

建 筑 项 目

横琴岛文化广场

建设地点：广东 珠海
建筑面积：128万平方米
设计时间：2011年

珠海十字门与南澳门隔海相望，项目位于新区中央商务区核心区域，旨在打造该区域顶级酒店、办公、商业配套，采用三江一湾边的盛开莲花和横琴之门两个概念指引深化方案，将形成区域地标的效果。

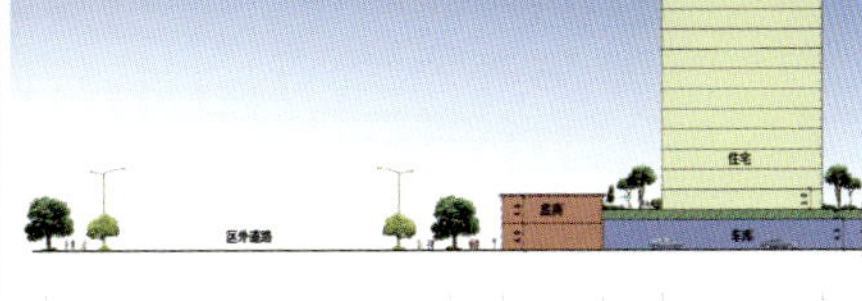
剖面一
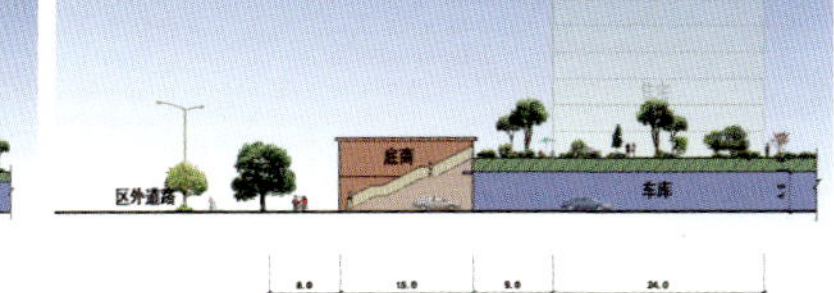
剖面二

连云港凌州广场

建设地点：江苏 连云港
建筑面积：110万平方米
设计时间：2010年

运用绿色科技打造生态城市综合体，建造高科技视觉演艺中心，打造城市的名片。

James Wang Design Associates

建筑项目

齐河东盟城

建设地点：山东 德州
建筑面积：37.7万平方米
设计时间：2010—2011年

1.集星级酒店、酒店式公寓、温泉水疗等功能为一体的现代化度假休闲中心。
2.安静、舒适、高档的独具风格的居住社区。
3.变化丰富的滨水生态商业空间。
全力打造集休闲、度假、旅游、会议、商业为一体的高端生态新城！

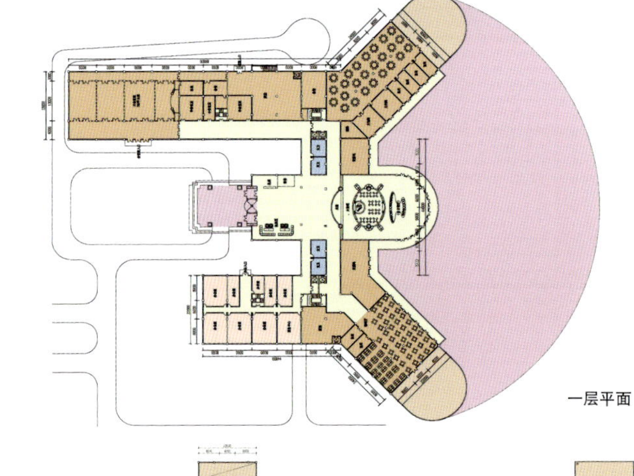

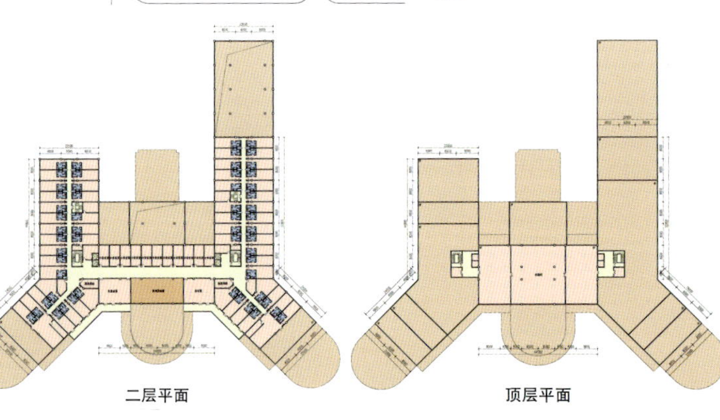

山东莱芜白马庄园

建设地点：山东 莱芜
建筑面积：26 000平方米
设计时间：2010年

酒店位于规划用地中西部，占地面积33 300平方米。如何利用基地的湖面景观为该酒店设计的主导因素，使酒店的房间尽可能多地面向水景，经过反复推敲比较，确定"H"形向心围合的平面可达到观海面的最大化。游客在酒店内能充分感受到水景的气息。建筑之间采用高大、通透、开放的外围护结构，最大限度地使各个方向的景观渗透到中央景观带，这一极具度假酒店的做法，通风效果理想，并使整个建筑群更为通透，最大限度地利用了水面资源的景观优势。

本酒店的建筑风格富有托斯卡纳风情，层层叠退，既体现度假酒店的特征，同时又与周围的自然环境协调。干净、简洁的立面设计元素与屋顶大大的挑檐活泼中又不失大气。立面文化石与防石涂料相结合，效果朴实，表达了休闲建筑的韵味。大面积连续的阳台增加了建筑的休闲情趣，亲切宜人、纯朴自然。

James Wang Design Associates

建筑项目

大连西郊公园高尔夫别墅

建设地点：辽宁 大连
建筑面积：38 300平方米
设计时间：2010年

立面设计采用纯粹的南加州风格，强调整体风格的统一与单体建筑的丰富变化。将古典建筑中繁复的细节加以提炼，保留古典的比例、尺度，在材料和细部处理上加以优化。展现出情趣高雅、稳重大方、返璞归真的人文主义建筑气质，能够体现目前这部分收入客户人群的价值取向。整个小区强调景观与建筑的结合，景观小品与植被的精巧搭配。低矮围墙作为道路与别墅前院之间的分隔，界定出公共绿化与私家园林的分界，视线可以穿透，园内的景色若隐若现，丰富了空间的层次。别墅私家庭院也根据功能的不同加以划分，配以不同的绿化植被。建筑本身的元素可以自然地延伸至庭院中，与小品相结合，成为景观建筑。

北京平谷阿凯迪亚度假村

建设地点：北京
建筑面积：8 000平方米
设计时间：2010—2011年

建筑设计突出“自然性、文化性、高效性、现代性”的设计原则。旨在打造一个既满足政府机关会务要求，也满足社会服务需求，兼顾国际会议、新闻发布各类会议功能的现代化、多功能的会议中心。

会议中心地块突出依湖、自然的整体设计的原则，通过良好的城市设计理念，合理地处理与周边现有山地环境的空间关系，把培训中心的建筑设计融入整体的城市肌理中去。

- 培训中心融入自然、开放的会议中心理念。
- 把本项目建设成富有平谷历史文化特色的会议中心。
- 合理特色的会议布局，提高会议中心的运作效率。
- 现代会议中心的设计理念。

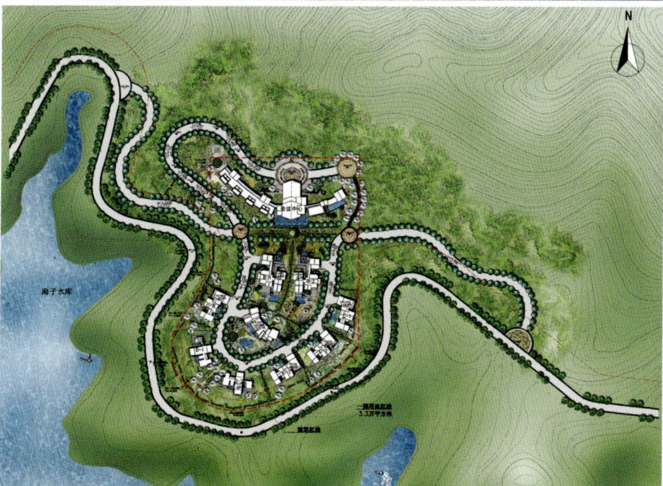

James Wang Design Associates

建筑项目

长春华瀚净月公馆

建设地点：吉林 长春
建筑面积：22万平方米
设计时间：2010—2011年

利用坡地创造了三维的台地，形成在立体空间上的错落、完全的人车分流，充分利用地下空间，为土地创造更大的使用价值。

大连凤凰谷

建设地点：辽宁 大连
建筑面积：11.147万平方米
设计时间：2010年

1.凸显基地依山就势的独特地理条件，打造高品质的居住社区。

2. 尊重每一位用户享有的优质居住权，以“均好性”的思想整体构思、精心布局，保证建筑的通风、日照及景观。

3. 采用法式的建筑风格，注重细节和概念，气势宏大，富有贵族气质。

James Wang Design Associates

建筑项目

海南邦溪镇山湖城

建设地点：海南
建筑面积：64万平方米
设计时间：2011年

项目充分考虑自身拥有的山、湖、江、温泉、水库、森林等多种复合型生态资源，建设以休闲为主题的新生态特色小镇。

焦作新城城市综合体

建设地点：河南 焦作
建筑面积：80万平方米
设计时间：2010年

运用绿色科技打造生态城市综合体，建造高科技视觉演艺中心，打造城市的名片。

James Wang Design Associates

建筑项目

天津翠金湖3期

建设地点：天津
建筑面积：240 000平方米
设计时间：2009—2011年

贯彻“以人为本”的设计思想，积极创造健康、宜人的居住环境，贯彻社会、经济和环境效益相统一的原则；休闲度假别墅建筑设计富有西班牙风情，注重整体环境质量，崇尚自然，强调健康，创造舒适宜人的居住环境。

James Wang Design Associates

建 筑 项 目

中建汤逊湖壹号

建设地点：湖北 武汉
建筑面积：52 363.8平方米
设计时间：2010—2011年

中建汤逊湖壹号作为欧洲建筑的典范法式建筑，格外注重细节和概念，气势宏大，富有贵族气质，是理性美的代表作。设计以法式风格为主旨，强调独特、鲜明的建筑风格特色。

——浪漫典雅的主题风格；

——造型严谨，颜色稳重大气，呈现出华贵气质；

——自然质朴的历史特色，建筑经典的重现；

——精心设计的细部节点突出建筑的质感。

室 内 项 目

北京民生金融中心—泛海集团总部

建设地点：北京
建筑面积：20 000平方米
设计时间：2010年

北京民生金融中心泛海集团总部，位于北京长安街东单路口东南角，使用面积21 850平方米。作为集团办公层，采用3种不同设计理念来演绎3个主要设计区域空间：标准办公层、高管办公层、会所。在设计上采用了现代和新古典元素相结合的手法，把古典融入设计之中，每个设计区域根据功能采用不同的材质，通过细节的处理，使整个设计达到尊贵、高品质的效果，体现出泛海集团自身的文化特征。标准层的设计，现代简洁，大气庄重，为员工提供了一个温馨的办公环境。高管办公层的设计则体现出高端、经典的设计理念，营造出一个既尊贵高雅，又富有活力，充满朝气、极具品位的办公场所。会所延续整体设计，通过现代的材料和古典建筑元素的对比，空间的凹凸变化及灯光气氛营造出一个温馨典雅、高贵自然的企业会所。

James Wang Design Associates

室内项目

廊坊君正红石庄园

建设地点：河北 廊坊
建筑面积：1 000平方米
设计时间：2011年

这是一处带有明显古典韵味的住宅项目，作为本案的售楼处，它需要表达出项目的价值感，但又不能过于张扬、辉煌而让人感到浮躁，于是节选了巴洛克时期的英式风格作为这个项目的设计语言。与意大利、法国等丰满繁复的形式完全不同，它在艺术风格上更加端庄、古雅，与本案的定位完全一致，用该风格来传达它的价值感和描述它的典雅。在这样的空间中驻足，会让人不自觉地感到一种品质的升华，这正是设计初衷。

北京门头沟城子新城天台山居住项目

建设地点：北京
建筑面积：1 500平方米
设计时间：2011年

天台山位于北京门头沟地区，是该项目开发商开发的高端品牌，华丽舒适，由于项目本身具有新古典的特点，为此在设计中以新古典为蓝本，增加了许多欧洲贵族的元素，体现出了高雅华贵的气质，另外还使用具有本地特点的大型装饰，从而衬托出一种中西合璧的时尚氛围。

James Wang Design Associates

室内项目

北京朝阳区太阳公元样板间

建设地点：北京
建筑面积：500平方米
设计时间：2011年

太阳公元是位于北京朝阳太阳宫地区的高端品牌项目，在建筑和室内的每一个细节中无不体现出该项目的高端品位，呈现给大家的是一个华贵的理想生活场景。设计中运用了新古典主义风格，在注重塑造效果的同时用现代的手法还原了古典的气质，使其具备了古典与现代的双重审美效果，达到一种完美的结合。

乌海君正花园别墅样板间

建设地点：内蒙古 乌海
建筑面积：400平方米
设计时间：2010年

美式风格是欧式风格的一种延续，一直以来颇受众人的喜爱与追求。美式风格从简单到繁杂，从整体到局部，精雕细琢，都给人一丝不苟的印象。在进行乌海别墅艺术创作的时候，设计人员一方面保留了材质、色彩的大致风格，同时又摒弃了过于复杂的肌理和装饰，简化了线条和造型。在设计上选取了大量的美式风格的元素及手法，这来源于特定人群的生活或者生活预期，使人产生一种很亲切的感觉，已达到交心的效果。

加拿大诺杰建筑设计事务所

ROGGEO DESIGN ASSOCIATES INC.

诺杰建筑设计事务所是一家综合性的国际建筑工程设计咨询公司，为各种类型的城市开发项目提供全方位的建筑工程咨询服务，包括市场研究、规划及方案设计、初步设计以及施工图设计和工程管理等咨询服务。

诺杰建筑设计事务所自1993年11月18日在加拿大多伦多创立以来，在全世界范围内参与过许多重大项目的咨询服务，业务范围遍布加拿大、美国、日本、欧洲、中国。特别是近十年，诺杰建筑设计事务所凭借其多年国内外的设计经验参与了多项国内外的重大项目。借助众多的设计精英团队、严谨的执行力、国际惯例思维，在高速发展的国内市场环境中，创造出一套完整的、综合的、独特的建筑工程解决方案。

诺杰建筑设计事务所中国区总部设在北京，在成都设有分支机构，在上海、天津、深圳设有合作机构。为遍布各地的城市开发项目提供高品质的市场研究、商业策划、规划及方案设计、建筑工程设计、工程管理等方面的专业咨询服务。诺杰建筑设计事务所高度本地化运作以及中、英双语设计团队的优势，使我们和客户之间的沟通更加顺畅，信息反馈更加便捷，从而大大提高了设计效率及质量水准。

诺杰建筑设计事务所承诺用智慧、诚信和专业服务来满足客户的需求和愿望。用国际化的水准灵活应对市场挑战，尊重客户、尊重团队，在和谐的氛围中打造最完美的设计作品。

联系人：卢莲珍 Cynthia LU
地　址：北京市朝阳区东三环北路霞光里18号佳程广场A座16D单元
邮　编：100027
电　话：+86-10-84400606
传　真：+86-10-84400062
邮　箱：info@roggeo.com
网　址：www.roggeo.com

1

2

1–2 北京望京华彩国际公寓

建筑面积：约17万平方米；项目类型：住宅、商业；项目周期：2006—2008年。华彩国际公寓地处朝阳区望京广顺北大街与望京北路交汇处的东南角，是望京乃至整个北京市区内的高品质公寓，项目整体分成南北区两大部分，由公寓和北区星级商务综合楼组成。110～160平方米的舒适型住宅为主力户型，是具备豪宅品质的纯居住型高档公寓与服务式公寓相融合的社区。

3–5 河北石家庄裕华区德鸿中央悦城

建筑面积：约22万平方米；项目类型：住宅、商业；项目周期：2009—2011年。A地块建筑设计上采用简约的设计风格，却体现出公共建筑的复杂性与趣味性。结合公共入口大堂及围合式休闲广场，创造出一种通透的室内外空间联系。B地块为商业居住用地。平面设计上强调灰空间的设置，为商户提供大尺度露台空间，最大限度地体现其商业价值。

6–7 内蒙古呼和浩特如意开发区东岸国际

建筑面积：约43.7万平方米；项目类型：别墅、公寓；项目周期：2007—2010年。项目致力于打造一个与众不同的舒适型高雅居住社区。通过两条大峡谷的创造将园林空间有机整合，大自然的美尽显其间。合理规划各个功能组团的居住建筑，强调人与自然、人与建筑、人与文化的互动，在呼和浩特东河岸边建立起一座高尚和谐的欧陆风情之城。

AGENCE C&P ARCHITECTURE

C&P(喜邦)国际建筑设计公司

C&P(喜邦)国际建筑设计公司集合了中国、法国、美国三国优秀的设计人才，在民用建筑设计及工程咨询方面表现出色。公司的业务范围涵盖亚洲、欧洲、美洲，设计内容包括商业、酒店、办公、文化、教育、体育、医疗、居住及城市规划等，深得客户好评并多次获奖。

C&P(喜邦)国际建筑设计公司在中国创建时就本着以人为本、中西结合的设计理念，将西方先进的设计理念及工程技术与中国深厚的文化底蕴相结合，积极探索，勇于创新，致力于设计绿色环保、实用美观的建筑。多年来丰富的实践和经验形成了C&P（喜邦）独特的设计风格，也赢得了大批客户的佳评与持续合作。随着公司业务范围的不断扩展，工程项目从中国、法国、美国三国扩展至亚洲、欧洲、美洲的其他国家，业务种类愈加广泛。

C&P(喜邦)国际建筑设计公司于1995年在中国成立厦门公司，并于2002年将公司总部设于北京。海纳百川有容乃大，公司优秀的建筑理念、先进的工程技术、融洽的工作氛围、丰富的设计业务吸引了大批优秀的建筑师加入，设计团队不断壮大，人才济济。随后，C&P(喜邦)国际建筑设计公司在中国又陆续成立了天津公司和南京公司。国内事业部与法国、美国事业部横贯亚、欧、美三大洲，集合三地人才，融合中国的古典自然、法国的浪漫自由、美国的现代科技，兼容并蓄，各展所长，做精品建筑，实现心中理想。

C&P(喜邦)国际建筑设计公司将努力开拓、积极进取、充分发挥中西方建筑文化交流之桥梁作用，为中国建筑的进步与发展贡献力量。

喜邦理念：

我们不浮躁。

我们不哗众取宠。

我们不做华而不实的建筑。

——我们拒绝平庸！

建筑是庇护，设计是艺术，建筑设计是生活和艺术的融合。无论初始如何喧嚣纷舞，经岁月销蚀终将归于平静，尘埃落定，更显设计的精髓和建筑的本质。用心设计每个项目，专注完善每处细节，全力把握每次机会，做出被记住、被喜欢的作品，是公司同仁不懈的追求；用专业设计求解生活答案，是喜邦全体坚持的信念；坚守热爱设计的心，让我们不断进步。过去、现在、未来，喜邦将在"设计和专业"的路上坚定地走下去。

服务宗旨：

诚信为本、热忱服务、不惧挑战、高效高质，是我们一贯的宗旨。

相信建筑比人类的寿命更长久，希望用设计点亮平淡的生活、用专业能力实现关于建筑与生活的美好梦想，是喜邦人对自己和客户的承诺。

C&P International Architectural Design Co.,Ltd. concentrates experienced professionals from China, France and America, with excellent performance in civil architecture design and also engineering consultations. The company's business covers Asia, Europe, America, and the projects include all types of buildings such as commercial, hotel, office, cultural, education, sports, medical, residential and urban planning etc. C&P has won a good reputation within the industry and a lot of design awards.

From its foundation in China, C&P emphasizes the human-oriented principle and integrates the oriental and occidental design conceptions. It explores actively to design the green, practical and beautiful building. The abundant experience of C&P has formed its unique style and won the customers' appreciation. With the development of the company, its business extends to other countries besides China, France and America.

C&P International Architectural Design Co.,Ltd. was established at Xiamen in 1995, and then set up its headquarters at Beijing in 2007. The company attracts many excellent architects and the design groups grow very fast, with the foundation of new branches at Tianjin and Nanjing. The offices in China, France and America take advantage of the divers strong points of the professionals from the three continents to promote the architecture design.

C&P will play its role more actively as the bridge between China and West, by contributing to the progress and development of China's building industry.

C&P Philosophy

We are not blundering.

We don't seek popularity with claptrap.

We don't make gewgawish buildings.

——We reject mediocrity!

The building is a life shelter while the design is an art, so the architecture design is an integration of life and art. The time will fade the ostentation and demonstrate the essence of the architecture. Attention on every project, every detail and every opportunity to achieve satisfactory works is our aim; to find the best solution for life according to professional design is our will. It is the enthusiasm on design that makes us advance continuously. From the past to the present we have covered a long road and we will go on the way in the future by sticking to the professional design.

Service Principle

Honest, enthusiast, no fear of challenge, seeking for high efficiency and high quality design are the principles that we always stand on.

Believing that building exists longer than man, we hope to lighten the life with excellent design and help you to realize your beautiful dreams with our professional ability. That's the commitment of C&P to itself and to its customers.

北京公司
地址：北京市东城区安定门东大街28号雍和大厦D座1102-1106室
邮编：100007
总机：+86-10-87561616
传真：+86-10-84195290

天津公司
地址：天津市南开区红旗南路彩虹花园52号
邮编：300042
总机：+86-22-23555271
传真：+86-22-23555270

厦门公司
地址：厦门市东渡路51号裕成大厦B座1101室
邮编：361013
总机：+86-592-3895590
传真：+86-592-3895596

南京公司
地址：南京市鼓楼区石头城6号文化产业园01栋3层H套
邮编：210013
总机：+86-25-83720007
传真：+86-25-83720770

法国公司
地址：里昂，马赛街66号
邮编：69007
总机：0033 4 7869 2121
传真：0033 4 7858 8242

美国公司
地址：明尼苏达州明尼阿波利斯市东南2街212号323室
邮编：55414
总机：651-983-8897
传真：651-222-9057

Beijing Office
Add: Room 1102-1106, Building D, Yonghe Plaza, No.28 Andingmen East Avenue, Dongcheng District, Beijing
P.C.:100007
Tel: +86-10-87561616
Fax: +86-10-84195290

Tianjin Office
Add: No. 52 Rainbow garden, Hongqi South Road, Nankai District, Tianjin
P.C.:300381
Tel: +86-22-23555271
Fax: +86-22-23555270

Xiamen Office
Add: Room 1101, Building B, Yucheng Plaza, No. 51 Dongdu Road, Xiamen
P.C.:361013
Tel: +86-592-3895590
Fax: +86-592-3895596

Nanjing Office
Add: Room H, 3F, Building 01, Cultural Industry Park, No. 6 Stone City, Gulou District, Nanjing
P.C.:210013
Tel: +86-25-83720007
Fax: +86-25-83720770

France Office
Add: 66 Rue De Marseille Lyon
P.C.:69007
Tel: 0033 4 7869 2121
Fax: 0033 4 7858 8242

USA Office
Add: 212 SE 2nd St, Suite 323 Minneapolis, MN
P.C.:55414
Tel: 651-983-8897
Fax: 651-222-9057

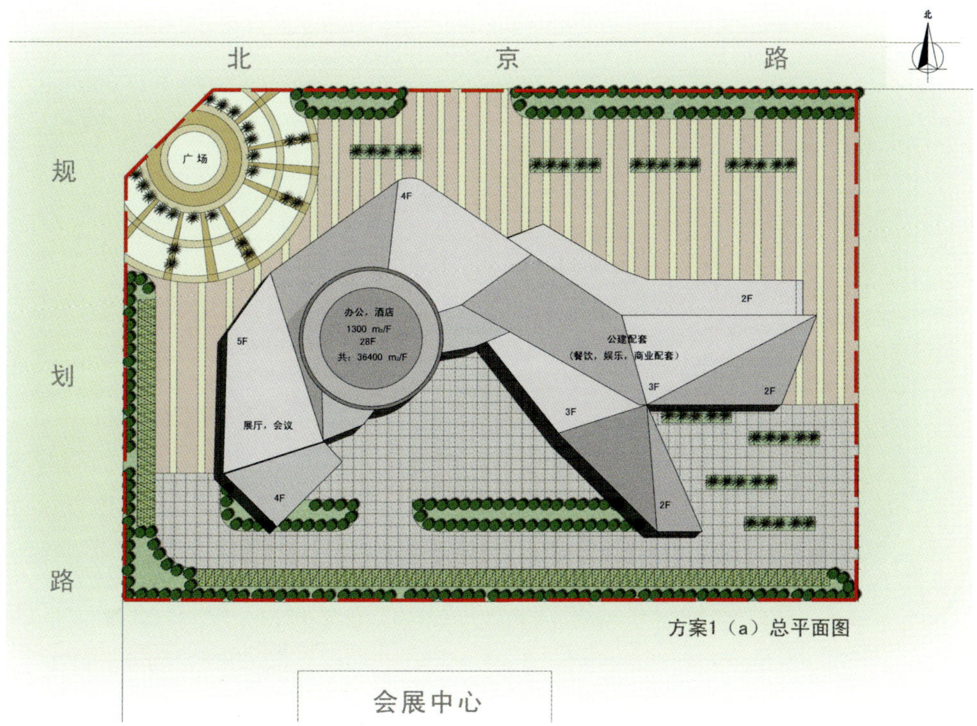

方案1（a）总平面图

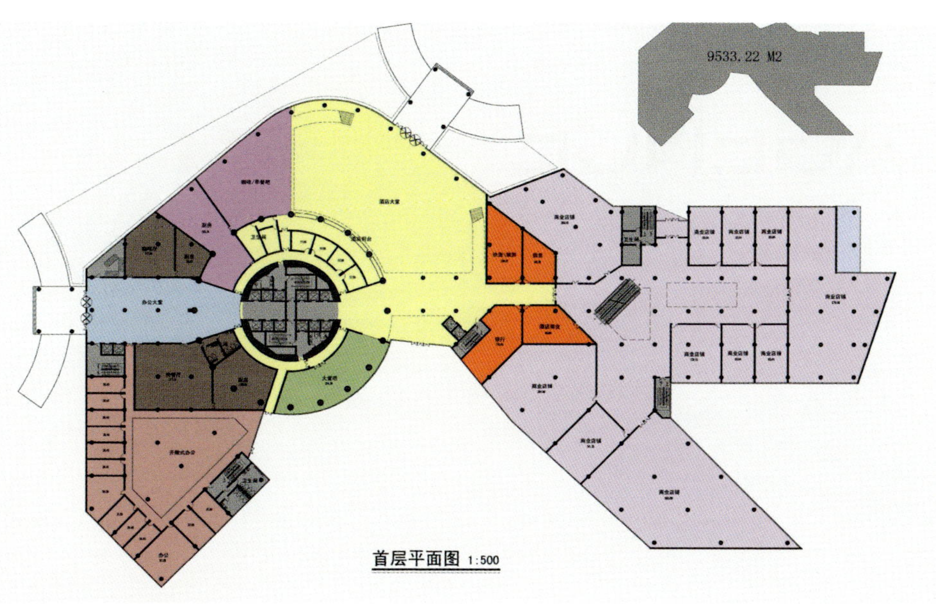

首层平面图 1:500

沧州管业大厦五星级酒店

建设地点：河北 沧州
委 托 方：东塑集团
用地面积：30 000.00平方米
建筑面积：72 068.00平方米
容 积 率：2.5
设计单位：C&P（喜邦）国际建筑设计公司
设计时间：2010年
主要设计人员：樊 斌、兰 剑、李梦瑶、杜 宁

关注建筑的功能性，引申管道“流通、交汇、聚集”的内涵，建筑造型简单巧妙，超百米楼高蕴含行业领袖的气质，白色的外墙简洁且明快，与周围的建筑和谐共存，是未来沧州的标志性建筑。

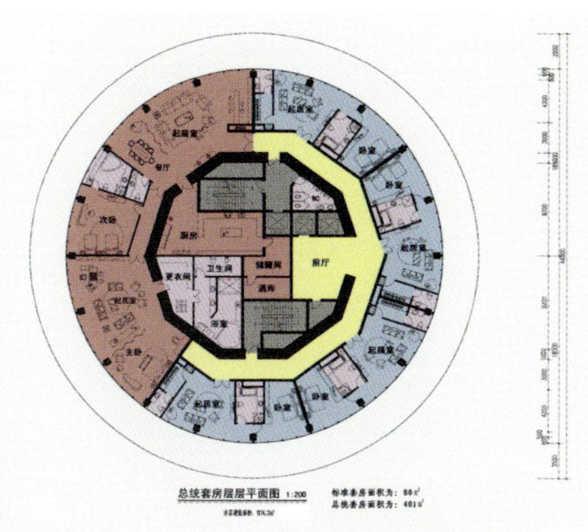
总统套房层平面图 1:200

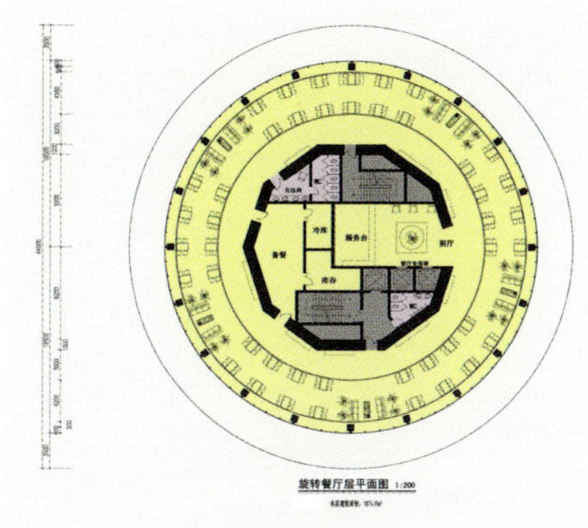
旋转餐厅层平面图 1:200

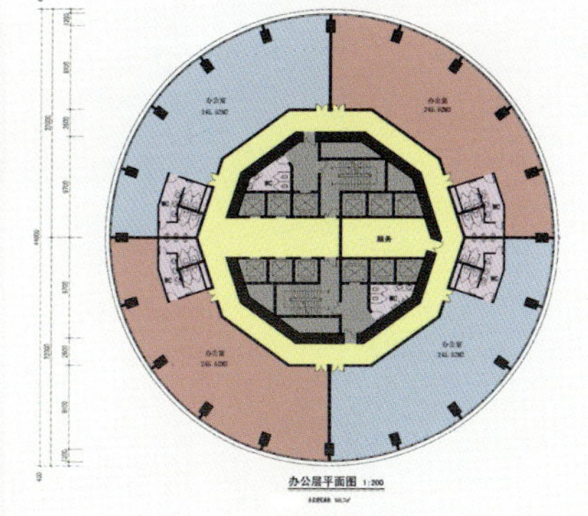
办公层平面图 1:200

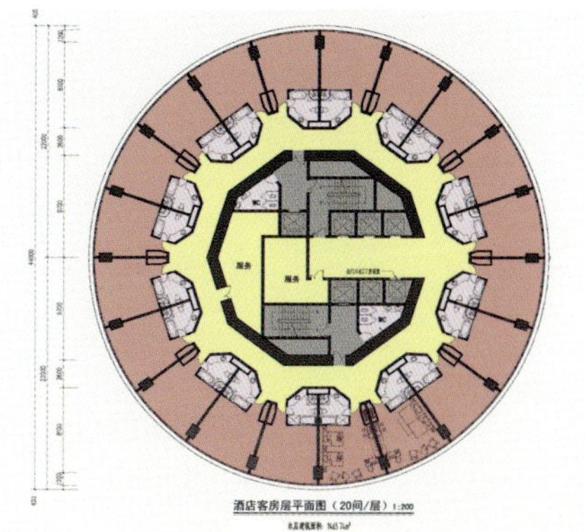
酒店客房层平面图（20间/层）1:200

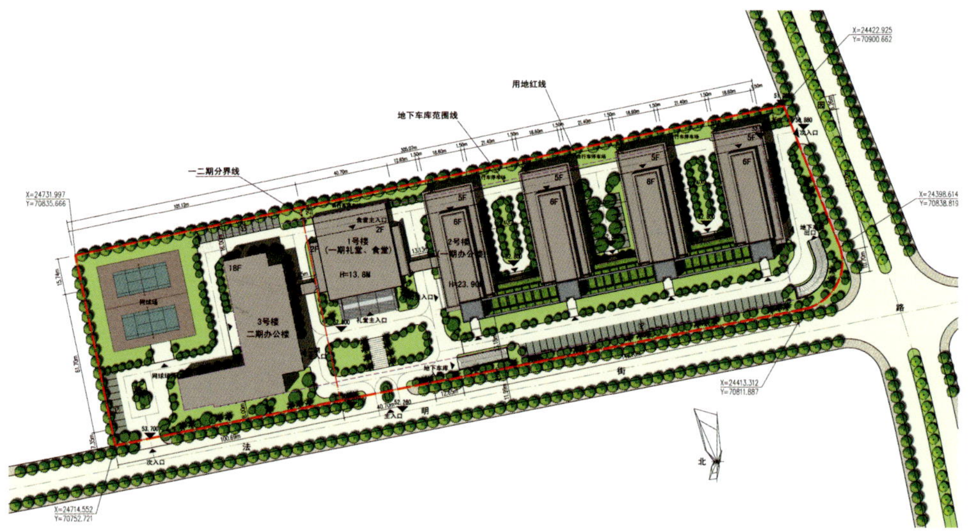

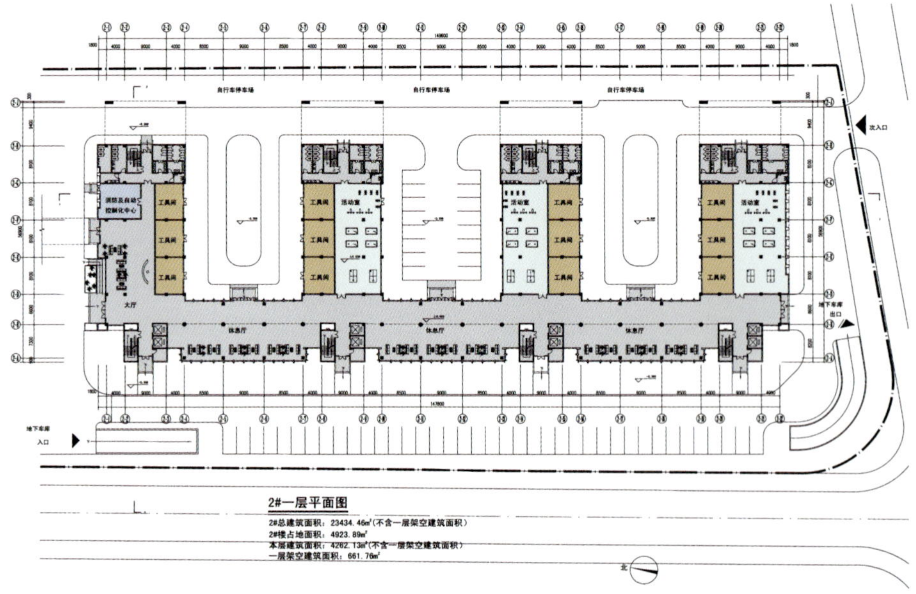

2#一层平面图

2#总建筑面积：23434.46m²(不含一层架空建筑面积)
2#楼占地面积：4923.89m²
本层建筑面积：4262.13m²(不含一层架空建筑面积)
一层架空建筑面积：661.76m²

金华电业局生产基地

建设地点：浙江 金华
规划总用地面积：27 976.00平方米
总建筑面积：65 801.63平方米
容 积 率：1.69
建筑密度：29.70%
绿 地 率：30%
设计单位：C&P(喜邦)国际建筑设计公司
设计时间：2010年
主要设计人员：魏黎华、林 蔚

该项目主要由修试、送电、变电与超高压、运输、计量等几个工区组成，担负着金华电业局重要的生产功能。

设计理念

强调办公建筑的特性，体现电力部门的行业形象。
提倡绿色办公，创造舒适宜人的办公环境，同时注意节能减排、生态环保。
优化布局，发挥多部门集约化办公的优势，强调办公高效化、集成化。
关注办公环境的均好性，利于各个部门独立运行管理。
以人为本，创造个性化的办公空间。

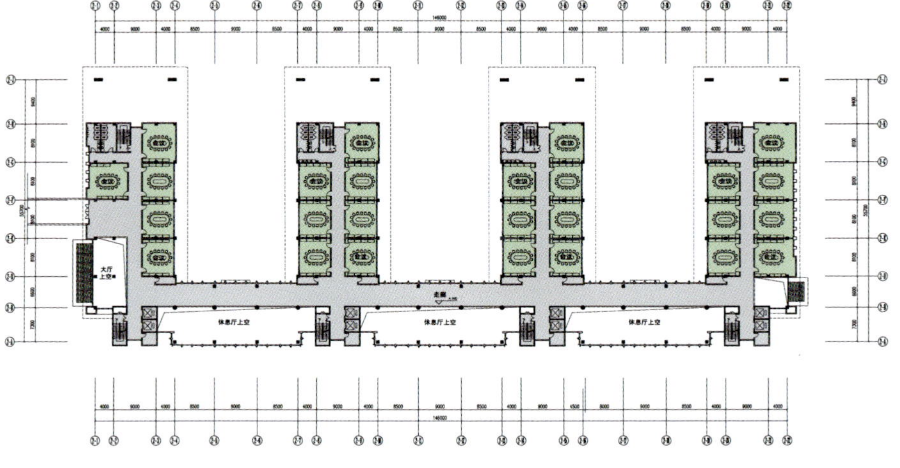

2#二层平面图

本层建筑面积：3358.49m²

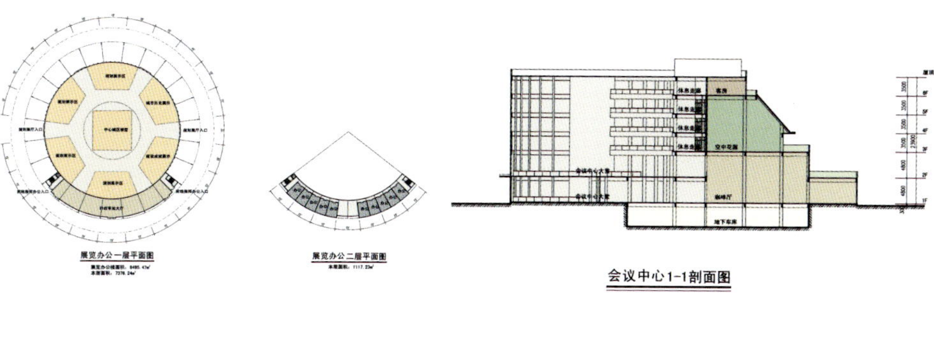

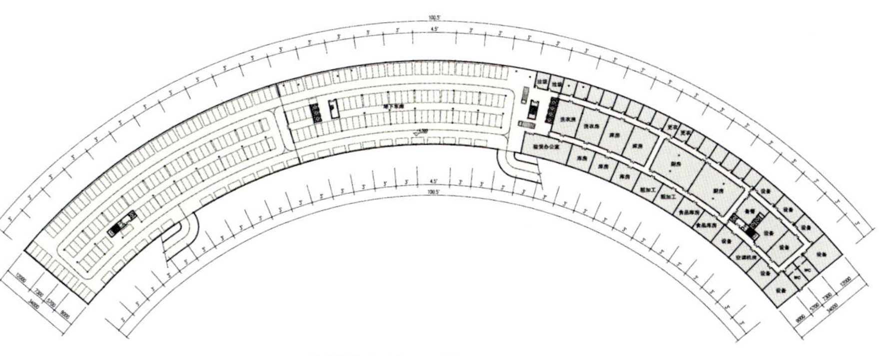

平潭综合实验区海峡两岸会展中心

建设地点：福建 平潭
总用地面积：273 400.00平方米
总建筑面积：75 021.21平方米
设计单位：C&P(喜邦)国际建筑设计公司
设计时间：2010年
主要设计人员：拉里·凯勒、魏黎华、伍红英

平潭综合实验区海峡两岸会展中心作为海峡西岸经济开发的前期启动项目，位于福建海坛岛中部偏南，西接环岛路竹屿口连接线，南临305国道，与规划中的滨海新城I区临水相望，与新区融为一体，将打造为平潭的标志性建筑。

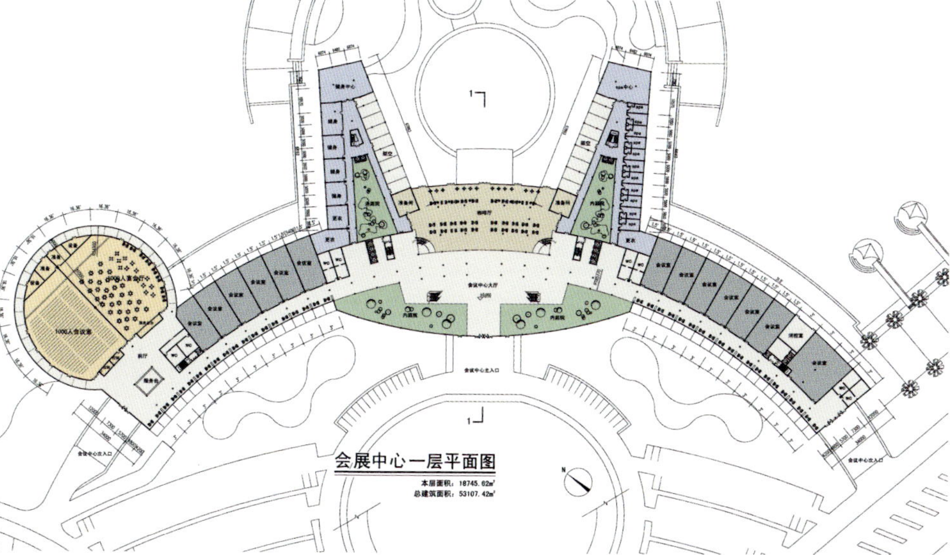

天津凤凰墅

项目位置：天津
用地面积：66 295.00平方米
建筑面积：32 470.00平方米
容 积 率：0.47
建筑密度：22.69%
绿 化 率：47.26%
停 车 位：262辆
设计单位：C&P(喜邦)国际建筑设计公司
开发单位：天津市房信集团
主要设计人员：樊 斌、丁晓波

本案借鉴“五大道”风貌区域特征，建构易于识别的环境的同时，将风格融合英式的院落和有机建筑，应用本土材料，形成传统创新的新地域主义风格和特有的文化内涵。

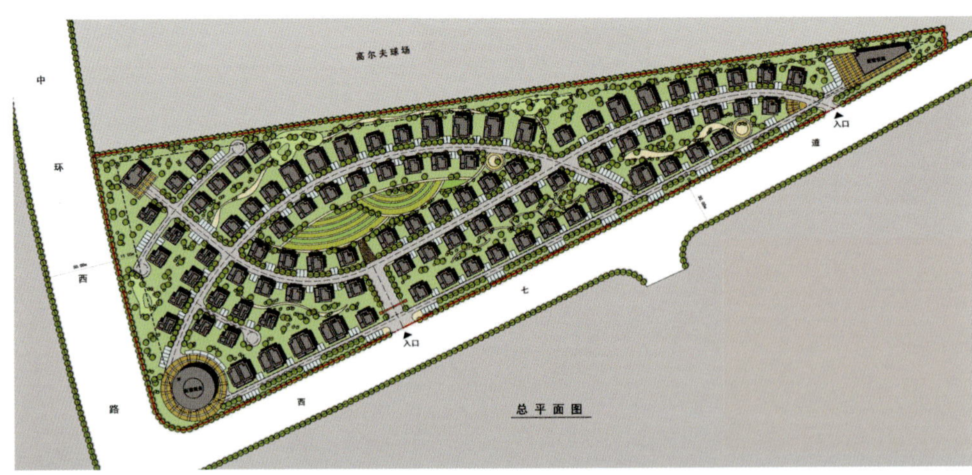

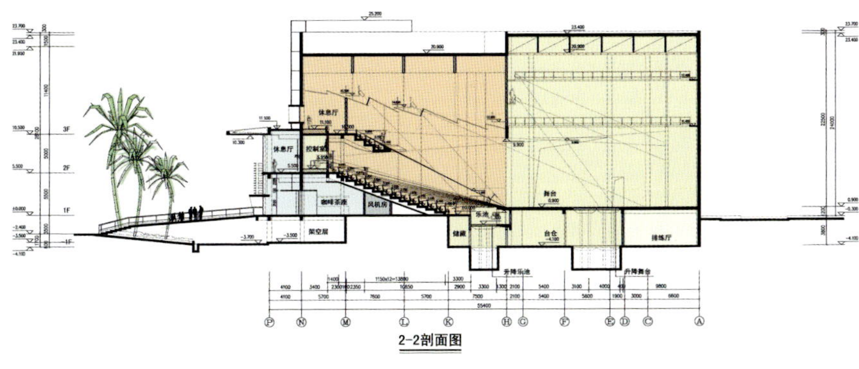
2-2剖面图

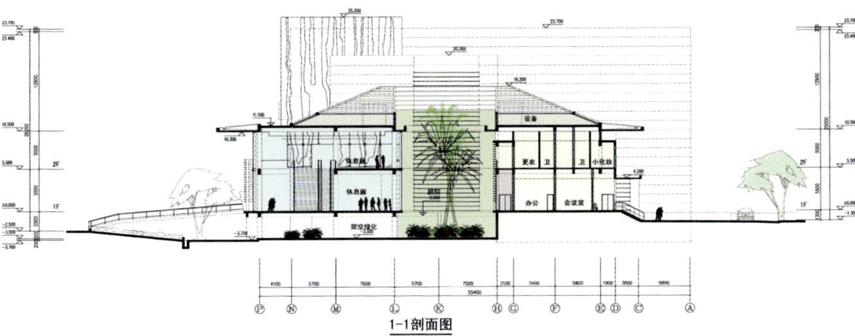
1-1剖面图

厦门小白鹭文化艺术中心

曾获奖项2004年国际投标一等奖
建设地点：福建 厦门
总用地面积：12 815.27平方米
建设用地面积：12 169.63平方米
总建筑面积：7 298.70平方米
地上面积：5 976.0平方米
地下面积：1 322.7平方米
建筑密度：29.5%
容 积 率：0.5
绿 地 率：36.2%
观众席位数：500座，池座：450座，楼座：50座
停车位：41个
设计单位：C&P(喜邦)国际建筑设计公司
开发单位：厦门市艺术学校
主要设计人员：魏黎华、胡朝虹

在小白鹭文化艺术中心的设计中，通过虚实的对比使封闭的演出厅与面向海滨通透开敞的公众大厅有机组合，使建筑恰如其分地融入其周边的海滨环境之中。

在一个巨大而简洁的屋面覆盖下，演出厅、剧院的公众区域及剧院的配套功能区域通过不同色彩的体块得以自我展现，又不失个性地展示其公共建筑的形象风格。

中国航空综合体技术研究所实验与检测中心

建设地点：北京 怀柔
用地面积：117 429.00平方米
建筑面积：128 418.00平方米
容 积 率：1.30
建筑密度：29.80%
绿 化 率：38%
停 车 位：699
设计单位：C&P(喜邦)国际建筑设计公司
设计时间：2010年
主要设计人员：樊 斌、兰 剑、范洪涛

绿色——借助雁栖开发区的丰厚生态资源，打造园区内部绿色环境，建筑对绿地敞开，人被绿色环绕，产业园区不再是冷冰冰的厂房，而是一个生机盎然、充满趣味的科研空间。

人文——坚持以人为本，赋予最细致的人文关怀，充分考虑人员的工作和生活需求，将办公、研发、培训、服务等功能进行最合理的分布，使地表建筑和中心景观绿地、休闲景观步行道有机结合，打造具有鲜明个性的办公空间。

科技——引入可持续发展的设计手法，充分尊重基地的自然条件，在总体规划上合理布局，在建筑设计中引入太阳能利用、遮阳板、水循环使用等技术手段，打造低污染、低噪声、低碳环保的世界一流现代化园区。

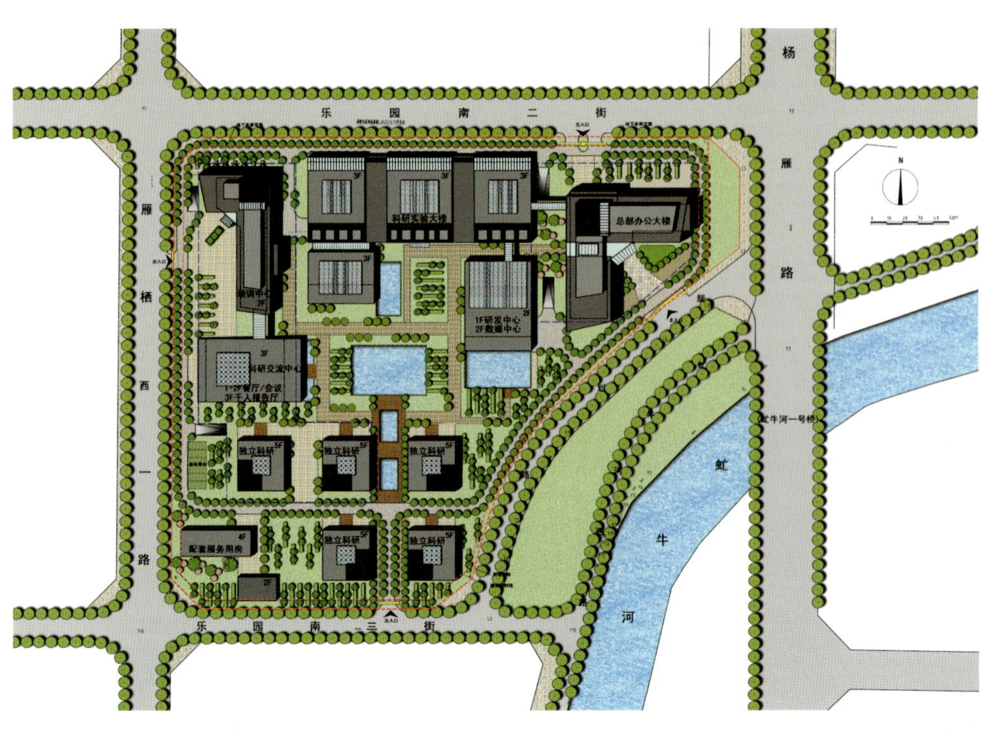

怀柔科研中心 剖面 Huai Rou Research and Development center Section

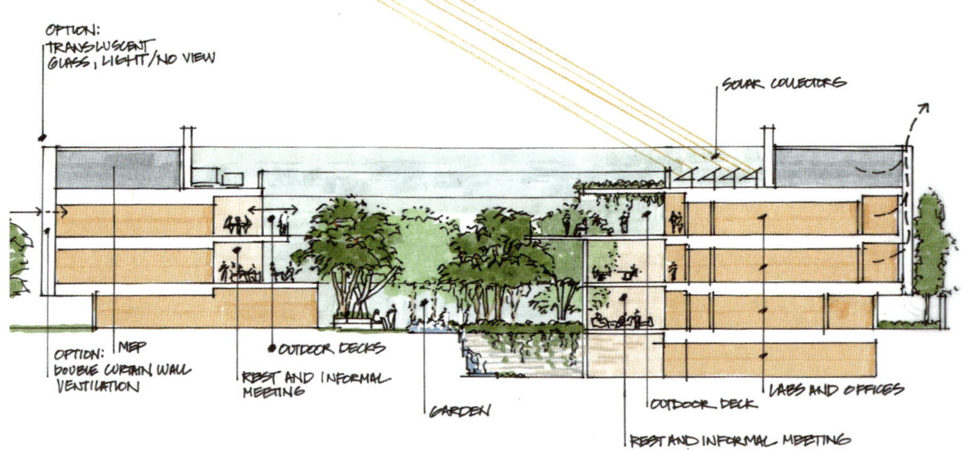

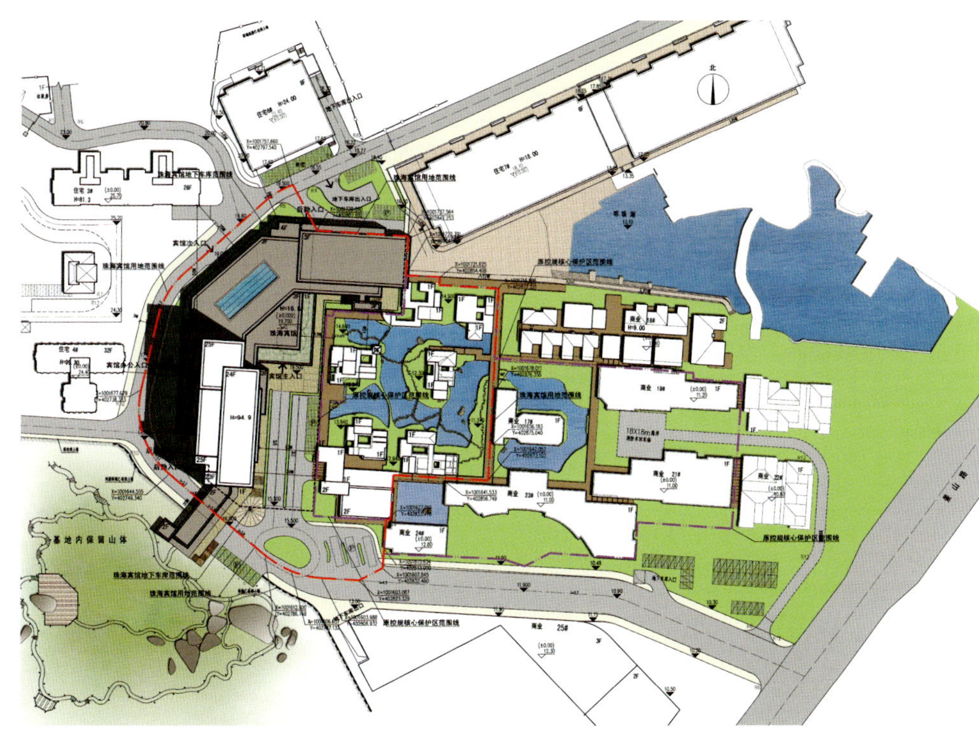

珠海万豪大酒店

建设地点：广东 珠海
规划用地面积：14 365.95平方米
总建筑面积：45 557.49平方米
规划用地面积：14 365.95平方米
容 积 率：2.45
建筑密度：54.87%
绿 地 率：31.20%
设计单位：C&P(喜邦)国际建筑设计公司
设计时间：2009年
主要设计人员：魏黎华、林 蔚

珠海宾馆项目位于珠海市吉大商业区景山路西侧，项目用地北临石景山，东北临九洲城，东面为珠海渔女风景区，南临吉大中心区，距离海岸线直线距离不足一千米，自然景观资源丰富且交通便利，工程总用地面积14 365.96平方米。

本项目是原政府接待酒店——珠海宾馆的改建项目，保留东侧原有底层园林式建筑，拆除原西侧建筑，建为高层宾馆，与保留建筑共同形成新的宾馆建筑群。

保证珠海宾馆及其周边建筑形象的和谐统一，强化其作为珠海特区城市印记的地标形象。

严格控制高层建筑的立面及造型设计，保护周边自然环境，烘托富有特色的岭南园林空间形象。

建筑色彩、材质要充分与珠海宾馆保留建筑的内部环境相协调，延续其原有宜人、精致、舒适的环境。

建筑立面的处理应保持视觉的整体性，获取耐人寻味的视觉效果。

注重新旧结合的建筑风格的过渡与演变，体现时代气息。

考虑到珠海的南方滨海城市的气候特征与建筑基地条件的限制，结合立面造型，设置遮阳镂空栅格，使建筑具有可持续性，更加天然舒适。

加拿大迈敦建筑师事务所
Metropolitan Architects Inc.

迈敦建筑师事务所（Metropolitan Architects Inc.）成立于加拿大多伦多市，是在加拿大拥有一定资质的专业建筑师事务所，多次在加拿大获得设计奖项。2009年，迈敦在中国上海成立了分公司，即迈敦（上海）建筑设计咨询有限公司。

设计的项目包括酒店、会所、住宅、写字楼和各种综合体项目，尤其擅长高端定制的温泉度假类产品或功能较复杂的项目。自上海分公司成立以来，经过不断的尝试和努力，已经和国内一些一流的设计公司形成了较为稳定和高效的战略合作关系，因此可以根据业主的需要，提供包括从规划到室内和景观设计的一站式服务，确保了设计构思的完整性和各工种配合的高效性，也给业主节省了这方面的时间和成本。作为一个已完成本土化的境外设计公司，迈敦可以做到真正的中西合璧，在将欧美先进的理念和技术应用在国内项目上具有得天独厚的优势。同时，迈敦还可利用与国际一线品牌酒店集团的良好合作关系，为有意引进境外高端酒店管理的国内开发商提供这方面的咨询和牵线搭桥。

在加拿大的业主包括国际著名酒店集团，如万豪酒店（Marriott）、希尔顿酒店（Hilton）、四季酒店（Four Seasons）及知名房地产开发商。迈敦已成为目前在加拿大同时获得国际两大主要酒店集团，即Marriott（万豪）和 Hilton（希尔顿）认可并推荐的少数几家事务所之一。进入中国市场后，利用在加拿大的多年实践经验，将侧重点放在了高端温泉度假类产品以及木结构建筑两个方面。木结构作为一种绿色低碳的结构形式在中国已开始本土化的进程，设计的汤泉公馆区木结构会所及别墅是中国目前规模最大的现代木结构项目。

设计理念：建筑是有生命的，作为人造物质环境的一个基本组成元素，始终不停地影响甚至改变着人们的生活，好的建筑除具有物质功能外，还应具有某些精神层面的品质。作为建筑师，力求设计有生命力的建筑。经营理念是：提供高附加值的服务，即利用我们中加两国的双重文化背景以及与国际一线品牌酒店管理集团的长期合作关系，为业主带来一般传统设计机构所不能提供的价值。

Founded in Toronto, Canada, Metropolitan Architects Inc. is an award-winning professional architectural firm, fully licensed to practice architecture in Canada. In 2009, the firm set up its branch in Shanghai, China.

Our expertise includes the design of hotels, luxury clubs, single and multi-family residences, office buildings, and mixed use projects, and in particular, high-end spa hotels or clubs and complex projects. As we have successfully established firm and effective relationships with a number of excellent local design firms in various disciplines since the Shanghai office became operational, we are able to provide our clients with one-stop service including planning, landscape design, and interior design, as required, generating multiple benefits of integrated design, seamless coordination, and great convenience to our clients in saving hassles of managing different design firms. As a localized foreign practice, we enjoy a true marriage of Chinese and Western backgrounds, and have unique advantages in applying the Western theories or technologies in Chinese projects. We can even provide consultancy or make initial connections for our clients who are interested in importing international hotel management, using our long-term relationships with prestigious global hotel groups.

Our clients include famous global hotel groups such as Marriott, Hilton, Four Seasons, and a number of major developers. Metropolitan Architects is one of the few architectural firms recommended by both Marriott and Hilton, the two most prestigious hotel groups in the world. Since the firm entered the Chinese market, supported by our years of experience in Canada, we have focused our practice on two specific areas: high-end spa facilities and wood construction. As a low-carbon structural form to reduce energy consumption, wood construction has started its process of localization in China, the Hot Spring Town Wood Frame Clubhouses and Villas we designed is so far the largest wood frame project in China.

Our vision is that buildings should have lives, and as the basic components of the built environment, they slowly influence and eventually transform our life. A good building should possess certain spiritual qualities, in addition to its physical functions. As architects we are committed to the design of buildings that have lives. On the marketing side, our approach is to provide value-added services, supported by our dual background of both Canada and China and our long-term relationship with prestigious global hotel groups. With no additional charge, our clients have the opportunity to receive extra values above and beyond a traditional design firm can offer.

上海
地址：上海市普陀区长寿路1118号悦达国际大厦A栋20E
邮编：200042
电话：+86-21-62522006-608
传真：+86-21-62522006-606
邮箱：shiying19@hotmail.com

多伦多
地址：62 Glen Echo Road, Toronto Canada M4N 2E3
电话：00416-386-1614
传真：00416-386-16140
邮箱：shiying19@hotmail.com

安徽会馆

湖南会馆

安徽会馆

广东会馆

浙江会馆

重庆会馆

上海会馆

北京会馆

山西会馆

中国商务会馆区（一期、二期）

建设地点：河北 霸州
建筑面积：38 600平方米
用地面积：120 340平方米
设计时间：2009年
项目备注：直接委托

中国商务会馆区项目位于河北省霸州市，是集居住、商务、会所、休闲、度假等多项功能于一体的综合性项目。设计试图以文化为切入点，探索高端商务会馆的一种新方向，即把现代五星级酒店的设施、当地的温泉资源和传统民居文化元素三者融合在一起，营造一种具有浓郁人文背景与地域特色的高端会馆类型。按照国际高端温泉度假酒店的模式对会馆的设施功能进行规划和设计的同时，根据当地具有丰富的温泉资源的特点，重点打造温泉水疗功能，最大限度地保留了各地传统民居文化的独特韵味。尽管会馆的建筑风格很多，但它们都有一个共同的侧重点，就是对院落的塑造，各种院落大小不一，各具特色。项目分期建设，目前进行至第二期，一二期总共开发会馆65栋，含350平方米、600平方米及2 600平方米等几种不同规格，每栋会馆都按其冠名省市的传统民居形式进行设计，整个会馆区建成后，将成为国内目前规模最大且种类最全的中国建筑之窗。

汤泉公馆区木结构会所及别墅(一期、二期)

建设地点：河南 许昌
建筑面积：59 400平方米
用地面积：91 670平方米
设计时间：2011年
项目备注：直接委托

汤泉公馆区木结构项目位于河南省许昌市鄢陵县陈化店镇西侧，311国道以北。本项目是自现代木结构作为一种主要建筑结构形式进入中国以来国内迄今规模最大的木结构项目。中国人对木结构其实并不陌生，木结构的应用在中国历史悠久。现代木结构延续了传统木结构的所有优点：保温、舒适、防震，同时又应用现代技术克服了传统木结构的缺陷，具有防火、防潮、隔音，以及缩短工期和耐久性等方面的显著优点。另外，木结构相对于混凝土结构和钢结构要更为低碳和环保，用于建材的木材皆可再生、循环使用和回归自然。以在北美的实践经验为基础，结合国内的建筑规范和施工工艺对部分节点进行了优化和调整，在设计层面上为木结构在中国的本土化做了不少有益的工作。同时，根据目前国内市场对于现代木结构的认知和接受程度，决定为本项目注入与一般轻型木结构房屋不同的特色，即在重点部位采用局部重木结构，暴露梁柱和屋架，以最直观的方式展示木结构的特色和魅力，以达到较强的视觉感染力和说服力。

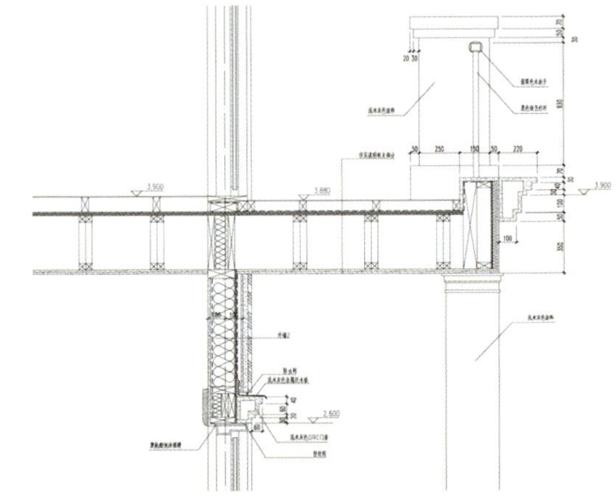

加拿大迈敦建筑师事务所 METROPOLITAN ARCHITECTS INC.

东方伟业城市综合体

建设地点：内蒙古 包头
建筑面积：285 500平方米
用地面积：32 469平方米
设计时间：2010年

本案毗邻包头市政府，是一个由商场、5A写字楼、酒店式公寓、公寓和住宅组成的功能复杂的城市综合体。规划设计的主要考虑如下：

（1）西高东低。沿钢铁大街建筑西高东低，高层塔楼在东南角向后退让，充分尊重邻近的市府和阿尔丁广场，符合政府规划远景。

（2）三点一线。东南角商业主入口、中部下沉式广场和西北角商业次入口三点一线，形成一条商业轴线沿对角线贯穿整个地块，加上东西向的步行街，将人流有效引入地块内部，从而解决了本案由于方形地块带来的商业纵深较大的问题。商业主入口放在东南角直接面对阿尔丁广场，通过造型独特的采光中庭形成强大的磁力，将大量人流吸入地块。

（3）动静分区。一条贯通东西的步行街在带动商业的同时也自然而然地将整个地块分成南北两个区域，北区以住宅和公寓为主，南区则以商场和写字楼为主。

博泰商住综合体

建设地点：内蒙古 包头
建筑面积：263 000平方米
用地面积：76 590平方米
设计时间：2011年
项目备注：直接委托

项目主要功能含住宅(含商品房和回迁房)、办公楼、集中商业及沿街商业。设计要点如下：

(1) 一点一线三区。“一点”指地块东南角的点式办公楼，设计成地标性建筑；“一线”指贯通地块南北的一条线形的水系，它由南段的曲折小溪向北逐渐扩大成一片较为广阔的水面，形成了小区的中心景观区和绿肺；“三区”由集中商业、回迁房和商品房组成，其中回迁房和商品房两个区域则由水系自然划分而成。

(2) 建筑风格。项目整体为现代风格。其中，住宅部分通过对弧形屋面以及暖色面砖的适当运用，营造出柔和舒适的感觉；商业和办公楼的设计则通过现代的元素和手法进行对地标性的塑造。

(3) 地上停车库。地块西北侧由于靠近临近地块的住宅导致利用率不高，设计师化不利为有利，在这个位置放置低层地上停车库的同时设置屋顶绿化和休闲平台，使这部分地块得到有效的利用。

加拿大迈敦建筑师事务所 METROPOLITAN ARCHITECTS INC.

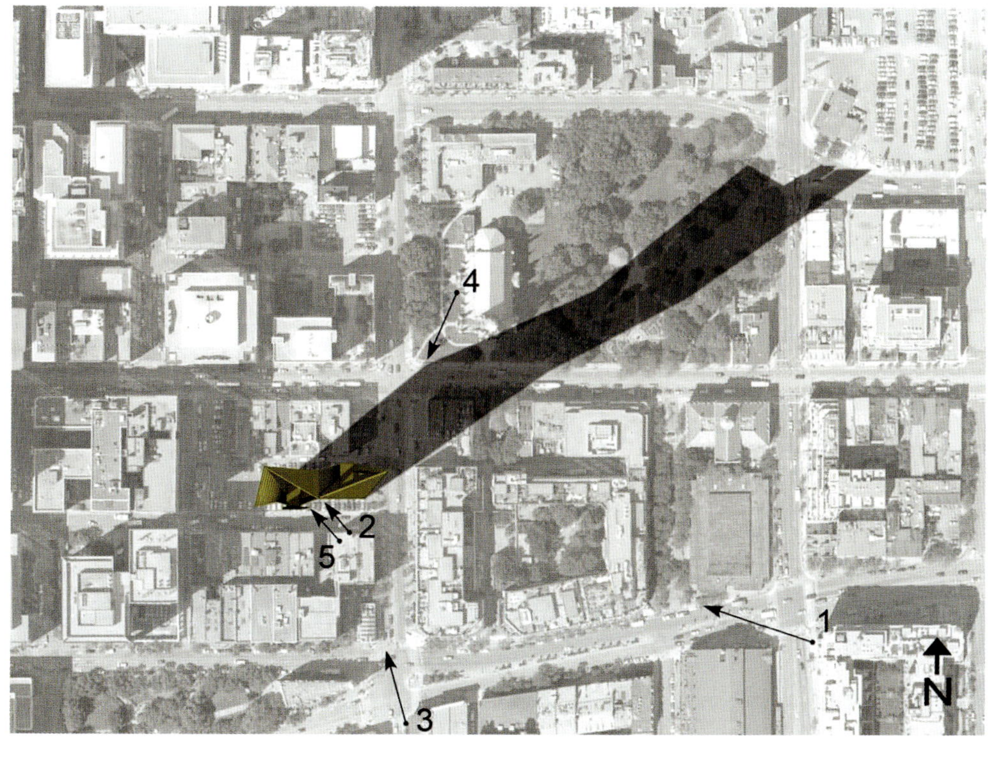

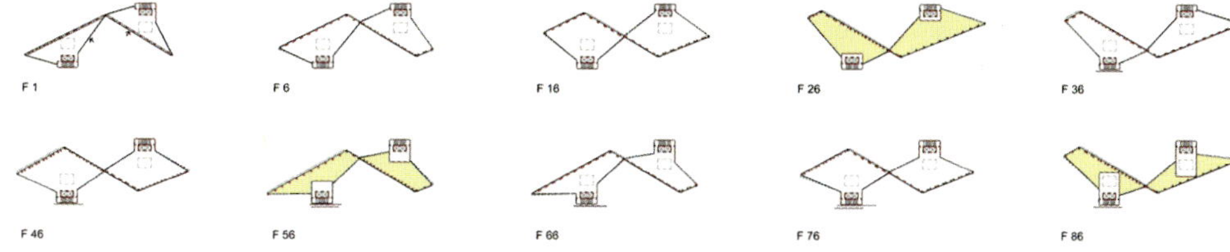

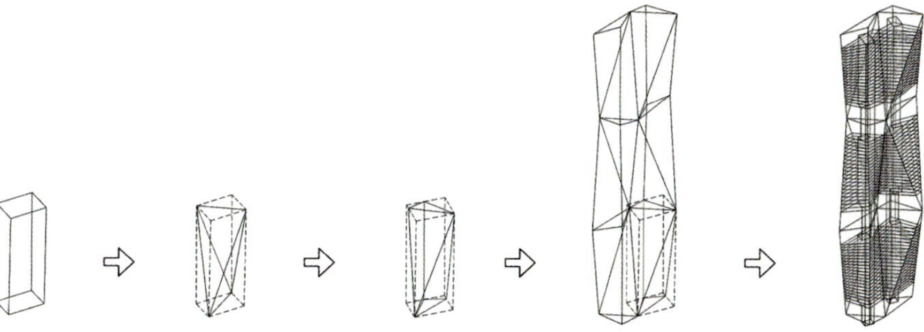

空中的多面体 —— 超高层概念设计

建设地点：加拿大 多伦多
建筑面积：69 300平方米
用地面积：2 300平方米
设计时间：2009年

多伦多，就像许多其他的国际大都市，在市中心有着不少露天的小停车场。这些散落各处的地块通常面积很小，因而很难用来建造传统的高层楼房。设计师尝试运用多面体的独特几何特性寻找在这类狭小地块上建造超高层的可行性，从而解决传统方法解决不了的问题。设计的几何原形来自一个简单的四面体，它由四个三角形组成：自然界最稳定的几何形状。将它稍加变化，即把它的两条对边换成两个三角形，就变为一个稍复杂一些的由四个三角形和两个扭面组成的多面体。将多个这样的多面体竖向叠加再横向重复，就可以构成一个巨型的空间体了。由于它大部分的面都由三角形组成，这个空间体的结构稳定性将远高于传统的由四边形构成的高层建筑，却有着相对较低的用钢量。另外，它动态的造型对城市街景的贡献也是不言而喻的。

加拿大迈敦建筑师事务所　METROPOLITAN ARCHITECTS INC.

1

4

2

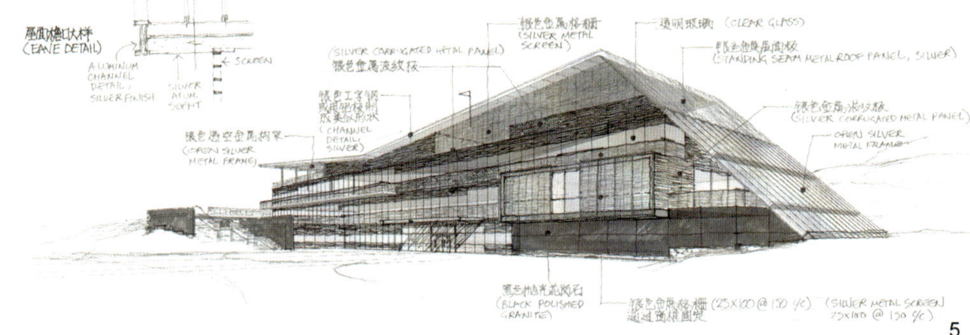

5

3

1. 枫丹白露住宅小区
   山东 高青

2. 淄博住宅小区
   山东 淄博

3. 中国网络文化创意中心
   北京

4. 别墅方案
   河南 许昌

5. 山地会所概念
   四川 重庆

6. 通信公司总部办公楼概念
   加拿大 多伦多

6

加拿大迈敦建筑师事务所　METROPOLITAN ARCHITECTS INC.

美国KAS设计有限公司
深圳凯斯筑景设计有限公司

KAS DESIGN COMPANY is a comprehensive architectural design company, whose business covers planning design, architectural design, landscape design, interior design and architectural photography etc. With headquarters in Hawaii, KAS has its Shenzhen office in Chegong Miao, Futian District.

KAS has been engaging in the combination and exploration of planning & architecture, architecture & landscape, architecture & interior after establishment. In KAS opinion, to make excellent works, must have better understandings of all the aspects mentioned above and combine the professional fields together to achieve interdisciplinary and integrated design. When facing clients with farsighted thinking and high sense of responsibility, KAS always sticks to "Be Honest & Creative, Keep Improving". Being developed with steady steps, KAS takes projects all over the country and cooperates with many developers, such as Vanke, Gemdale, Excellence Group, Galaxy Group, Fuchun Orient, Pearl River Investment, Horoy, Poly, L'sea Group, Poly and R&F Properties etc. Designers from USA, England, Australia, Philippine and Japan etc. collaborate well with the local high-efficient and creative team, creating high-quality works approved and admired by the society.

KAS设计有限公司是一家综合性的建筑设计公司，其业务范围涵盖规划设计、建筑设计、景观设计、室内设计以及建筑摄影等。总部设在美国夏威夷，深圳公司位于深圳市福田区车公庙。

KAS成立以来，一直致力于规划与建筑、建筑与景观、建筑与室内之间的融合与探索。KAS认为，一个好的作品，必须将以上几个方面很好地理解，并通过将各个专业领域融合在一起的方式，来实现整体性、跨学科、一体化设计的效果。KAS一直遵循诚信可靠、精益求精、富有创意、有远见的思维和高度的责任感。

KAS成立以来，稳步发展，承接的项目遍布全国各地，与万科、金地、卓越、星河、富春东方、珠江投资、鸿荣源、广州利海、保利、富力等著名的房地产开发商有过合作，并由多名美国、英国、澳大利亚、菲律宾及日本的规划师、建筑师、景观设计师、室内设计师参与了从方案到施工阶段的配合，共同创造出一个个高品质，让客户满意，并得到社会各界认可和赞赏的作品。

KAS DESIGN COMPANY
HEADQUARTERS
250 KAWAIHAE
HONOLULU, HAWAII 96825, USA
Tel: 001-808-927 8139

SHENZHEN COMPANY
Room 11B, Yunsong Tower,
ChegongMiao. Futian, Shenzhen,
Guangdong 518040, P.R.C
Tel: +86-755-8835 0900
Fax:+86-755-8835 0400
Website: http://www.kas.cn
E-mail: kas@kas.cn

美国KAS设计有限公司
美国夏威夷檀香山
Kawaihae大街250号
邮编：96825
电话：001-808-927 8139

深圳凯斯筑景设计有限公司
中国广东省深圳市福田区车公庙云松大厦11B
邮编：518040
电话：+86-755-8835 0900
传真：+86-755-8835 0400
网址：www.kas.cn
邮箱：kas@kas.cn

a. 保利花卉市场项目 建设地点：广东 广州 设计时间：2010年 竣工时间：预计2012年

b. 郑州清华大溪地 建设地点：河南 郑州 设计时间：2009年 竣工时间：2010年

a. 郑州清华大溪地　建设地点：河南 郑州　设计时间：2009年　竣工时间：2010年

b. 京基天涛轩　建设地点：广东 深圳　设计时间：2009年　竣工时间：2010年

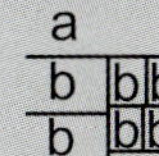

KAS DESIGN COMPANY

a. 利海郑州雁鸣湖温泉度假酒店　建设地点：河南 郑州　设计时间：2010年　竣工时间：2011年

a. 广州万科柏悦湾　建设地点：广东 广州　设计时间：2009年　竣工时间：2010年

b. 苏州万科国际广场　建设地点：江苏 苏州　设计时间：2008年　竣工时间：2010年

a. 天津顺驰太阳城	建设地点：天津	设计时间：2010年	竣工时间：2010年
b. 深圳阳基天御山（售楼处）	建设地点：广东 深圳	设计时间：2010年	竣工时间：2010年

a|a|a|a
b|b|b

a. 深圳阳基天御山（夏威夷） 建设地点：广东 深圳 设计时间：2010年 竣工时间：2010年

b. 深圳阳基天御山（新古典） 建设地点：广东 深圳 设计时间：2010年 竣工时间：2010年

ORIGINAL VISION

ORIGNAL PLANNING ORIGINAL ARCHITECTURE ORIGINAL INTERIORS

Hong Kong 22/F, 88 Gloucester Road, Wanchai, Hong Kong
Tel: + 852-2810 9797 Fax: + 852-2810 9790

Phuket 393 Moo 1, Srisoontorn Road, Cherngtalay Subdistrict Thalang District, Phuket 83110, Thailand
Tel: 0066-76270755 Fax: 0066-76270757

DIRECTORS

Adrian McCarroll
B.Sc., B.Arch (Hons), M,Sc.
HKIA, RIBA.
ARB (UK)

Stephen Gorton
BA (Hons) Arch, Dip Arch, ARB.
RIBA

Ken Leung
Bachelor of Built Environment
(Interior Design)

ASSOCIATE DIRECTORS

Jamie Jamieson
B. Arch, RIBA

Prasert Chanhom
B. Arch, M. Arch, ASA 3719

SELECTED CLIENTS

Asia Properties Asia Capacity Exchange Asia Island Homes Axiom Investment Management Baker & McKenzie Bovis Lend Lease Castaway Bay Café Piatti Chesterton Petty Cheung Kong Holdings Ciena Clifford Chance Double Star Café FCC Cambodia East Asia Properties Elite Model Management Frontline Clothing Ltd. Hyson IFM (Asia) Ltd. Indochina Assets Ltd. Internet Media House iRegent Jaspas Lee Marie Manhattan Holdings Midas Mövenpick Mundo Latino Nopawong Construction Co. La Rose Noire Lotus Asset Management Olivers Penta Finance Asia Ltd. PCCW PCPD Pepperonis Pinnacle Asia PT Padma Ohm Raimon Land Samsara Salem Attitude Sedgwick Richardson Sergio Tacchini Sino Land Starcore Asia Time Inc. Time Life Urban Entertainment Wilson Parking WSP XS Media Zac's Zuellig

COMPANY DATA

Established	创立年份	1992
HKIA member	香港建筑师学会会员	3
Professional	专业人员	17
Support	辅助人员	10

ORIGINALVISION

Creativity is not a linear process. It involves establishing then understanding a matrix of information and references which must be viewed in parallel to recognise the opportunities that lie within.

创作，从来不是直线进行的。在过程中，必须搜罗各式资讯及参考素材，然后深入剖析，并纵观彼此关系，方能发觉潜藏良机。

Many skills and disciplines are involved in creating elegance and beauty in the built environment. Appropriate orchestration of that expertise is as essential as the spark of innovation.

要缔造高雅优美的建筑环境，更需要运用不同技巧及专业知识。能把个中智慧发挥得恰到好处，其实与激发创意一样，至为重要。

ORIGINALVISION

A N S
ARCHITECTS & PLANNERS

ANS国际建筑设计与顾问有限公司

中国总部：上海市静安区西苏州路71号6楼 200041 电话：+86-21-51697511 传真：+86-21-62668470 邮箱：info@ansarchitects.com

ANS国际建筑设计与顾问有限公司是一家起源于澳洲，并通过其在澳洲的合作伙伴进行全球推广及专业服务的国际建筑设计与顾问有限公司。于1997年进入中国市场并于上海建立上海爱恩斯建筑设计有限公司，并相继于武汉、天津、成都设立了分公司，其中还包括拥有甲级设计资质的北京彩恩建筑设计有限责任公司。主要业务包括：城市规划、建筑设计、室内设计、景观设计，同时提供房地产项目前期工程的咨询顾问服务。ANS公司力图通过设计作品形式和功能的完美结合，成为生活和行为模式的创导者。公司的承诺是不断创造革新方案、国际设计标准，并以客户的满意为最终结果，就像ANS所代表的含义：ANSWER，公司一直致力于将国际化的视野融入本土化的充分理解，从而为客户带来最完美的解决方案。

ANS International Design & Consulting Pty. Ltd., is a multi-disciplinary international design company originally founded in Australia. We offer expertise services in the fields of urban planning, architectural design, landscape design and Interior design, as well as pre-project consultations during the early stages of a project. ANS established its first Chinese studio in Shanghai in 1997. Due to its success in its fields of expertise, ANS has since expanded its operations to include branch offices in Wuhan, Tianjing, Chengdu, partner studios in Australia and an A class architectural licensed studio in Beijing . ANS is dedicated to become a leading innovator and creator of the built environment with the goal of enhancing the qualities of life and lifestyle of the end users by combining form, function and art throughout its creative process. We endeavor to exceed client expectations with our innovative designs in every project. ANS WE R=ANSWER. Our international designers' fresh and innovative perspectives coupled with our localized knowledge offer our clients THE ANSWER to their question.

2010上海世博民营企业联合馆

武汉东湖开发区服务中心行政楼设计

DONGHU OFFICE BUILDING, WUHAN

工作内容：方案设计
设计团队：Joe Lau、Gabriel Gonzalez、张　海、顾甲天、熊　曦
建设地点：湖北 武汉
用地面积：400 000平方米
建筑面积：124 290平方米
时　　间：2010年

东湖公共服务中心考虑了东部的国际俱乐部和酒店，将其作为整体设计的一部分，提出了一个更加综合的无缝集成的总体规划。

为了着重体现基地的自然特质，建筑形态有机退后设置，将功能与形式和谐统一。管委会大楼与市属分局行政办公楼矗立于设置公共功能的基座之上，形成一个穿孔式的景观平台，促进自然采光、通风。建筑中不同的部分集合于一个综合体中，正如豆荚中的圆豆。我们以此简单的图示作为灵感的来源，创造出一个均质的、容纳不同功能的建筑。

我们的目标是设计一座现代化、高效率的建筑设施，为最终用户和公众提供一个愉快有效的服务环境。一条环绕基地四周的景观道路，连接了综合体的主要出入口，包括南边的迎宾及北部服务区域，与此同时把办公楼串联了起来。

大部分来客可以从建筑的南部进入，游览独特的空间，即周边环境和湖泊形成的前院。通过公共空间环绕的中心庭院，最后将视线收束在高耸连绵的山岳和天际一线形成的背景。

湖景与公园区域将作为一个舒适的环境场所服务于公众。其中的湖滨步行道，起自高新大道与规划路的拐角处，形成一条优雅的弧线延伸开来，最终以一座跨湖桥梁连接起东部的国际俱乐部与酒店。

天津滨海新区新港船厂改造综合文娱区城市设计

XINGANG SHIPYARD CULTURAL AND ENTERTAINMENT PRECINCT, TIANJIN

工作内容：城市规划、城市设计
设计团队：Joe Lau、熊曦、王燕霞、李敏、Gabriel Gonzalez、唐琛
合作设计：James Murray – Tandem Design Studio
建设地点：天津
用地面积：123.04公顷
建筑面积：1 815 254平方米
时　　间：2010年

项目地块目前是造船厂区，区内丰富的工业遗存、多变的岸线边界，一系列的码头、干船坞、水闸、浮桥等，为我们的设计带来了灵感。设计用新的空间和建筑设计语言进行诠释，唤醒人们对该区域历史的记忆。

“水域边缘”具有两方面的含义：一方面是处于天然的海河和人工化的天津港之间的临界位置；另一方面，寓意呈楔形交的城市与景观。

设计采用新的几何秩序和新的连接，塑造新的蓝（水）、绿（景）、红（活动）、灰（建筑），将滨水文化娱乐区与腹地最大限度地进行沟通互动。

为了设计新的城市方案，我们采用新几何结构。新几何结构把自身与直角梳状水域边缘和新街道、开发区网络区分开来。新几何结构采用三角网格，在位置和规模上充分利用历史优势，借机嵌入新的建筑砌块当中。三角测量是测量船岸之间距离的常用方法。选择性使用三角网格已经在此地创造了新的海岬和河床网络。新网格具有自己的边界，建立行人和自行车连接、风景线、绿色走廊和引入地下深水。近水和公园更有价值，精心策划水和景观引入方案，为这个令人兴奋的新文化和商业方案开辟了多种自然建筑面，成为活跃的水域边缘。

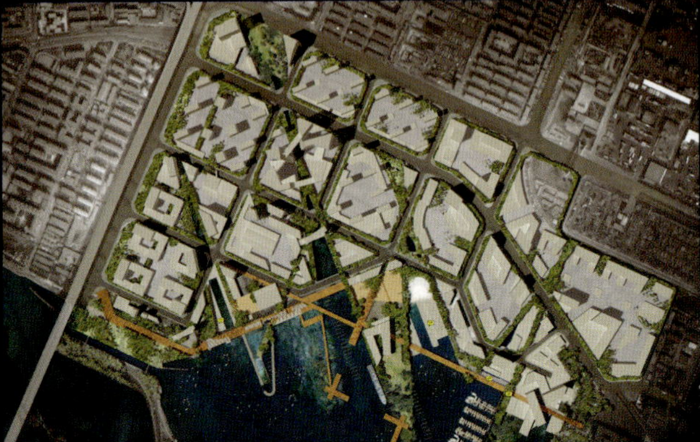

上海申江十六铺码头二期设计

16 PIER PHASE II PROJECT, SHANGHAI

工作内容：方案设计
设计团队：Joe Lau、张海、陈加、顾甲天、Gabriel Gonzalez、唐琛、熊曦
建设地点：上海
用地面积：9 880平方米
建筑面积：10 665平方米

十六铺码头二期项目，北起东门路，南至复兴东路，西起外马路，东至黄浦江的狭长地带，总占地面积约为9 880平方米，设计为黄浦江游码头、并作为一期的延续与拓展。

设计首先满足的是旅游码头高效的功能。将登船流线与下船流线完全分开，可以满足最大的人流量，并且将下车集散、安检、候船、检票、登船的空间设计在同一平面上，以避免高差对人流量的影响，从而达到高效。

其次，码头位于黄浦江S形弯道上，是黄浦江畔唯一能将外滩和陆家嘴两岸景色尽收眼底的最佳景观点，所以在码头的二层设计了6 938平方米的花园平台。此平台与一期的平台相接通，提供了一个可以饱览黄浦江两岸景色的绿地公园，成为游客们流连忘返的旅游胜地和周边居民的河滨休憩区。

如果滨江平台不能体现十六铺的文化氛围，那只是景观材质的堆砌。现代的是当下的，文化的是永远的。旅途本是收获，我们希望人们通过十六铺码头二期的时候可以收获十六铺的精神、文化的精髓，以及理解十六铺作为水上门户对上海城市发展的价值——千里外滩路，海上十六铺。文化平台除了复写十六铺历史文化氛围，传达十六铺的精神以外，也应该体现现代化公共空间的休闲性和轻松感——提供多层次、多角度的黄浦江观景方式。新十六铺和历史十六铺的有机结合可以激发人们对十六铺的情感，升华十六铺对于上海这座繁华大都市的价值。

天津万豪国际酒店

MARRIOTT HOTEL, TIANJIN

工作内容：建筑设计（方案设计、扩初设计）
设计团队：蔡 磊、张剑堃、石 岩、刘 珂
建设地点：天津
用地面积：7.2公顷
建筑面积：90 000平方米
时 间：2008年

项目位于天津市中心城区，处于滨水道与紫金山路交口。地处天津政府接待区域地块内的西南侧，北侧、东侧紧邻大面积自然水景，周边地区是天津市的政治、文化、展览中心，地理位置优越。

项目在总图摆位上采用折线的形式，最大限度地展现东北侧的景观，沿街立面完整、大气，突出整体严谨、稳重的气质，北侧则结合景观，以及中庭、餐厅等功能，运用了曲线元素，在不失大气的基础上充分考虑了景观因素。

主要功能为政府接待用五星级宾馆，包括500套高档客房、2套总统套房、大宴会厅、多功能厅、高档会见厅、中餐厅，以及各种风味餐厅、酒吧、室内泳池、室内网球馆、健身康体中心等。

成都大魔方

BIG CUBE, CHENGDU

工作内容：建筑设计（方案设计）
设计团队：宓立军、钱　阳、简川淞、颜美华
合作设计：现代集团 华东院
建设地点：四川 成都
用地面积：11.89公顷
建筑面积：910 323平方米
时　　间：2008年

该案所处的天府新城位于成都中轴线南段。在城市总体规划中，该段轴线定义为科技商务中轴线，是整个城市向南发展的先导区和核心区。

设计针对场地特征，将大魔方的三种基本功能中的商业布置于西侧主要交通方向，而居住区则在东侧沿河布置，使用西动东静的基本模式，符合商业综合体的分区特点。在沿快速路方向形成有魅力，有特色的工建化界面，而在沿河的区域，则将优美的自然景观渗透引入。商务及酒店功能由于需要直接的对外交通，因而布置于用地的外侧，与商业区和居住区协调安排。

在空间组织上，以大魔方城市舞台为核心组织建筑的布置，高层建筑形成完整的内部空间，灵活的水系进行小空间的连接。集合空间节点的设置呈现多核心布局，不规则的建筑形体互相有机结合，辅以空中连廊体系的设置，强调了商业综合体特有的活力感和戏剧性。核心广场建筑向城市空间敞开，以功能划分为空间组织的导向，突出了"门"的概念。居住区则发散布置，与商业区相对独立，闹中取静，建筑分层而列，既形成了富有韵律感的节奏空间，又有效地使得城市景观得以渗透进来。

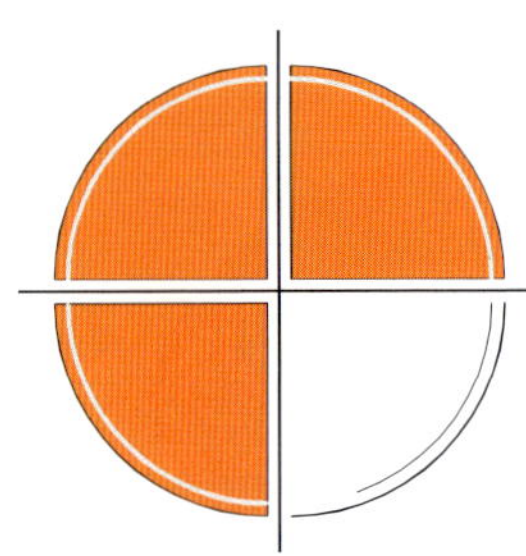

澳大利亚道克设计咨询有限公司

DECO-LAND DESIGNING CONSULTANTS (AUSTRALIA)

澳大利亚道克设计咨询有限公司(DECO-LAND DESIGNING CONSULTANTS，AUSTRALIA)于21世纪初在澳大利亚悉尼创建。自道克设计(DECO-LAND)进入中国以来，已在深圳与北京设立了分公司，在成都与西安设立了办事处，迅速成为国内颇具影响力的设计机构；至今公司的分支机构在中国的设计师已超过80人，在中国市场的设计面积已累计超过了2 000万平方米。

道克设计的核心团队吸纳了来自澳洲、欧美与亚太地区的优秀设计师，对土地利用、投资咨询、规划设计、建筑设计、景观园林、室内设计以及施工管理都具有丰富的经验。公司的设计服务涵盖了规划与城市设计、城市标志性建筑设计、文化与商业建筑设计、住宅设计、城市景观设计、室内设计等诸多领域。公司依托国际化背景，拥有国际及本土的多元化的专业设计团队，形成了独具一格的创作和服务特色，公司致力于为客户提供便捷、高效，更具成本效益的专业服务。道克设计力图为新世纪建筑和城市所面临的新问题提供新的解决办法，从广阔的城市视角和特定的城市体验中解读建筑的内涵。

道克设计的主旨是：从理想把握现实，在现实中调整思想；能在每一个项目中发现其中关键性的问题，然后提出一整套创造性的解决问题的方案，同时秉承了现代主义先驱的设计理念，认为建筑是使生活变得更加美好的动力之一。公司力图通过广泛的实践来发展和完善其理念，每个成员都在以主动参与的心态加入到项目设计中，坚信团队合作的精神将产生最佳的作品，最终以完善的设计成果展示给业主。我们的设计成果不仅是一种完善的建筑形式和风格，更是一种对项目完善的解决方案，使普通的开发产生诗意的生活和富有内涵的文化意念，借国际视野，以中国哲学思想精髓内涵服务中国地产，使项目在先进的思想、构思、信念、技术和材料等组织设计中提升价值。

近年来，道克设计积极参与国内多项重要的建设项目和规划设计，从2001年至今，公司多次在大型项目的国际招标里中标，近年来更荣获多项国际、国内大奖，并担任多个开发机构和建筑媒体的顾问理事单位。

● **Sydney**
Barker Street Kingsford NSW2032, Australia
Tel:00612-93983753 Fax:00612-93266818

● **Shenzhen**
广东省深圳市天安数码城天吉大厦A座
Tel:+86-755-83869932 Fax:+86-755-83869959

● **Beijing**
北京市朝阳区远洋国际中心C座
Tel:+86-10-58081092 Fax:+86-10-59081094

DECO-LAND DESIGNING CONSULTANTS, AUSTRALIA was established in Australia Sydney at the turn of the century. Since DECO-LAND entered into China, the branch companies have been established in Shenzhen and Beijing, and the branch offices have been established in Chengdu and Xi'an, which have become the most influential design organizations in China promptly; up to the present, there are more than 80 designers in the branch organizations of the company in China and the design area on Chinese market are accumulated to exceed 20 million m^2.

The core team of DECO-LAND has recruited outstanding designers who come from Australia, EU and America as well as Asian and Pacific regions and have rich experience in land utilization, investment consultation, planning & designing, architectural design, landscape design, interior design as well as construction management. The design services of DECO-LAND cover many fields, such as planning & urban design, design of urban landmark building, cultural and commercial architectural design, design of dwelling houses, urban landscape design and interior design, etc. Relying on internationalization background, we have international and national diversifying professional design team, which has formed distinctive creation and service feature, and the company devotes itself to supplying convenient, efficient and cost-effective professional services for customers.

DECO-LAND tries to supply new solutions for the new problems which are faced by architecture and cities in the new century and tries to read the meaning of architecture from the extensive urban visual angle and given urban experience.

The major idea of DECO-LAND is: grasping the reality from the ideal and adjusting thinking in reality; being able to find out critical problems in every project, and then being able to propose a whole set of creating solution proposals, and at the same time taking orders of pioneers of Modernism design philosophy, considering that architecture is one of the power which make life become finer. The company tries to develop and perfect its philosophy through extensive practice, every member is taking part in project design with the mentality of active participation, and firmly believing that spirit will produce the best works and will show the perfect design achievements to owners at last. Our design achievements are not only a kind of perfect building type and style but also a kind of solution to perfect project, which make common development generate poetical life and connotative cultural idea, and based on global vision, it serves China Property with the connotation of Chinese philosophic thought essence, making the project improve its value in the design of advanced thinking, conception, belief, technique and materials, etc.

In recent years, DECO-LAND has actively taken part in many important construction projects and planning & designing in China. From 2001 to now, the company has won bids in the international bidding of many large-scale projects, has been honored many international and national big prizes in recent years, and has worked as the consultation & direction unit of many development organizations and architectural medias.

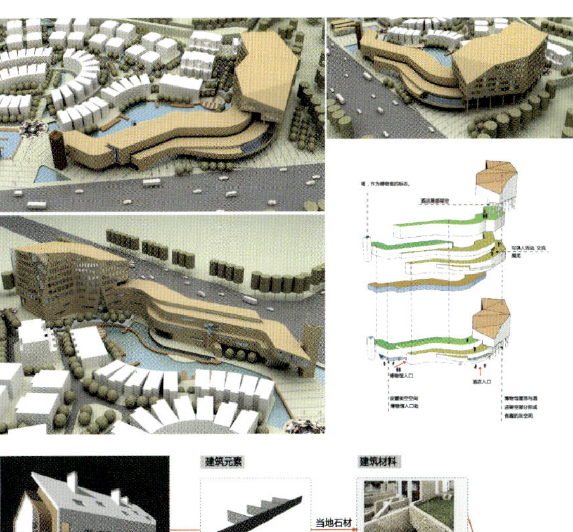

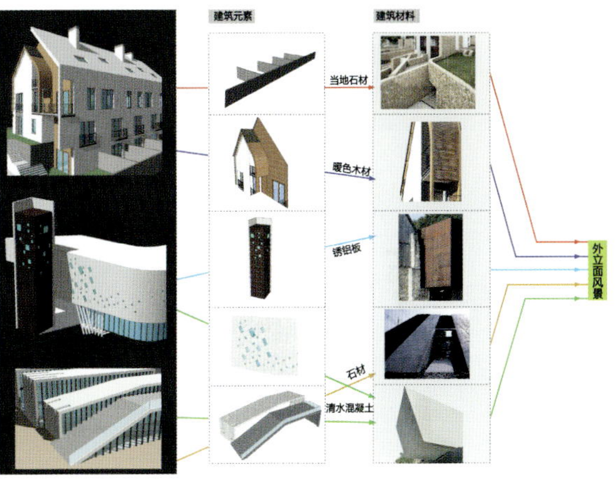

贵阳花溪项目

该博物馆“并不是一个容器，而是一个艺术品营地”。在这里，走廊和天桥相互叠加和连接，创造出一个具有生机的动感空间。尽管该建筑的功能清晰、平面组织合理，但寻求空间的灵活使用性仍是该设计的主要目标。空间的连续性设计避开了大量的墙体划分和干扰，为建筑内的多样动线和临时展示提供了良好场所。进入博物馆的中庭，混凝土弧形墙、悬浮的黑色楼梯和吸收自然光线的开敞天花，这些建筑的主要元素映入眼帘。借助这些元素，“力求创造出多视点和分散几何体的新型空间流动性，以此来征现代生活的纷杂动感。”

● 龙湖·成都牧马天堂悠山郡（一期）

建设地点：四川 成都
建筑密度：38.51%
绿 化 率：35.00%
总用地面积：137 680.66平方米
总建筑面积：126 213.09平方米

本项目在城市的位置，与大都市的距离、交通，以及景观环境中，使其具有第二居所的特质，也可兼作休闲度假居所。

项目用地山川起伏，壮丽恢弘，山丘、森林、草原、溪流，一派可忙里偷闲的氛围，是营建贵族式慢生活的难得之处。

如果一定要在地球上找一个地方来类比，首先想到的是英伦乡村。

英国人对于乡村天生情深意浓，对自然之美、乡村乐趣与劳作的喜爱，作为一种贵族式的休闲生活方式恰得其意，与当今都市人崇尚健康自然舒缓的生活态度又相切合。

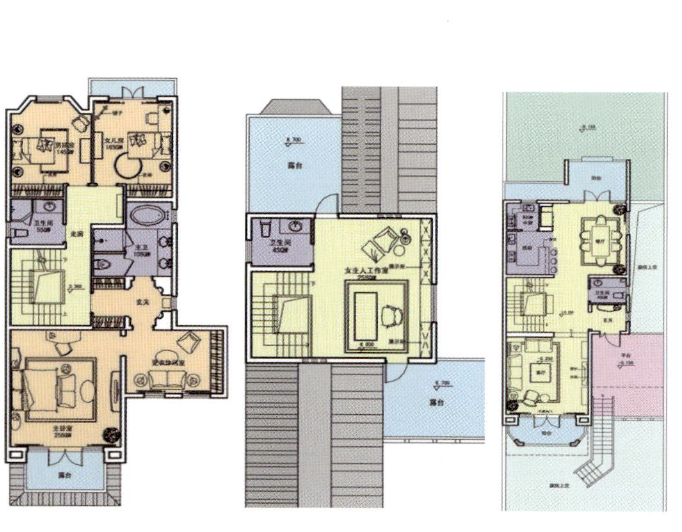

南通·海门东恒盛国际公馆

建设地点:江苏 海门
用地面积:82 735平方米

项目概述

本项目地块位于海门市南部新区南进轴线上，用地面积约82 735平方米，与市政府隔北京路相望，与东恒盛国际大酒店为邻。

用地范围交通便利，景观资源丰富。北临东海南路，西为城市干道张謇大道，南面为北京路，东面为天然河道。得天独厚的地理位置，让这里势必成为海门高品质景观住宅社区的新标杆。

设计理念

我们始终坚持"建筑服务于人，服务于城市"的理念，运用最新的设计思想，为住户创造便捷、舒适、安全、优美的21世纪生活新模式，为投资者的发展创造新的高度，同时为城市创造一个"梦想"。

我们对该项目寄予厚望，期望建设成为一线景观豪宅，将地段价值潜力放到最大，并在居住模式、空间形态与环境创意上树立新标杆，提升该地域地块的整体价值。

苏州·太湖波罗密山庄

建设地点：江苏 苏州
总用地面积：64 476平方米
总建筑面积：25 809.52平方米
容积率：0.2
绿地率：40%
建筑密度：10%

设计思路

项目位于苏州市金庭镇大龙山西北山坡，面临太湖，北眺飘渺山。用地面积64 476平方米。西山金庭镇是苏州市未来城市发展规划的重要地区之一，位于姑苏城西南35千米。

喧嚣、拥挤、浮躁，身处城市丛林中的都市人在城市繁华面具下越来越感觉到压抑和窒息。远离城市，到湖边、到山谷、到大自然中去享受宁静，拥抱自然，放松心情。面朝太湖，春暖花开。立于长三角都市"后花园"，打造高端休闲、娱乐、度假为一体的太湖"伊甸园"。

"Paramita"是梵语，音译为波罗密多，梵音"波罗"汉译"彼岸"，"密"译"到"，"多"是语尾的拖音，译"了"。译成汉文合起来是"到彼岸了"。城市之外、太湖之滨、群岛之间、梦想之地，彼岸就在大龙山脚下……

合理开发

结合地缘优势，发挥经济效益和实现销售客户群最大化，考虑地形特点和观景视线的规律，打造不同规模的户型和建筑形式。

在满足建筑容积率为0.2的情况下，本案秉持"产品价值最大化，景观面利用最大化"的设计原则，并结合地形现状将本地块分为六级别墅用地和一温泉会所用地。

单体平面空间构思

结合中式传统庭院建筑的蜕变和传承，设计出现代新中式度假型建筑。

户型平面设计模式主要采用三种建筑形态。

（1）围合式。传承中式传统建筑特点，采用具有静谧私密的内天井模式，并吸取现代生态建筑和环境相融合的设计特点，采用大面积的观景露台和落地窗，亲近自然，享受现代品质生活。

（2）半围合开放式。在扩大景观视野的同时，兼具传统中式建筑布局特点，保证别墅具有开阔通透的建筑格局。

（3）退台式。地势较陡，依坡而建，减少交通面积，充分利用地形特点创造功能空间利用最大化。

会所入口

AUROS ESPACES CONCEPTION (FRANCE)

法国AECF颐朗联合建筑设计有限公司

法国AECF建筑设计有限公司上海事务所
上海：上海市杨浦区大连路(海上海)970号1308室
邮编：200092
电话：+86-21-65909515
传真：+86-21-65909526
邮箱：YL_AECF@163.COM

法国AECF建筑设计有限公司厦门事务所
厦门：福建省厦门市思明区莲前东路123号加州商业广场嘉盛建设11楼
邮编：361004
电话：+86-592-2292893
传真：+86-592-2292892
邮箱：YL_AECF@163.COM

法国AECF建筑设计有限公司巴黎事务所
巴黎：155 Rue du Fbg Saint-Denis Paris
邮编：75005
电话：0033-179242012
传真：0033-179242013
邮箱：YL_AECF@163.COM

颐朗联合设计是立足于法国巴黎，联合众多欧洲资深设计师的联盟事务所。拥有建筑、室内、园林景观、规划、高尔夫专项设计资质。其设计团队将以独特的创意、丰富的专业知识和对中国文化的深刻理解为中国客户提供优质的服务。

颐朗联合设计是以欧洲设计师为设计核心、以中国设计师为市场引导，并以国内设计院为技术依托而形成的针对中国市场的中欧合作的设计师团队。颐朗可以为中国市场提供各类专业的设计服务，并与中国设计师紧密联系，使欧洲原创的设计理念在中国更有可实施性和持续性。

颐朗认为建筑师作为改变城市形态的带头人，除了为人们提供先进、专业的设计理念外，还应该对今天人们的生活模式和社会模式进行深入的思考，并对当今城市在未来的可持续发展提供可实施性的理论和建议，成为改变人们生活的先行者。

颐朗设计服务建立在设计资源和设计管理完全共享的平台上，坚持〝多元文化、多重表达〞的设计理念，以国际水准提供专业服务的客户宗旨，与每一位关注它的朋友共同发展。

YL-ACEF is a Paris-based association of senior designers from Europe. It's qualified for architectural, interior, gardening & landscaping, planning, golf specific designs. This design team will provide high quality services to Chinese customers by its unique creation, rich expertise and profound understandings toward Chinese culture.

European designers are the core of this team, while Chinese designers are the market guide, and this team is a Sino-European cooperation based on Chinese design institutes, oriented to Chinese market. It provides all kinds of professional design services for Chinese market, and closely links with Chinese designers, to make European original conceptions more feasible and sustainable in China.

YL-ACEF believes that architects, as leaders of urban reformers, will provide advanced professional design theories, and additionally reconsider modern lifestyle and social model, and propose viable theories and suggestions to future sustainable development of cities, and become pioneers of life reforms.

YL-ACEF design services are based on a fully shared platform of design resources and design management, where the design philosophy of "multi-cultures and multi-expressions" is performed to take care of every friend for joint development, following the customer tenet of "international level professional services".

AECF Shanghai
Shanghai Office: Room 1308, 970 Dalian Road (Hai Shang Hai),
Yangpu District, Shanghai
P.C.: 200092
Tel: +86-21-65909515
Fax: +86-21-65909526
E-mail: YL_AECF@163.COM

AECF Xiamen
Xiamen: Floor 11,Jiasheng Construction,Jiazhou Mall,123
Lianqian East Road, Siming District, Xiamen, Fujian
P.C.: 361004
Tel: +86-592-2292893
Fax: +86-592-2292892
E-mail: YL_AECF@163.COM

AECF Paris
Paris: 155 Rue du Fbg Saint-Denis Paris
P.C.: 75005
Tel: +86-033-179242012
Fax: +86-033-179242013
E-mail: YL_AECF@163.COM

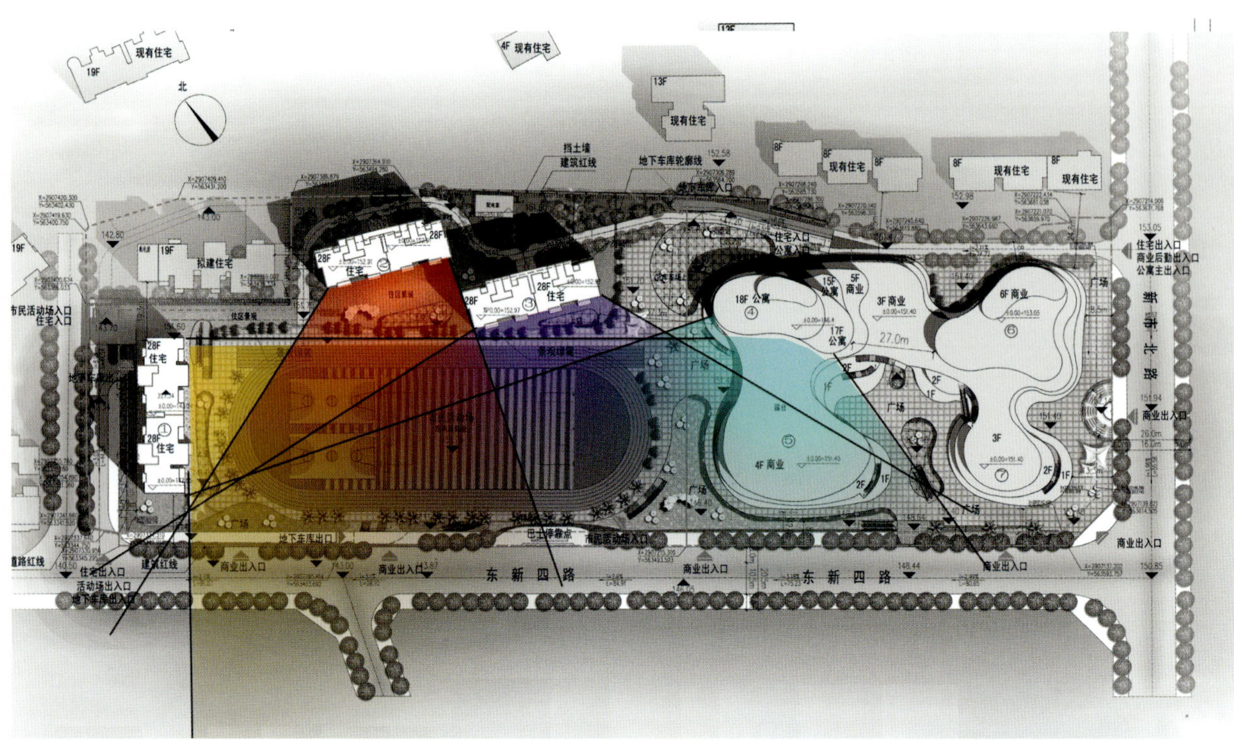

三明市列东商业

主创人员：刘智敏、巴学天、丁 旋
建设地点：福建 三明
建筑面积：162 000平方米
设计时间：2011年
项目状态：方案报批阶段

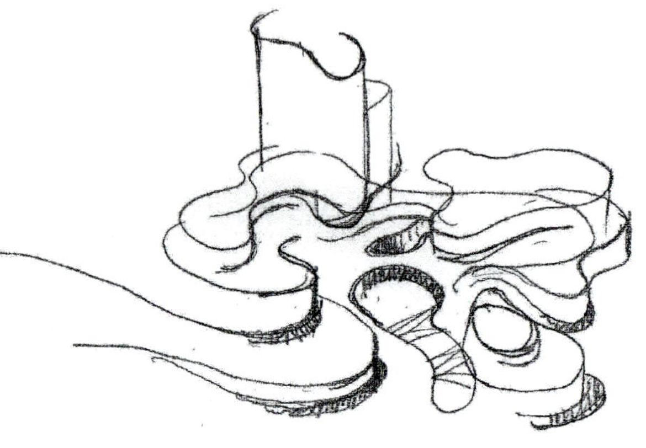

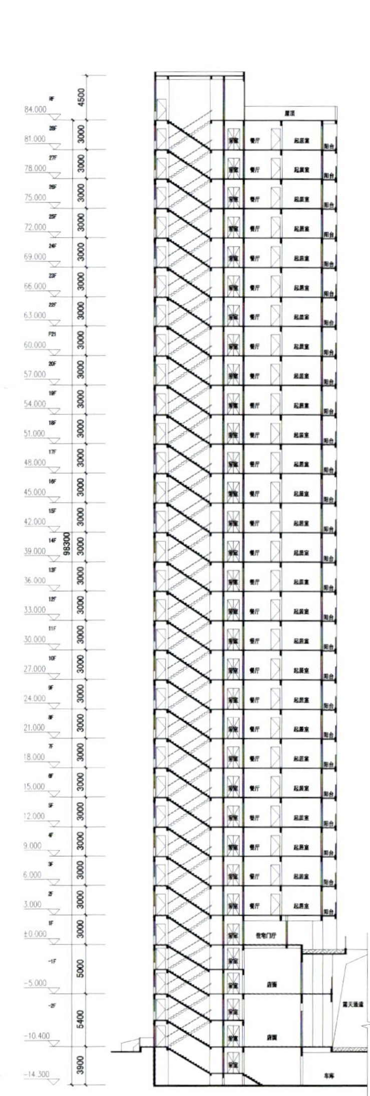

厦门长泰泛华酒店

主创人员：巴学天
建设地点：福建 漳州
建筑面积：31 000平方米
设计时间：2010—2011年
项目状态：施工建设中

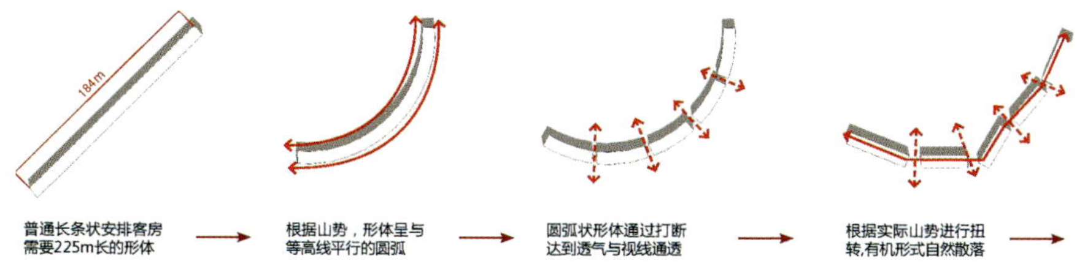

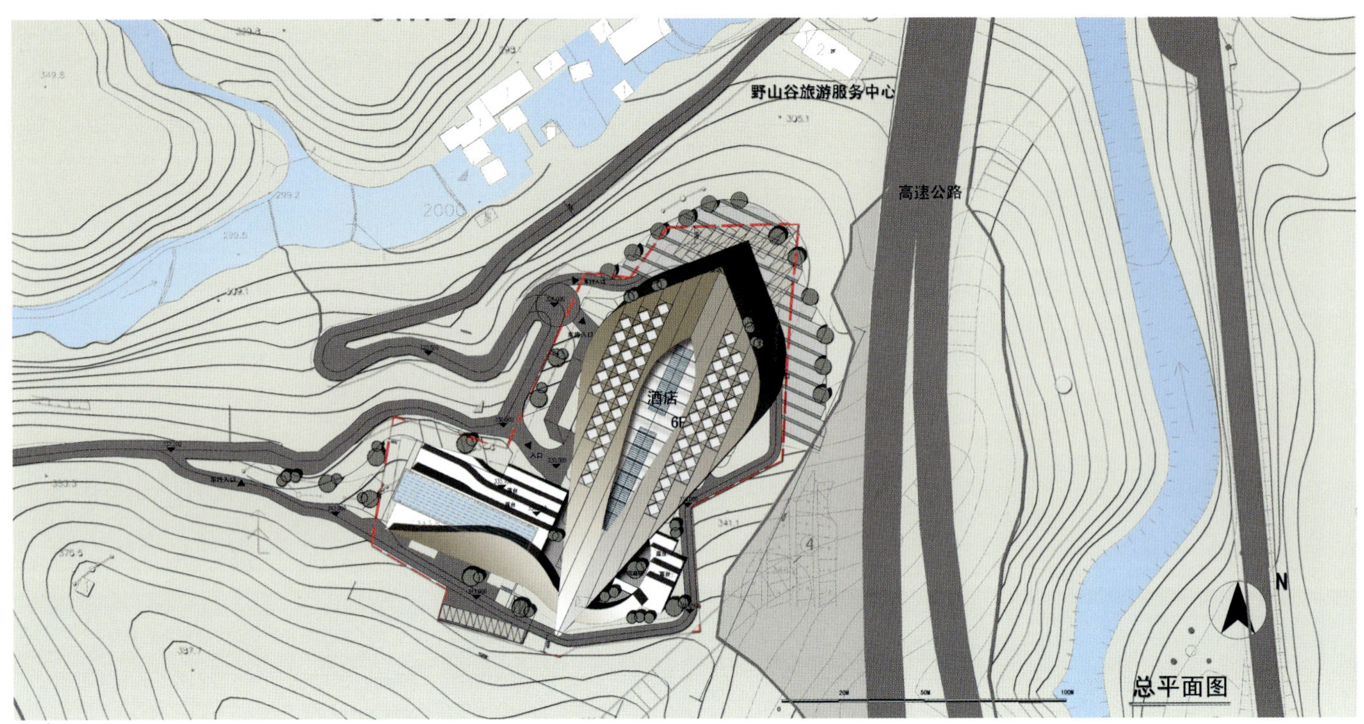

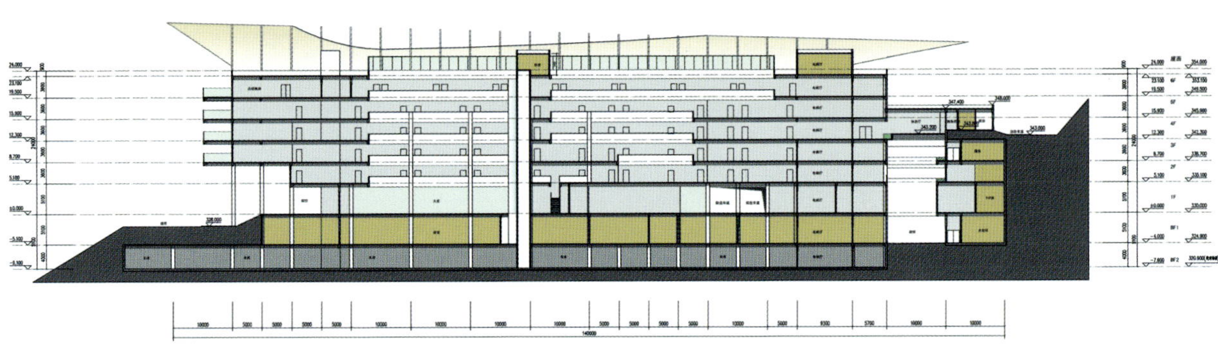

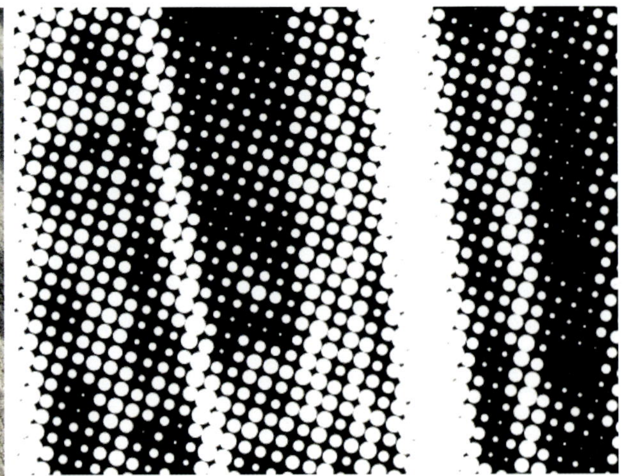

野山谷酒店

主创人员：巴学天、金虎星
建设地点：福建 厦门
建筑面积：36 600平方米
设计时间：2011年
项目状态：方案设计阶段

御龙湾项目

主创人员：刘智敏、朱 盛、徐耀鑫
建设地点：福建 晋江
建筑面积：275 550平方米
设计时间：2010年
项目状态：施工图报批阶段

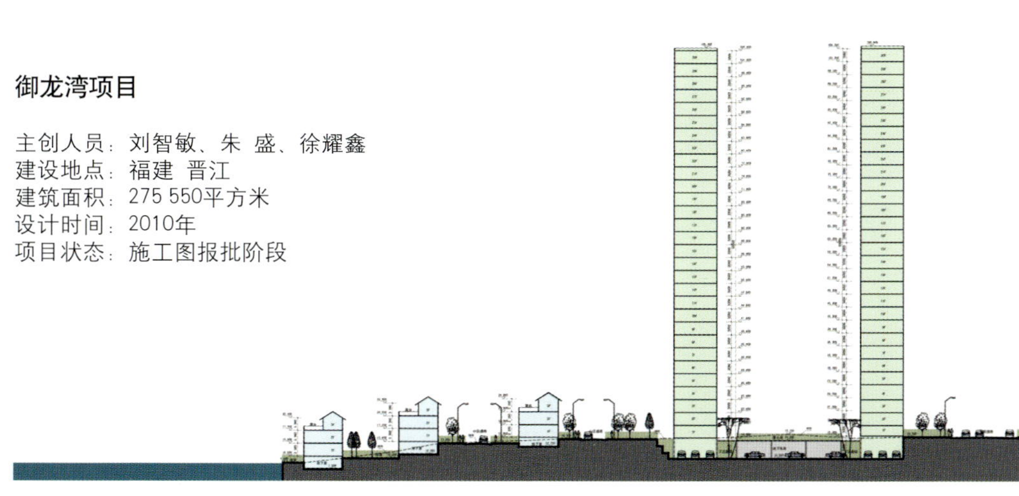

土地价值性

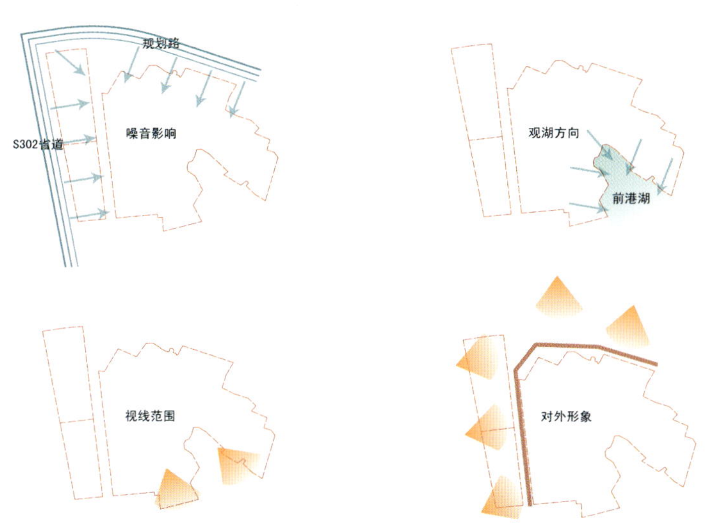

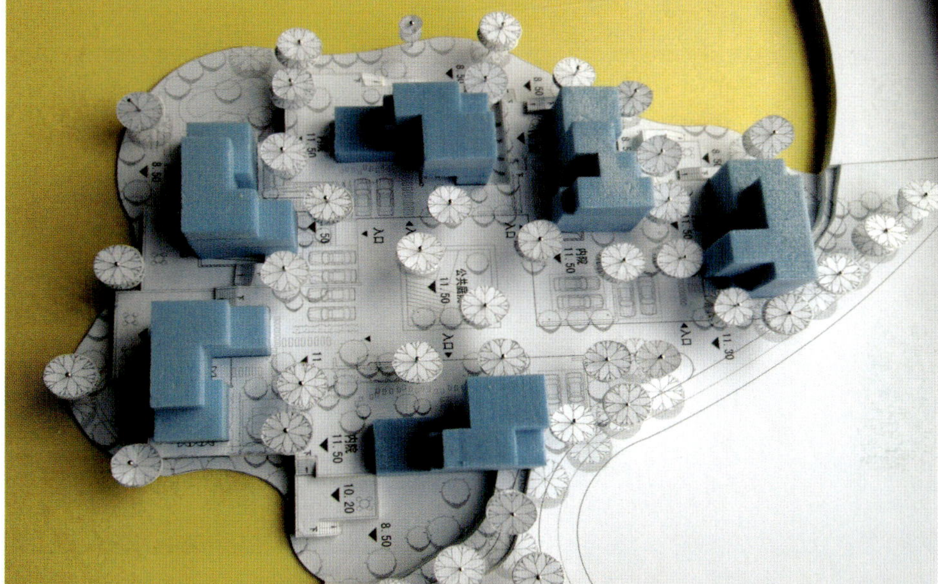

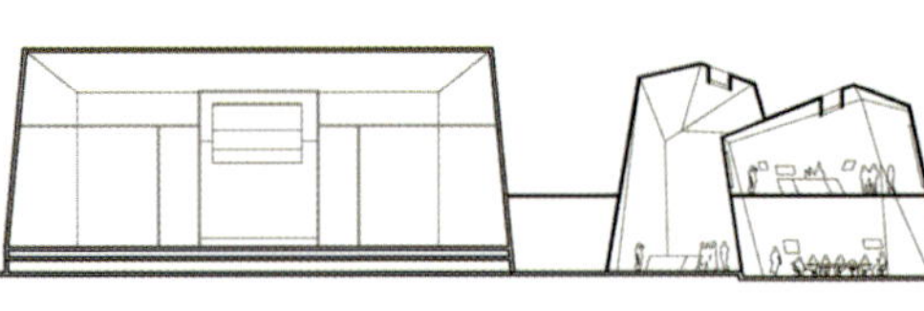

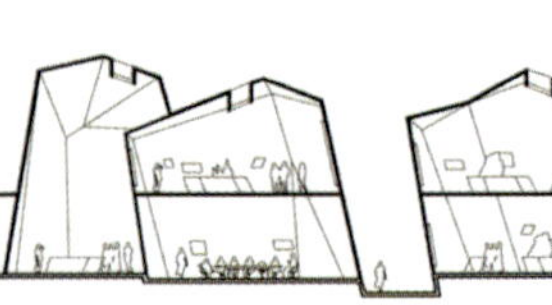

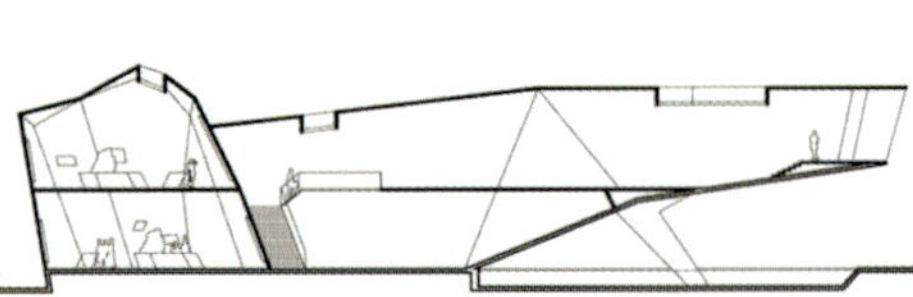

1-1 剖面图

厦门漳州火山岛博物馆

建设地点：福建 厦门
建筑面积：4 500平方米

本案位于厦门漳州火山地质公园内，设计概念把抽象的折线变化作为本案的基础元素，运动的轴线上有古堡休闲中心、林进屿和南碇岛，建筑的造型为几块不规则切面的多面体，寓意为富有动感的火山岩。

建筑造型表现火山岛特色，轻巧、简洁、通透，为园区标志性景点。同时充分利用地势，创造高低错落的室内外空间，室内空间灵活多变，在参观路线上设计布置丰富的观测点，使游客在参观展品的同时，又能观赏景区内的景观。特别是室内多变的空间顶部采光效果，更使游客仿佛置身于火山内部，产生奇妙的心理体验。

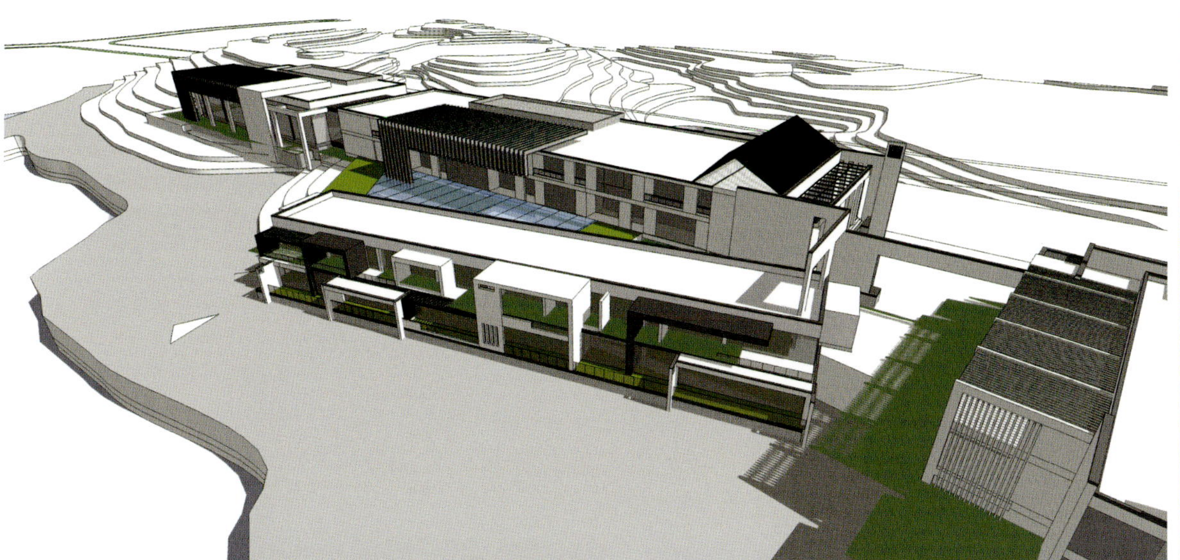

笋山竹海旅游度假区

主创人员：金虎星、巴学天、丁旋
建设地点：安徽 广德
建筑面积：40 000 平方米
设计时间：2011年
项目状态：方案报批

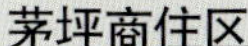

茅坪商住区

主创人员：刘智敏、朱 盛、徐耀鑫
建设地点：福建 三明
建筑面积：210 000平方米
设计时间：2011年
项目状态：方案报批

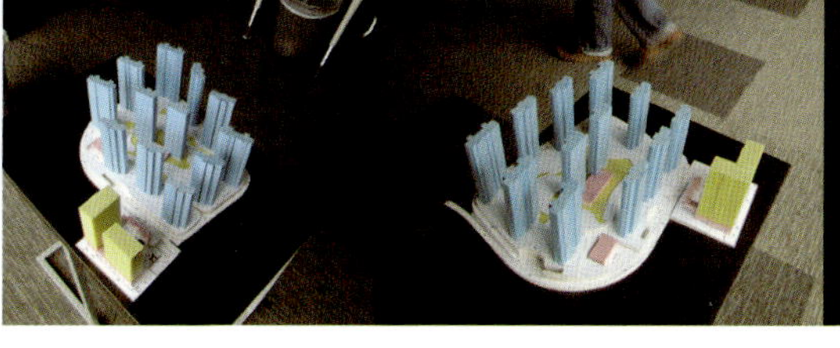

Aedas

1 交通运输 | 广深港高速铁路香港段西九龙总站（中国香港）

- 2010 Cityscape Awards for Architecture in the Emerging Markets Award
  Winner-Tourism, Travel &Transport Future
- 2010 Cityscape Awards for Architecture in the Emerging Markets Award
  Highly Commended-Commercial/Mixed Use Future
- 2010 World Architecture Festival Award
  Winner-Future Project-Infrastructure
- 2010 World Architecture Festival Award
  Category Winner of Future Projects
  Infrastructure Category Winner of Future Projects-Competition

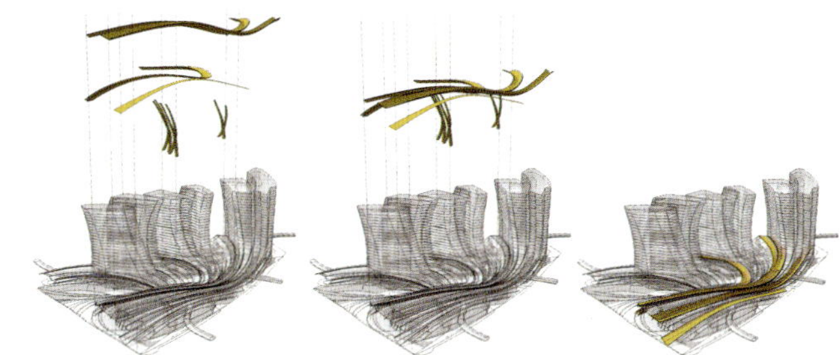

1

2

3

4

6

7

8

9

Aedas

10

11

12

12

2 研发项目 | Valeo Thermal Systems（美国底特律）
- 2002 American Institute of Architects Honor Award
- 2000 Business Week/Architectural Record Awards
- 1999 AIA NEW York State Excellence In Design Award
- 1998 AIA NEW York Chapter, Distinguished Architecture Award

3 体育设施 | 多哈体育城
国际足联2022年世界杯体育场（卡塔尔多哈）

4 酒店项目 | 皇冠假日酒店（中国惠州）

5 办公项目 | Boulevard Plaza （阿联酋迪拜）

6 住宅项目 | DAMAC Ocean Heights I（阿联酋迪拜）
- 2005 Bentley Design Award for Best Architecture

7 综合项目 | 北京北辰综合发展项目（中国北京）
- Cityscape Awards for Architecture in the Emerging Markets
- 2010 Highly Commended-Commercial/Mixed Use Built
- Asia pacific Commercial property Awards
- 2010 The Architecture Award (Mixed Use) Asia pacific
- 5 Star-The Architecture Award (Mixed Use)

8 城市规划 | 香港西九龙文化中心（中国香港）
- 2006 MIPIM Architectural Review, Future Project Awards
- 2004 Cityscape Architectural Review Awards, Commendation in Master Planning Category

9 零售项目 | 仁和春天广场（中国成都）

10 教育设施 | 香港城市大学深圳产学研大楼 （中国深圳）

11 交通运输 | 杭州萧山国际机场二期（中国杭州）
- 2010 AIA Hong Kong Chapter,Merit Award for Interiors

12 文化设施 | 市政、文化及商场综合发展项目（新加坡）
- 2010 MIPIM Architecture Review Future Project Awards Mixed Use-commendation
- 2009 CNBC Asia Pacific Commercial Property Awards The Architecture Award Singapore
- 2009 CNBC Asia Pacific Commercial Property Awards The Architecture Award Asia Pacific
- 2009 Intl Property Award-Finalist,Architecture Award
- 2007 Design Competition Award-1st Prize

S A I N A S I A

Sainasia 西班牙建筑设计事务所

西班牙Sainasia建筑设计事务所

（西班牙建筑设计在亚洲）

地址：中国上海市静安区江宁路420号和一大厦15E
Add: 15E Heyi Mansion, NO. 420 Jiangning Road, Jing'an District, Shanghai, China
邮编：200041
电话：+86-21-62712489 传真：+86-21-62712894
E-mail: shanghai@sainasia.net Http://www.sainasia.net

天津昆明路小学现代化设计方案

天津市河北路310、312号设计方案

西安咸阳国际机场旅客过夜用房工程方案

季梦滨
Benito Jiménez González

合伙建筑师
Architect Partner
医疗建筑专家
Specialized in Healthcare Architecture
政府机关建筑专家
Specialized in Government Buildings
公共住房专家
Specialized in Public Housing

费 尔
Horacio Fernández del Castillo

合伙建筑师
Architect Partner
翻新与修复大师
Master in Rehabilitation & Refurbishment
城市规划大师
Master in Urban Planning
博物馆，剧院，音乐厅
Specialized in Museum,Cultural Buildings and Concert Hall

裴立柯
Enrique Prieto Catalán

合伙建筑师
Architect Partner
城市规划专家
Specialized in Urban Planning
楼宇翻新专家
Specialized in Rehabilitation
节能专家
Specialized in Sustainability

PAN-PACIFIC DESIGN GROUP LTD. (CANADA)
SHANGHAI PAN-PACIFIC ARCHITECTURAL DESIGN CO., LTD.
SHANGHAI PAN-PACIFIC SPACE-WORK ASSOCIATES

泛太平洋设计集团有限公司（加拿大）
上海泛太建筑设计有限公司
上海泛巢建筑设计事务所

泛太平洋设计集团有限公司（加拿大）

泛太平洋设计集团有限公司（加拿大）1994年成立于加拿大多伦多，并于1995年在上海成立设计分支机构。公司设计范围包括城市规划及建筑设计，包括酒店、办公、商业文化设施、住宅、学校、体育场所等，并重视发展城市景观设计、建筑设计与室内设计的衔接与渗透。

公司依托国际化背景，以国际化加本土的专业化团队，形成自己的创作和服务特色。在创作过程中，公司全面认知项目的土地价值、开发定位、文脉延续、艺术感知及心理体验，以开放、理性和积极的姿态，去寻求创作中的每一个机会和最佳答案。公司还重视实施过程的技术控制和提升，逐步发展了一套适应于当代市场多样变化的服务体系。

公司成立至今完成了上海浦东新区接待中心、九寨天堂国际会议度假中心、德隆集团总部大楼、成都新国际会展中心和上海烟草集团科教中心等国内许多重大工程的设计，并赢得了建设部全国人居经典综合金奖、建设部建筑形态金质奖、建设部创新风暴中国建筑设计示范住宅单项奖、上海市优秀住宅小区工程设计项目一等奖、上海市优秀勘察设计二等奖、上海市住宅小区优秀规划奖等近30个设计奖项。2004年12月至2005年12月，公司获得由中国住交会委员会等颁发的"2004年度中国建筑二十大品牌影响力规划建筑设计事务所（公司）奖"及"2005年度中国二十大品牌影响力规划建筑、景观设计公司，2006年获得法国建筑设计师协会、国际房地产商协会和亚洲房地产峰会组织委员会等颁发的"2006年度亚洲建筑规划设计机构100强"以及由中国建筑规划设计师协会等颁发的"2007引领中国建筑规划设计十大品牌设计机构"等奖项。

泛太平洋设计集团有限公司（加拿大）于2001年1月成立上海泛太建筑设计有限公司，并于2005年9月成立上海泛巢建筑设计事务所。

上海泛太建筑设计有限公司
上海泛巢建筑设计事务所

上海泛太建筑设计有限公司成立于2001年1月，于同年获得上海市首批颁发的建筑专业专项设计资质，2006年改为建筑工程设计乙级资质；上海泛巢建筑设计事务所注册于2005年3月，同年获建设部建筑设计专项甲级资质。均是泛太平洋设计集团有限公司（加拿大）在中国境内设立的全资控股企业。公司承接从规划、建筑方案到施工图全程所有专业的设计工作。

公司以高标准的设计与服务建立起了市场知名度和信誉度，在此基础上，公司着眼全国市场，在江苏、浙江、四川、山东、吉林等近20个省份开展了一系列项目，并在成都成立了分支机构。目前，公司有设计人员近120名，专业涵盖建筑、结构、机电、室内设计和景观等，其中国家一级注册建筑师25名、外籍建筑师10名、建筑学博士2名、硕士25名，均具有一流的设计能力和工程协调能力。自2003年起，公司在各年度民用建筑设计市场民营设计企业二十强排行榜中排名均位于前列。并且自2001年以来，每年都有数个设计作品获得国家及地方评比的奖项。

随着社会经济与技术的发展，公司致力于前所未有的人性化的建筑创造，并通过我们的双手实现梦想。

PAN-PACIFIC DESIGN GROUP LTD. (CANADA)
SHANGHAI PAN-PACIFIC ARCHITECTURAL DESIGN CO., LTD.
SHANGHAI PAN-PACIFIC SPACE-WORK ASSOCIATES

PDG was established in Toronto, Ontario, Canada in 1994. Since the opening of the Shanghai Office in 1995, PDG has become one of the most active design firms in the Chinese market.
Inspired by international trends of modern design, PDG offers its expertise in architectural design, urban planning, landscape design and interior design. With over 100 professionals from all over the world, PDG combines Chinese and western philosophy in its commitment to excellence under the company's mission, that is, to maximize the value of the project for the clients, users and society in general through design.
Based on the specific project, our team works closely with the clients to achieve a full understanding of their facility needs and site conditions. The application of creativity to solve challenging problems, the pursuit for excellence and the spirit of the team collaboration are the means for our success.
In 2001, Shanghai Pan-Pacific Architectural Design Co., Ltd., which was licensed by the Chinese government and in 2005, Shanghai Pan-Pacific Space-work Associates was incorporated in Shanghai and thus became the legal base for PDG to work in China.
On recognition for this excellence on both professional design and service, PDG has been awarded numerous national awards both from the government and from professional associations.

加拿大温哥华
地址：2403-1250 Hasting Street, Vancouver, BC, Canada V6E 4T7
电话：001–(604) 899 2639
传真：001–(604) 899 2692

中国上海
地址：上海市浦东新区张江高科技园区伽利略路11号伽利略商务公馆6单元
邮编：201023
电话：+86–21–31265800
传真：+86–21–50814700
邮箱：ppddg@ppddg.com
网址：www.ppddg.com

Vancouver, Canada
Add: 2403-1250 Hasting Street, Vancouver, BC, Canada V6E 4T7
Tel: 001–(604) 899 2639
Fax: 001–(604) 899 2692

Shanghai · China
Add: Unit 6, No. 11 Galileo Business Mansion, Jialilue Road, Zhangjiang High-Tech Park, Pudong, Shanghai
T.P.: 201023
Tel: +86–21–31265800
Fax: +86–21–50814700
Email: ppddg@ppddg.com
http:// www.ppddg.com

成都医科总部产业园

项目性质：办公楼
占地面积：125 877平方米
建筑面积：356 712平方米
设计时间：2010年
项目状态：方案

郑东新区办公商业项目

项目性质：办公楼
占地面积：12 475平方米
建筑面积：108 500平方米
设计时间：2010年
项目状态：方案

绿地长沙金融中心

项目性质：商业
占地面积：7.53公顷
建筑面积：1 083 022平方米
设计时间：2010年
项目状态：方案

重庆珠江实业太阳城商业

项目性质：商业
占地面积：33 235平方米
建筑面积：11 7720平方米
设计时间：2011年
项目状态：方案

大连育马场

项目性质：住宅
占地面积：67 485平方米
建筑面积：221 074平方米
设计时间：2009年
项目状态：方案

上海东航金叶苑

项目性质：住宅
占地面积：7.53公顷
建筑面积：1 083 022平方米
设计时间：2010年
项目状态：方案

上海东航金叶苑位于上海市徐汇区WS5地块滨江商务区内。WS5地块与世博区隔江相望，是未来又一滨江CBD。项目定位于徐汇滨江商务区高品质大型国际化社区，其高端定位将进一步提升商务区的品质形象。东航金叶苑占地约25公顷，地上面积约50万平方米，其中居住建筑面积33万平方米，商业办公17万平方米。整体项目分为五个地块，中间三个地块为住宅用地，东西两个地块为商业办公用地。居住地块组团式布局，强调大间距和均好性，人车立体分流，地下车库高标准的停车配置，地面通过景观的营造和强化自然与人文、休闲和尊贵感相结合的特色。建筑风格定位于"上海新式老洋房"。

上海堪称装饰艺术建筑的圣地之一，高层住宅采用装饰艺术风格成为最佳选择，而低层住宅则采用了意式和法式的"老洋房"风格。

浙江仙居大卫世纪城

项目性质：住宅、商业、酒店
占地面积：120 000平方米
建筑面积：250 000平方米
设计时间：2009年
项目状态：在建

本项目旨在创造一个有活力的城市滨水空间，有机组合商贸、文化娱乐休闲与公共空间环境、教育展示、居住、特色历史、游憩码头设施、酒店会务住宿等七类功能。传承历史文脉并创造具有地方传统特色的建筑空间形态，塑造宜人的游憩空间，融入自然的滨水建筑景观界面，打造城市新型滨水居住区。

由于基地被道路分开成相对独立的五部分，因此在设计中使所有的建筑既相对独立，又以连贯的整体形态组织起来，使它们所围合的空间更具动感与吸引力。此外，在形成城市空间的同时，努力使空间不形成封闭的角落，内部空间、外部空间和城市空间互相渗透交融在一起，给人一种开放的感觉。空间的使用者会感到它比实际大，因为空间穿透建筑并与街道空间融合，同时较自由的建筑形态也就意味着整个地区是有机、自然、可成长的。在设计中平衡建筑物高低的关系，以达到经济性和空间质量的平衡，让各个地块建筑的高低错落的形式组合起来，既有百米的标志性建筑，又有亲切宜人的低多层建筑。在空间上，在城市道路界面、城市滨水界面上，与地块的自然地形巧妙结合，给人丰富和动感的形象，最终形成复合城市滨水空间。

WHI International

ADELAIDE阿德莱德
MELBOURNE墨尔本
BANGKOK曼谷
SEATTLE西雅图
GUANGZHOU广州
SHANGHAI上海
BEIJING北京

WHI International建筑设计集团是澳大利亚知名的综合性设计机构。WHI提供完整综合的创造性设计服务，设计范围包括规划、建筑、室内、景观、图文等专业；专注于设计创意、优良服务以及市场为导向的设计定位，作品形态涵盖了城市规划、旅游度假区及住宅区规划、高级酒店、度假村、高档住宅、别墅、商业、办公楼、综合性公共建筑等。

WHI International is an Australia-based multidisciplinary design practice and a leader within the architectural field.

Services

WHI International aims to provide innovative design services based on market needs. This diversified cultural background and our career experiences make for a professional design practice, which has successfully provided designs for numerous projects.

Scope

Our design covers urban design and master planning of large scale urban areas, tourist zones and residential areas. We also specialise in architectural, interior, and landscape designs for high quality hotels, resorts, commercial, retail, villas, residential buildings and public buildings.

SHANGHAI 上海

E-mail:shanghai@whiint.com
TEL：+ 86 21-62996297 , 53930250
FAX：+ 86-21-62998336
ADD：上海市西康路928号静安创展大厦413室
Rm.413, Chuangzhan Building, 928 Xikang Road Jing'an District, Shanghai

GUANGZHOU 广州

E-mail: guangzhou@whiint.com
TEL：+86-20-38303025 , 38303020
FAX：+86-20-38303021
ADD：广州市天河区华庭路4号富力天河商务大厦611室
Rm. 611, Tian He Fu li Business Mansion No.4 Hua Ting Rd Guangzhou

http://www.whiint.com

.Planning
.Landscape
.Arctecture
.Consulting

ARIF-international, Ltd.

加拿大埃瑞弗建筑规划设计机构
埃瑞弗（上海）工程咨询有限公司

ARIF-international,Ltd具有多年海外设计机构的项目规划设计及服务经验，有着来自北美工作经历的专业团队，主要从事城市规划与设计、风景区与旅游度假专题策划、城市公共空间与园林绿地规划设计、城市开发与地产全程策划等综合专业服务。公司为客户的各类投资开发项目提供"策划咨询（Consulting）—规划设计（Planning）—实施策略（Enable）"的全过程服务。

公司凭借其成熟的专业实力，国际化的视野，卓越的技术团队，多年项目操作经验以及对中国文化理念和国情的深刻理解，使规划设计充满了活力与创意，并能够适应不同地区和项目类型的多重需求。公司的理念来自于对自然规律、对人类发展、对城市演进的感悟与理解，来自于对历史人文地理的关怀与尊重，来自于对科学技术的熟稔与掌握。公司坚持走可持续发展道路，凝聚东西方文化、规划理念与设计手法的结晶，融会贯通、求实创新。

ARIF-international, Ltd. with years of overseas project planning, design and service experience, has professional team with work experience from the North American, mainly engages in urban planning and design, scenic and tourist holiday special planning, urban public space and green space planning and design, urban development and planning, comprehensive real estate full professional service. The company provides customers with all types of investment and development projects in the "consulting – planning – implementation" of the entire process of service.

The company relies on its mature expertise, international perspectives, excellent team, multi-year project experience, as well as on Chinese culture and philosophy of understanding, which make planning and design full of vitality and creativity, and has the ability to adapt to different areas and multiple types of projects. The company's philosophy on nature, human development, comprehension of the evolution of the city and to understand, is from the historical geography of caring and respect for people familiar with the science and technology. Company insists on the road of sustainable development, cohesion and occidental culture, planning, concept and design practices in the Crystal, holistic, and innovation.

地址：上海市四平路773号金大地商务楼811室
电话：+86-21-61435019 / 61435018
传真：+86-21-61435020
网址：www.ARIF-international.com
邮箱：bcy1107@126.com
manager@arif_international.com

Add: Room 811, No. 773 Siping Road, Shanghai Golden Land Business Building
Tel: +86-21-61435019 / 61435018
Fax: +86-21-61435020
http://www.ARIF-international.com
E-mail: bcy1107@126.com
manager@arif_international.com

宁波阳光海湾启动区控制性规划
Ningbo Sunshine Gulf Promoting Region Consulting Planning

委托单位：宁波阳光海湾发展有限公司
设计单位：加拿大埃瑞弗建筑规划设计机构
（合作单位：加拿大MYP工程咨询公司）

项目说明
Interpretation

以凤凰山为核心形成"一心八片"结构："一心"，凤凰山观景区；"八片"，活力小镇综合区、生态山林度假区、主题公园综合区、海上游艇度假区、鸟岛生态保护区、悬山生态体验区、仙人度假公园、景观围坝区。

规划总用地为2 174公顷，城市开发建设用地为982.92公顷。其中，居住和公建开发用地为434.5公顷。

设计目标：

依托基地山海资源，以滨海度假加运动休闲为创造阳光海湾的契机，打造宁波高端休闲度假新名片。

规划理念：
Planning philosophy

1．整合式规划：承上启下式的规划设计，多层次、多专业的整合。
— 规划中整合景观设计的构思，为景观空间的打造预留场地。
— 规划中整合建筑的外部空间，创造独特的场所感。
— 规划中整合前期策划和后期运营的考虑。

2．原生态的发掘：保护和修复原始生态条件，发掘山水人居的文脉肌理。
— 森林山居。
— 溪流湖畔。
— 海景别墅。
— 山地运动。
— 幽谷花田。

3．家园感塑造：通过道路交通和管理的组织，界定出优质的生活空间感。
— 构建社区主干道、组团道路和入户道路的三级过渡系统。
— 设置进入园区主入口、进入组团和进入私家花园的三级管理系统。

4．节点的引导：节点区域的建设到位与否关系园区品质的高低。
— 门户形象的打造。
— 商业空间及会所周边。
— 道路景观的设计。
— 核心景观体系的建构。

总平面图
Master Plan

规划构架
Planning Framework

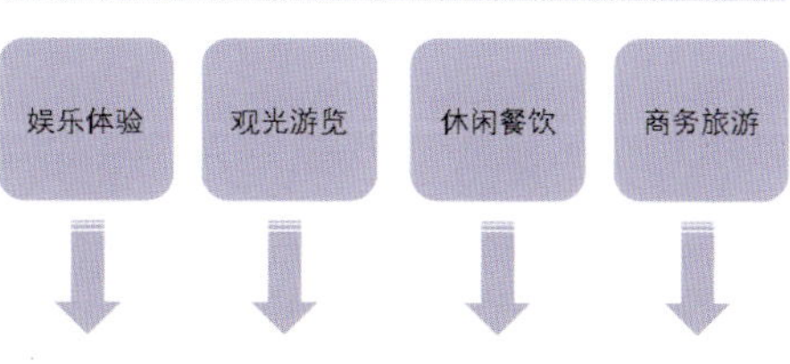

成功要素

- 核心吸引物与周边开发结合
- 提供多样化的旅游体验
- 宜人的步行系统与外部交通
- 优美的环境景观与开放空间
- 令人印象深刻的娱乐体验

项目开发方向

- 综合考虑娱乐体验、观光游览、休闲餐饮、会议会展和生态环境的元素
- 保留原有的生态资源与旅游优势，进一步提升质量至领先水平
- 开发主题印象鲜明、能提供多重体验的休闲娱乐项目设计完善的内外交通系统
- 创造足够的开放空间，提升周边开发的环境质量

上海国际研发总部景观设计概念方案
R & D Center Shanghai International

委托单位：上海祁连房产开发总公司
设计单位：加拿大埃瑞弗建筑规划设计机构

项目说明
Interpretation

本项目地处宝山区大场镇，西邻上海市沪太路，南侧为上大路，东侧为告泾河。基地共分为两期，北侧为一期，南侧为二期，一、二期总计用地面积12.5公顷。

设计理念
Design Philosophy

“水滴”表现出园区内以“合作”—“共享”—“海纳百川”为主题，以五组“水滴”形的建筑组成，象征了科技研发的独立和融合，契合了上海的文化特征。景观理念：能量漩涡则展现了动漫产业的崛起、势能。

从创意之眼为起点，表征IT、动漫、生物科技等高新产业崛起，发展势能的汇聚，并由此向各个园区挥洒发散的景观规划概念与空间结构。

效果图
Perspective Drawings

概念分析
Concept Analysis

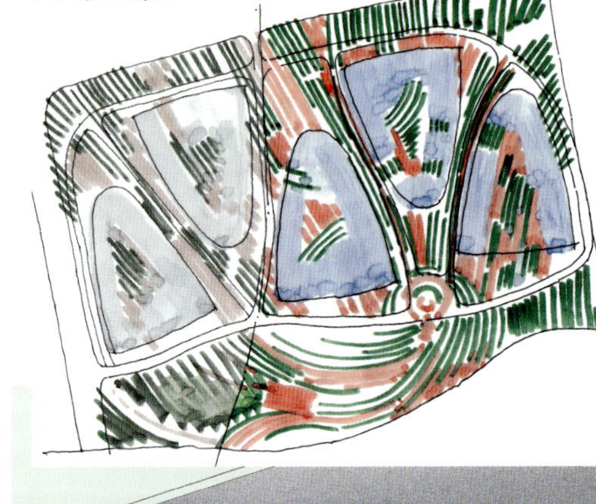

总平面图
Master Plan

效果图
Perspective Drawings

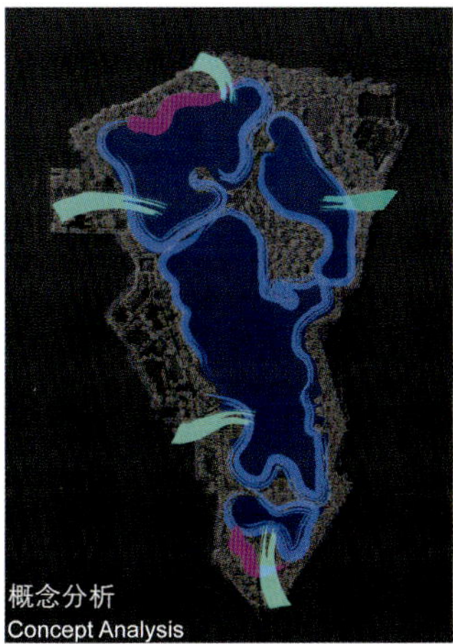

概念分析
Concept Analysis

湖南省常德市滨湖公园生态环境整治工程
Changde Lakeside Park Landscape Design

委托单位：常德市风景园林绿化管理局
设计单位：加拿大埃瑞弗建筑规划设计机构

项目说明
Interpretation

滨湖公园位于常德市江北城区中心区，朗州路以东，洞庭大道以南，光荣路以西，建民巷以北，公园周边被水榭花城东城、南城、西城、北城所环抱，老护城河从公园游乐区穿园而过。

公园规划总面积27.34公顷，且是以水景为特色的大型综合性城市公园，园内有铁经幢等重要文物，有三观亭、水榭、假山等主要景点，园中湖面开阔、景色优美。

设计理念
Design Philosophy

“仁者乐山，智者乐水”。山让人持重，水让人轻盈，山水合一，才是完美的风景。根据常德“多湖多水”的特征，我们提炼出水的四种形态“源”、“涓”、“澜”、“泓”，通过这四个元素来诠释一个城市公园的魅力。现代、简约、活力，具有城市肌理的延续，城市空间的共享，城市文化的交融。

总平面图
Master Plan

天津静海南片区示范区规划设计
Tianjin Jinghai South Area Demonstration Area Planning and Design

委托单位：天津市松江生态产业有限公司
设计单位：加拿大埃瑞弗建筑规划设计机构

项目说明
Interpretation

基地隶属天津市静海县，总占地面积约合5公顷。地块西侧为旭华南路，是通向县城及县政府的主要干道；南侧为静海的南外环，南外环与高速公路相通，交通十分便利。

规划理念
Planning Philosophy

发展之城——天津静海新兴的发展中心
对于地块的重新定位在于为其带来更新、更快的发展。地块的发展不仅提升其自身的价值，并且能带动周边地块的发展。
活力之城——现代时尚、充满活力的魅力之心
依托周边行政、文化、产业功能，面向年轻的高素质阶层，打造富有人气的城市区域，规划有商业、居住、休闲、办公、娱乐等多功能之间的复合，形成24小时都充满生机的现代化区域。
宜居之城——人文融合、绿色平和的宜居之地
依托基地良好的生态环境，创造环境优良、人文荟萃、生态宜居的高档生活场所，功能和谐：强调复合互动的功能组织，实现不同功能区块之间的分解与合作，打造复合社区。以吸引年轻、高素质的人群为诉求，建设静海高素质人群集聚区，打造静海居住品牌。

总平面图
Master Plan

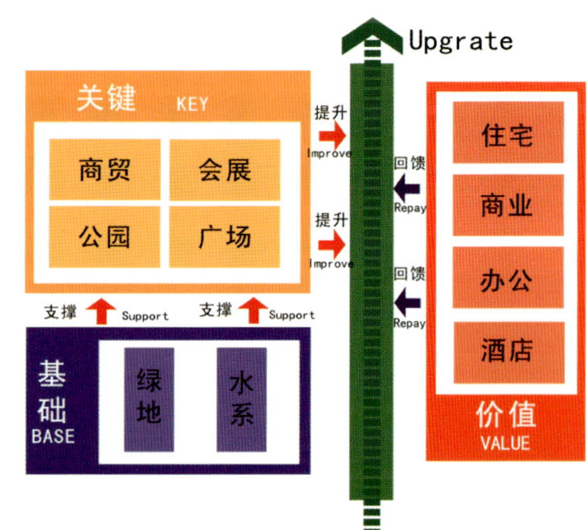

规划构架
Planning Framework

洛阳龙门一号景观规划设计
First Standard Dragon Gate Landscape Design

委托单位：河南洛阳天地辉煌置业有限公司
设计单位：加拿大埃瑞弗建筑规划设计机构

项目说明
Interpretation

规划区域位于洛阳古都，邻近龙门石窟。区域交通十分便利，东侧依畔伊河水系。

景观设计充分利用外部自然条件，引伊河水入基地内部，并与生态花园相结合，形成丰富变化、灵动活泼、水绿交融的景观效果。

前期规划深入挖掘了地域文脉，建筑样式体现了“唐韵之风”。因此，景观风格延续了这一独特的文化特色，将“唐风”元素融合到景观艺术中。

规划理念
Planning Philosophy

——突显唐风古韵

为体现“古韵”的人文环境，景观设计结合建筑风格，塑造具有“唐风”的艺术氛围，让居民能感受到历史的印记。

——营造人性场所

景观设计充分考虑到各类居民的不同需求，策划了一系列健身、娱乐、游戏、聚会的活动设施，为居民带来了丰富多样的户外体验。

——建构优美生境

以“水”为脉，以“绿”为底的景观为社区住户提供了清新宜人的生态系统。植物搭配的合理性把这种景观效益最大化，使得处于此处的居民有远离喧嚣、回归自然的感觉。

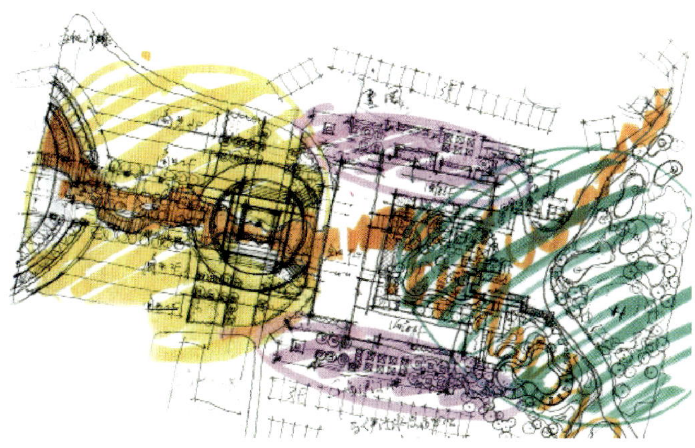

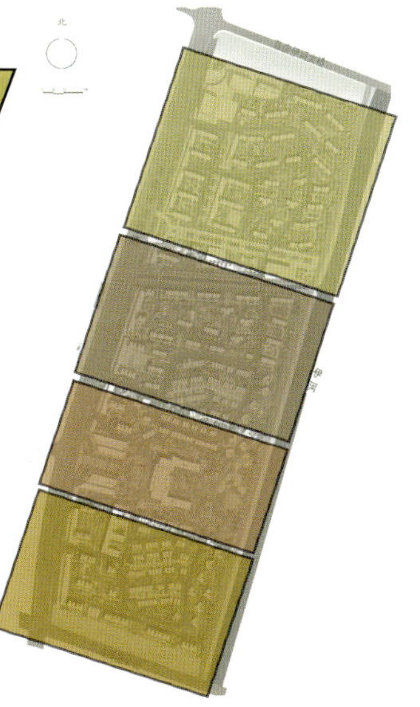

花 溪 流 园

景观特色：潺潺溪流的水景、四季相异且争奇斗艳的花景。

枫 林 逸 园

景观特色：收放有序的水景、各类色叶且茂密的林景。

古 韵 谐 园

景观特色：活力四射的商务酒店聚集区彰显“唐风古韵”的雅趣。

香 泽 和 园

景观特色：安静平和的水景、各类独具康疗特色的植物绿景。

草店伊河大桥
洛栾快速通道
伊河
环城高速公路

景点名称

1. 商业核心景观
2. 花蝶园
3. 养心曲水
4. 彩溪园
5. 芸溪岛
6. 引蝶广场
7. 中式合院花园区
8. 高尔夫景观
9. 酒店庭院花园区
10. 酒店核心景观
11. 老年活动中心花园

唐山凤凰新城
Fenghuang New Town,Tangshan

北京保利新茉莉
Poly New Jasmine, Beijing

ADS 建筑·规划设计事务所
ADS ARTEC DESIGN STUDIO architecture & urban planning

地址

北京代表处：
海淀区车公庄西路甲19号华通大厦396室
中国 北京100044 邮箱：ads_china@163.com
电话：+86-10-68700548 传真：+86-10-68700545

Beijing Office:
300 Huatong Bldg. No.19A Chegongzhuang West Rd.
Beijing China 100044
Tel: +86-10-68700548 Fax: +86-10-68700545
E-mail: ads_china@163.com

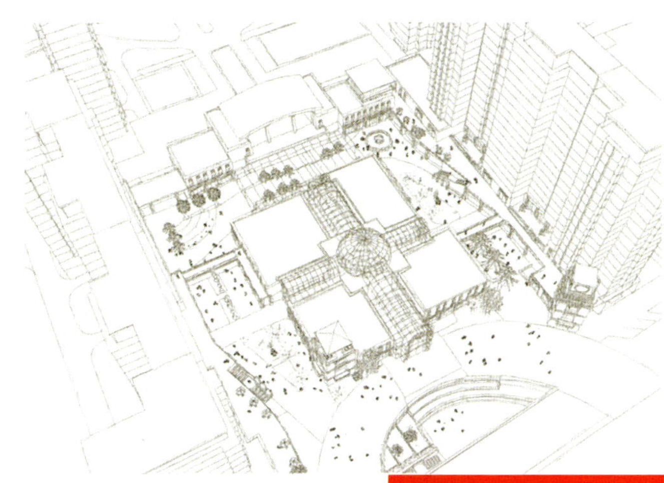

ARTEC DESIGN STUDIO architecture & urban planning

北京首开璞瑅二期
Capital Development Boutique Living Phase II, Beijing

简介

ADS建筑·规划设计事务所创立于加拿大安大略省，公司在北京和上海两地设有办事处。近几年来在国内十几个城市参与了许多公共建筑与住宅项目的规划及设计工作，其中多个项目成为当地的代表性作品，并获得了许多优秀设计奖项。在设计过程中，不仅应用最先进的规划设计理念和方法，而且十分强调为业主提供国际水准的专业化全程服务。公司非常重视与业主间的充分沟通，尤其在概念设计阶段。只有在更深的层面上理解业主的需求和项目本身，才能做出既能充分体现客户意愿，又能完整表达设计创意的方案。

Artec Design Studio (ADS) is an architecture & urban planning firm registered in Ontario, Canada. Its local representative offices locate in Beijing & Shanghai. It has been actively designing many public buildings and housing projects across China in recent years. Many of them are award-wining projects and become local landmark. In our design concept, we emphasize on in-depth communication with our client. Understanding the requirement and characteristic of each project enable us to address our clients' needs better while maintaining the artistic integrity of the design.

北京通州
Tongzhou, Beijing

ADS 建筑·规划设计事务所
ADS ARTEC DESIGN STUDIO architecture & urban planning

公司近年主要项目一览表：

环渤海地区

01 保利 · 垄上别墅（北京）
02 保利 · 西山林语（北京）
03 慧城 · 君山别墅（北京）
04 泰禾 · 红御别墅（北京）
05 鲁商 · 蓝岸丽舍（北京）
06 华润 · 风景翠园（北京）
07 润泽 · 润泽庄园（北京）
08 政泉置业 · 金泉广场（北京）
09 保利 · 香槟花园（北京）

ARTEC DESIGN STUDIO architecture & urban planning

北京霄云路
Xiaoyunlu, Beijing

10 城开 · 北京香颂（北京）
11 首开亿信 · 紫芳园四区（北京）
12 首开亿信 · 顺义商业住宅（北京）
13 公安大学高级警官培训中心（北京）
14 大成 · 日坛商业街（北京）
15 华润 · 凤凰城二期（北京）
16 保利 · 茉莉公馆（北京）
17 绿城 · 通州项目（北京）
18 武清区文化广场（天津）
19 保利 · 上河雅苑（天津）
20 保利 · 香槟国际花园（天津）
21 绿城 · 大连项目（大连）
22 天泽锦程 · 红屿云上别墅（北戴河）
23 金海湾 · 森林逸城（北戴河）

合肥万城南山郡
Wancheng Nanshan County, Hefei

ADS 建筑·规划设计事务所
ADS ARTEC DESIGN STUDIO architecture & urban planning

公司近年主要项目一览表：

长三角地区

24 保利·十二橡树庄园 （上海）
25 万科·提香别墅 （上海）
26 万科·新里程（上海）
27 招商·依云郡 （上海）
28 久事·西郊花园二期 （上海）
29 万里雅筑 （上海）

天津首创京津同城
Capital Jingjin City, Tianjin

北京润泽庄园
Runze Manor, Beijing

北京东方普罗旺斯
Eastern Provence, Beijing

ADS 建筑·规划设计事务所
ADS ARTEC DESIGN STUDIO architecture & urban planning

公司近年主要项目一览表:

其他地区

36 上海城开·托斯卡纳 （长沙）
37 德普·美洲故事 （长沙）
38 泰禾·住宅项目 （福州）
39 新长城·龙湖花园 （郑州）
40 佳达利·阳宗海凹子山项目 （云南）
41 海悦天伦 （海口）

ARTEC DESIGN STUDIO architecture & urban planning

42 科尔沁大酒店 （通辽）
43 君地・君地天成 （长春）
44 亨泰伟业・大漠高尔夫（榆林）
45 万城・南山郡（合肥）
46 湘银・金色阳光（株洲）
47 合盛・兆瑞广场（唐山）
48 隆鑫・杨柳河（成都）

天津中澳游艇城
Zhong'ao Marina City, Tianjin

ADS 建筑·规划设计事务所
ADS ARTEC DESIGN STUDIO architecture & urban planning

公司近年主要项目一览表：

长三角地区

24 保利 · 十二橡树庄园 （上海）
25 万科 · 提香别墅 （上海）
26 万科 · 新里程（上海）
27 招商 · 依云郡 （上海）
28 久事 · 西郊花v园二期 （上海）
29 万里雅筑 （上海）

天津中澳游艇城俱乐部
Zhong'ao Marina City Club, Tianjin

天津中澳游艇城
Zhong'ao Marina City, Tianjin

ARTEC DESIGN STUDIO architecture & urban planning

30 银都名墅 （上海）
31 浙电 · 九龙坞别墅 （富阳）
32 绿城 · 桃花源 （杭州）
33 地税局办公楼 （杭州）
34 在水一方 （苏州）
35 宝华 · 海滨庄园 （上虞）

ADS 建筑·规划设计事务所
ADS ARTEC DESIGN STUDIO architecture & urban planning

公司近年主要项目一览表：

海外项目

49 MNSK学院总体规划（印度）
50 HSC医疗中心/盛捷服务公寓（吉隆坡）

ARTEC DESIGN STUDIO architecture & urban planning

英国UK.LA太平洋远景国际设计机构

U.K PACIFIC LONG-RANGE PLANNING & DEVELOPING DESIGN CONSULTANT LTD.

中国境内公司联系方式：
地址：江苏省南京市奥体大街128号奥体名座大厦F座10楼整层
电话：+86-25-84739678
传真：+86-25-87763798
网址：www.UKLADESIGN.com
邮箱：UKLA2000@126.com

Contact Method (China):
Add: Fl 10, No. 2 Building, Aotimingzuo Mansion, 128, Aoti Street, Nanjing, Jiangsu
Tel: +86-25-84739678
Fax: +86-25-87763798
Http: //www.UKLADESIGN.com
E-mail: UKLA2000@126.com

英国UK.LA太平洋远景国际设计机构（UK.LA PACIFIC LONG-RANGE PLANNING & DEVELOPING DESIGN CONSULTANT LTD.）是一家英国专业设计公司，旗下设计机构覆盖英国，以及中国香港、南京、上海和郑州等地。公司致力于城市与建筑的功能规划和空间设计，业务涵盖城市规划、建筑设计、景观设计、建设工程咨询等诸多领域。在设计的各个阶段，公司秉承“专业”、“创新”的宗旨，不断在设计作品上精益求精。近年来，公司在实践中所展示的创造性能力、先锋的设计理念和不懈的探索精神得到了公众和学术界的广泛认可，是江苏规模最大的境外设计机构之一。

UK.LA的核心优势来源于其独特的多专业、多文化和国际型的主设计师群与设计团队。公司云集了一批不同年龄、不同背景经历的设计师，他们带来了创新的设计理念和手法、先进的设计管理经验和各种文化的精华。多种文化的融合和碰撞，形成了UK.LA独树一帜的创作风格，凭借其对自然、文化和经济环境的高度责任感和洞察力，UK.LA形成了“适用、和谐、创新”的原则，在设计理念和技术手段上不断追求更高的境界。

UK.LA的核心优势还来源于对卓越设计的不懈追求和坚韧探索，无论是在南京总建筑面积达2 300 000平方米的华欧国际友好城的规划中，还是在郑东新区意大利格拉姆集团的超高层双塔格拉姆国际中心的建筑设计中，公司都充分分析设计任务中每个要素的重要性和整体性，其根本宗旨是实现设计最优化，将城市设计、建筑设计、景观设计、生态设计交汇融合，自始至终将创造性、和谐性和超越性完美统一，最终提供给客户一个完整的、卓越的设计方案。

世纪之交的中国，社会、经济、城市与建筑等各个方面均发生着巨变，资本、技术和思想领域都进行着前所未有的交流，城市空间和建筑作品必须适应这一潮流。在走向全球化的同时，必须发扬自身文化的特点，发掘本土文化的魅力。UK.LA的团队在结合世界先进理念的同时融入中国本土的建筑元素和语汇，UK.LA公司一直重视与开发商和政府建立和发展伙伴关系，准确把握市场脉搏，在业务不断扩展的过程中以创新的设计和优质的服务，赢得了众多客户的信赖与支持。

UK.LA PACIFIC LONG-RANGE PLANNING & DEVELOPING DESIGN CONSULTANT LTD. is a British professional design company, with branches in the UK, and Chinese cities such as Hong Kong, Nanjing, Shanghai and Zhengzhou. It is committed to functional planning and space design of cities and buildings, and its business covers urban planning, building design, landscape design, and construction advisory among many others. In every stage of design, this company follows the tenet of “specialization” and “innovation”, constantly seeking for the best work in design. In recent years, it demonstrated remarkable creativity, leading design philosophy and unremitting adventure spirit in practice, widely recognized by the public and academic community, and became one of the most sizeable foreign design organizations.

The core advantage of UK.LA comes from its unique multi-specialty, multicultural and international chief designers' group and design team. It gathers a group of designers at different ages and from different backgrounds, who bring forward innovative design concept and approach, advanced design management experience and the essence of various cultures. The integration and clash of various culture form the unique creative style of UK.LA, and with high responsibility and insight into natural, cultural and economic environment, UK.LA establishes the innovation principle of “application, harmony and innovation”, constantly seeking higher accomplishment in design philosophy and technical approach.

The core advantage of UK.LA also comes from its unremitting pursuit and diligent exploration in excellent design, no matter in the planning of Hua'ou International Friendship Town (floor area 2,300,000 m^2) in Nanjing, or in the architectural design of high-rise twin-tower Cram International Center (sponsored by Italian Cram Group) in the Zhengzhou East Development Zone, the company makes full analysis on the importance and integrity of every elements of the design task, for the ultimate purpose of optimal design, to integrate urban design, architectural design, landscape design, biological design, and unite creativity, and provides customers with harmonious and transcendent design services.

In the new century, China is undergoing tremendous transformations in the society, economy, cities and architecture, where capital, technology and ideas are exchanged in unprecedented conditions, while urban space and building works have to follow this tide. In the way to globalization, we should also promote our own cultural characters, and develop local cultural charms. The UK.LA team combines world-leading ideas with Chinese local building elements and lexicology, and meanwhile, this company is paying great attention to partnership with developers and local governments, touch market pulse, and in business expansion, win general trust and support from vast customers, by creative design and high-quality services.

主要获奖情况

2010年 中国最具业主满意度设计机构
2009年 中国最具影响力境外设计机构
2009年 国际建筑设计创意企业金奖
2007年 亚洲建筑规划设计奥斯卡国际风尚大奖
2007年 中国建筑规划设计诚信百强品牌机构
2006年 江苏十强诚信品牌设计机构

Major Awards

2010 China Owner's Most Satisfaction Design Institution
2009 China's Most Influential Foreign Design Institution
2009 International Architecture Design and Creative Enterprises Gold Award
2007 Asia Architecture Planning and Design Oscar International Fashion Awards
2007 China Architecture Planning and Design Integrity Hundred Brand Institution
2006 Jiangsu Top Ten Integrity Brand Design Institution

1-3 红太阳·乌江新城规划设计

建设地点：安徽 和县
项目业主：江苏红太阳房地产开发有限公司
总用地面积：700万平方米
总建筑面积：538万平方米

乌江新城项目位于安徽和县乌江镇，北靠乌江，东临被喻为"黄金水道"的长江，是八百里皖江第一镇，处于南京"一小时都市圈"核心层。乌江是千年古镇，镇内有千年老街、3A级国家风景区霸王祠等名胜。

作为上百万平方米的大型项目，整个规划将地块划分成三大功能片区，即生态居住区、商贸旅游区以及湿地观光区。规划结合安徽文化，恢复了江南水乡河街并行的传统格局，将老街改造成水街，形成一系列的水上旅游环形线路。这样一个通透的集旅游、观光、休闲、养生、度假、会务、交易、居住为一体的徽派社区，必将符合人们对中国传统文化的需求和在城市寻根的心理，同时还营造出原生态山水豪门的生活气势。

1-3 Planning and Design for Red Sun • Wujiang New Town

Construction Location: Hexian, Anhui
Project Owner: Jiangsu Red Sun Real Estate Development Co., Ltd.
Total Site Area: 7,000,000 m^2
Total Building Area: 5,380,000 m^2

The project of Wujiang New Town is located in Wujiang Town, Hexian County, Anhui, adjacent to Wujiang River on the north, the "Golden Watercourse" - Yangtze River on the east, the No. 1 town along the 400 km Wanjiang River, located at the core layer of Nanjing's "1-hour Metropolitan Circle". Wujiang is an ancient town with one thousand years of history, boasting millennium streets and a great many places of interests such as Overlord Temple – a National 3A Scenic Area.

As a large-scale project covering a land of over 7 million square meters, the land is divided into three key functional areas in the overall plan, i.e. eco-residential area, trade and tourism area and wet-land sightseeing area. The planning combines the cultures of Anhui to restore the traditional pattern of southern Chinese riverside towns – with streets stretching along the river, and reconstruct the old streets into water streets to form a series of loop routes for cruising tourism. Such transparent Anhui-style community will integrate tourism, sightseeing, recreation, health preservation, holiday, conference, trade and residence, and it will meet people's needs of traditional Chinese culture and the psychology of seeking the root of the town, and meanwhile, it makes a luxury life style in the primitive landscape.

5

1

4-7 海南儋州双联滨海住区规划设计

建设地点：海南 儋州
项目业主：海南儋州双联置业有限公司
总用地面积：101万平方米
总建筑面积：88万平方米

规划区位于海南儋州白马井镇，距海口市135千米，西临大海，具有得天独厚的景观优势。地块分为居住板块与度假板块两个独立的功能板块。

高档滨海社区与海上度假区通过一条横贯东西的中央景观大道连接为一体，由东向西的序列依次为商业步行街—会所—景观大道—湿地公园—沙滩游乐区—内港游乐区—百米海景酒店。

在整个项目规划建设过程中，我们积极倡导生态节能和低碳环保的设计理念，推广生态技术工程的应用，力争使本项目成为儋州首席"低碳零排放示范社区"。在进一步保护自然生态环境系统的同时，提升居住品质和舒适度。

4-7 Planning and Design for Danzhou Shuanglian Coastal Residential Community, Hainan

Construction Location: Danzhou, Hainan
Project Owner: Hainan Danzhou Shuanglian Real Estate Co., Ltd.
Total Site Area: 1,010,000 m^2
Total Building Area: 880,000 m^2

The planned area is located in Baimajing Town, Danzhou, Hainan, 135 km from Haikou City, adjacent to the sea on the west, boasting unique landscapes. The land is divided into two independent functional areas, i.e. residential area and holiday resort area.

The top-grade coastal residential community and the sea resort area are linked into a whole by the central sightseeing avenue stretching from the east to the west, the layout (from east to west) is: shopping street – clubs – sightseeing avenue – wetland park – beach recreation area – inner harbor carnie – 100 m sea-view hotel.

During the planning and construction of the entire project, we actively advocate the ecological, energy-efficient, low-carbon and environment-friendly design concept, and promote the application of ecological engineering. We aspire to build the project into the first "low-carbon and zero-emission model community" in Danzhou, and improve the quality and comfort of residence besides further protecting the ecological and environmental system of the nature.

6

2

3

4

7

1–5 江苏新沂市窑湾古镇规划设计

项目业主：新沂市政府
建设地点：江苏 新沂
总占地面积：100万平方米

窑湾古镇是目前苏北地区在京杭大运河滨水古镇中保存最完好的一个。它形成于春秋战国时期，明清时期达到鼎盛。古镇有发达的水系，分别是大运河、后河以及护城河，形成独特的半岛形态。

窑湾古镇滨水景观规划在京杭大运河2013年申报世界文化遗产的背景下展开，本次规划中充分考虑了窑湾古镇在京杭大运河遗产中的地位，在保护窑湾特有的"镇有前后河，城在两湖中"的滨水古镇形态的同时，打造中国"苏北第一运河古镇"。

我们在总平面规划中设计了三条水轴、四大功能区、十三组景观节点。

景观规划着力打造的窑湾新"十三景"，旨在将窑湾三条河的沿线打造成展示窑湾历史文化的重要景区，并通过景观步道紧密相连，形成窑湾穿越时空感的如翡翠项链般的沿河景观带。

1-5 Planning and Design of Yaowan Ancient Town, Xinyi, Jiangsu Province

Project Owner: Xinyi Municipal Government
Construction Location: Xinyi, Jiangsu
Total Site Area: 1,000,000 m^2

Yaowan Ancient Town is the best preserved waterfront ancient town along Beijing-Hangzhou Grand Canal in Northern Suzhou Area. It came into being in the Spring and Autumn and Warring States periods and reached its Peak in Ming and Qing Dynasty. The Ancient Town has sophisticated river system, namely the Grand Canal, back river and the moat, forming a unique peninsula shape.

Yaowan Ancient Town waterfront landscape planning was carried out under the background of declaring for the World Historical and Cultural Heritage in 2013. In this planning the position of Yaowan Ancient Town in Beijing-Hangzhou Grand Canal Heritage—forging "First canal ancient town in Northern Suzhou" of China in the meantime of protecting Yaowan's unique shape of waterfront ancient town of "town with both front and back river, city between two lakes".

Our site planning has designed 3 water axles, 4 big functional areas and 13 groups of landscape nodes.

The landscape planning puts forth efforts on forging Yaowan's new "Thirteen Sceneries", aiming at making the areas along the three rivers of Yaowan into significant scenic spot representing Yaowan's historical culture, and through close connection with landscape footpaths, forming Yaowan's emerald necklace through the veil of time.

3

4

1

2

5

6

8

6-8 Zhengzhou Tsinghua Yijiangnan Construction Planning and Design Construction

Construction Location: Zhengzhou, Henan
Total Site Area: 1,460,000 m^2
Total Building Area: 860,000 m^2

This project locates at the juncture of Huiji District of Zhengzhou and Guangwu Town of Xingyang, With Yellow River Landscape Area in the east, the ruins of Two Hegemons' City in the north, and Guangwu Town Government 3 kilometers to the west. The transportation of the whole district is very convenient. The total site area of this district is 1.46 million square meters. In the middle of this district lies a long river traversing from west to east. In the district the plants are exuberant and distribute naturally, forming a great greenery condition. Three high-tension cables slant through the middle area.

This project constructs ideal sustainable community and provides living space of higher leveled public health and better quality for the residents. At the meantime of creating good in and out door space state of the construction, it emphasizes creating a good in the adapting to the requirement of development of market economy and urban morphology, making the planning more flexible and adaptable.

Making use of the mountain's momentum, incorporating with the terrain and river, Tsinghua Yijiangnan Subdistrict sets an ecological wetland of 320 thousand square meters within the boundary of 1.46 million square meters. Under the environment of lacking water in Zhengzhou, Tsinghua Yijiangnan pays attention to the utilization of natural resource, builds along the slopes and rises and falls according to the mountain terrain, which fuses the construction style and the natural landscape together and realizes the resort of "scenery in the house, house in the scenery". It is precious due to the location and the river, and it owns the sky, the earth, the river and the scenery. According to the terrain, the subdistrict sets multiple architecture forms like waterfront independent villas, semi-detached villas, rowhouse villas, garden houses, medium height buildings, resort style apartments, wetland wooden house etc., and even manor-style single island villas, which represent dignity.

6–8 郑州清华·忆江南建筑规划设计

建设地点：河南 郑州
总用地面积：1 460 000平方米
总建筑面积：860 000平方米

本案处于郑州惠济区与荥阳市广武镇交界处，东临黄河大观景区，北依汉霸二王城，西部3千米处为广武镇政府。整个地块的交通十分便捷。地块总用地面积为146公顷。地块中部有一条自西向东流的长河横贯东西，地块内植物茂盛，并且呈自然式分布，绿化条件优越，中间地区有三条高压线斜穿。

本案旨在建设理想的可持续社区，为居民提供更高水准的公共式健身空间和更优质的生活空间。方案在创造良好的建筑室内外空间形态的同时，注重打造良好的社区人文环境，以适应市场经济和城市形态发展的需要，使规划具有弹性和应变能力。

清华·忆江南小区依山就势，结合地形及河流，在146公顷的范围内设置了32万平方米的生态湿地。在郑州水资源稀缺的环境下，清华·忆江南小区规划重视对自然资源的利用，建筑依坡而建，随山势起伏，将建筑风格与自然景观统一融合，实现了"景中有房，房中有景"的胜境。因地而贵，因水而贵，有天、有地、有水、有景。小区结合地形设置了滨水独立别墅、双拼别墅、联排别墅、花园洋房、小高层、度假酒店式公寓、湿地木屋等多种建筑形态，更有庄园式独岛别墅，彰显尊贵。

1

2

3

1–3 南京红太阳·旭日上城

建设地点：江苏 南京
项目业主：江苏红太阳集团
总用地面积：100万平方米
总建筑面积：200万平方米

南京红太阳·旭日上城地处中国东南部交通要地——南京长江大桥北端。依托长江三角洲区域经济强劲发展的大背景，借助"南京市发展长江两岸，建设跨江发展的国际化大都市"的发展战略，将建成一个集购物、度假、旅游、休闲、娱乐、文化为一体的综合性、标志性商业城及一个国际化的大型居住区。项目为南京市"跨江发展，两岸齐飞"的重要领航项目，项目规划中的旭日上城将成为中国乃至全球城市综合开发的一大亮点。

规划区位于南京市（浦口区）南京长江大桥北侧，整个旭日豪庭项目的最东端地块。地块东西长约581米，南北长约468米，北邻柳州路东沿线，东依南浦路，南滨丽江河，基地西侧为规划市政道路。

本设计构想主要依据基地临水之特征，以中国江南水乡之风貌，结合托斯卡纳人居之理念，以塑造一个具有住（居住）、游（休闲）、创（文化）特点的"生活理想国"，为南京市浦口区提供一个悠闲的、舒适的、崭新的居住环境，同时依据基地和城市文脉，旭日上城也将成为建筑与公共开放空间相结合的独特典范。

1-3 Nanjing Red Sun· Rising Sun Top Town

Construction Location: Nanjing, Jiangsu
Project Owner: Jiangsu Red Sun Group
Total Site Area: 1,000,00 m^2
Total Building Area: 2,000,000 m^2

The project is located at the traffic hub of southeast China,the north end of Nanjing Yangtze River Bridge. Under the background of the rapid development of Yangtze Delta Area and based on Nanjing's development strategy "To seek development on both banks of Yangtze River and build an international metropolis across Yangtze River", a comprehensive business town integrated with shopping, tourism, recreation, entertainment and cultural facilities and a large international modern residential community will be built. It's a pioneering project for Nanjing's "trans-Yangtze development" strategy. The planned Rising Sun Top Town will become a spotlight of comprehensive urban development in China or even in the world.

The planned area is located at the north side of Yangtze River Bridge in Pukou District, Nanjing. Sun Rising Top Town sits on the east end of the land. The land extends 581 m from east to west and spans 468 m from south to north, with east line of Liuzhou Road on the north, Nanpu Road on the east, Lijiang River on the south, and the planned municipal road on the west side.

The design, based on the geographic features (sitting on the riverside) and combining the style of South China water town and Tuscany's concept of residential buildings, aims at creating a "Utopian land of life" integrated with residential, tourism and cultural elements, and providing an easy, comfortable and fresh living place in Pukou District. Meanwhile, Sun Rising Top Town will become a model in respect of integration of architecture and public open space based on its location and the cultures of the city.

4

4 洛阳定鼎路以西A、C地块规划设计

建设地点：河南 洛阳
项目业主：洛阳增凯置业有限公司
总用地面积：60万平方米
总建筑面积：158万平方米

本项目位于洛阳市道北地区，背靠邙山，与洛阳火车站、洛阳机场交通联系便捷，区位优势明显。基地内部有占地4.3公顷的文化公园，未来将成为本区居民重要的休闲娱乐资源。

规划以城市文脉为线索，延续地方传统文化，形成与地方文化相融的个性品质，营建文化社区；打造中高档社区，营造现代舒适、归属感强的社区环境；秉承低碳环保设计理念，建设可持续性社区；通过合理富有创新的设计，秉承以人为本、自然和谐的设计理念。

4 Plot A, Plot C Planning & Design, West of Dingding Road, Luoyang

Construction Location: Luoyang, Henan
Project Owner: Luoyang Zengkai Property Co., Ltd.
Total Site Area: 600,000 m^2
Total Building Area: 1,580,000 m^2

This project is located in Daobei District, at the foot of Mt. Mang, near Luoyang Railway Station and Luoyang Airport, and this geography is obviously advantageous. In the project base is a Cultural Park (land area 4.3 ha.) which will become an important leisure and entertainment resource of local residents in the future.

The planning follows the cultural veins of Luoyang City, to succeed its local traditional cultures, and form a characteristic integrated with local cultures, to create a cultural community; to build a mid-high end community, create a modern, cozy and habitable environment; to follow low-carbon environment-friendly design idea, build a sustainable community; to follow the human-based, natural and harmonious design philosophy through reasonable and creative design.

1–4 郑东新区龙子湖湖心岛

建设地点：河南 郑州
项目业主：郑州市人民政府
总用地面积：99万平方米
总建筑面积：101万平方米

郑东新区龙子湖位于该区东部龙子湖区中部，是一个水面为120万平方米的人造湖，周围环绕12所高校和一块高校预留地。规划确定的12所高校建筑面积约365.1公顷，在校生人数约14.48万人。湖心岛为龙子湖中心岛，为科研中心和可共享资源用地。规划用地范围北至校园六路，南至校园七路，西起龙子湖纵贯一路，东止龙子湖纵贯三路，总用地面积约100万平方米。

湖心岛结合文化、休闲、娱乐、购物等设施，匹配大学城的特定功能，创造良好的科研环境，塑造绿色的教育园区的同时，体现了郑东新区“共生城市”和“新城代谢城市”的规划概念。湖心岛的规划建设目的在于“以学兴城”、“以城促学”的良性结合，提升城区整体活力，形成以大学为纽带，辐射周边地区，集教育功能、产业功能和生活服务为一体，文化氛围浓厚、科技产业发达、服务体系完善、优秀人才汇聚、生态环境良好的优化区域，最终建成一个发展的智力支持密集区、文化科技积累与传播的中心区、生态环境良好的新世纪现代化城区。

1-4 Zhengdong New Area Longzihu Lake Island

Construction Location: Zhengzhou, Henan
Project Owner: Zhengzhou Municipal People's Government
Total Site Area: 990,000 m^2
Total Building Area: 1,010,000 m^2

Located in Eastern Longzihu lake New District, the Longzihu Lake is in central eastern area, a man-made lake covering 1,200,000 square meters, surrounded by 12 universities and a university set aside land. Plan identified 12 construction area of about 365.1 hectares university, in the number of about 14.48 million students. Lake Island is Longzihu Center Island, as research centers and shared land resources. Planned land area is in the north to campus bus No. 6 line and the south of campus No.7 line , West from Longzihu No.1, east to Longzihu Longitudinal No.3, with a total land area of about 1,000,000 square meters.

Lake island has combined culture, leisure, entertainment, shopping and other facilities, to match the specific function of University City, creating a good environment for scientific research, and shape the green zone for education; meanwhile it also reflects the planning concept of Zhengdong New Area, i.e. "Symbiotic City" and "Metabolic Metro City". The purpose of its planning and construction is lied on the virtuous combination of "let university strive the city", "Better city better education", to enhance the overall vitality of urban areas, formed by the University as a link, radiating surrounding areas, integrating educational, industrial and health services into one, optimizing to being with strong culture atmosphere, developed science and technology industry, sophisticated service system, outstanding talent gathering, and good regional environment. All in all, it is aimed to develop into a community intelligence-intensive support area, accumulation and spread center of culture and technology, and a modern city with well environment in the new century.

5–7 东湖宾馆规划设计

建设地点：河南 郑州
项目业主：郑州九华房地产开发有限公司
规划用地：1 330 000平方米
水面面积：460 000平方米

本项目定位为“具有典型中原自然环境特征和文化特征，具有极佳自然生态环境，能够代表河南省的政务形象和生活品质的健康休闲、会议接待的度假胜地”。地块规划分为星级度假酒店、体检中心、温泉养生、艺术文化展示、度假别墅、中原美食街、生态湿地、高尔夫练习场、水上游览、生态果园、球类运动、滨水活动、湿地观赏、园林绿化等多个功能区块，充分体现了该区集休闲健身、文化娱乐、度假旅游于一体的综合功能。规划以园林绿化、高尔夫练习场、生态湿地、生态果园等大量绿色“面”域为背景，以精致的、古典建筑风格的、不同功能的建筑单体为“点”缀，以水体沿岸步道及迂回弯曲的道路为“线”条，联系各个功能组成部分，使该案形成一个佳境宜人的旅游胜地。

5-7 Donghu Hotel Planning and Design

Construction Location: Zhengzhou, Henan
Project Owner: Zhengzhou Jiuhua Real Estate Development Limited Company
Planning Areas:1,330,000 m^2
Waters Areas: 460,000 m^2

This project is oriented as "healthy leisure and conference reception resort with typical natural environment and cultural characteristics of Central Plains, and excellent natural ecological environment, which can represent government image of Henan Province". The sections are arranged as the following: multiple functional sections such as star level resort hotel, physical examination center, warm spring regimen, artistic culture exhibition, resort villa, Central Plains food street, ecological wetland, golf practicing, superaqueous tourism, ecological orchard, ball games, waterfront activities, wetland enjoyment, gardens greenery etc., which fully represent the district's comprehensive functions combining leisure and fitness, cultural entertainment, resort and tourism as a whole. The planning takes abundant green "surface" area like garden greenery, golf driving ranges, ecological wetland, ecological orchards etc. as the background, lying out construction monomers of delicate classical style with different functions as "spots", with the coastwise footpaths and roundabout roads and "linear" paths, which connect each functioning organization parts, and make this project form an enjoyable and pleasant resort.

1–4 翰荣·历史文化庭院长廊

建设地点：河南 郑州
项目业主：郑州翰荣管理有限公司
总用地面积：7.5万平方米
总建筑面积：2.5万平方米

本项目位于郑州惠济区与荥阳市广武镇交界处，东临黄河大观景区，北依汉霸二王城，距西部三千米处为广武镇政府。整个地块交通十分便捷。

地块所处位置环境良好，建筑全部设计为低层建筑，以中国历代庭院文化建筑为原型，强调突出时代文化特色，创造了一条中国庭院建筑发展历程的文化长廊。

1-4 Hanrong – Historic and Cultural Court Gallery

Construction Location: Zhengzhou, Henan
Project Owner: Zhengzhou Hanrong Management Co., Ltd.
Total site area: 75,000 m^2
Total building area: 25,000 m^2

This project is located in the boundary between Huiji District of Zhengzhou City and Guangwu Town of Xingyang City, west to Yellow River Landscape Area, south to the ruins of Two Hegemons City, and 3 km east to the office building of Guangwu Town People's Government. Traffic is very convenient in this land plot. The plot is located in good environment, where all buildings are designed in low-storey ones, taking the prototype of historic courtyard building, to highlight the cultural feature of this epoch, and create a cultural gallery of Chinese courtyard building history.

5 郑州格拉姆国际中心

建设地点：河南 郑州
项目业主：意大利格拉姆财团、罗马市政府
项目规模：12万平方米

郑州格拉姆国际中心是受意大利格拉姆财团和罗马市政府委托，郑州唯一一家家外商独资开发企业独立开发的商业地产项目。该项目现为中原第一高楼。

意大利是欧洲文明的摇篮，河南是中国黄河文明的发源地，在我们的设计中，希望通过意大利和中国文化的交融与碰撞，以及建筑语言的融汇来塑造现代、高效，具有时代特色，体现中意两国文化的地标性建筑。

建筑造型层层收进，寓意"芝麻开花节节高"，顶部造型宛如钻石般熠熠生辉，成为郑州东部中央商务区内一颗耀眼的明珠。

5 Zhengzhou Clam International Center

Construction Location: Zhengzhou, Henan
Project Owner: Clam Consortium of Italia, Government of Rome
Project Scale: 120,000 m^2

Zhengzhou Clam International Center is the only one commercial real estate project developed by a foreign enterprise in Zhengzhou. The project is entrusted by Clam Consortium of Italian and Government of Rome, and it's currently the No. 1 high-rise in Central China.

Italia is the cradle of European Italian and Henan is the birthplace of Yellow River civilization of China. The design aims at creating a modern and high efficient landmark building embodying Italian and Chinese cultures by integrating the architectural languages of the two nations.

Each floor of the building recesses inward like "a gingili growing taller and taller", the crests of the twin towers shining like two diamonds. The project has become a pearl of CBD in Zhengdong New Area.

2
3
4

5

1–4 连云港昌圩湖瑞丰天成

建设地点：江苏 连云港
项目业主：连云港瑞丰房地产开发有限公司
总用地面积：11.3万平方米
总建筑面积：9.5万平方米

基地位于城市中轴线花果山大道与港城大道交汇处东北方向，雄踞城市发展中轴和生态居住核心，周边市政配套完善，坐观国家4A级花果山景区，拥揽30万平方米城央最美的生态湿地——昌圩湖，可谓“近可揽水，远可观山”，拥有不可替代的自然景观优势。

本项目力求打造具有幽雅、宁静的生活气氛的英式小镇，塑造优美现代、健康生态、舒适宜人方便、极具特色的社区环境。

1-4 Changwei Lake Ruifeng Tiancheng, Lianyungang

Construction Location: Lianyungang, Jiangsu
Project Owner: Lianyungang Ruifeng Real Estate Development Co., Ltd.
Total site area: 113,000 m^2
Total building area: 95,000 m^2

The base is located in the northeast of the intersection between Huaguoshan Avenue and Gangcheng Avenue, on the central axis and bio-living core of the city, with complete surrounding municipal utilities. It overlooks the AAAA-grade Huaguoshan Scenic Spot, embraces the most beautiful biological wetland – Changwei Lake in the middle of the city, covering 300,000 m^2, so it is undoubtedly “waterside, mountainside”, owning irreplaceable landscape advantages.

This project seeks to build serene, tranquil living town of English style, to build a beautiful, modern, healthy, biological, habitable community with featured environment.

1

2

3

4

5-8 Lianyungang Crystal Palace Architecture

Construction Location: Lianyungang, Jiangsu
Project Owner: Lianyungang Umikawa Properties Ltd.
Total Planning Area: 22,467.8 m^2

The project base is located southwest of the planning East River New City of Lianyungang, China's well-known crystal capital. This project, East China Sea Crystal Palace, named after the Journey to the West as the case name, being a bond to cultural relics. In the design, the southwest corner of the building is presented in the typology of "Dragon" plane, and its facade is made of glass mosaic forms of polygons, implied crystal and "Dragon scale" metaphorical sense; in the middle of the its body we use whole image of linking dragon head, dragon body, and dragon tail through, to generate the "Dragon gate" space pattern.

Dual-enclosed structure is adopted to create an inner and outer ring layout, with a strong sense of direction and consumer guide, integrated business and leisure with the individual region cultural icon, like city landmarks in Shanghai "New World", Nanjing "1912", Hangzhou "West Lake" and so on. Modeling techniques and reasonable creative layout of the land contribute it a novel beautiful "pearl" in the East River New City of Lianyungang.

5

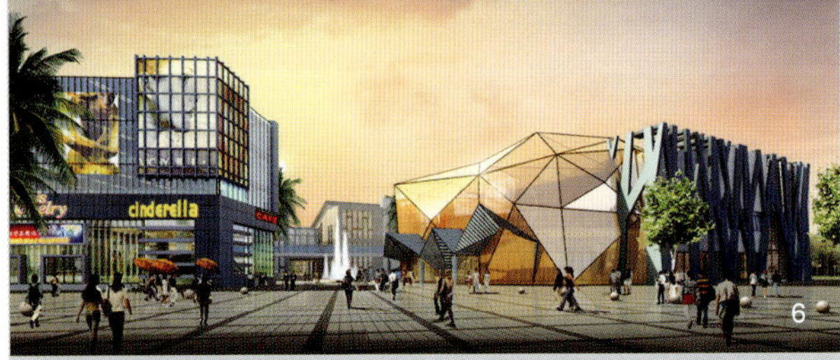
6

5–8 连云港水晶宫建筑规划设计

建设地点：江苏 连云港
项目业主：连云港海川置业有限公司
总规划用地：22 467.8平方米

本项目基地位于规划中的东河新城西南侧，连云港为中国知名水晶之都，本项目以东海水晶宫为案名，以西游记传说为文化纽带。在设计过程中，西南角的建筑以“龙”形平面的形态展示，其立面形式采用多边形的玻璃拼接，隐含水晶之喻义的同时，又有龙鳞附体的感觉。在中间部位我们采用了连接体，将龙首、龙身、龙尾联系于一体，在空间形态上产生了“龙门”之意。

设计采用双围合结构，形成内外环布局，具有强烈的方向感和引导性，使休闲商业与各地区域文化相结合，成为富有城市特征的地标，类似于上海“新天地”、南京“1912”、杭州“西湖天地”等。现代的造型手法和合理的创意布局，将把地块打造成连云港东河新城新颖靓丽的“明珠”。

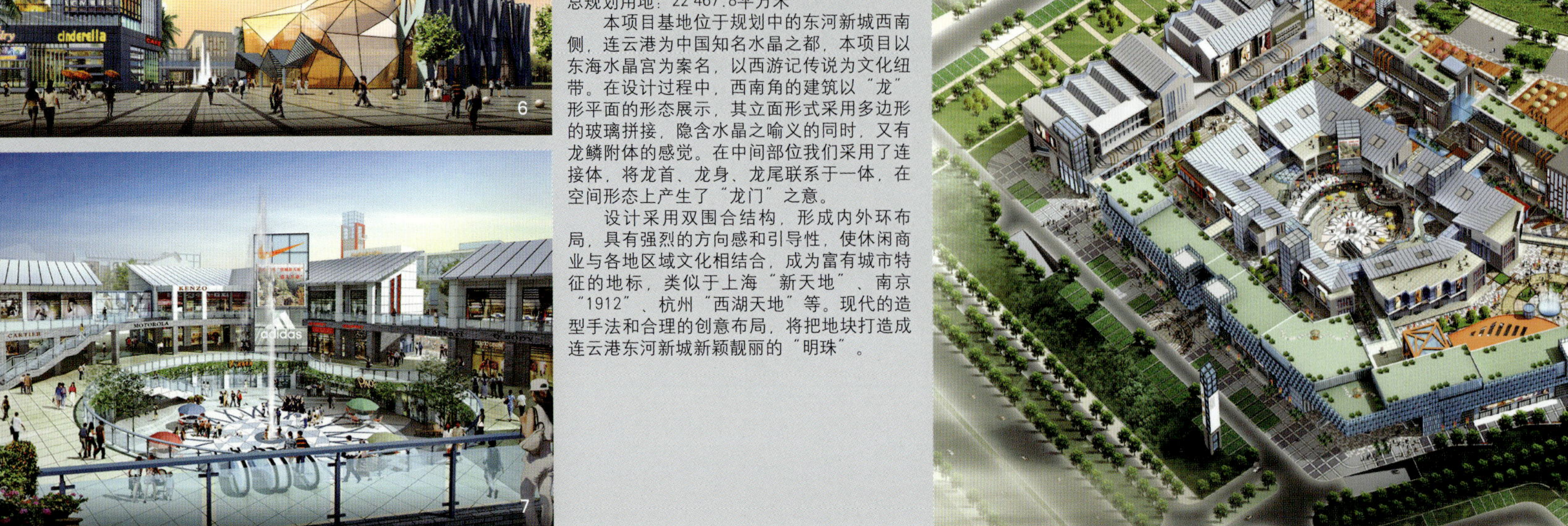
7
8

1 郑州一生缘·翡翠谷建筑规划设计

建设地点：河南 郑州
项目业主：郑州一生缘置业有限公司
总用地面积：117万平方米
总建筑面积：77万平方米

本案作为郑州远郊一个具有高度竞争力的项目，以其优美的景观为依托，在葱葱河畔，雕琢出一个个精致的组团。其最大的特点就是隐映在河湾林丛之中，与自然融为一体。

翡翠谷涵盖多种产品形态，有联排住宅、双拼住宅、花园洋房、坡地洋房、多层、小高层、高层以及酒店公寓等多种住宅形态，除此之外，还有商业街、假日酒店、温泉会所等多种商业形态。规划整体呈现两边高，中间低的态势，全面体现自然水景生活，以舒适、自然、休闲作为样板标准，使舒适的生活记忆与休闲记忆永留在翡翠谷。

翡翠谷在注重空间设计的基础上，加强了对个人空间的重视和关注，保证了主体景观的突出与个性生活空间的延伸，力争做到整体景观与个人生活的融合统一。

1 Zhengzhou Yishengyuan • Jadeite Valley Architectural Planning and Design

Construction Location: Zhengzhou, Henan
Project Owner: Zhengzhou Yishengyuan Properties Ltd.
Total Site Area: 1,170,000 m^2
Total Building Area: 770,000 m^2

As a highly competitive project at the suburb of Zhengzhou City, this project relying on the beautiful landscape of river bank carves out a set of exquisite composition. It is especially characterized by its location in the bay, among the forest cluster, integrated with nature.

A variety of products co-exist in the Jadeite Valley, including town houses, semi-detached house, garden houses, sloping houses, multi-layer, medium high-rise, high-rise, hotel apartments, etc. In addition, there are also various business forms such as commercial street, holiday hotels, clubs, hot springs clubs. Generally, it presents a higher on both sides of overall trend, explaining natural water life in an all-round way, running for comfortable, natural, recreational standard, so that only the comfortable and leisure memories are stayed in the Jadeite Valley forever.

This space-oriented Jadeite Valley enhances the emphasis and focus on personal space to highlight the main landscape and extend individuality living space and also strives to achieve the integration and reunification between overall landscape and personal life in the best way.

1

2

2–4 清华·大溪地项目规划设计

建设地点：河南 郑州
项目业主：郑州清华园房地产开发有限公司
总用地面积：198万平方米
总建筑面积：264万平方米

该项目位于河南省郑州市与荥阳交界处，地处郑州西大门——中原西路两侧，路况良好，交通出行十分便利，周边交通网络便捷。

规划设计了大型水上游乐设施、温泉浴场、酒店客房、餐饮娱乐以及大型商业等多种设施，强力打造中原地区特色鲜明的"吃、住、游、购、娱、休"一条龙的特色休闲娱乐基地以及中原休闲娱乐顶级强势品牌。

该方案本着可持续发展，以人为本，社会环境、生态环境、建筑环境和谐统一，远近结合，保持弹性，坚持规划的可操作性为原则，力求打造缤纷、清新、气息浓郁、配套完善的都市近郊生活氛围，使这里成为城市精英圆梦的地方。

2-4 Tsinghua • Tahiti Project Planning and Design

Construction Location: Zhengzhou, Henan
Project Owner: Zhengzhou Tsinghua University Park Real Estate Development Co., Ltd.
Total Site Area: 1,980,000 m^2
Total Building Area: 2,640,000 m^2

This project is located in boundary of Zhengzhou City, Henan Province and Xingyang, near Xingyang side, where both sides are West Road of Zhongyuan, West Gate, Zhengzhou city. There are good road conditions, convenient traffic travel, expedite traffic network around.

Its planning has designed various forms of product function including large water amusement facilities, Spa pool, hotel rooms, dining, entertainment, Shopping Mall, etc., striving to create a one-stop recreation base featuring in "food, housing, travel, shopping, entertainment, rest " and top entertainment brand with great influential in Central Plains.

This project is in line with sustainable development as well as people-oriented tenet; to realize the harmony among the social environment, ecology as well as built environment; to adjust as the immediate and long term development as well as maintaining flexibility; to adhere to the principle of manipuility; to create a fun suburban life atmosphere with various ancillary facilities. This will be the place where the dreams of urban elites come true.

3

4

ArchLong Design Group

美国朗基建筑设计有限公司

朗基建筑设计有限公司是一家国际性的设计机构，在中国上海、北京、美国加利福尼亚州设有办公室或联络机构。

公司提供建筑设计、城市规划、景观设计、室内设计、房地产策划咨询等多方面的服务内容。从2000年起，朗基作为独立的项目公司，已完成了许多重要的工程设计。2002年朗基合并ARCH1，正式成立美国朗基建筑设计有限公司（ArchLong Design Group）并对外承接项目，成立至今已经承接了包括都市规划、城市公共建筑、大型居住社区、公共景观等近百项各类设计项目，业务范围从北美、中东遍及中国的上海、天津、重庆、苏州、杭州等地。

在专业领域，朗基一直致力于将设计理论的研究和建筑实践紧密结合的理念，深信“设计创造生活”，并将这一理念积极地贯彻在设计的每件作品中。

朗基的国际设计团队提供先进的设计方法、理念和灵感，同时本地的优秀人才能够全面地把握项目方向并最终完成具有丰富表现力的建筑作品。在项目发展过程中，始终充分理解客户的需求与意图，并积极寻求客户的充分参与，同时建立了有效的沟通与服务体系，使项目在专业领域和市场价值间达到精确的平衡。

朗基在全球有超过100名的建筑师、结构工程师、设备工程师、景观建筑师及其他专业的技术人员提供专业化的服务。在上海有超过50人的设计团队，有三个设计项目部，其中两个提供建筑及规划设计，一个提供景观设计。我们建立以总经理、首席建筑师、技术总监和项目总监为领导的综合管理团队，在项目进度和设计创意、质量方面提供全方位的保障。

基于每个项目不同的背景、预算及目标，朗基提供与之相适应的服务：充满创意的构思、专业的技术支持、严谨的后期服务。以下专业领域的实践，我们具有特别丰富的经验和竞争优势：

· 城市设计
· 行政办公建筑
· 教育与文化设施
· 高级住宅
· 酒店及度假设施
· 商业及多用途建筑
· 高科技研发中心及工厂
· 公共景观

同时，朗基将继续关注业内的最新趋势，提供在建筑及环境设计领域更加全面的服务，使每个项目都成为具有独特表现力的设计作品。

ArchLong Design Group is a Shanghai based international architectural office. It has practiced in architecture, landscape, urban planning, urban design and interior design in many places such Beijing, Shanghai, Toronto, Berlin, Mid-East and etc.

ArchLong Design Group is providing services in architectural design, urban planning & design, landscape, interior design and real estate consulting. Starting at 2000, the office, LGL Architects, has completed many important projects. At 2002, LGL Architects combined with Arch1 Design. ArchLong Design Group was founded.

As professions, ArchLong Design Group always concentrates on the architectural practice. We believe that a successful combination of the international design methods, ideas and technologies with the excellent local practice team will develop out brilliant projects. We highly believe that "Design creates Life".

In the progress of each project, ArchLong Design Group will always have a good coordination with the client. The successful established coordinating channel has had our projects kept a good balance between the architectural request and marketing request.

地址： 上海市长宁区长宁路350号日旭商务中心4层
邮编： 200042
电话： +86–21–52381556
传真： +86–21–52381556–802
邮箱： ArchLong@archlong.com

Add: Floor 4. No.350 Changning Rd, Changning District, Shanghai PRC
P.C.: 200042
Tel: +86–21–52381556
Fax: +86–21–52381556–802
E-mail: ArchLong@archlong.com

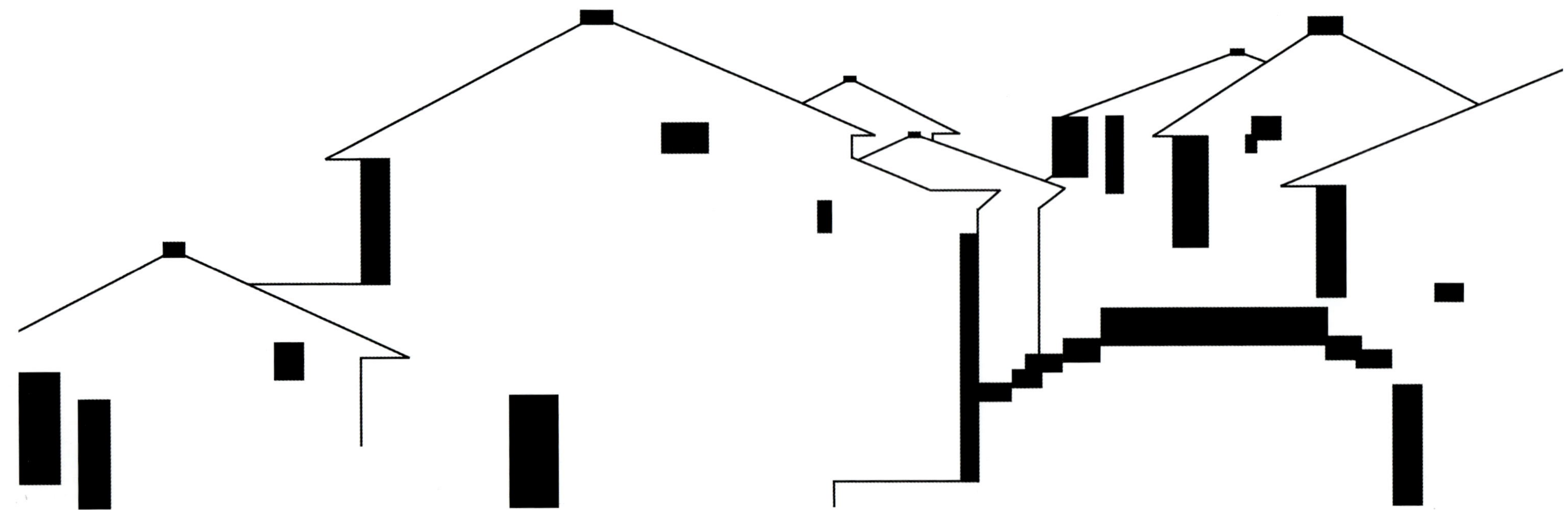

尹山湖水街

总用地面积：503 768.92平方米
总地上建筑面积：79 103.39平方米
总地下建筑面积：112 426.94平方米
容积率：0.16

该项目是继金鸡湖商业综合体之后的第二个苏州市的大型商业综合体。

内部建筑具有大型歌剧院、IMAX影剧院、大型图书馆、商业、文化等全方位的配套设施。整体规划采用下沉式设计，结合多层次的景观设计，并全部采用屋顶花园，完全实现低碳、环保、自然绿色等理念，随环境的变化而变换建筑形态，建筑成为环境的填充物，并与之共存。

天山苏州研发中心

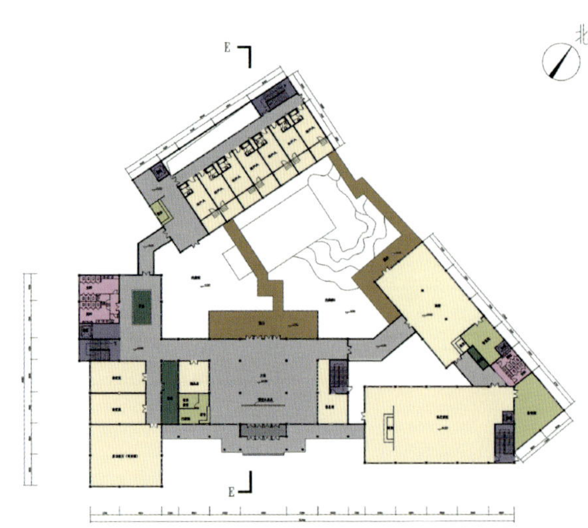

涵碧晓筑

总用地面积：32 000.11平方米
总建筑面积：14 531平方米

该项目位于苏州吴中太湖度假区内，东南侧环拥太湖水景，城市交通便利，环境资源优厚。

涵碧晓筑是吴中集团继环秀晓筑之后设计的第二个精品酒店。设计采用院中院、湖中湖、外景含内景再含户景，一环扣一环的设计理念，将整个酒店做到公共和私人空间、动和静、公开和隐蔽绝对分离。

整个酒店总共有48间高级客房，按六星级标准建设，内外全部采用苏州传统风格，将苏州文化展现得淋漓尽致，必将成为苏州太湖一大景区亮点。

南门坝商业街区

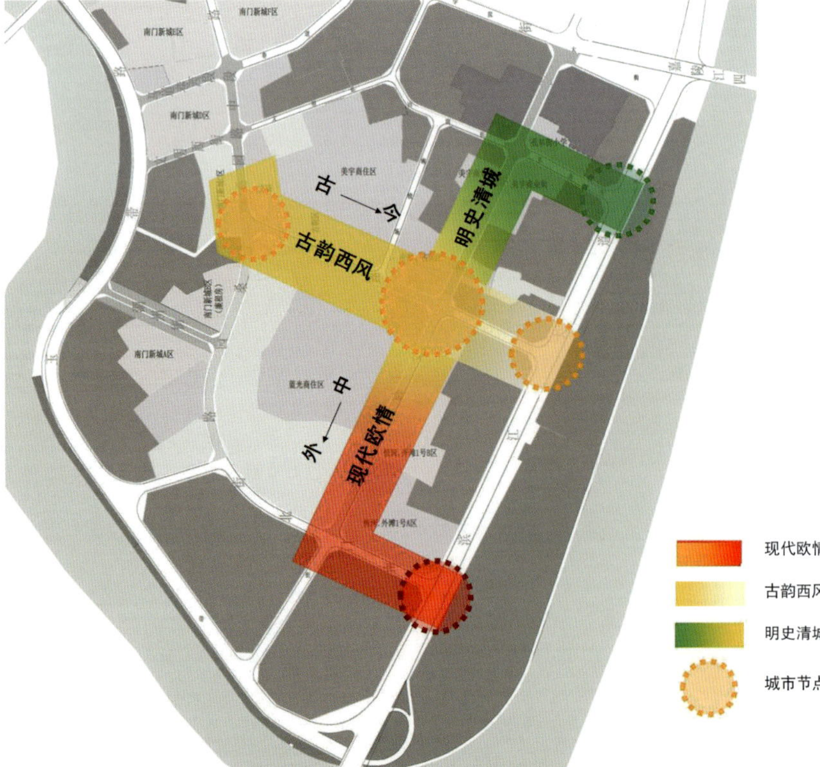

科技园水街

总用地面积：141 925.88平方米
总建筑面积：100 075.35平方米
容积率：0.7
覆盖率：20%
绿地率：44%
停车位：450辆

项目用地位于苏州吴中旺山科技园区，南侧仅靠吴中大道，西侧临近旺山山体，苏旺河、旺山河位于地块之中，总用地面积约为141 925.88平方米。东南侧吴中科技创业园一期已局部建设完成。城市交通便利，拥有自然水景、山景，环境资源优厚，将成为景观优越的现代化科技园区。

木渎胥江景观

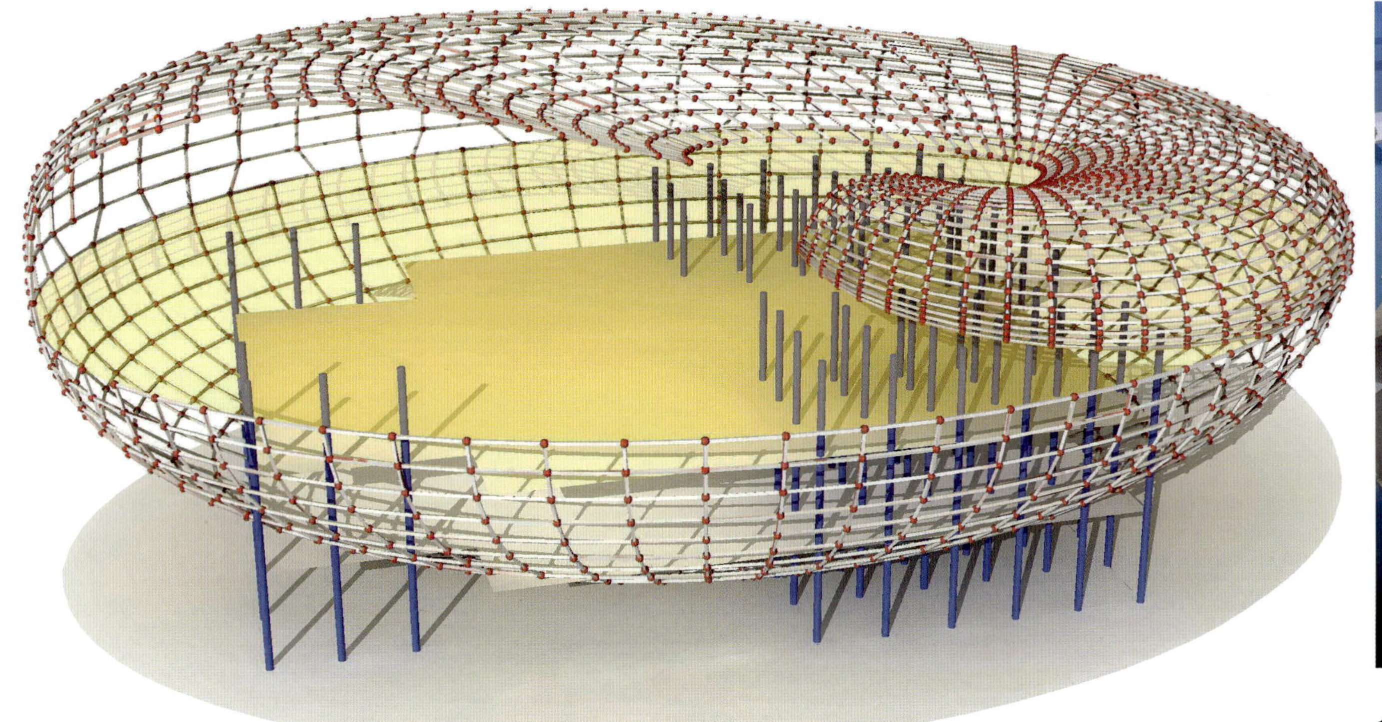

釜山歌剧院

总用地面积：32 000.11平方米
总建筑面积：14 531平方米

方案的构思为“漂浮双贝”，寓意一对漂浮于海岛的贝壳，它们伫立在海岛上，互相对望，生根于动感十足的城市空间，仿佛在齐力合奏出海洋的交响乐。

“双贝”起伏流畅的线条、引人入胜的轮廓与海水交相辉映，充满了生命力和浪漫激情。再通过类比的手法把主体建筑与环境景观关联起来，以制造一种嵌入式的效果。将景观的元素渗透到建筑形体和空间中，以动态的建筑空间和形式、模糊边界的手法形成功能交织，并使之有机相连，从而实现空间的持续变化和形态交集。将建筑内部、外部直至城市空间看做是城市意象的、不同的但连续的片段，通过切割与连接，使建筑和城市景观融合共生。

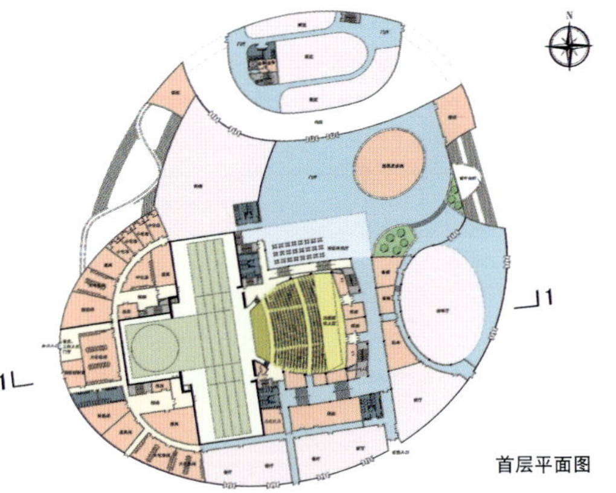

首层平面图

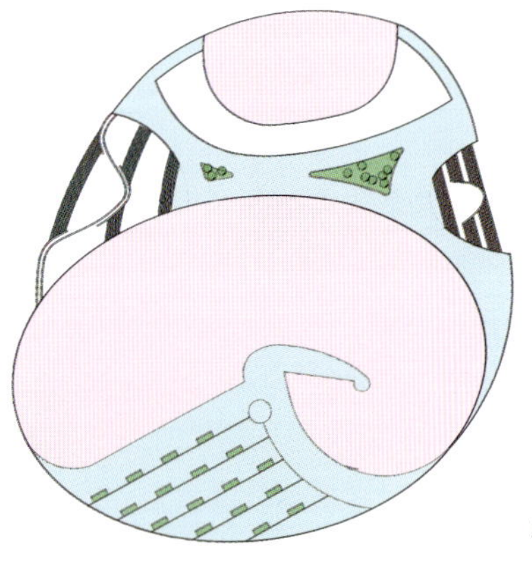

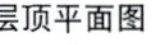

层顶平面图

ARCHITECTURE
PLANNING
LANDSCAPE
INTERIOR

Joseph Wong Design Associates, Inc. (Jwda) is a firm that has provided outstanding architecture and planning since its formation in 1977.
As a firm, we are proud of our reputation for on-time and on-budget delivery of architectural services and believe budget delivery of architectural services and believe that our client's success is the key to our success. Evidence of this commitment is the fact that most of our projects in the last ten years have been with repeat clients.

美国JWDA建筑设计事务所全称Joseph Wong Design Associates，成立于1977年，注册地为美国加利福尼亚州。JWDA于1993年设立中国事务部，进入中国市场，并以一流的设计与服务，在中国赢得了客户与市场的认可。1997年在上海设立代表处，为中国客户提供优质的设计服务。2005年11月，正式注册成立了上海骏地建筑设计咨询有限公司，2007年成立了深圳骏地建筑设计有限公司。

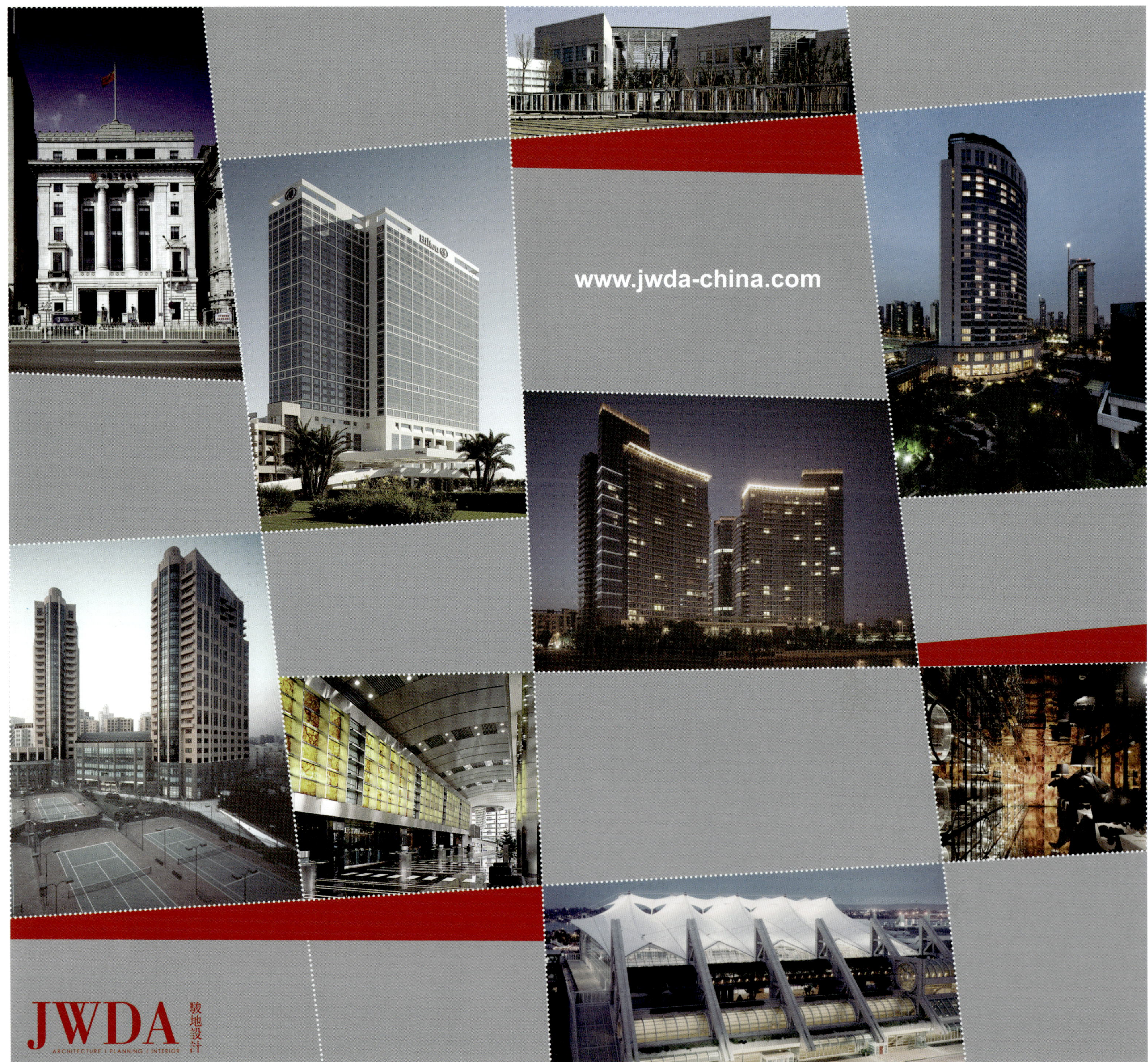
www.jwda-china.com
JWDA 骏地設計
ARCHITECTURE | PLANNING | INTERIOR

Hainan Crabapple Bay Resort Hotel

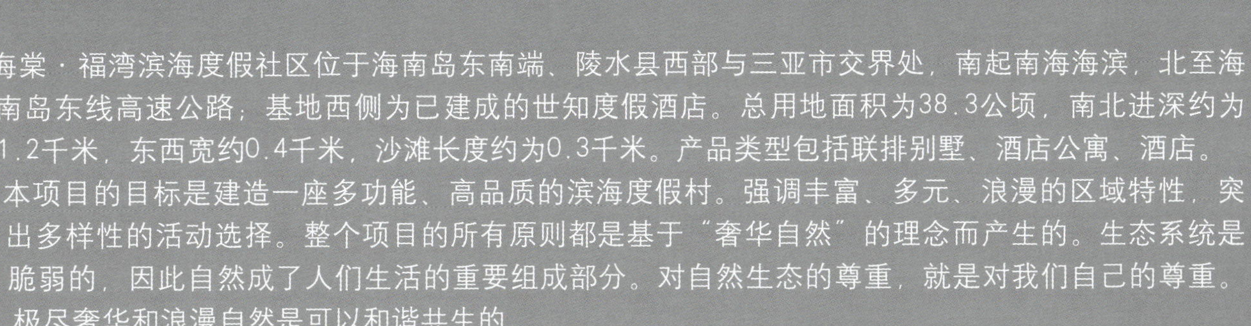

The Crabapple Bay Resort takes advantage of the natural beauty of Hainan Province. It is located at the border of Lingshui and Sanya. This tropical site rests its southern border on the South China Sea shore. The north side is bordered by East Line Express Highway. Designed as a California-Spanish style resort, this 38 hectare property combines luxury quality and natural environments. The Crabapple Bay Resort is comprised of a townhouse, a service apartment and a five-star hotel, making it a true holiday destination in Southern China.

海棠·福湾滨海度假社区位于海南岛东南端、陵水县西部与三亚市交界处，南起南海海滨，北至海南岛东线高速公路；基地西侧为已建成的世知度假酒店。总用地面积为38.3公顷，南北进深约为1.2千米，东西宽约0.4千米，沙滩长度约为0.3千米。产品类型包括联排别墅、酒店公寓、酒店。

本项目的目标是建造一座多功能、高品质的滨海度假村。强调丰富、多元、浪漫的区域特性，突出多样性的活动选择。整个项目的所有原则都是基于"奢华自然"的理念而产生的。生态系统是脆弱的，因此自然成了人们生活的重要组成部分。对自然生态的尊重，就是对我们自己的尊重。极尽奢华和浪漫自然是可以和谐共生的。

对小城镇的塑造就是对生活模式的营造，度假作为一种休闲方式不应该是对传统生活模式的辅助，而是一种新的生活态度；度假村的生活构成本项目的设计主线。外立面设计采用了充满阳光气息的加州特征的西班牙风格，演绎出简朴、大气而精致的地中海度假村建筑特征。

Hotel Indigo

JWDA was retained to provide design services for the newest Hotel Indigo, located in downtown, San Diego. The 12-storey hotel consists of 210 rooms and suites with a roof terrace and pool/SPA facilities. The indigo brand, developed by Intercontinental Hotels Group, seeks to appeal to travelers tired of bland "beige box" hotels. Each hotel in the line adheres to the franchise essentials while reflecting the region of their location. The "branded boutique" hotel offers an accessible and unique experience for the technology and design savvy customer where guests can feel at home.

JWDA作为这座位于圣地亚哥市中心最新的英迪格酒店的设计者，在这座12层高的酒店中设置了210间客房和套房，并配置了天台、泳池和温泉设施。

英迪格作为洲际集团旗下的品牌，主要针对那些厌倦了枯燥乏味、千篇一律"米色盒子"的酒店客人。

该品牌下的每一座酒店在坚守连锁精神的基础上都尽量体现当地特色。这种"品牌化精品酒店"的路线通过提供科技便利和精细的设计细节，为客户提供了亲和的、独特的、如家般的酒店体验。

Shanghai Zhangjiang Entrepreneur Industrial Center

We set up 4 major principles to guide our projects: 1) Harmony with Environment; 2) Effectiveness and Efficiency; 3) Rhythm in Landscape; 4) Pleasant from Openness. Our architectural design highlighted the ideas of "Ecological, Communicational, and Flexible" industrial spaces, which enriched the spatial dynamic and intimacy. Combined with effective yet efficient supporting campus facilities and landscaped venue, Zhangjiang Entrepreneur Industrial Center encouraged interactions between occupants and the natural environment. Ultimately, this project promoted the interactions between cultural and creative industry and its surrounding areas, at the same time, increase the influenced and appeal of a work/living style space.

上海张江创业产业中心——我们把生态和谐、高效便捷、极富层次和开放舒适作为项目的四个原则，在设计上突出“生态性、交流性、灵活性”的理念，使园区呈现更加丰富和亲切的空间形态；同时结合便捷实用的功能配套设施和景观活动场所，力求创造园区使用者与环境和谐共生，对文化创意产业和周边地块具有影响力和吸引力的工作生活空间。

Z-O-A is an international design firm registered in the US. Our offices are located in Seattle, Denver, and Beijing. We help our clients create innovative and timeless places. The firm's practice is characterized by a combined emphasis on design excellence and high standards of client service.

The design service we provide planning & urban design, and architectural design services. Project types include sports & entertainment, hospitality, corporate, mixed use & residential, and retail and commercial buildings. Our current practice borrows from both suburban and urban backgrounds, and both Eastern and Western cultures. Site and building design are conceived of simultaneously—exterior space and interior space developed as dual elements in an overall composition—landscape and urban design elements helping to connect the building with its local physical context. Each project design is intended to contribute to the quality of its neighborhood / context as well as fulfill the needs of its own program. We have strong experience in sports, entertainment, commercial, residential, and mix-use development projects. Our projects are located globally including North America, Europe, Asia, and Africa.

Z-O-A是一家在美国注册的建筑设计事务所.事务所分别在美国西雅图、丹佛和中国北京设有办公室，作为国际化的建筑设计事务所，我们力求为业主提供最优质的设计和一流水准的服务，建造出不同凡响、跨越时空的建筑。

公司的设计服务包括:建筑设计、规划、城市设计及室内设计。我们的设计融入了东西方思想文化的精华，且非常注重人的生活环境品质，以设计创造和改善环境及人文生态为主旨，我们的设计是在考虑到与其周边社区的整体和谐与共生,充分满足我们业主要求的基础上，达到最高环境品质、景观与建筑，整体与局部的交融，有效地使业主的投资发挥出最大的价值，我们的设计涉及内容广泛的设计领域，在办公、体育、娱乐、商业、文化、综合体、居住建筑方面有着丰富的设计经验，公司主创设计师的设计曾多次荣获美国建筑师协会奖及中国建筑学会等奖项。公司参与的设计项目遍及世界各地。优质创新的设计和服务使我们赢得了业内广泛的赞誉。

1 “创意山东”城市文化综合体，中国 济南
2 75 Emerson, Denver，美国 丹佛
3 3131 Zuni, Denver，美国 丹佛
4 阳光文化中心，中国 济南
5 包头体育中心，中国 包头
6 烟台海滨城市综合体，中国 烟台
7 8 Norris Design Office, Denver，美国 丹佛

Z-O-ARCHITECTURE

www.zo-architects.com

1

5

4

3

2

设计服务内容	Design Service:
建筑设计	Architectural Design
规划、城市设计	Planning, Urban Design
室内设计	Interior Design
环境设计	Environmental Design

设计项目类型：	Project Types:
体育、娱乐建筑	Sports, Entertainment
商业、文化建筑	Commercial, Cultural
住宅建筑	Residential
办公建筑	Office
城市综合体	Mix-Use Development

北京办公室：
海淀区车公庄西路19号
华通大厦 901室，邮编：100044
电话：+86–10 6879 0020
传真：+86–10 6879 0021
xzhang@zo-architects.com

Seattle Office:
4920 156th Avenue SE
Bellevue, WA 98006
Tel: 001-206-973 7607
xzhang@zo-architects.com

Denver Office:
1101 Bannock Street
Denver, CO 80204
Tel: 001-303-623 2330
Fax: 001-303-623 2335
market@zo-architects.com

www.zo-architects.com

4

3

2

5

6

Z-O-ARCHITECTURE

1. 竹文化博物馆，中国 重庆
2. Bonfils Stanton, Denver，美国 丹佛
3、5. 中建大厦，中国 南京
4. 段店旧城改造，中国 济南
6. 烟台丽景海湾度假酒店，中国 烟台

1

映時設計

TIME MIRROR International

美国Time Mirror International 映时设计公司，以美国洛杉矶为基地，并积极在中国开拓市场，不断探求设计艺术与经营管理的最佳平衡点，力求使每个项目均获得最大的成功。把不断为业主提供更好的设计服务作为公司服务的总目标，除在设计上求新外，更在项目管理上求精。永远站在客户的立场上去寻求最佳并有创意的解决方法。

创新是我们灵感的来源，作为设计师，我们倍感欣慰的是总能找到目标与价值的平衡点，使每个不同的客户均享受到优质服务。

多年来我们的设计项目及规模也在不断地开拓发展。迄今为止，我们的设计已涉及各个领域，包括酒店、写字楼、住宅、商业等不同的设计项目。

设计公司的宗旨是在设计及服务上精益求精，为客户营造出有创意的设计作品。

American Time Mirror International Design Company, based in Los Angeles, USA, develops China market actively. The company constantly seeks the best balance between design art and operating management, trying to make each project successful. It makes the improving services continuously provided to the clients as the general goal of the company's services. It not only pursues innovation in design, but also pursues refinement in project management. It always seeks the best and creative solution from the stand point of the clients.

As designers, what makes us gratified and content is the reaching of the balanced point between the object and value all the time, which can let every different client enjoy excellent service.

For the past many years, our design project and scale have also continuously expanded and developed. So far, our design has already covered a lot of fields, including the design projects of hotel, office building, residence and business, etc. The aim of TMI Design Company is to seek for greater perfection in design and services and provide creative design works for the clients.

地址：北京市朝阳区朝阳公园西路19号佳隆国际大厦B座601
电话：+86–10–65390171
传真：+86–10–65390176
邮箱：tmioffice@163.com
网址：timemirrordesign.com

Add: Room 601, Tower B of Jialong International Building,
No.19 Chaoyang Park West Road, Chaoyang District, Beijing
Tel: +86–10–65390171
Fax: +86–10–65390176
E-mail: tmioffice@163.com
Http://www.timemirrordesign.com

碧海方舟别墅

位置：北京
面积：1 500平方米

项目坐落于北四环的高尔夫球场旁，享有得天独厚的视野景观和交通便利性。业主有丰富的国外生活经验，所以对别墅的功能布局、生活方式及设计细节，提出许多的想法。

别墅共计4层，包含一层的客厅、餐厅、起居厅、客房、书房，二层的男孩女孩房、主卧室，三层和室、主卧室及图书厅，以及地下室的家庭影院、桑拿健身房等。设计风格的主轴为现代法式。八角厅书房采用山纹樱桃木妆点墙饰面，凸显庄重的氛围。设计时尽量避免在墙面使用过多石材，以免造成压迫堆积感，仅在挑高2层的客厅局部使用。家庭厅、早餐厅和开放厨房是居家的核心，色调以米色为主，空间虽不是很高，却更能表现居家的温馨气氛。三层充分利用屋脊的优势，做成特殊造型的吊顶，空间更显恢弘，配上樱桃木饰面的整体墙板和实木地板，重现了业主国外生活的法式古堡风格。

佳隆君悦餐饮会所

位置：北京
面积：2 400平方米

佳隆御园餐厅位于朝阳公园西南侧，北拥朝阳公园，南瞰林荫大道，东面是凤凰卫视新居。

项目定位为高级的粤菜餐厅，共有14间不同设计形式的豪华间，包括一个VIP包间。VIP包间面积近100平方米，挑高4.5米，面对朝阳公园景观，有专属的电梯直达门厅，包间内各种商务及休闲功能齐备，餐区足够24位宾客宴会享用。

设计师对于房间的布局和宾客、服务动线做了深入的研究，达到开放、私密、合理的三重目的；为调和主入口及双向通道的问题，设计师以倾斜的椭圆厅作为接待大厅，让动线更流畅。走廊引领宾客到达A区及B区不同风格的房间，C区是两套豪华间，拥有独立的门厅通道，各以红色与灰色为设计的主要色调。

东侧临街面是开放的酒吧区，也可以作为多功能包间的前厅使用。原建筑结构的"M"形斜柱，也被设计师保留下来，利用这一特殊的结构造型组成了带状的展示柜。

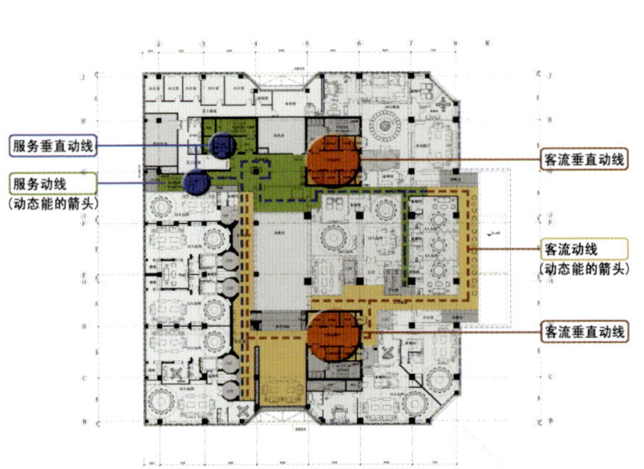

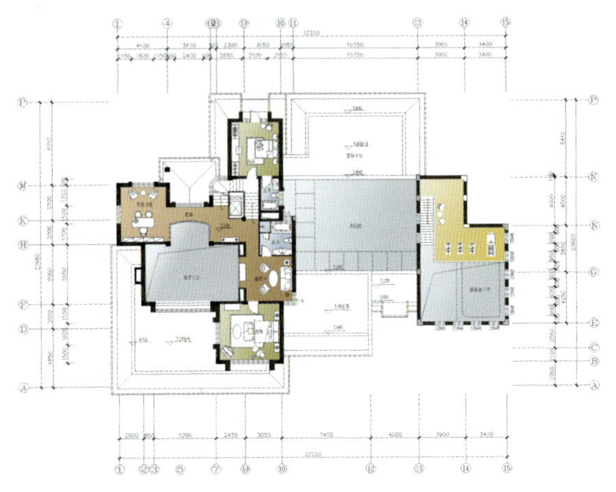

名都园别墅

位置：北京
面积：1 000平方米

项目坐落于北京顺义别墅区，原建筑约600平方米，加建后面积约1 000平方米。加建的面积主要为增加别墅居住的功能空间，如游泳池，健身区，开放厨房及家庭影院。

受限于原建筑体的高度和结构形式，设计中设计师花费许多工夫在细节空间的调整，尺寸的拿捏，期望在有限的空间创造出合理舒适的尺度，甚至运用了空间缩放的对比方式，以此塑造出相对宽阔和相对高耸的空间感觉。同时，由于新旧建筑体的联结组合，造成空间之间的立面是连续的，而高度却相差很大；这样的重叠与错落，增加了设计的难度，但是也让设计师有开启另一番独特设计的可能性。

许多有趣的设计的确是由受限的条件演绎而来，如一层客厅与起居厅中间的鱼池与天窗、客厅的海水鱼缸、挑高的开放厨房，以及主卧室的天窗，都因此演变而来。中式环廊与庭院在这个装饰艺术风格的设计案中，让人眼睛一亮，让古朴的牌楼、碑石巧妙地融合在建筑群体之中，完整呈现出一个具有个性的别墅。

澳大利亚AUD建筑设计公司（上海）
上海工程勘察设计有限公司（二所）

澳大利亚AUD设计有限公司（上海）是一家在澳洲注册的专业建筑、装潢、景观设计公司，是与上海工程勘察设计有限公司联合发展的设计单位。

公司承接有大中型住宅小区、商业广场、办公楼、各类学校等项目，在国内外有一定知名度。20世纪90年代，上海的《解放日报》专访该公司的一篇文章中就将公司比喻为“设计界跑出的一匹黑马”。

公司现有国家一级注册建筑工程师、国家一级注册结构工程师多人，并具有国家甲级设计资质。

公司愿与各业界共同创设未来新空间。

Australia AUD Architectural Design Co.,Ltd.(Shanghai) is a professional construction, decoration, landscape design company registered in ANZAC, which is a united company of Shanghai Engineering Survey and Design Co.,Ltd.

We have undertaken large and medium-sized residential area, commercial plazas, office buildings and other projects and we have received popularity at home and abroad. In 1990s *Shanghai Liberation Daily* reported our company in special interviews as a "dark horse of designing industry".

There are so many National Class A Registered Architects and National Class A Registered Structural Engineers in our company.

We would like to invite all elites hand in hand to our bright future.

地址：上海市武宁南路318号2楼
邮编：200042
电话：+86-21-62323145-201
传真：+86-21-62311424
邮箱：No.2@saec.cc

Add: 2#, Wuning South Road 318, Shanghai
P.C.: 200042
Tel: +86-21-62323145-201
Fax: +86-21-62311424
E-mail: No.2@saec.cc

1—3 恒盛鼎城
Glorious Top City

4—6 河南许昌塔湾社区
Henan Xuchang Tower Bay Community

规划用地面积
Planned Land Area: 84,804 m^2
建筑面积合计
Total Building Area: 287,992 m^2
总建筑面积
Total Floor Area: 316,912 m^2
机动车位
Motor Vehicle Spaces: 594

7—8 南京六合
Nanjing Liuhe

总用地面积
Total Land Area: 131,946 m^2
容积率
Floor Area Ratio: 1.56
建筑密度
Building Density: 19%
绿化率
Green Rate: 39%

9 桃浦西路公建
Taopu West Road Public Buildings

10

11

12

13

10—13 兰州留学人员创业园 彭家坪产业研发基地

Lanzhou Overseas Students Pioneer Park, Pengjia Ping Industrial R & D Base

14

16

15

17

14—15 顾村公园二期陈广路入口服务区项目

Gu Village Park Phase Two, Chenguang Road Entrance Service Area Project

用地面积
Land Area: 7,100 m^2
建筑面积
Building Area: 4,160 m^2

16—17 顾村公园二期奇石园项目

Gu Village Park Phase Two, Stone Garden

用地面积
Land Area: 4,300 m^2
建筑面积
Building Area: 3,000 m^2

18

19

20

18—20 新江湾城20、21号地块项目

New Riverside City Site No. 20 and 21 Project

地上总建筑面积
Total Floor Area on Ground: 74,359 m^2
地下总建筑面积
Total Floor Area under Ground: 86,359 m^2
容积率
Floor Area Ratio: 2.3

21—23 上海思博学院

Shanghai Si Bo Institute

用地面积
Land Area: 6,000 m^2
建筑面积
Building Area: 25,000 m^2
地下面积
Ground Floor Area: 5,000 m^2

21

22

23

ZPLUS 法国普瑞思建筑规划设计公司

地址：天津市河西区黄埔南路万顺温泉花园D座4F
电话：+86-22-28011388 / +86-22-28011288
传真：+86-22-28133933
邮箱：puruisi001@163.com / zhou_xiangjin@eyou.com
网址：www.zplus-fr.com

Add: 4F D Building, Wanshun Hot Spring Garden, south Huangpu Road, Hexi District, Tianjin
Tel: +86-22-28011388 / +86-22-28011288
Fax: +86-22-28133933
E-mail: puruisi001@163.com / zhou_xiangjin@eyou.com
http:// www.zplus-fr.com

ZPLUS法国普瑞思建筑规划设计公司由欧洲资深建筑师、规划师、景观设计师和留学欧洲的中国学者及设计师组成，旨在结合西方先进理念、设计手法与技术支持，为客户提供一流品质的设计服务，塑造品牌、作品与魅力。

ZPLUS公司对于项目的考察与分析是全面而客观的，提出的解决方案是建立在丰富的经验基础之上的，既散发时尚美感、蓬勃想象与创造精神，又在实现层面上本着务实的态度，经济而有效地从整体到细节都予以精心推敲，保障了项目的实施。

公司业绩集中在高层公寓、住宅、写字楼、新区规划等领域，完成项目均已成为城市重点地标。近年来，公司的建成作品屡获殊荣，如国际经济贸易中心（ICTC）被评为2004年最具影响力、最佳销售业绩的A级写字楼。海河之子、赛顿中心、华门名筑荣获优秀社区奖，慧谷大厦、富力克拉公馆、松江水岸江南、渤海创智中心等著名项目都已建成，并得到业界的好评与瞩目。 新近中标的滨海新区两大项目——中国五矿旷世国际和浙江商会大厦体现了公司日益提升的专业水准和综合实力，为今后的发展打下了更坚实的基础。同时，空港以及团泊湖的几个项目则加强了我们与老客户的紧密合作，使公司业务又涵盖了星级酒店、大规模新城规划等范畴。住宅方面涉及中高档居住社区规划及建筑设计，如大学城英、法风格社区为25万平方米的高层建筑、联排别墅和洋房；华北城二期为38万平方米的以中高层住宅为主的新古典风格综合社区。

The Paris-based company ZPLUS is composed of senior architects, plannes and landscapes both from Europe and China, with international working background, the team aims at combining the most up-dated technology, design method and concept to provide reliable and time lasting services, and to create brand, masterpiece and charm.
ZPLUS gives each problem a full-range and indiscriminate investigation, with the design ideas and representations in the most efficient way integrating imagination, creation and technology, the realization of the projects is based on the knowledge gained through longtime experiment and readjustment.
The main works are high rise apartments, housing, offices which are all spectacular landmarks in the important area of the city. New development zone urban plan is also included in the company's project category.
Over recent years, many completed works have won special honors. For example, International Economy and Trade Center (ICTC) was appraised as the most influential and marketable class-A office building in 2004. The Naga Center along HAIHE, Centown in the city center, and Huamen Building won the Excellent Community Award; Huigu Building, R&F Kela Mansion, Songjiang Riverbank South China, Bohai Intelligence Center and other famous projects have been completed, winning general recognition and good reputations. The Company has won the bids for two new projects in Binhai New District – China Minmetals Business Center and Zhejiang Chamber of Commerce Building, representing the company's increasing expertise and overall strength, and laying solid foundation for future development. Meanwhile, several projects in Airport Zone and Tuanbo Lake strengthens our close cooperation with existing customers, enabling our business to cover star-level hotels, large scale new urban planning etc.
Residential business involves mid-high end residential community planning and architectural design, such as university town of English and French-style communities featuring 250 000 m^2 high-rise buildings, row villas and townhouses; North China Town Phase II of 380 000 m^2 neoclassicism comprehensive communities mainly of mid- to high-rise residential buildings.

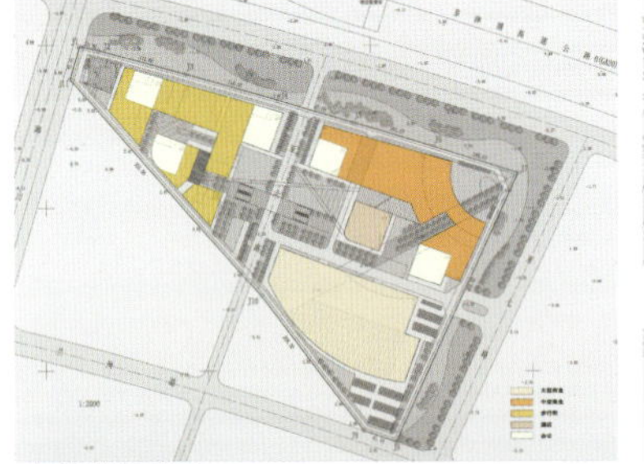
功能平面图

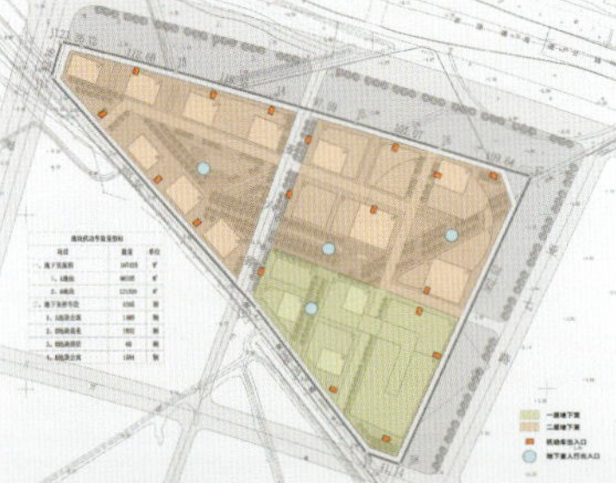
机动车平面图

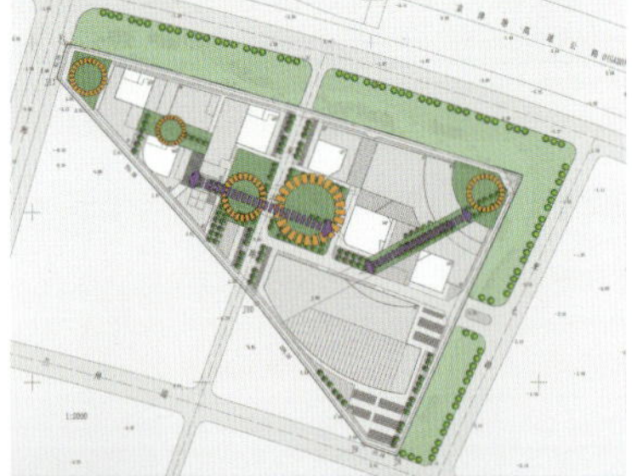
景观平面图

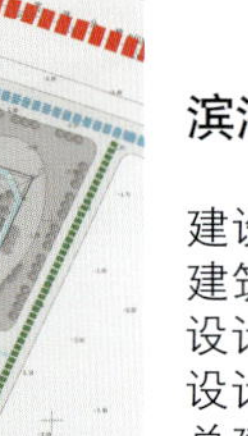
交通平面图

滨海新城

建设地点：天津
建筑面积：800 000平方米
设计时间：2010年
设计公司：ZPLUS法国普瑞思建筑规划设计公司
总建筑师：刘顺校
开 发 商：滨海新城建设发展公司

区域位置及规划设计

滨海新城位于天津京津塘高速公路收费站以南、中环线以西，本工程地上建筑面积约为58万平方米，地下建筑面积约为22万平方米，共80万平方米。

该项目交通主要出入口方向位于东侧和北侧。

方案巧妙利用三角形的地界将高层分为三个一组的两大组团，沿高速公路一侧展开，余下的基地一角布置大型综合商业，围合出内部的中空庭园，形成有序的、高品质的都市群体风貌。为印证当代经济技术发展特征，建筑外立面采用多种材质，呈现出多种肌理，形成亮丽而又挺拔的建筑形态。规划中注重连续空间的塑造，关照人的行为及空间使用的便利，以丰富和谐的环境为人们提供一个功能齐全、安全舒适、充满艺术气息的场所。

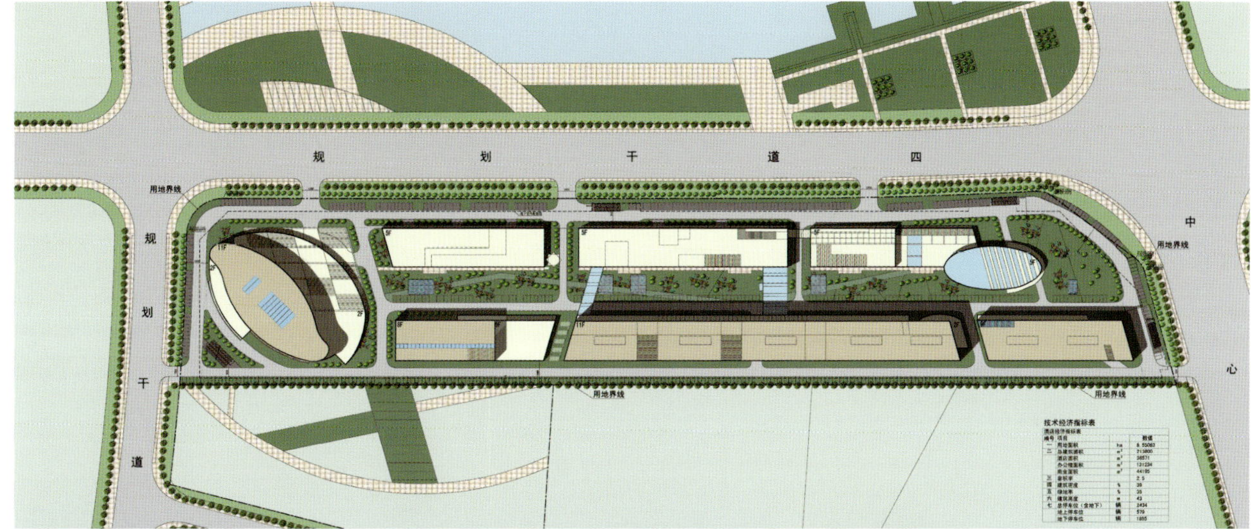

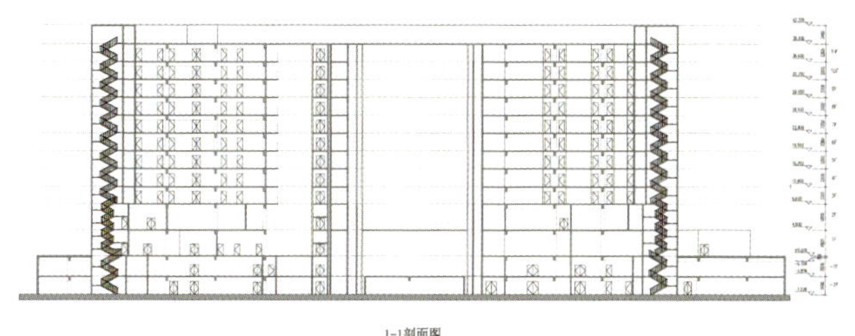
1-1剖面图

2-2剖面图

天津空港经济区融和广场及五星级酒店

建设地点：天津
占地面积：85 508.3平方米
设计时间：2005–2009年
竣工时间：2010年
设计公司：ZPLUS法国普瑞思建筑规划设计公司
总建筑师：刘顺校、周湘津
开 发 商：万顺置业有限公司

面对30多万平方米的建筑体量、600余米长的街区景观，如何把握这一新的都市片段，使其不容忽视的巨大影响力具有持续积极的作用，让设计师具有强烈的社会责任感和使命感。

设计师以开放的心态、现代的材料及手法精心打造作品，雕刻着种种主题，演绎着种种细节，试图让新材料、新形式的大量积累与爆发，给人万花筒般的视觉体验，浓缩出一种新的场所和时代精神。

这600米长的现代建筑诗篇，在日积月累的精雕细刻中，在长达5年的时断时续的更改中，终于呈现。它在2011年空港最受欢迎的十大建筑评比中，名列首位。

公司简介 BACKGROUND

吕邓黎建筑师有限公司创立于1983年，以香港为基地，国内分公司分别设于上海及重庆。

公司的宗旨是坚持以创新精神与稳健周全的专业态度，服务不同顾客和满足项目的需求。多年来，吕邓黎建筑师有限公司积极投身建设香港，设计的项目包括各类规模的住宅、办公楼、酒店、商场、厂舍、学校、公共房屋、宗教建筑及生态环境等。通过长时间的实践，奠定了一个优质服务信誉的基础。

随着中国经济的起飞，吕邓黎在20世纪80年代中期开始踏足国内，在各大都会如北京、上海、重庆、成都及广州等地，先后参与完成多个建筑及室内设计项目。尤其在作为全国经济龙头的上海，已完成的项目包括瑞安广场、城市酒店、东方巴黎、东方剑桥、东方曼克顿及瑞虹新城等。以上项目都以优秀的质量完成，并获得多个奖项，瑞安广场更荣获鲁班奖的殊荣。

吕邓黎建筑咨询(上海)有限公司的成立，更标志着公司对积极建国建港的抱负，秉承公司的宗旨，通过与国内专业单位共同努力，为广大业主创造更美好的生活环境。

Lu Tang Lai Architects Ltd. originated in Hong Kong and started off as a small owner-managed design studio in 1983.

Realizing the rapid economic transformation and growth of China in the last decades of the twentieth century, Lu Tang Lai Architects Ltd. sought to contribute to the development and redevelopment in China by actively participating in building and environmental design projects on the Mainland as early as the mid eighties, focusing in shanghai in the early nineties. To increase competitiveness and enhance services, the Shanghai and Chongqing Offices were put into operation in 2002 and 2004 respectively.

Over the years, the practice maintains a steady growth in sizable architectural projects ranging from offices and commercial developments, hotels, service apartments, high-class residential developments, schools and institutional buildings to master planning in Hong Kong, Mainland China and S.E. Asia. Its reputation for providing personalized and quality services remains unchanged.

奖项 AWARDS

上海瑞安广场
Shui On Plaza, Shanghai
- 鲁班奖
- 白玉兰优秀建筑奖

上海瑞虹新城一期
Rui Hong Xin Cheng Phase 1, Shanghai
- 上海市优秀住宅评选优秀房型奖
- 上海市优秀住宅评选科技奖
- 上海市"四高"优秀小区
- 上海市优秀住宅评选优秀住宅银奖

上海瑞虹新城二期
Rui Hong Xin Cheng Phase 2, Shanghai
- 詹天佑住宅大奖小区优秀规划设计奖
- 全国人居经典方案大赛住宅组环境金奖
- 最受欢迎楼盘综合金奖

上海东方曼克顿–世纪豪庭
Century Metropolis, Shanghai
- 上海市优秀住宅工程单体设计项目二等奖

广州西门口广场一、二期
Westmin Plaza Phase 1 & 2, Guangzhou
- CNBC Asia Pacific Commercial
- Property Awards 2009
- 5-star award for
- "Best Mixed Use Development China"

重庆市化龙桥雍江苑
The Riviera, Chongqing
- Residential Building
- 4-star Performance Award 2008

香港皇冠假日酒店
Crowne Plaza, Hong Kong
Asia Pacific Commercial Property Awards
in Association with Bloomberg Television 2010
- The Best Hotel Construction & Design Hong Kong
- 5-star Award
- The Architecture(Leisure & Hospitality)Hong Kong
- 5-star Award
- Best Interior Design Hong Kong
- Highly Commended Award

香港骏逸峰
The Morrison, Hong Kong
- HKIS Property Marketing Award 2008
"The Best Environmental Design Award"

香港总办事处 HONG KONG HEAD OFFICE

香港柴湾永泰道60号柴湾工业城第一期2001室
Room 2001, Chai Wan Industrial City Phase 1
60 Wing Tai Road, Chai Wan, Hong Kong

Tel: +852–2521 0681
Fax: +852–2524 3319
E-mail: info@ltlarch.com
Website: http://www.ltlarch.com.hk

上海办事处 SHANGHAI OFFICE

上海市延安西路1566号龙峰大厦15楼C室
邮编：200052
15C Long Life Mansion,
No. 1566 Yanan West Road, Shanghai 200052

Tel: +86–21–3363 0103
Fax: +86–21–3363 0120
E-mail: info.sh@ltlarch.com.cn

重庆办事处 CHONGQING OFFICE

重庆市青年路3号时代豪宛A座3004室
邮编：400010
Room 3004 Tower A, Timesquare
No.3 Young Road, Chongqing 400010

Tel: +86–23–6371 3593
Fax: +86–23–6371 3592
E-mail: info.cq@ltlarch.com.cn

1. 上海瑞安广场
Shui On Plaza
Shanghai

2. 上海城市酒店
City Hotel and Apartment
Shanghai

3. 成都中汇广场一期及二期
Central Point Plaza
Phase 1&2
Chengdu

4. 广州越秀区瑞安广州中心
The Centrepoint
Yuexiu, Guangzhou

5. 广州西门口广场一期及二期
Westmin Plaza Phase 1&2
Zhong Shan Road
Guangzhou

6. 成都中环广场
Plaza Central
Chengdu

7. 香港铜锣湾皇冠假日酒店
Crowne Plaza
Leighton Road, Hong Kong

8. 香港大新金融中心
Dah Sing Financial Centre
Hong Kong

9. 香港观塘鸿图道工业大厦
Light-Industrial Building
Hung To Road, Kwun Tong,
Hong Kong

10. 香港长沙湾永康街工业大厦
Light-Industrial Building
Wing Hong Street,
Cheung Sha Wan, Hong Kong

11. 香港湾仔捷利中心
Jubilee Centre Phase 1&2
Fenwich Street, Hong Kong

12. 香港葵涌政府海关大楼
Kwai Chung Custom House
Hong Kong

13. 香港元朗皇冠车行新检定中心
Crown Motors Pre-Delivery Center
Yuen Long, Hong Kong

14

15

14. 广州恒宝华庭
Heng Bao Garden
Guangzhou

15. 上海康城一期
Shanghai Cannes, Phase 1
Xin Zhuang, Shanghai

16–17. 上海东方曼克顿
Century Metropolis
Xuhui, Shanghai

18–19. 上海瑞虹新城一期
Rui Hong Xin Cheng, Phase 1
Shanghai

20. 上海瑞虹新城二期
Rui Hong Xin Cheng, Phase 2
Shanghai

21. 上海瑞虹新城三期
Rui Hong Xin Cheng, Phase 3
Shanghai

22–24. 重庆化龙桥雍江苑
The Riviera
Hualongqiao, Chongqing

18

16

17

19

20

21

22

23

24

25

26

27

28

30

31

29

25．香港司徒拔道别墅项目
Residential Development
Stubbs Road, Hong Kong

26．香港赫兰道别墅项目
Residential Development
Headland Road, Hong Kong

27．香港蒲岗村道汇豪山
The Forest Hills
Po Kong Village Road,
Hong Kong

28．香港上水御皇庭
Royal Green
Sheung Shui,Hong Kong

29．香港上水御景峰
8 Royal Green
Sheung Shui, Hong Kong

30．香港湾仔骏逸峰
The Morrison
Wanchai, Hong Kong

31．香港油塘四山街商住发展项目
Proposed Composite
Development
Sze Shan Street,
Yau Tong, Hong Kong

lwk&partners architects 梁黄顾设计

www.lwkp.com www.lwkyh.com.cn

梁黄顾建筑师（香港）事务所有限公司（以下简称"梁黄顾"）由梁鹏程先生与两位合伙人于1986年成立，总公司设于中国香港，在深圳、广州、上海、沈阳、成都等城市，以及澳大利亚等国家设有分公司和办事处。目前"梁黄顾"在香港和内地的设计团队达到300多人，拥有香港注册建筑师近30人，国内注册建筑师、工程师20多人，有多年建筑专业经验的设计人员200多人。

2009年，梁黄顾通过公司并购成立超过300人并具备甲级建筑设计资质的建筑设计公司"深圳市梁黄顾艺恒建筑设计有限公司"。

"梁黄顾"拥有20余年的建筑设计经验，为中国、韩国及中东等国家和地区的客户在城市设计、大型居住区规划、大型城市商业综合体、轨道交通与物业发展、住宅小区等各大领域提供综合建筑设计服务和景观设计服务。通过与国内外大型蓝筹房地产商合作，"梁黄顾"在过去20余年中，在中国的30个城市完成100余个总建筑面积超过2 000万平方米的项目，并多次荣获"詹天佑优秀住宅小区金奖"、"联合国全球人居环境最佳社区奖"、"中国建设部中国城市标志性楼盘奖"等大奖。

"设计创新"与"高品质服务"一直是"梁黄顾"的企业文化与核心价值。"业精于勤、行成于思"，"梁黄顾"将在全新的规模与平台上，成为更为国际化，更熟悉本土文化的建筑设计公司，并在城市设计、大型居住区规划、大型城市商业综合体、轨道交通与物业发展、住宅小区等各大领域，为客户提供规划、总平面、建筑方案设计、施工图设计全阶段的综合建筑设计服务。

In 1986, IWK & Partners Architects (HK) Limited ("IWK & Partners") was founded by Mr. Liang Pengcheng and another two partners, the head office of which is in Hong Kong, China, with branch companies and offices in Chinese Cities Such as Shenzhen, Guangzhou, Shanghai, Shenyang, Chengdu and Countries such as Australia. At present, the number of designers of IWK & Partners in Hong Kong and mainland has reached over 300, with almost 30 Registered Architects in Hong Kong and more than 20 Registered Architectural cngineers, as well as over 200 experienced designers.

In 2009, with over 300 designers, IWK & Partners established Shenzhen IWK & Partners Architectural Design Co., Ltd by merging, a company with Class-A Architectural Design Qualification.

With over 20 years of architectural design experience, in all main fields such as urban design, planning of large-scale residential area, large-scale urban commercial complex, rail transportation and property management development, residential community etc. "IWK & Partners" has provided integrated architectural design and landscape design services for customers of China, R.O. Korea and Middle East countries.

"Innovative design" and "high-quality service" have always been the culture and core value of "IWK & Partners". Under the spirit of "diligence creates excellence, thinking leads to success", with a brand-new scale and platform, "IWK & Partners" will become a more international, yet more localized foreign architectural design company, which provides integrated architectural design services of planning, general layout, project proposal design, construction drawing design for customers, in main fields of urban design, planning of large-scale residential area, large-scale urban commercial complex, rail transportation and property management development, residential community etc.

获奖项目

荣获"BCI亚洲2011年香港十大最杰出建筑设计公司奖"
荣获"2010第七届中国人居典范·金牌建筑设计机构"
荣获"2010第七届中国人居典范·最佳城市别墅设计方案金奖"（观湖园）
荣获"2010第七届中国人居典范·最佳建筑设计综合金奖"（重庆弹子石长嘉汇项目总体规划设计）
荣获"2010第七届中国人居典范·最佳建筑设计方案金奖"（海口天利三星级酒店和酒店式公寓及写字楼项目）
荣获"2010年中国十大城市豪宅"称号（中海城南1号）
"2010中国土木工程詹天佑奖住宅小区优秀建筑奖"（中海龙湾半岛）
"2010中国土木工程詹天佑奖住宅小区优秀建筑奖"（中海 银海一号）
"2010国际地产奖（亚太区）最佳建筑奖"（中海杭州钱塘山水）
"2010年中国土木工程学会詹天佑优秀住宅小区金奖"（中海杭州钱塘山水）
"2009年香港环保建筑协会白金级别大奖"
"2008年韩国Kimpo Gochon，Hillstate Villa，优秀设计大奖"
"2008迪拜棕榈岛44桥国际竞赛大奖"
"2007年韩国Yong-In Sang-Hyun，Hillstate Villa，优秀设计证书"
"2004年国际房地产大奖"（香港浅水湾道117号Grosvenor Place）
"2004年香港房屋署大奖"（香港石荫村5期）
"2003年香港建筑师协会银奖"（香港九龙升悦居）
"2009年中国土木工程学会詹天佑优秀住宅小区金奖"（苏州熙岸花园）
"2008年中国土木工程学会詹天佑优秀住宅小区金奖"（苏州湖滨1号）
"2008年中国土木工程学会詹天佑优秀住宅小区金奖"（苏州星湖国际）
"2008年联合国全球人居环境最佳社区奖"（苏州半岛华府）
"2005年中国建设部中国城市标志性楼盘奖"（苏州湖滨1号）

Awards and Projects

Won the award of "BCI Asia Year 2011 – Hong Kong Top 10 Excellent Architectural Design Companies"
Won the award of "Year 2010 The 7th China Human Habitat Example – Gold Prize Architectural Design Institution"
Won the award of "Year 2010 The 7th China Human Habitat Example – The Best Urban Villa Design Conception Gold Prize" for Greenlake Garden
Won the award of "Year 2010 The 7th China Human Habitat Example – The Best Architectural Design General Gold Prize" (General Planning Design of Dangzishi Changjiahui Project, Chongqing)
Won the award of "Year 2010 The 7th China Human Habitat Example – The Best Architectural Design Gold Prize" (Tianli 3-Star Hotel and Apartment Hotel and Office Building Project, Haikou)
2010 China Best Ten Luxury Apartments (Zhonghai-Chengnan No.1)
2010 Tien-yow Jeme Civil Engineering Prize For Residential Community (Zhonghai-Longwan Peninsula Community)
2010 Tien-yow Jeme Civil Engineering Prize For Residential Community (Zhonghai-Yinhai No.1 Community)
2010 International Property Awards (APAC) For Best Building (Zhonghai-Hangzhou Qiangtan Shan Shui Community
2010 Tien-yow Jeme Civil Engineering Gold Prize For Residential Community (Zhonghai-Hangzhou Qiantang Shan Shui Community
2009 BEAM Society Platinum-Class Awards
2009 Good Design Award Korea Kimpo Gochon, Hillstate Villa,
2008 International Competition Awards Dubai Palm Islands 44-span Bridge
2007 Good Design Certificate Korea Yong-In Sang-Hyun, Hillstate Villa
2004 International Property Awards (Hong Kong Repulse Bay No.117 Grosvenor Place)
2004 Hong Kong Housing Authority Awards (Hong Kong Shiyin Village Stage-V)
2003 Hong Kong Institute of Architects Silver Awards (Hong Kong Kowloon Sheng Yue Ju)
2009 Tien-yow Jeme Civil Engineering Gold Prize For Residential Community (Suzhou Xian Garden Community)
2008 Tien-yow Jeme Civil Engineering Gold Prize For Residential Community (Suzhou Hubin No.1 Community)
2008 Tien-yow Jeme Civil Engineering Gold Prize For Residential Community (Suzhou Xinghu International Community)
2008 U.N. International Awards For Livable Communities (Suzhou Peninsula Huafu Community)
2005 China Ministry of Construction City Landmark Building (Suzhou Hubin No.1)

梁黄顾建筑师（香港）事务所有限公司
www.lwkp.com

Add: 15/F, North Tower, World Finance Centre, Harbour City, Kowloon, HongKong
Tel: +852-25741633
Fax: +852-25724908

深圳市梁黄顾艺恒建筑设计有限公司
www.lwkp.com.cn

地址：广东省深圳市福田区深南大道4019号航天大厦21楼
邮编：518048
电话：+86-755-82031633
传真：+86-755-82031623

THE ONE

项目位置：香港 九龙
用地面积：3 125.6平方米
建筑面积：37 507.2平方米
设计时间：2010年

THE ONE是一座位于香港尖沙咀闹市中的大型综合购物中心，全幢大厦共29层，总高度171米，是亚洲最高的商场建筑之一。

其设计概念的重点是为该区提供一个全新的、悠闲的舒适的健康生活环境，并在有限的空间融入多种元素，糅合室内与室外、消闲、娱乐、休憩及购物于一体，塑造全新形式的建筑。以"形式紧随功能"为原则，设计师把建筑内部划分成数个区域，而每个区域的外部则由设计为不同形状的玻璃幕墙包围及区分，并达到不同的功能要求。设计师特意在16层设置了一整层空中花园，可以远离地面交通的废气，让游人能切身感受格外清新的自然环境。最高的5层楼预留做高级食府之用，提供各国美食，每层均能远眺整个维多利亚港的景色、香港岛的海岸线以及两岸时尚多变的建筑。香港美景配合国际美食，更加凸显出香港与国际接轨及其在国际上的地位。

Located in downtown area of Hong Kong Tsim Sha Tsui, the ONE is a large-scale shopping mall. As tall as 171 meters with 29 floors, it is one of the tallest buildings in Asia.

The design idea focuses on providing brand-new leisure and comfortable healthy living. By bringing different elements into the limited space and integrating indoor, outdoor, leisure, entertainment, recreation and shopping as a whole, the design creates a new-form of building, as its appearance shows. According to the design principle of "form follows function", the indoor space is divided into several areas by designers, while the outside of each area is surrounded and differentiated by distinctive glass walls for different functions. Designers specially design a whole-floor hanging garden at L16, from where, tourists can stay away from traffic exhaust gas and enjoy a particularly fresh natural environment. The highest five-storey are especially reserved for high-level restaurants. On every storey, tourists can overlook the complete view of Victoria Harbor, coastlines of Hong Kong Island, fashionable and various buildings on each side. Beautiful view of Hong Kong combining with international cuisine specially highlights Hong Kong's international standards and its position in the world.

成都国际金融中心

项目位置：四川 成都
用地面积：54 900平方米
建筑面积：763 718平方米

项目位于成都市红星路，红星路不仅是该市的主要街道之一，更是成都充满活力的城市中心。项目开发包含3个办公性质的塔楼和一个酒店，并由5层高的商业裙房加以连接。成都国际金融中心的设计理念是使其成为一种触媒，促进该区域周边的再生与重振。随着城市金融、文化和娱乐的新中心的出现，这个开发项目将给该地区带来新的商业机会。

Located on Hongxing Road, one of the main streets in Chengdu, China, the project is in Chengdu's vibrant city center. The development has three office towers and one hotel, all on top of a 5- storey commercial podium. The intention with the design of the Chengdu IFC is to be a catalyst through which the regeneration of the surrounding area will occur. As the city's new center for finance, culture and entertainment, this development will bring in new commercial business to the area.

太古汇

项目位置：广东 广州
用地面积：49 000平方米
建筑面积：446 000平方米

该地块位于中国广州天河区，项目位于天河路以南，天河东路以东，N1路以北，正在建设的地铁3号线石牌桥站以南，地铁1号线体育中心站300米以西。开发项目是两站之间的地段包括地下连接部分。 该地块用地面积为4.9万平方米，建筑面积44.6万平方米，其中地上建筑面积28万平方米。开发包括4个主要部分：① 40层办公楼的一号塔楼和28层高的办公楼的二号塔楼，分别位于基地东南部的专卖店商场与位于基地西南部的酒店B座；② 基地东北部的五星级酒店A座和一座在基地西北角的文化中心；③ 地下部分，其中包括一个3层地下停车场和4层上下交通区域；④ 位于3层和4层的绿化广场。设计旨在四角布置酒店、文化中心和办公部分，从而增加城市视觉廊道，同时消除每栋塔楼的视觉遮挡。

The site is located in Tianhequ, Guangzhou, on the northwest side of the intersection, bound by Tianhelu to the South, Tianhedong Road to the east and N1 Road to the north. The adjacent lot to the south is the Line 3 Railway Shipaiqiao Station under construction and 300 m to the west is the Line 1 Railway Tiyuzhongxin Station. The development is to connect the stations to the basement levels.The site area is 49,000 m^2 with a total floor area of 446,000 m^2 including a GFA above ground of 280,000 m^2. The development includes 4 major elements: 1) a 40-storey office tower 1 and a 28-storey office tower 2/boutique hotel B at the southeast and southwest corners respectively; 2) a 5-star hotel A at the northeast side and cultural centre at the northwest corner; 3) the below ground area including a 3-storey basement car park and 4-storey loading/unloading area; and 4) the green plaza at levels 3 and 4. The design intent is to locate the cultural centre, hotels and office towers at the four corners to increase visual access for the city, and at the same time to eliminate the visual barrier for each tower.

东大街时代豪庭

项目位置：四川 成都
用地面积：70 747平方米
建筑面积：560 000平方米

成都东大街综合开发项目占地70 747平方米，包括12个住宅楼，一座180米高的办公楼，一座五星级酒店，一座酒店式公寓塔楼和一个1 860平方米的购物中心。该项目基于其市中心的重要地位，设计上力求重新定义成都的城市中心，并实现与现有的城市肌理的有机结合。商业建筑和6层的居住建筑均靠近基地中央的绿化公园，进而创建一个私人的、安静的、内向的开放空间，其中布置了令人放松的功能空间，如咖啡馆。位于基地的西北角，可以直接俯瞰府河的办公楼被设计为城市地标性建筑。便捷的交通流向与开放的景观环境是该开发项目的特色。酒店和高级公寓位于主要道路的北边，这样使其拥有城市多层建筑群上空的开阔视野，更多的购物人流将通过新增的地铁站来到这个新的购物目的地。

The Chengdu Dongdajie Comprehensive Development occupies a 70,747 m^2 site and includes 12 residential towers, a 180 m tall office building, a five-star hotel, a serviced apartment tower and a 20,000 sqft shopping mall. The project, in recognition of its downtown prominence, is designed to redefine Chengdu's city centre as well as achieve integration with the existing urban fabric. The commercial building and six residential towers enclose a green park in the middle of the bigger lot, creating a private, quiet and hip open space in which to relax and perhaps stop for a coffee. Designed as a landmark, the office tower addresses the prominent northwest corner of the site and commands views of the Fu River directly below. Efficient access and open views are characteristics of the development. The hotel and serviced apartment building address the main road to the north, where they are oriented to capture sweeping views over the mid-rise rooftops of the city. Circulation of pedestrians to this new shopping destination is enhanced by an underground connection to the MTR.

重庆弹子石

项目位置：重庆
用地面积：359 504平方米
建筑面积：1 826 589平方米

项目位于重庆南岸区，地处长江东岸，毗邻长江、嘉陵江两江交汇处。项目用地属于中央商务区的核心配套区，设计师根据“现代城市、国际社区、人文空间、生态环境”的发展理念， 以长江景观资源为依托，充分利用沿江面，使得更多的住宅能看到江景。住宅布局以点式住宅为主，打开空间，减少对江景的遮挡，形成通廊。以“老街、滨江商业、滨江广场”为发展主轴，打造城市名片，服务城市，住宅临靠老街及滨江商业布置，从而提升住宅的开发价值及自身品质。

人们可以通过项目的两条步行线路，从城市周边步行到江滨游憩。在高差较大的地方会布置自动扶梯、楼梯，同时也设置有风雨连廊，为人们遮风挡雨。

项目内的绿化有城市绿地，能在此眺望长江与滨江公园。三种不同形态的绿化构成了整个住区点、线、面的立体绿化体系。项目还可以享用两江景观，从沿江面看，楼栋起伏有致、主次分明，而沿江面楼层高度的变化呈现波浪形的高低起伏态势，蜿蜒有致，正是江城意趣。从南滨路往弹子石路纵深，建筑高度由“18层到32层再到48层”由低往高地变化。低的灵秀，高的挺拔，自低往高，恰似山城威势。

This project is located at the south of Chongqing, along the east shore of Yangtze River, adjacent to the conjunction of Jialing Jiang. The development is at the core of the CBD, incorporating urban living, international community, human and eco living into a master planning of the area. All apartments are designed in custard along the river to maximize viewing of the famous scenery and creating this impeccable scenic picture as part of the décor of the area. Neighboring to the Bin Jiang commercial area and old streets, the development stands out as an integrated self-contained residence and of greater value.

Pedestrians can access Bin Jiang by walking along 2 covered corridors, facilitated by with escalators, stairs for added convenience.

Greenery and buildings in the area are sophisticatedly planned to capture views from Yangtze River and Bin Jiang Park. When viewing along the shores, there are layers of greenery, smooth curve formed by buildings in different heights, echoing with the long and flowing Yangtze River and its hilly terrain.

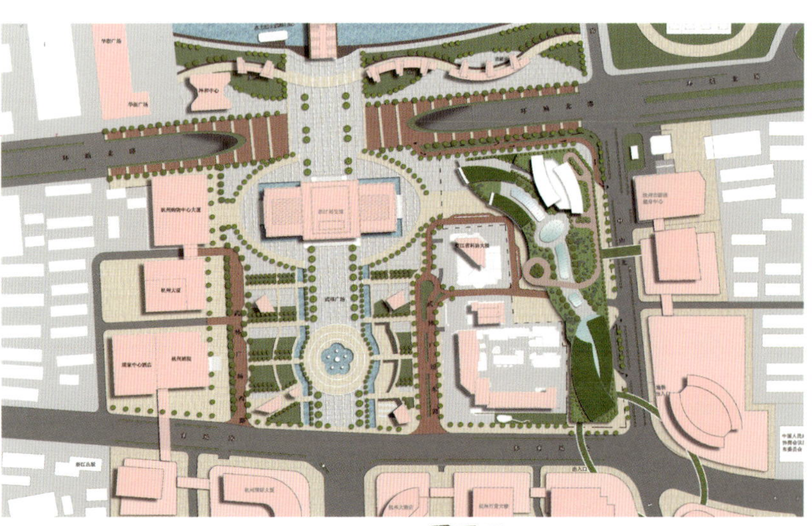

杭州武林广场

项目位置：浙江 杭州
用地面积：25 000平方米
建筑面积：125 000平方米

武林广场位于杭州中央商务区，与杭州地铁1号线直接连接。武林广场是一个混合功能的商业中心，它的明显标志是建立在7层基座上的双塔，它还包括4层的地下层。其用地面积约2.5万平方米，建筑面积12.5万平方米。设计思想是将香港岛中环中心城区的“都市绿洲”概念引入到杭州的中央商务区。

Wulin Plaza is located in the Central Business District of Hangzhou with a direct connection to the Hangzhou Metro Line 1. It is a mixed-use commercial centre with the signature double towers above a 7-storey podium, which includes 4 levels of basement floors. The site area is approximately 25,000 m², with a total GFA of 125,000 m². The design concept is to bring the idea of the Central downtown area of Hong Kong Island, an "Urban Oasis", to the Hangzhou CBD.

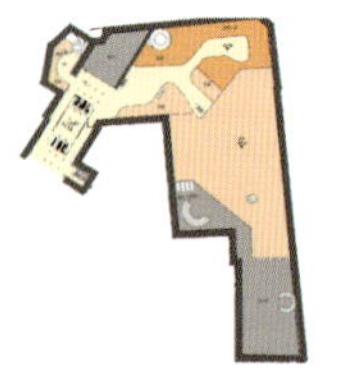

地下二层-12.00

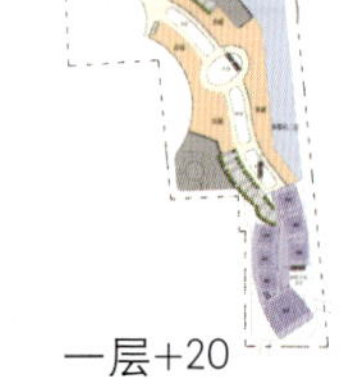

地下一层-5.00

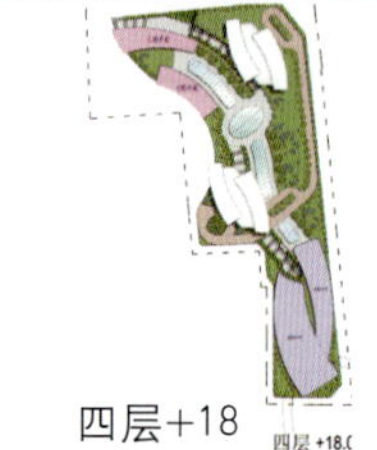

一层+20

四层+18

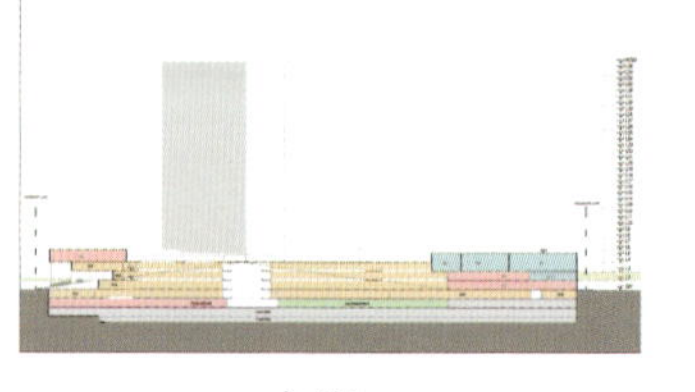

立面

天津金茂广场

项目地点：天津
用地面积：25 000平方米
建筑面积：73 424平方米

该项目包括数幢住宅塔楼、1座酒店式公寓和独立的5层高的商业综合体。建筑形体大胆采用了变形的三角形体量，以突出其重要的城市中心地位。

This project includes residential towers, one serviced apartment block and a separate 5-storey commercial complex. The prominence of its location at the city centre calls for a more daring architectural approach manifested by the variation of triangular geometry.

武汉时代广场

项目位置：湖北 武汉
用地面积：17 474平方米
建筑面积：184 541平方米

该项目是一项综合开发项目计划，包括4个住宅塔楼、一座酒店式公寓和一家五星级酒店。武汉时代广场，旨在通过开发成为周边地区的视觉焦点，它可能没有纽约的洛克菲勒中心那样的显赫地位，但该项目巨大的体量反映的肯定是渴望成为一个城市中心地段标志性的建筑群。建筑群的高度从28层到56层，4个住宅塔楼直指天空，它们在形式上追随鼎盛时期的芝加哥学派的建筑风格。其中较高的两个塔楼间间隔的连接体，形成两个建筑物之间的空间与视觉上的联系。前景两个较矮的塔楼成为整个开发项目的重点，入口处采用穹顶提示主入口，同时也是有着戏剧性变化的不同建筑语言立面构成的衔接过渡。4层高的商业裙房灵感来自文艺复兴时期的拱廊的启发：每侧入口两旁采用3层连续拱圈，裙房的灯光在夜里强调了水平线条效果的同时，形成了富有韵律感的雕塑效果。

This project is a comprehensive development which includes four residential towers, a serviced apartment building and a five-star hotel. Wuhan Times Square is designed to capture attention from both near and far. It may not have the footprint of New York's Rockefeller Center, but in terms of massing the project certainly aspires towards the same iconic status on a prominent city centre site. Ranging in height from 28 to 56 storeys, the four residential towers soar towards the sky, their subtly articulated forms harking back to the heydays of the Chicago school. The two taller towers are linked at intervals, creating physical as well as visual connections between the two buildings. In the foreground, the two shorter towers flank the development's focal point and a dome marks the grand entrance as well as a transition to a dramatically different language and articulation of the facade. The four-storey commercial podium draws its inspiration from the arcades of the Renaissance, with three rows of flat arches spreading out on either side of the entrance. Uplighting emphasizes the podium's horizontality and orchestrates a sculptural effect at night.

沈阳金港大厦

项目位置：辽宁 沈阳
用地面积：7 105平方米
建筑面积：125 635平方米

本建设项目地块位于沈阳中心地区，毗邻太原街市级商圈，项目周边有众多大规模的百货公司、购物广场，具有100多年的历史为浓厚，也是沈阳人逛街必去之处，地块的交通住优越，通达性好，为商业气氛奠定了基础。项目周边有数间星级酒店，休闲娱乐设施发达，如剧场、游泳馆等，吸纳大量全国各地的业主及外籍人士入住，为太原街商圈及本项目提供更多的消费人群。项目开发拟建设成一流的甲级办公楼与产权酒店公寓，提升太原街商圈的商务档次，构建一个高档次、时尚、繁华的新都市空间，并成为该区域的标志性建筑。

This hotel project, consisting Grade A offices and hotel apartments, is strategically situated in the heart of Shenyang, neighboring Taiyuanjie commercial district, famous for its numerous big-scaled department stores and shopping malls which have over 100 years of history of presence in the region. Infrastructure inside this district is well developed and highly accessible which makes it a definite choice for business development. Besides other international hotels, presence, the area is facilitated with leisure and recreational amenities like opera house and swimming complex. Domestic and international tourists gather and spend in the area. With the completion of this project, it will upgrade the area to a high-end, trendy and prosperous era and become the icon of the district.

沈阳青年大街(金廊)综合商住项目

项目位置：辽宁 沈阳
用地面积：532 000平方米
建筑面积：2 539 318平方米

沈阳市青年大街(金廊)综合商住项目，地处沈阳市中心地带。项目不但位于市内两条地下铁路的交汇点，更是沈阳市的商业主轴“金廊”的重心位置所在。本项目含住宅区、高密度及多功能商业区两大部分。多功能商业区设有酒店、办公大楼、高层公寓及大型商场，主要沿沈阳市的商业主轴“金廊”两旁设置。各大楼外形极富时代感，还与金廊一带的商业悠闲的气氛融合。

蜿蜒的半开放式商业长廊连接着大型商场，把东西方向的人流引入项目内的金廊轴线。另外商业区内以不同主题的特色广场联系，把人流导向于项目的各个部分。最后，两个住宅区设置在项目的东西两面，以绿化及公建做缓冲。配合中央园林水景，使住宅区在享有交通与购物的方便之余，也不失宁静悠闲的居住环境。

Shenyang Youth Street (Golden Corridor) integrated commercial and residential project, located in the heart of Shenyang city. The project is not only located in the city's intersection of two underground railways, it is a commercial spindle center of Shenyang "Golden Corridor". The project contains the two parts, residential area, high-density and multi-functional commercial area. There are hotels, office buildings, high-rise apartments and shopping malls in the multi-functional business district, mainly set along the axis of Shenyang's business "Golden Corridor". The appearance of the buildings is very contemporary, integrated with the commercial leisurely atmosphere of Golden Corridor.

Winding and the semi-opened commercial corridor connects large shopping malls, leads the east and west direction of the crowds to the axis of the Golden Corridor. Different theme characteristics of square links in the downtown, the crowds are directed to the various parts of the project. Finally, two residential areas set on both sides of the project, the green and public buildings to make the buffer. Central garden with water features, enjoy the traffic in the residential area with shopping convenience, but also an quiet yet relaxed living environment.

法博国际（香港）规划建筑设计有限公司

FABER INT'L (HK) LAYOUT CONSTRUCTION DESIGN LIMITED

法博国际（香港）规划建筑设计有限公司是一家国际化企业，从事规划、建筑、景观及工程咨询的专业方案设计公司。2002年进入国内市场发展，成为一支异军突起的生力军，公司发展迅速，成果丰硕，在全国建立客户网络及良好口碑，并且成为了湖南城市学院研究生实习基地。由主创设计师王军民领衔，致力于将国际先进的设计理念与本土优秀的传统文化相结合，科技与环保相结合，技术与关怀相结合，以“创意无限、造就经典”为宗旨，创造高效、宜人、经济、可持续发展的个性化作品。公司坚持研究精品化的实践方向，不断创新，充分利用香港与内地相通的文化理念以及创意资源优势，使客户的要求可以从我们的专业设计和优质服务中得到充分满足。公司旗下聚集来自国外及各全国各地的优秀设计师，对于传统文化及地域特征具有深入了解，同时具备丰富的工作经验和高度的敬业精神，使项目设计一贯保持优良水平，设计技术精良，服务周到细致，为开发商提供建筑、园林景观设计、设计咨询、项目策划等全方位服务。

Faber Int'l (HK) Layout Construction Design Union Company Limited is an international professional-scheme design company engaged in planning, architecture, landscape and engineering consulting. Since 2002, it entered domestic market and developed into a new force. With rapid development and fruitful achievements, the company has built customer network and good reputation, and has become training base of Hunan City College for postgraduates. Led by Chief Architect – Wang Junmin, the company is dedicated to combine international advanced design ideas and outstanding local traditional culture, technology and environmental protection, technology and concern. With the purpose of "Infinite Creation, Creating Classic", the company creates efficient, pleasant, economic and sustainable development individualized works; it adheres to the practice direction of researching competitive products, continuous innovation and taking full advantage of similarities of cultural ideas and creative resources between Hong Kong and mainland, so that customers' requirements can be fully met from our professional design and quality service. Outstanding foreign designers and designers from all over China join the company. They have in-depth understandings of traditional culture and geographical features. With excellent design skills and considerate services, they provide full-service of architecture, landscape design, design consulting and project planning for developers.

地址（香港）：香港九龙尖沙咀漆咸道南67-71安年大厦1104室
电话：+852–36583890
邮箱：wjm1413@163.com
网址：www.hkfabo.com

地址（深圳）：广东省深圳市福田区八卦四路中浩大厦18楼G/F
电话：+86–755–25927276
传真：+86–755–25923916
网址：wjm1413@163.com

Address (HK): Room 1104 Oriental Centre, Chatham Rd South No. 67-71, Tsim Sha Tsui, Kowloon, Hong Kong
Tel: +852–36583890
Email: wjm1413@163.com
Website: www.hkfabo.com

Address (Shenzhen): G/F 18F Zhonghao Building, Bagua Fourth Road, Futian District, Shenzhen, Guangdong
Tel: +86–755–25927276
Fax: +86–755–25923916
Website: www.hkfabo.com

桂林市临桂新区 · 创业大厦

安墩温泉小镇概念规划方案

项目位于惠州市惠东县安墩镇，地理位置优越，交通方便。

经典开发模式定位：结合用地现状与安墩社会经济条件，通过温泉与其他产业的嫁接以达成1＋1>2的效果，把安墩温泉小镇开发模式定位为温泉加运动游乐加旅游地产。通过对用地的现场踏勘调研、现状地势与水系的分析以及对项目的战略定位，将地块的功能区划从南向北依次布局为：温泉度假区—运动休闲区—旅游地产区。

温泉度假区：围绕湖面、水系支流、温泉泉眼的开发利用打造三个品牌文化，即多元化的温泉泡浴文化、独具特色的温泉饮食文化、新潮时尚的居住文化。

运动休闲区：包括高尔夫休闲区、山地赛车区。旅游地产区：依托周边丰富的旅游资源开发，融旅游、休闲、度假、居住为一体的置业项目。

桂林市临桂新区·创业大厦及市民广场设计方案

本案基地位于临桂新区中心区，东临新中路，西接西城大道，南临万福路，北为规划道路。总用地面积约19公顷。

设计理念及构思：新旧文化的交融、“天圆地方”的经典。

功能结构：“一轴”为南北向延伸贯穿市民广场、创业大厦的景观核心轴线。

“二区”是指山水大道以北，以创业大厦为主体的商务办公区和南面的市民广场休闲活动区、集会区。商务办公主要功能为商务办公、会议接待以及后勤服务。市民广场布置艺术馆、会展中心、科技馆、文化宫等大型公建，以丰富市民的精神文化生活。

“三节点”是指市民广场南面大气磅礴的主入口广场节点，衔接南北二区的交通广场节点，以及轴线延伸与北面山体形成的背景节点。

福龙湾规划建筑方案

本案基地位于湖南省衡阳市雁峰区，东临鄱白路，西接白沙大道，北接幸福河，南临李家塘路，白沙大道为双向四车道，是衡阳市雁峰区最重要的道路之一。项目交通便利，生活配套设施完善，生活氛围极为浓厚，同时区位条件优越。项目规划用地面积95 503平方米，东西长约580米，南北宽约30米，用地形状不规则，地块地势整体较为平整，内部有少数低洼水塘，规划总建筑面积约321 930平方米。

“龙踞福湾”——根据现状地形，将居住组团设计成一个有机整体，使其具有连续性。地形呈不规则形状，犹如一条盘龙盘踞在此，“龙”象征团结整合、利泽天下、天人和谐的精神。因此本次规划设计不论是从形式上，还是从功能上均按照功能整合的原则进行设计。

海南中部家居建材市场规划方案

海南中部家居建材市场位于屯昌县东北片区内，北邻兴业路，西至昌盛北路，东至环东路，南接北干路。规划总用地面积95 171.5平方米。呈基本规则的长方形，地势平坦。

总建筑面积120 852.76平方米，一期总建筑面积58 132.2平方米，分五大区：洁具城、家电城、陶瓷城、油漆城、铝材城。现A栋、E栋、M2、M3、M4、M15、M16、M7、M8建筑主体已完工，B栋、C栋、D栋、M1、M5、M6建筑正在施工。二期总建筑面积62 720.56平方米，为新规划建筑，包括M9、M10、M11、M12、M13、M14、M17、M18。市场分四大区：家具城、灯饰城、木地板城、五金机电城。市场内商品品种齐全，规划贯穿“一站式”购物理念。

衡阳步行街规划设计

本项目位于衡阳市中心城区，南临解放大道、北靠常胜西路、东接蒸湘北路、西为静园路。其中，解放大道、蒸湘北路为衡阳市交通干道，常胜西路为次干道。地理位置十分优越。

规划总用地面积11.68万平方米。地块南北长约550米，东西宽约250米，呈不规则形状。

该项目可建成建筑面积约52万平方米，并以住宅开发为辅、商业开发为主的开发模式。

惠州好益康休闲国际会议中心

本项目基地位于惠州市惠阳区以北，距离惠州市区约27.5千米，距离惠阳区政府约8千米，距离深圳市约25千米，地理位置优越，淡水河从基地外围绕行经过，惠澳铁路从基地东北侧经过，西侧为120米宽的惠南大道，总用地面积约82公顷，可建成建筑面积约100万平方米。基地用地构成主要为土壤条件较好的平原，内部有若干水塘，东侧地形有一定高差。基地地块依山傍水，风景优美，是打造高尚国际会议中心与高档休闲中心的极佳地点。我方由2009年开始设计，现方案已经政府部门通过，场地正在平整中。

长沙市芙蓉区马王堆小学

长沙市芙蓉区马王堆小学位于长沙市古汉路88号，是一所全日制小学，于2010年9月正式落成开学，办学规模24个班。学校占地面积15 312平方米，校舍面积8 564平方米，绿化面积6 490平方米。学校拥有1个60米直跑道，1个200米环形跑道田径场，2个标准化篮球场和1个绿茵足球场。校园内设计了林荫道，大小花坛分布在道路场馆周围，绿树花草掩映其中，6个特色宣传橱窗伫立在校园醒目处，整个校园宛如幽雅宁静的公园。

学校秉承“骏驰汉道，硕通经纬”校园文化设计理念，以厚重的汉文化为立足点，结合现代学校需要，营造了一个温馨、人文、童趣、崇德、富有底蕴的育人场所。

钦州市北美国际商贸城规划设计

该项目位于钦州市区东南部，市行政中心与保税港区的中心地带，金海湾东大街南面，扬帆南大道两侧。

规划设计构思：通过分析周边已有城市资源，项目的实施体现了“以人为本”的原则，实现人车分流。

规划结构：实现院落—组团—小区的空间关系变化。

道路交通系统：小区车行道与城市道路形成环路，不直接进入院落，把机动车对住户的干扰降到最低值。

景观绿化系统：本方案采用了组团、中国院落式景观布局方式。

公共服务设施：公共服务设施大部分设置在用地中部沿南面城市道路一侧，这样既能保证小区住户自身的需要，又能服务社会。

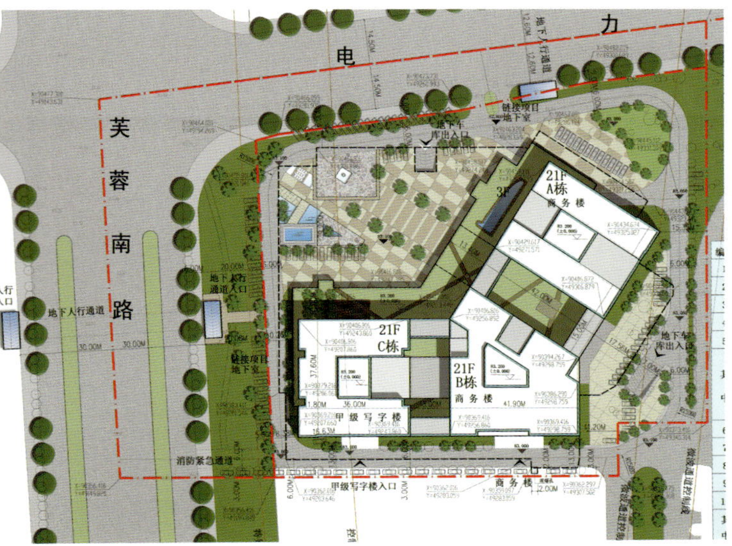

心星国际商务中心建筑设计方案

本项目位于长沙市天心生态新城中心区域，北靠电力路，西临芙蓉南路，地处天心区区治大院西北角，紧靠湖南省政府机关大院。

随着长、株、潭三市融城步伐的加快。本区域将是长沙市乃至湖南省政治文化及经济中心。

湖南心星国际商务中心项目规划总用地面积25 381平方米，是集两栋商务楼和一栋甲级写字楼于一体的现代化商务中心。拟建地上总面积约91 388万平方米。

合作单位（深化设计）：湖南省建筑科学研究院。

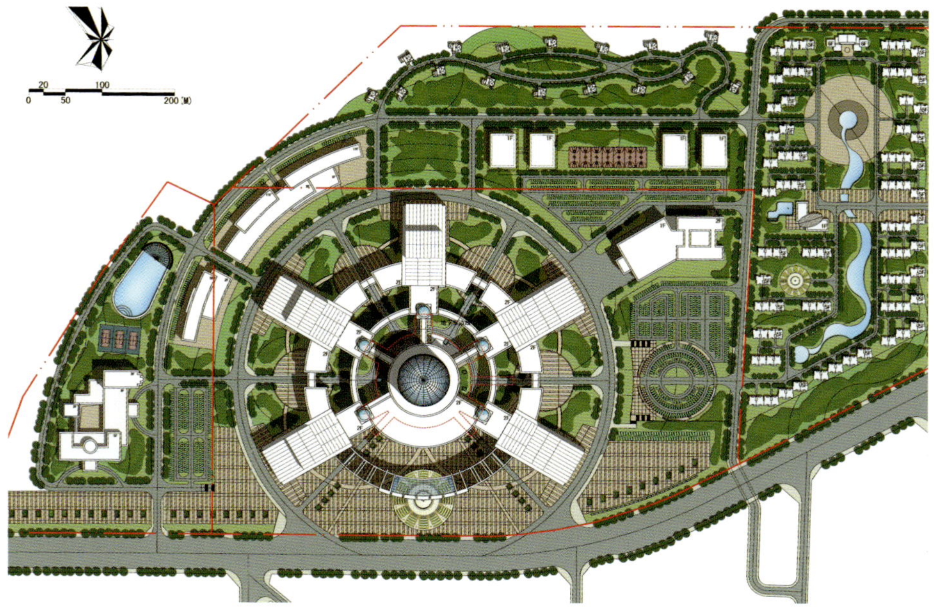
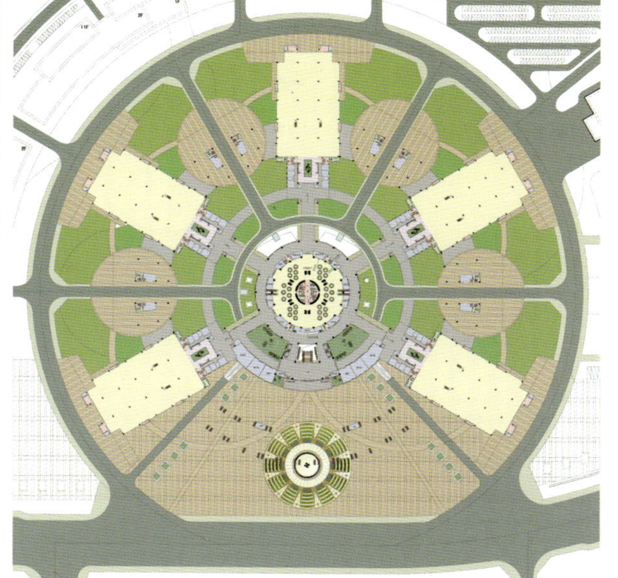
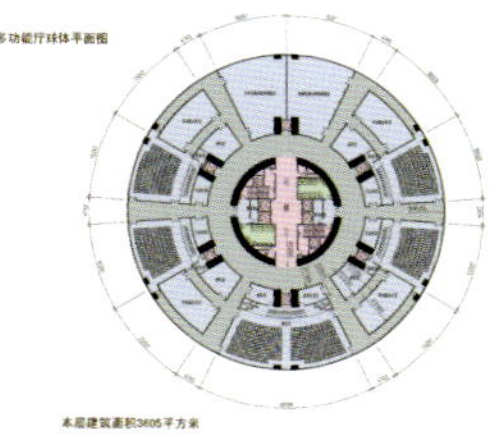
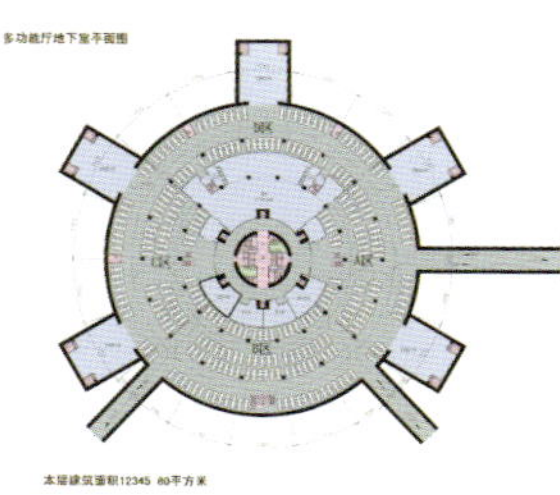

新疆国际会展中心

新疆国际会展中心在设计中，整个建筑由内到外呈现出高低错落和几何化的秩序感，设计上主要利用这种高差变化和不断咬合的结构关系，营造出一种节奏感，使建筑的主体部分更加突出，视觉效果更加强烈，对与建筑自身的展示功能也起到了很好的烘托作用。在细部处理上，我们引入了很多有着浓厚地域文化特色的建筑语言，强调了新疆的地域文化特征，整个建筑将现代风格与新疆的地域文化完美地融合在一起，使整个会展中心浑然一体，建筑形象简洁且明快。

仰天湖

规划总用地面积：338.57公顷
仰天湖水面面积：103.60公顷
陆地面积：234.97公顷
道路面积：44.43公顷
规划总建筑面积：200.14万平方米
建筑容积率：0.85（不含水）
绿地率：61.5%（不含水）
规划停车位：6737个

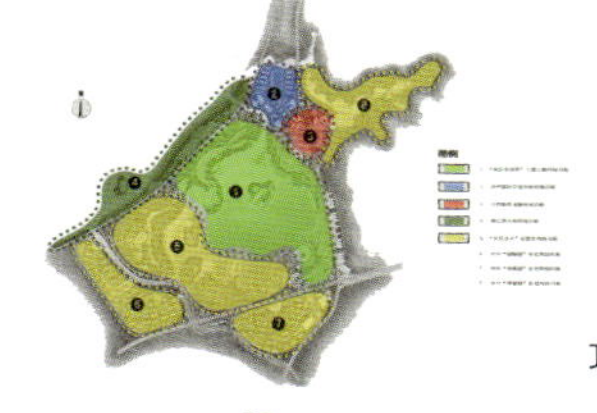
功能布局图

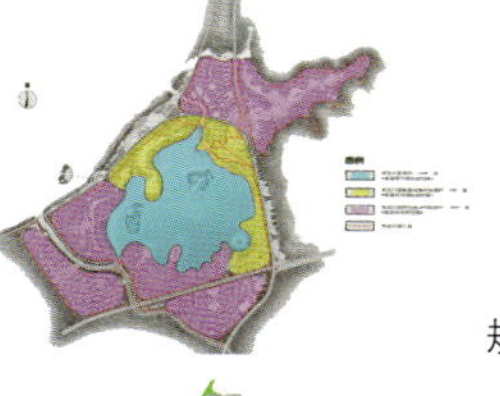
规划范围分析图

景观分析图

602A, Innocentre
72 Tat Chee Avenue
Kowloon Tong
Hong Kong

Tel +852–2863 0800
Fax +852–2528 2226

www.ida-hk.com
info@ida-hk.com

Principal
Winston Shu

Directors
Jochen Tombers
Ana Shu
Dick Mak

Associate Director
Pusey Chan

Associate
Ed Peter
Albert Lau

Beijing office
Pusey Chan

Tokyo office
Yasuyori Yada

India office
Ritesh Agrawal

Integrated Design Associates Ltd. (IDA) aims to provide a high calibre research, design and planning consultancy service for clients and projects of wide ranging levels of complexity and challenges in Hong Kong China, mainland China and the world. By combining original, innovative thinking and design excellence our goal is to add qualities and values to all the projects we undertake. Our philosophy is always clear and unequivocal–our efforts must make a difference for all our clients, irrespective of size and scale of the project.

Our approach to design takes an integrated consideration of sustainability, environmentally as well as functionally, cost effectiveness, energy efficiency, spatial awareness, user well-being and architectural aesthetics. We believe good design must address the appropriate use of technologies, engineering, humanity and ergonomics. We adopt a rational design process, which is often analytical and typically highly focused. Our in-depth technical know-how, design expertise and understanding of the construction process have produced many successful projects which are extremely advance in engineering, acutely budget driven and being carried out in very tight time constraints.

综汇建筑顾问有限公司旨在为中国香港、中国内地及国际客户的各种项目提供高水平的设计及策划咨询，将创新独到的理念融入一流设计之中，从而提高每个项目的质量及价值。我们的理念十分明确：无论项目规模大小，均竭尽所能使客户能从中获益。

我们的设计综合考虑环境及功能的可持续性、成本效益、能源节约、空间观感、人本因素及建筑美术等要素，相信出色的设计必须与科技、工程、人文和功效学互相配合。设计中采取理性的设计流程，高度集中分析，加上丰富的技术知识、专业的设计意见和对建筑流程的充分了解，成功完成了多个项目，无论在工程质量、资金管理和时间掌握等方面都表现出色。

1. Parkview Green, Beijing, China _ 1, 2, 3
2. One LaSalle, Hong Kong, China _ 4, 5, 6
3. Hanimaadhoo International Airport, Maldives _ 7, 8
4. Super Tower, Beijing, China _ 9, 10, 11
5. Cameron, 33 Cameron Road, Hong Kong, China _ 12
6. Hyderabad International Airport, India _ 13, 14, 15

1. 中国北京侨福芳草地：1，2，3
2. 中国香港喇沙利道1号：4，5，6
3. 马尔代夫Hanimaadhoo国际机场：7，8
4. 中国北京商业中心超高层大楼： 9，10，11
5. 中国香港金马伦道33号办公大楼：12
6. 印度海得拉巴国际机场：13，14，15

1

2

3

4

5

6

7

9

10

8

11

12

13

14

15

16

7. Male International Airport, Maldives _ 16, 17, 18
8. ESF Private Independent Schools, Hong Kong, China _ 19, 20, 21
9. New Delhi General Aviation Terminal, India _ 22, 23
10. Nortel Wangjing Campus, Beijing, China _ 24, 25, 26
11. ISA Technology Park, China _ 27
12. Ho Man Tin MTR Station, Hong Kong, China _ 28, 29
13. Jinan International Airport, China _ 30
14. Heliport, Hong Kong, China _ 31
15. Residential Projects _ 32, 33, 34

7. 马尔代夫马里国际机场：16，17，18
8. 中国香港愉景湾英基学校协会小学及中学：19，20，21
9. 印度新德里商务客机中心：22，23
10. 中国望京科技园北电中国总部及研发中心：24，25，26
11. 中国天津ISA科技园：27
12. 中国香港港铁何文田站：28，29
13. 中国济南国际机场航站楼：30
14. 中国香港直升机航务中心及机坪：31
15. 住宅项目：32，33，34

17

18

19

20

21

22

23

24

25

26

28

27

29

30

31

32

33

34

香港华天国际建筑与城市设计有限公司
HongKong Witen International Limited

香港华天国际建筑与城市设计有限公司(Hong Kong Witen International LTD.)成立于2000年6月2日，依靠的是精益求精和富有团队合作精神的专业人员，推崇质量，追求卓越。为客户提供城市规划设计、建筑工程设计、景观设计和房地产发展顾问等服务。

公司于2008年7月被香港国际名牌理事会授予常务理事会员单位。

香港华天国际建筑与城市设计有限公司于2010年被授予〝中国城市规划与建筑设计行业最佳优秀设计机构〞称号。

Hong Kong Witen International Ltd. was established in June 2nd, 2000, relying on the excellent professionals full of team spirit, respecting quality, and pursuing excellence. The company provides urban planning and design, construction design, landscape design and real estate development consulting services for clients.

The company was awarded Executive Director of the Council of Hong Kong International brands member in July 2008.

"Time Architecture," the Council of the unit

In 2010, it was awarded title of the Best Excellent Design Agency in China Urban Planning and Architectural Design Industry.

公司地址：香港九龙弥敦道625号雅兰中心二期15楼1508室
电话：+852–67662969 / 30605049
传真：+852–30626606

深圳公司：广东省深圳市南山区艺园路133号田厦IC产业园3014室
电话：+86–755–26470085
传真：+86–755–86604392
邮箱：witen999@163.com

南宁公司：广西壮族自治区南宁市星湖路南二里6号4楼
电话：+86–711–5337586
传真：+86–711–5336675

烟台公司：山东省烟台市开发区万寿山路5号宝威科研楼667室
电话：15853592277 13828809897

Add: Room 1508, Floor 15, Phase II of Grand Tower, No. 625 Nathan Road, Kowloon, Hong Kong
Tel: +852–67662969 / 30605049
Fax: +852–30626606

Shenzhen Company: Room 3014 of IC Industry Park of Tian Sha, Yiyuan Road, Nanshan District, Shenzhen, Guangdong
Tel: +86–755–26470085
Fax: +86–755–86604392
E-mai: witen999@163.com

Nanning Company: Floor 4, No.6 of Nanerli, Xinghu Road, Nanning, Guangxi
Tel: +86–711–5337586
Fax: +86–711–5336675

Yantai Company: Room 677 of Baowei Research Building, No.5 of Yantai Development Zone, Shandong
Tel: 15853592277 13828809897

1

2

3

4

5

8

10

6

7

9

1 福临家园

地址：吉林 长春
面积：30万平方米

2–4 鹭园会所

地址：贵州 遵义
面积：4.5万平方米

5–7 温馨家园

地址：江西 赣州
面积：19.6万平方米

8–10 豪德翠山豪庭

地址：广东 深圳
面积：23万平方米

11 黄河文化水公园

地址：甘肃 兰州
面积：6.69公顷

11

12

13

14

12 六枝特区人民医院门诊急诊楼

地址：贵州 六盘水
面积：3.6万平方米

13–14 锦绣世纪花园

地址：湖南 长沙
面积：13.5万平方米

15–17 南昌十九中学

地址：江西 南昌
面积：13.1万平方米

18–20 祁阳鸿运大市场

地址：湖南 永州
面积：16万平方米

15

16

17

18

19

20

21

22

24

23

25

26

27

21–23 容桂会议中心

地址：广东 佛山
面积：3.6万平方米

24–25 容桂文化中心

地址：广东 佛山
面积：4.2万平方米

26–28 台湾群创创意产业区

地址：福建 龙岩
面积：50万平方米，一期18万平方米

29 前北流鸣玉泉规划

地址：山东 莱州
面积：68万平方米

30 坦州区人民政府

地址：广东 中山
面积：2.9万平方米

28

29

30

31

32

33

34

31–33 香格里拉古城旺角商业中心

地址：云南 迪庆藏族自治州
面积：28万平方米

34 银海迎宾馆

地址：广东 东莞
面积：5.4万平方米

35–36 星辉大厦

地址：山东 威海
面积：3.2万平方米

35

36

主要经济技术指标

总用地面积		341068.37M²（511.6亩）
总建筑面积		252854.51M²
其中	计容积率部分	199199.81M²
	不计容积率部分	53654.70M²
容积率		0.584
建筑密度		25.0%
绿地率		42%
停车位		1007 辆
其中	室内	643 辆
	室外	364 辆
总户数		793 户
其中	独栋别墅	1 户
	双拼别墅	172 户
	联拼别墅	296 户
	叠拼别墅	324户

37

38

39

40

41

42

37–41 玉龙山庄

地址：辽宁 阜新
面积：26万平方

42 宜春国宾馆

地址：江西 宜春
面积：3.87万平方米

43 月亮湾广场

地址：山东 蓬莱
面积：40万平方米

43

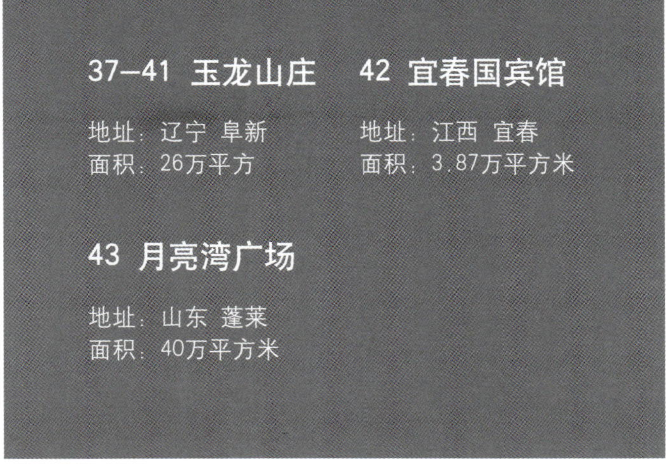

FRANKIE LUI STUDIO

吕达文设计所

www.frankielui.com　　E: mail@frankielui.com

吕达文设计所为香港汇创国际建筑设计有限公司的核心部分，皆于2007年成立，由Frankie Lui 吕达文先生和 Anita Lo 卢可欣女士创办，分别在香港与深圳设有办事处。

作为一所从事多专业的建筑设计事务所，我们追求卓越品质，力求创新。通过现代化设计去提升生活空间品质。我们的设计范围广泛，服务内容包括室内设计、建筑及城市规划，业务遍布香港及国内重点城市。设计所拥有丰富项目经验，尤其在商业、酒店、文化艺术、住宅等领域表现出色，被国内外业界高度认可。

2003年 Frankie Lui 吕达文先生获“香港建筑师学会年青建筑师奖”，两位创办人均收录在2005年香港建筑师学会出版的《2021》香港新进建筑师中。

2009年— “香港尖沙咀宝勒巷10号与深圳华侨城欢乐海岸生态展厅”获选为香港建筑师学会年奖《HKIA 2009》“商业组别”入选作品。

2010年— “香港京士顿国际学校”在第二届中国建筑传媒奖被选为最佳建筑奖入选作品。以及香港建筑师学会年奖《HKIA 2010》的入选作品。

Global Atelier Limited, established in 2007 by Frankie Lui and Anita Lo together with the core team - Frankie Lui Studio, works on a wide spectrum of projects in Hong Kong and other large cities in China. Global Atelier has offices in Hong Kong and Shenzhen providing interior design, architectural design and urban planning consultancy.

Our aim is the creation of possibility by quality design through contemporary approach. We have received high recognitions within professional industry and in commercial, hotel, cultural art and residential sectors.

The chief designer - Frankie Lui received Hong Kong Young Architect Award from Hong Kong Institute of Architects (HKIA) at 2003 and both founders were selected as one of the Emerging Young Architects by HKIA at 2005.

<2009> “10 Prat Avenue and OCT Happy Coast Pavilion” have been selected as shortlist projects for *HKIA 2009* Best Architectural Award - Commercial Group.

<2010> “Kingston International School” has been selected as a Shortlist Project for Best Architectural Award in CAMA China Architecture Media Awards 2010 and *HKIA 2010* Best Architectural Award.

吕达文 | 主创设计师
Frankie Lui | Design Director

美国纽约哥伦比亚大学建筑及城市设计硕士
美国绿色建筑协会LEED专业人员
中国香港注册建筑师
中国香港大学建筑硕士（优等荣誉）
中国香港大学建筑学士（一级荣誉）

M.Sc.(Arch & Urban Design)(Columbia U)
LEED AP
HKIA, HK Registered Architect
M.Arch.(Distinction)(HKU)
B.A.A.S(First Hon.)

GLOBAL ATELIER LIMITED
香港汇创国际建筑设计有限公司

Rm, 201-202, Carnival Comm. Bldg.
18 Java Rd. North Point, H.K.
Tel: +852-3106 3128
Fax: +852-2537 2811

9B, Shuisong Bldg, Terra 8th Rd.
Chegongmiao,Shenzhen
Tel: +86-755-3305 1505
Fax: +86-755-3305 1506

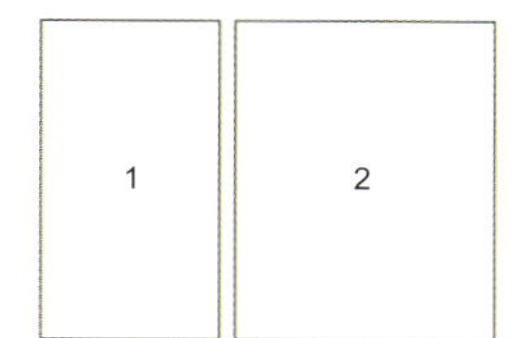

1. 香港尖沙咀宝勒巷10号垂直餐饮
Hong Kong “10 PRAT” Vertical Restaurant Building
2. 武汉南国中心城市综合体项目
Wuhan Lan Gold Commercial Complex
3. 深圳华侨城欢乐海岸生态展厅
Shenzhen OCT Happy Coast Exhibition Pavilion
4. 南京中航科技城——科技艺术中心
Nanjing AVIC Cultural & Technology Centre
5. 沈阳万象城商场室内设计优化
Shenyang “The Mixc” Interior Shopping Mall Enhancement Design
6. 沈阳万象城美嘉荟欢乐影城
Shenyang Megabox Cinema
7. 香港京士顿国际学校
Hong Kong Kingston International School

嘉柏建筑师事务所于2003年创立。基于公司不愿再延续一般传统的设计及进行类似生产线的工作模式，并坚持打开设计理念的新一页，集合一群充满活力和有志于创意设计的专业人才，在蜕变的环境中，以崭新的理念，配以敬业的精神，提供高质素及独特的设计，务求确立卓越的建筑典范并对社会负起应有的专业责任。

嘉柏成立至今已获得多家知名发展商对其独特设计的认同，并受委托参与多项不同类型的规划及建筑设计，包括万科房地产有限公司于深圳、佛山、东莞、珠海、成都、武汉、天津、大连等城市的项目；金地房地产有限公司于武汉及杭州的项目，招商房地产有限公司于深圳及天津的项目，厦门建发集团于厦门、福州及成都的项目；其他发展商如山东鲁能集团、广州侨鑫集团、广州时代地产控股有限公司、深圳联泰房地产开发有限公司、惠州金融街置业有限公司、香港新鸿基地产、香港九龙仓集团有限公司及位于越南的印度支那集团等。

Gravity Partnership was founded in 2003 with a vision to deliver innovation through non-conventional architectural designs and work processes to better serve society and the profession. This vision is shared by 40+ enthusiastic and motivated professionals who are passionate in their endeavors to create hith quality design, and who take every opportunity to explore new ways in achieving original and imaginative results.

Recognized for its distinctive design, Gravity Partnership has been invited by prominent developers to participate in prestigious planning and architectural design commissions, including Vanke Real Estate Co.,Ltd. projects in Shenzhen, Foshan, Dongguan, Zhuhai, Chengdu, Wuhan, Tianjin and Dalian; Gemdale Real Estate Co.,Ltd. projects in Wuhan and Hangzhou, China Merchants Real Estate Co.,Ltd. projects in Shenzhen and Tianjin, Xiamen C&D Real Estate Co.,Ltd. projects in Xiamen and Fuzhou; others such as Shandong Luneng, Guangzhou Kingold Group, Guangzhou Times Property Holdings Limited, Shenzhen Liantai Real Estate Co.,Ltd., Huizhou Financial Street, Hong Kong Sun Hung Kai Properties, Hong Kong Wharf (Holdings) Limited and Indochina Group in Vietnam.

嘉 柏 建 筑 师 事 务 所

g r a v i t y partnership ltd
architects & planners

w w w . g r a v i t y p a r t n e r s h i p . c o m

1–3 中国成都环球贸易广场 Chengdu International Commerce Center, China

总监
Directors

设计总监
Principal-Design

余啸峰
Frank Yu
B Arch(Pratt Institute) New York

项目总监
Principal-Project

王克江
Claude Wong
Registered Architect
AA Dipl.
RIBA
HKIA
AP (List of Architects),
PRC Class 1 Registered Architect Qualification

副总监
Associate Director

方正道
Solomon Fong
M Arch (HKU)
LEED® AP

廖国安
Roy Liu
Registered Architect
RAIA
HKIA
AP(List of Architects)
PRC Class 1 Registered Architect Qualification

白元卿
Won Paik
AA Dipl.
RIBA
ARB

主任设计师
Associate

汪皓
Wang Ho
Registered Architect
HKIA
PRC Class 1 Registered Architect Qualification

邓天齐
Tang Tin Chai

联络
Contact

香港铜锣湾电气道148号39楼
39fl, 148 Electric Road
Causeway Bay, Hong Kong
Tel: +(852) 3106 8711
Fax: +(852) 3106 0754
E-mail: studio@gravitypartnership.com
www.gravitypartnership.com

4–6 中国成都新鸿基悦城 SHKP Jovo Town Phase 1, Chengdu, China

7–8 中国佛山市顺德区中毅弘越中心 Zhong Yi Hong Yue Centre, Shunde, Foshan, China

9–11 中国深圳招商数据大厦 China Merchants Digital Building, Shenzhen, China

12–13 中国重庆中安翡翠湖一期B Enrich Jade Lake Phase 1B, Chongqing, China

14–16 中国厦门建发爱琴海项目 C&D Aegean Sea, Xiamen, China

17 中国天津招商钻石山 China Merchants Diamond Hill, Tianjin, China

18 中国厦门夏新电子城研发办公楼 Amoi Research & Design Headquarter, Xiamen, China

19–21 中国厦门海峡交流中心二期一号楼 Straits Exchange Centre Phase II Tower 1, Xiamen, China

22–24 中国深圳联泰梅沙湾 Liantai Meisha Wan, Shenzhen, China

嘉柏建筑师事务所
www.gravitypartnership.com

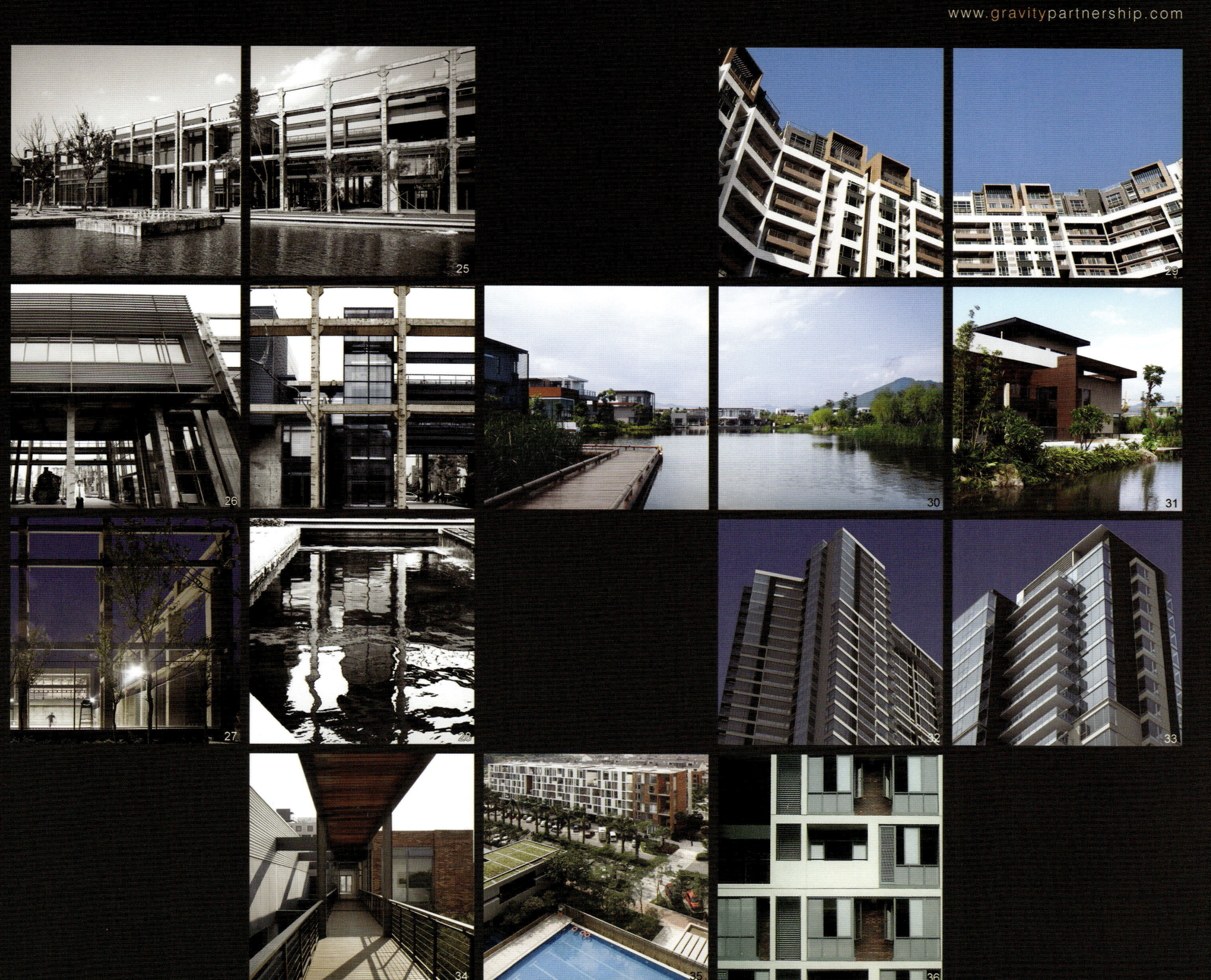

25–28 中国天津万科水晶城中心会所 Vanke Crystal City Sports and Recreation Facility, Tianjin, China

29 中国佛山万科顺德新城海畔 Vanke The Paradiso, Shunde, Foshan, China

30–31 中国福州建发领域 C&D Domain, Fuzhou, China

32–33 中国广州番禺时代珠江花园二期 Times Property Zhu Jiang Garden, Panyu, Guangzhou, China

34–36 中国东莞万科城市高尔夫花园 Vanke Golf Town, Dongguan, China

37–39 中国天津万科水晶城北入口公建 Vanke Crystal City North Entrance Apartment, Tianjin, China

40 中国海南三亚鲁能美丽城 Sanya Luneng Beauty City, Hainan, China

41 中国珠海市万科金域蓝湾 Vanke Paradiso, Zhuhai, China

42 中国深圳万科金域蓝湾 Vanke Paradiso, Shenzhen, China

43–45 中国深圳万科天琴 Vanke Vega Villa, Shenzhen, China

46–47 中国深圳万科东海岸三期及四期 Vanke East Coast Phase 3&4, Shenzhen, China

嘉柏建筑师事务所
www.gravitypartnership.com

48–49 越南河内IPH项目 Indochina Plaza Hanoi, Vietnam

50–52 中国成都东大街电视娱乐综合商场 Dongda Street Cinematique & Entertainment Mall, Chengdu, China

53–54 中国武汉金地国际花园 Gemdale International Garden, Wuhan, China

55–57 中国厦门建发国际大厦 C&D International Tower, Xiamen, China

58–60 中国东莞万科运河东一号商业区 Vanke East Canal No.1 Commercial District, Dongguan, China

65–67 中国天津招商钻石山会所 China Merchants Diamond Hill Clubhouse, Tianjin, China

68–69 中国福州建发领域公建 C&D Domain, Fuzhou, China

70 中国香港中文大学教学楼（方案竞赛） Teaching Building at Chinese University of Hong Kong (Competition)

71–72 中国香港中文大学晨兴书院 Morningside College, Chinese University of Hong Kong, China

73–74 中国成都四川航空广场 Sichuan Airline Plaza, Chengdu, China

嘉柏建筑师事务所
www.gravitypartnership.com

75–77 中国佛山市北滘文化中心 Beijiao Cultural Centre, Foshan, China

78–80 中国广州侨鑫从化养生谷别墅及水疗中心
Kingold Conghua Yangshenggu Villas and Spa Centre, Guangzhou, China

81–83 中国北京鲁能优山美地A区二期 Luneng Youshan Meidi, Beijing, China

84–86 中国重庆新天地 Xingtiandi, Chongqing, China

87–88 中国深圳梅沙海与建筑 Seashore Promenade, Meisha, Shenzhen, China

89–91 中国天津万科西青假日风景 Vanke Holiday Town, Tianjin, China

92–93 中国重庆鲁能星城十三街区 13th Avenue, Luneng Star City, Chongqing, China

94 中国深圳招商光明科技企业项目(方案竞赛) China Merchants Guangming Science and Technology Park, Shenzhen, China (Competition)

95–96 中国重庆鲁能星城十三街区 13th Avenue, Luneng Star City, Chongqing, China

97–98 中国佛山市跨G325国道乐从段人行天桥工程 Pedestrian Footbridge, G325 National Highway, Lecong, Foshan, China

嘉柏建筑师事务所

www.gravitypartnership.com

99–100 2010上海世界博览会香港馆概念设计比赛（亚军）
Concept Design Competition for Hong Kong Pavilion, World Exposition 2010 Shanghai (Second Prize)

101 香港·深圳城市/建筑双城双年展
Hong Kong & Shenzhen Bi-City Biennale of Urbanism/Architecture 2010

102–103 香港北角汀综合发展设计概念比赛（冠军）
North Point Harbour Conceptual Design Competition, Hong Kong (Champion)

104 第十一届威尼斯建筑双年展——香港在威尼斯 11th International Architectural Exhibition- Hong Kong in Venice

105 2008 WA–万科青年建筑师设计大赛 2008 WA-Vanke Young Architects Design Competition

106 香港·深圳城市/建筑双城双年展2010（策展组）
Hong Kong & Shenzhen Bi-City Biennale of Urbanism/Architecture 2010 (Curatorial Team)

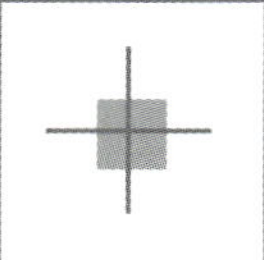

Joy Choi Arquitecta

蔡田田建筑师事务所

澳门南湾大马路325号昌辉大厦一楼AB座
Avenida Praia Grande No.325,
Edit, Cheong Fai 1°A e 1°B, Macau
Tel: +853–28752667
Fax: +853–28752544
E-mail: joyc@macau.ctm.net

蔡田田建筑师事务所是一家扎根于澳门本土的建筑设计顾问公司，成员来自世界各国的不同地区，多元文化和创作为澳门的项目在不同的领域牵领时代的气息，为澳门新生代建筑文化树立新的形象和品位。

该公司参与和提供的城区改造研究方案包括：历史建筑复修、保育和改建研究和实践；在建筑设计和室内设计项目中，公共建筑以教育建筑和社会设施为主，民用建筑中涵盖超高层住宅大楼、别墅住宅和商业建筑。其建筑顾问工作提供一条龙服务，包括规划评估、方案设计、政府报批、招评标及施工合约管理等。

该公司的理念是着力发挥澳门多元文化的优势，整合团队的专业职能，以有效的管理模式为客户提供高水平、具国际视野的设计作品。同时，致力于研究和开发澳门历史文化的特色，以创新的手法开拓更符合可持续发展的新建筑，使我们的客户获得最大的利益。

JCA is a Macau local architectural design consultancy company founded in 2001.We have architectural professionals from different parts of the word. Together we have formed strong design teams in serving different aspects of architectural projects, which acting a leading role of new architecture generation in Macau.

JCA takes important steps in enhancing urban research and design in Macau old town and Unesco heritage sites.There are projects in relation to restorations, revitalization and renovations for historical buildings and reconstruction projects in Heritage Protected Zones. New architecture projects known in the city are including high-rise office building, high-rise residential building, luxury private villas and numerous educational buildings. The practice provides full package of architectural consultancy services including urban design, architecture, interior design and post contract project management.

We aim to elaborate architectures with multi-cultural influences and international visions. Our team provides efficient management for our clients in all aspects of innovation and exploration in Macau to endeavor the most interest of client and community.

澳门氹仔圣善学校重建

澳门何东图书馆扩建

澳门圣罗撒女子学校英文部扩建

澳门圣罗撒女子学校中文部重建

澳门化地马女子学校扩建

澳门氹仔客商街13、17号

澳门交通局及工务局大楼

澳门氹仔出入境事务局设计投标

澳门迎泉住宅

澳门落环乡村马路别墅

公司资料 Company Data

成立年份	Established	2001
专业人员	Professional	15
协助人员	Support	5

主要专业人员表 Key Staffs

蔡田田 Joy, Choi Tin Tin
中国澳门注册建筑师，主理建筑师
Principal Architect

蔡子钰 Dominic, Choi Chi lok
中国澳门注册建筑师，英国注册建筑师，设计总监，合伙人
Design Director, Partner

高武洲 Joseph, Wu-zhou Gao
中国澳门注册建筑师，高级联营董事
Senior Associate

梁志伟 Tony, Leung Chi Wai
中国澳门注册机电工程师，高级项目经理
Senior E/M Engineer, Senior Project Manager

澳门土木实验室总部

澳门基金会总部复修

澳门轻轨第一期设计

澳门路氹城行人天桥设计

2010—2011
主要建筑设计作品
Main Architectual Design Works

Annual Review of Chinese Architectural Design Works

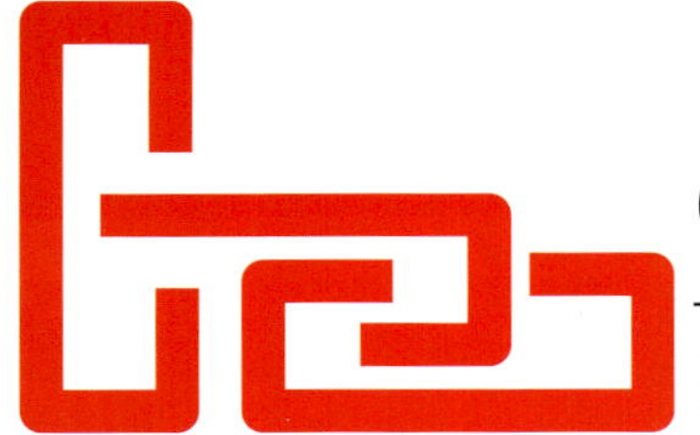

C.J Chen Architects

陈子弘建筑师事务所

陈子弘建筑师事务所(C.J Chen Architects)成立于2006年1月，中国台北。事务所主要的服务项目包含私人住宅、商业、办公、室内设计、混合使用内容以及相关的公共建筑等。除了实质的景观设计、建筑设计、室内设计服务外，事务所也提供概念设计，以前卫的设计理念配合数字科技创造出独特的空间感。

2007年，为了提升事务所的设计质量与水平，陈子弘前往纽约哥伦比亚建筑研究所进修学习，以吸取建筑设计上的形而上的养分。事务所的经营理念以满足顾客的需求为根本，并在过程中寻找创意设计的空间，透过团队的默契合作将空间设计呈现出来。

C.J Chen Architects is a 20-person international design firm offering services in architecture and interiors. Founded in 2006, we are committed to forgoing a greater connection between the built environment and the quality of life, and creating thoughtful and responsible architecture or “design product”, which is tied to a true story of place.

Focused on the unique and compelling retail and mixed-use destinations worldwide, the cohesive of sensibility of our practice is expressed in a customer-centric approach, an emphasis on communication and teamwork, and a dynamic and positive office culture.

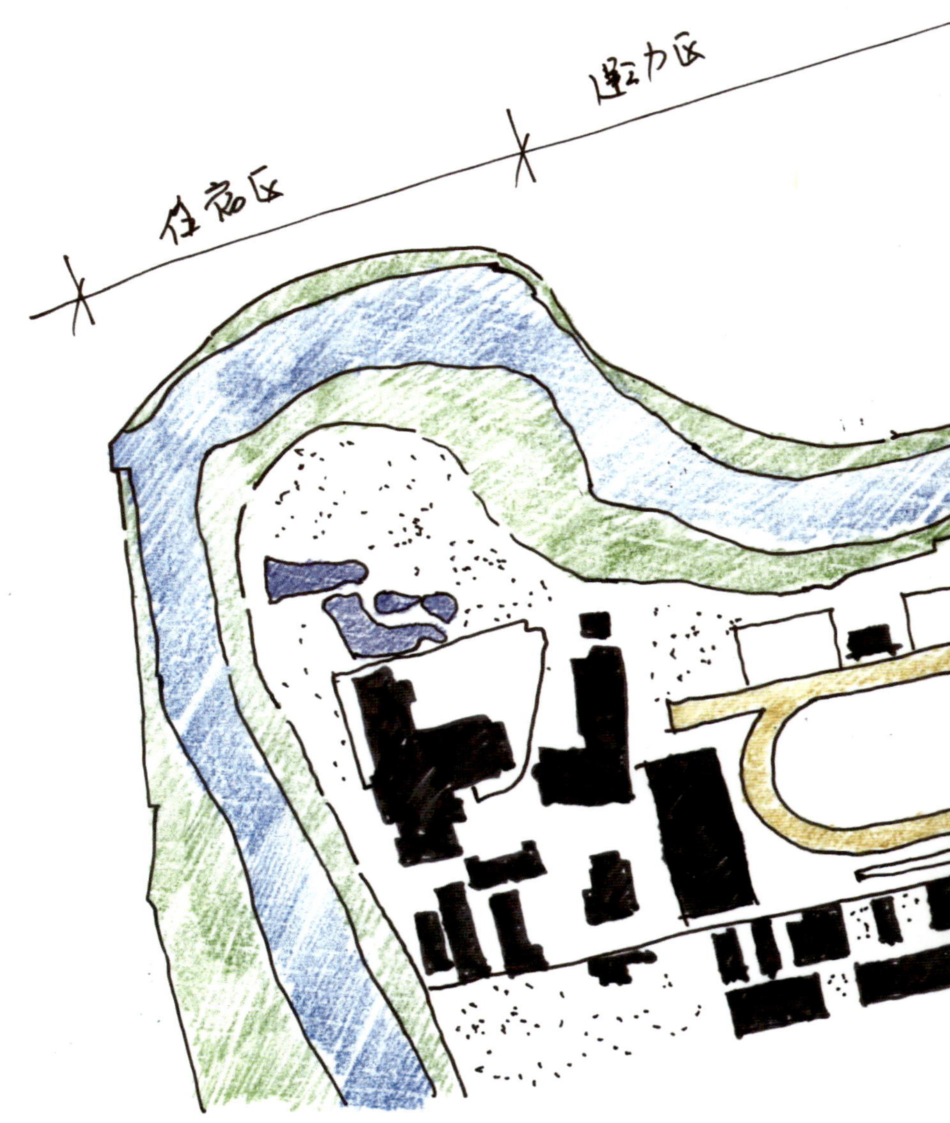

地址：台湾台北110基隆路二段51号3楼之7
电话：+886-2-27392075
传真：+886-2-27392327
网址：www.cjchenarchitects.com
邮箱：cj@cjchenarchitects.com

Add: 3F.-7, No.51, Sec. 2, Keelung Rd., Xinyi Dist., Taipei City 110, Taiwan, China
Tel : +886-2-27392075
Fax: +886-2-27392327
Http: //www.cjchenarchitects.com
E-mail: cj@cjchenarchitects.com

新北市双溪高中综合大楼

建设地点：台湾 台北
设计单位：陈子弘建筑事务所
设计业主：新北市政府
空间性质：学校建筑
主要结构：RC
工程造价：新台币21.5亿
设计时间：2009年

缘起

一万年前，上帝创造了亚当及夏娃。上帝为了启发人类的探索欲望，在西方设置了一份潘多拉的盒子用以隐藏许多科学的智能；在东方则埋藏一个名为月光宝盒的神物，直到庚寅年因为环保意识以及环境永续的概念抬头，在复育萤火虫的过程中，发现了传说中的月光宝盒，它就坐落在今日的双溪乡。

设计理念

双溪高中的特色在于承袭了所在地的绿色气候，萤火虫、铁马单车成为宁静乡村的写照。校园内的漆弹场、生态池、古典的水塔等诉说着校园古老的历史。

面对一个有历史情怀且具有低碳环保概念的校园，设计承袭月光宝盒的启示，将图书馆幻化成知识的萤火虫，随着顶楼的演艺厅的音乐演奏，渐渐开启了这个迷人的月光宝盒。

旧有的水塔，经过整理后在新建的建筑物上以圆形开孔示人，并与户外庭园设计、夜间照明设计结合起来。象征意义上，旧有水塔所产生的“孔洞”空间恰巧成为月光宝盒的钥匙孔，借助钥匙贯穿古今时空。

图书馆隐喻为知识的萤火虫，创造了良好的阅读氛围以及空间特色。同时将自然光线加以利用，让月光宝盒除了具有象征性的意义外，也能真正体现绿能建筑的特色。

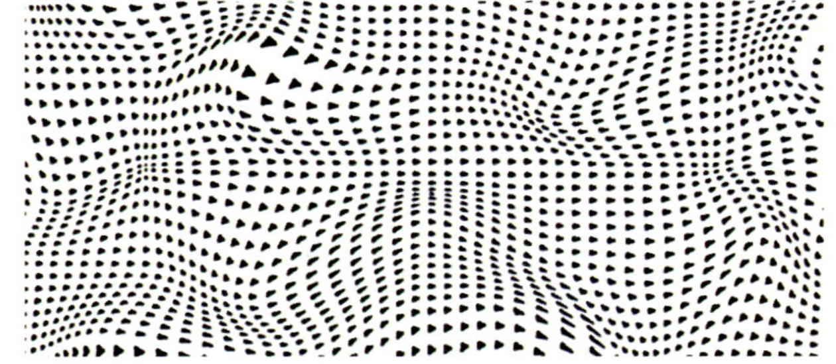

高雄县台铝厂房整建规划案

建设地点：台湾 高雄
设计单位：陈子弘建筑事务所
设计业主：高雄市政府
空间性质：工业厂房
主要结构：RC
建筑面积：1 035平方米
工程造价：新台币1亿元
设计时间：2007—2008年

本案位于经贸特区内，为旧时台铝厂房改建的展览馆。展览馆属于临时性展场建筑，日后即将与港边的永久性展览馆形成一、二号展览馆。基于高雄港都的都市特性，配合城市营销之概念，该案旨在借由展览馆和经贸特区都市魅力，塑造整体水岸风貌。因此，展览馆除了满足多样化的展场需求外，还需塑造另类的都空市间。

设计理念

1．后现代夜间都市灯笼

本案为旧台铝厂房改建而成，建筑物属于低矮扁平型。为了打造夜间都市魅力活动场所，构想上以灯笼之意象转化而成多孔型外墙，透过内部活动的灯光，周围空间所形成的趣味性空间氛围将重新定位展览馆的建筑个性。

2．绿能设计示范建筑

本案采用多孔型外墙，内部使用高效隔热玻璃，如此双层表皮的设计理念将白天的热能阻隔于建筑物墙体之间，通过风压以及浮力通风方式将热能排出。多孔型外墙为木质材料，内部玻璃框架系统使用钢构造，符合绿色建筑减废以及回收再利用需求。其他诸如太阳能光电系统、雨水回收、基地绿化、基地保水等诸多绿建筑手法亦实现在此案中。

3.Digital Fabrication 建筑组构方式

有别于传统的建筑设计、制造方式，本案采用客制化量身定做建筑物专属的构法以及工法，旨在彻底实践原有建筑的规则。制作上以现有国内的建材加以改良加工，以符合本案之功能、构造以及意境上的需求。本案试图重新思考材料、形式与建筑之间的关系，并探讨空间存在的价值。

4.建筑的自明性

本案尊重周边建筑纹理涵构，主要体量以矩形为构成单元，反映周边环境，但在内部展场动线以及空间视觉经验却是本案的另一设计重点。

较之于以往的外形大胆以追求建筑物的自明性，本案以"暖暖内含光"的方式，强调以视觉空间经验重新诠释的建筑物的自明性。

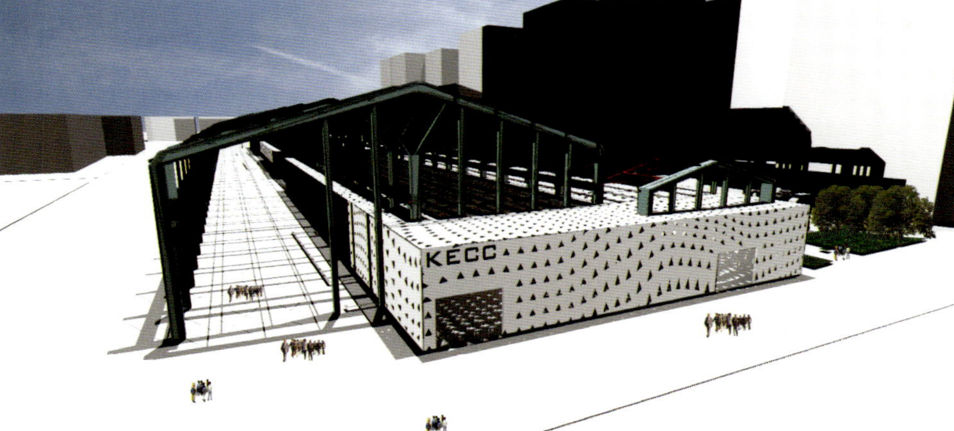
KECC

KECC

KECC

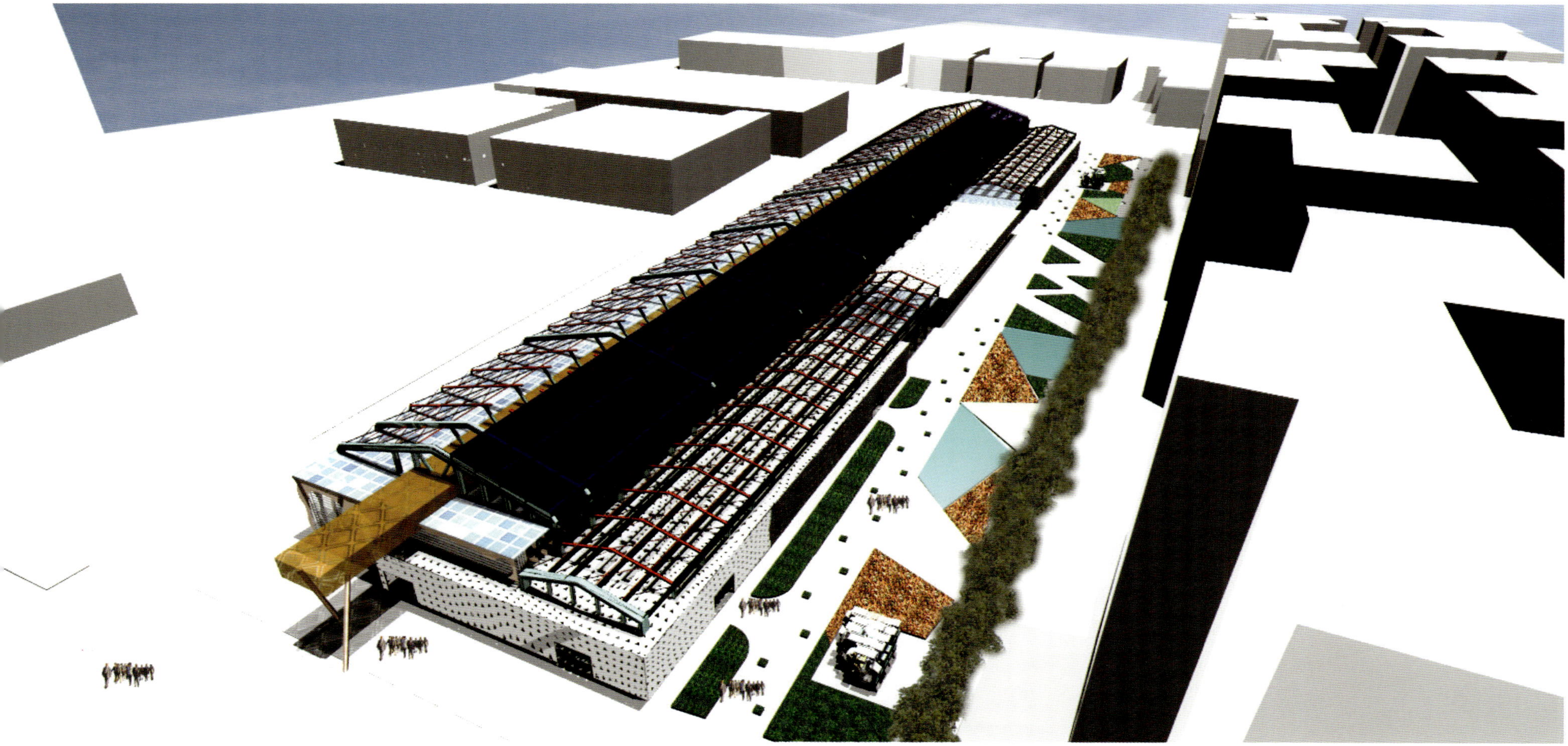

天一制药厂迁厂暨创意文化馆

建设地点：台湾 台南
设计单位：陈子弘建筑师事务所
设计业主：天一制药
空间性质：工业厂房、博物馆
主要结构：RC薄膜及SC
建筑面积：17 000 平方米
工程造价：新台币5亿
设计时间：2009年

制药工厂本质上属于功能建筑。药厂转型的契机在于加入许多人文、艺术文化的元素后，形成复合式建筑空间。以美术馆、博物馆的精神，让观光工厂变得人文化更有亲和力是该案的目标。

设计理念

能量：本案的设计核心，表现的方式以螺旋(Spiral)的形式呈现。地标建筑（Icon building）本身不是以怪异造型呈现，而是为制药厂量身定做的"深"绿建筑，以能量光塔作为柳营科学园区视觉地标。

博物馆空间文化：本案以博物馆的空间气质，运用光影美学，营造空间氛围。

多元的创意展览空间：利用螺旋参观动线兼具展售、展室的功能营造。

主题生态公园：将工业厂区塑造成主题公园，兼具教育以及景观功能。

绿建筑的应用：诱导式设计绿建筑应用内容包含浮力通风、自然采光、内流水微气候、壳隔热设计、结构轻量化。

惟桑之道：诊时弊，织锦衣，育英才，大哉矣！

桑葉機構

MULBERRY TEAM

上海桑葉建築設計咨詢有限公司

桑叶机构（MULBERRY Team）前身PDI，成立于2005年。于2011年成立桑叶机构。

上海桑叶建筑设计咨询有限公司隶属于桑叶机构，是一家以设计咨询、招商运营、投资管理为主的综合性公司。

运用国际化商业专业理念、专业资深团队、丰富的本土实战经验，结合中国经济发展状况，致力于国内商业地产的深度研究。向广大开发商、品牌商家以及机构投资者提供商业不动产投资、策划、定位、改造、招商、运营和管理等综合服务。

设计咨询：特别强调设计的“动态三宜”，即因时、因地、因人制宜。从规划设计初始就整合“市场可行性”、“文脉”、“环境”、“建筑艺术”，不断进行价值探讨，塑造独特魅力空间，为客户创造品质与回报的完美演出。提供从都市开发到建筑设计、从室内空间到室外景观的全方位解决方案，特别是在商业地产和都市复合型开发领域。

招商运营：专业于将项目策划、商业组合、定位战略与建筑设计的软硬件整合在一起，针对客户的不同开发阶段，提供适时正确的服务要求，坚持从市场及业主的角度出发，针对关键性问题提出解决策略和项目定位。对于不同地域和都市涵构，突出不同的设计主题构想，并为业主创造更大价值的空间产品，以保障整个开发案的有效发展，并完成创新和成功的执行要求。

已成功地完成了国内多个不同都市层级和各种经济发展阶段大小城市的商业及商业复合体建筑的设计，并得到客户的一致认可。
桑叶机构致力于服务全国客户，打造中国精品地产项目，创造独具魅力的、富有价值的建筑空间。

策划咨询 / 城市规划 / 建筑设计 / 商业景观 / 室内设计 / 商业策划 / 经营管理

上海桑叶建筑设计咨询有限公司/台湾赖育志建筑师事务所

地址：上海新华路543号新华大厦1号楼4F-G座
电话：+86-21-62817700
传真：+86-21-52300755
邮箱：mulberryteam@163.com
网址：http://www.pdi.net.cn/

河南郑州商都路项目

用地面积:39 793平方米
总建筑面积:131 063平方米
容 积 率:3.29

从基地分析来看，基地相邻的红星美凯龙以庞大的体量对基地形成巨大的压迫感，另一方面，基地目前仅有一小地块邻接商都路，这成为本地块的挑战所在。但从商都路往南看，基地的西北角反而成为基地最大的展示面，由此，裙楼、塔楼一体化的“巨构建筑”（Mega-structure）概念进入我们的设计思维里。以60米限高为顶点，将西面与北面筑起完整的连续体量，对内围合以向南获得足够的阳光，试图以西北角的巨构建筑体量成为区域地标的象征，沿逆时针方向从西北或北为出发点，以整体而富有节奏的造型手法绕行360°，形成丰富的协奏曲，同时向南创造一个超大的阳光温室，形成郑州罕有的“空中城市客厅”，成为可同时容纳庆典中心、酒店、会议及餐饮的聚客大堂，让此巨构建筑成为东建材区的“郑东之星”。

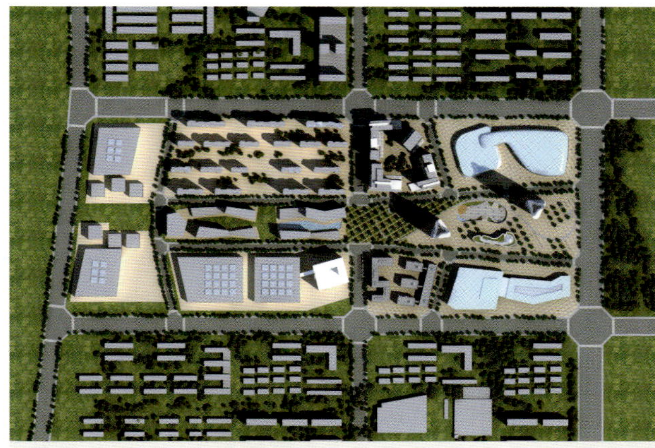

兰州金城国际商贸园区概念规划

建设地点：甘肃 兰州

一、基地

两栋超高层地标建筑、两个城市广场及城市森林构筑了一个气势恢弘的中央商务区.在规划中轴线上坐落着两栋遥遥相望的超高层建筑（超五星级酒店及甲级智能化写字楼），以满足其域内商务人士及中央商务区周边企业对高档商业办公场所、会议商务中心及配套酒店的需求。两栋超高层恢弘的外立面设计既符合地标建筑的王者风范，又不失地域时尚风格。

中央商务区由三个城市开放区域（市民广场、城市枢纽广场、城市森林），两个超高层建筑及文化会馆组成（功能分别为：超五星级酒店、甲级智能写字楼、金融中心及文化会馆）。城市枢纽广场与各功能区整合，使得大量人流通过本集散地后形成多层次的互动，打造兰州新区新的地标性中央商务区。

二、购物中心

定位：① 金城购物中心不仅是兰州商贸的引领者，更应该是兰州周边等商业的领跑者；② 目标消费者以年轻、中高收入的都市新贵为主；③ 价位多层次以满足不同消费者的需求；④ 整体商业以时尚、品质、品位、体验与娱乐为主；⑤ 业态多元化，充分体现综合形态的娱乐、购物、餐饮等体验。

三、高层办公区域

甲级智能化写字楼，可满足其域内商务人士及中央商务区周边企业对高档商业办公场所、会议商务中心及配套酒店的需求。

四、超智能化的区域型总部办公大厦

该总部办公大厦尽显尊贵地位及品位。

五、商住混合区

30万平方米的居住区配套社区商务，以步行商业内街的规划手法将特色餐饮、休闲咖啡、简易西餐串联，并同时为城市物流区及东街坊城市森林提供休息、餐饮配套。

社区商业规划追求强烈的街道街区感，以合理的街道尺度充分考虑人的需求，将水面、绿化、景观元素融入其间，亲切的景观结合震撼的现代手法赋予该区域时尚灵动的空间。

六、居住区

社区内组团关系明确，围合感强。在组团内营造亲切宜人的空间与尺度，有可识别性和归属感。居住区配有集健身、娱乐为一体的社区会所。

七、城市物流区

在这里，预计年人流量将超百万。本区域将以厂商直销、优惠零售的商业模式，服务来自多个不同城市及民族的群体。该区域的配送、批发、零售取代了传统的物流形式。借助于这些大盒子的设计，对配送、批发、零售进行重新组合，将以往的配送、批发、脏乱环境的理念进行重新塑造。

八、城市家居区

整合兰州地区现有家装、家居资源，推广新材料及新的生活家居趋势互动。

马鞍山佳山公园东侧麓山地块

项目地点：安徽 马鞍山
设计单位：上海桑叶建筑设计咨询有限公司
总占地面积：33 366平方米
总建筑面积：93 426平方米
容 积 率：0.98
建筑密度：0.35
设计时间：2010年

本案总用地面积95 333平方米，产品包括坡地合院联排住宅、半岛水岸双拼住宅和东侧18层高层住宅、会所以及沿街商业等多种类型。在规划设计中，最大限度地实现了建筑与山水的和谐交融。巧妙地遵从山地的地形，顺从流水的体态，慢慢地舒展在如画的山水间。

从茂盛的森林到幽静的竹林，从层层跌落的清泉到四季常青的草地斜坡，本案的大格局是开放的、自由的，而层层叠叠、宛如珍珠般的水景又将低层住宅区域与高层住宅区域巧妙分隔，让私密性得到有效保证。

整个小区的超低容积率带来更开阔的视野、更充裕的采光，让视线更加无拘无束，让心灵有更大的舒放空间。

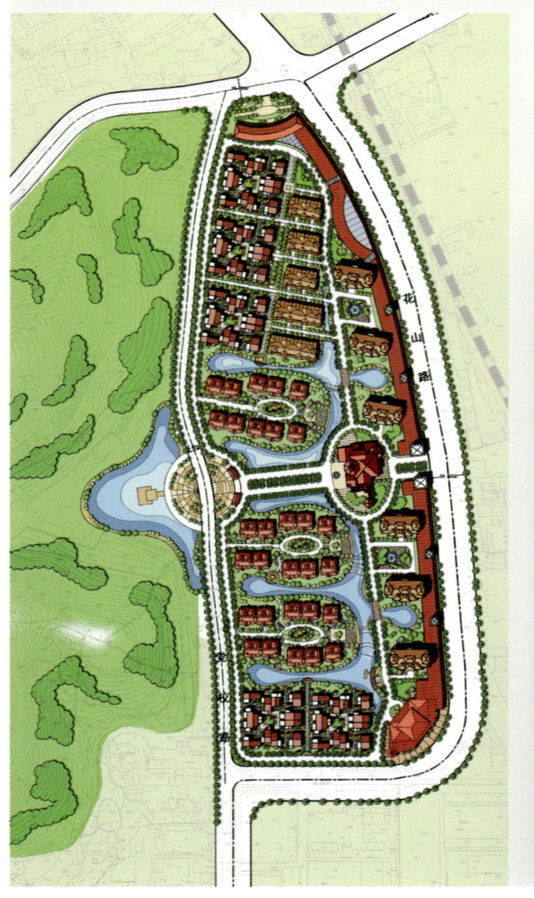

从问"为什么"开始

好的商业项目会成为一个城市的代名词、名片，瑞安有城市名片吗？
在不同点中业态定位如何结合城市环境找到准确点？

一、项目定位出发点

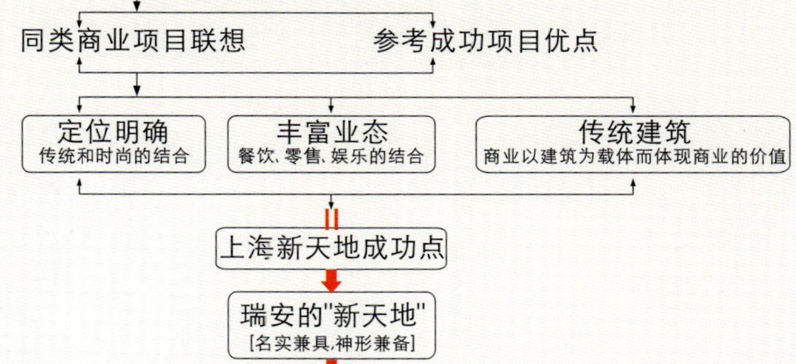

- 玉海广场和新天地有着很多相似点
  我们认为玉海广场同样可以成为瑞安的城市名片，瑞安的新天地。

二、商业定位设想

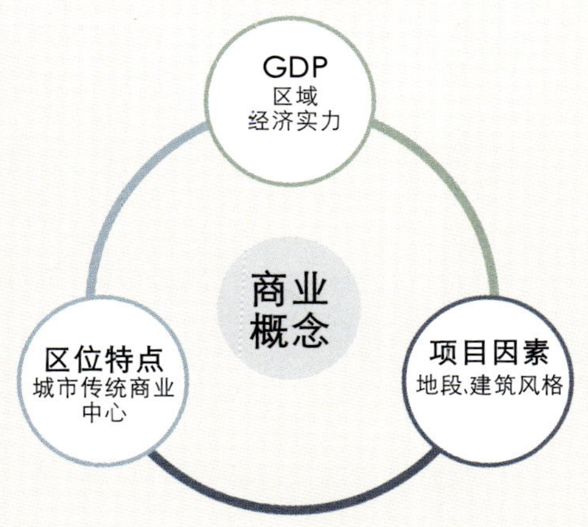

1.商业主题定位

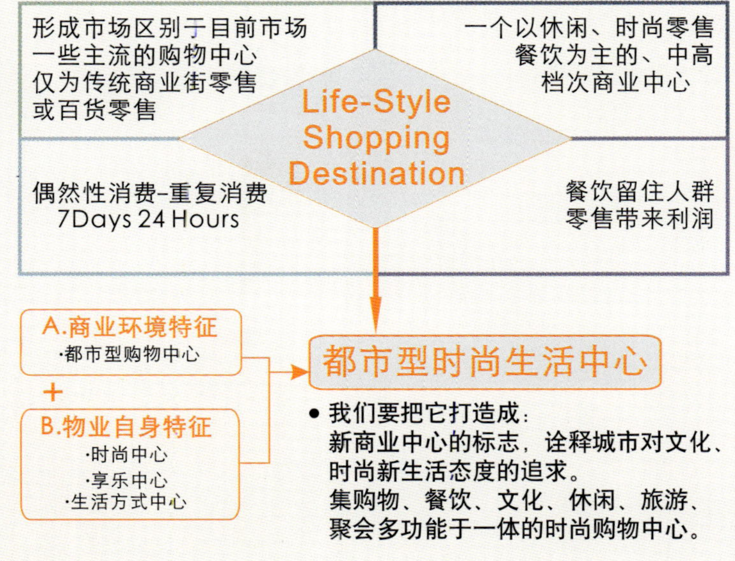

- 我们要把它打造成：
  新商业中心的标志，诠释城市对文化、时尚新生活态度的追求。
  集购物、餐饮、文化、休闲、旅游、聚会多功能于一体的时尚购物中心。

2.商业业态定位

玉海尚街

时尚-购　时尚-饮　时尚-享　时尚-玩

四大主题功能
对时尚核心概念的体现

- 主力店目标是决定项目成功的基本因素

港瑞新玉海·国际街区策划案

项目地点：浙江 温州
总用地面积：21 258.55平方米
建设用地面积：20 966.69平方米
容积率：1.31

三、商业营销推广设想

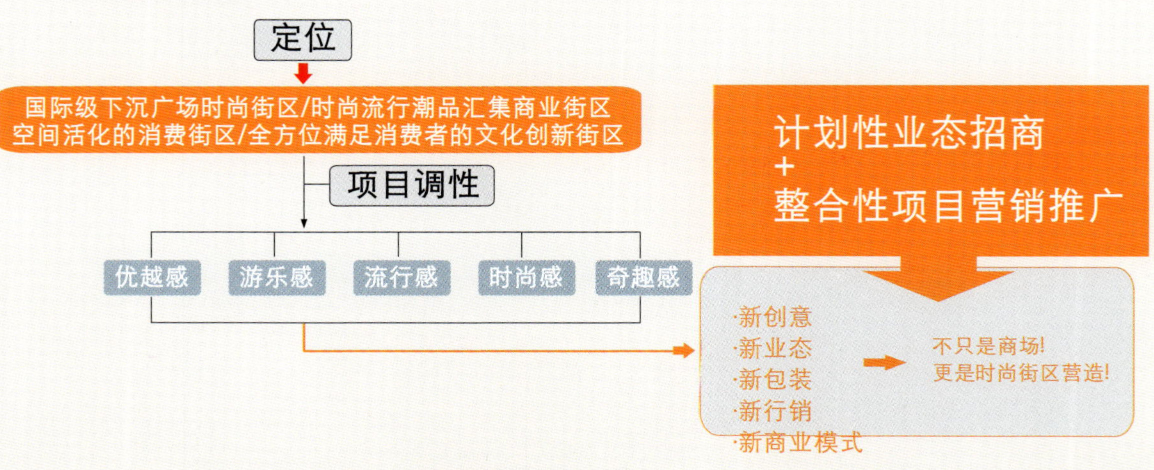

商丘市行署地块

项目位置：河南 商丘
总用地面积：183 895平方米
总建筑面积：595 339平方米
容 积 率：3.24

设计意图突破原有的城市文脉，以若干不规则线条活跃城市“新的成长”。“新的成长”是多维度的，在形体上将二维的跃动往上转至第三维度，直接延伸至天空，想象天际线与天空灵活地对话。在“城市关怀”和“以人为本”的前提下，打造和谐社区，设置了由东向西的“城市中央公园”，为市民提供休闲、运动、活动场所。

两旁的商务住宅区也留设大型中心绿地，与“城市中央公园”共同形成新的“城市绿肺”。城市中央公园的最西边为双塔，两栋百米高的5A级写字楼和五星级酒店，形成巨大的双塔门户，塑造新的城市地标。5A级写字楼的裙楼为大型购物中心，其主力店为百货公司和大型超市，引入国际品牌名店；五星级酒店的裙楼为大型餐饮宴会厅和国际会议中心，满足城市发展新的需求。力求以“量子跃”的“质变”来带动城市的活力与新一轮的成长，让绿色生态介入城市的有机成长。 建筑物采用现代高科技的节能手段和符合自然生态的阳光通风原则，充分做到“低碳”、绿色节能。

最终以“城市中央公园”为核心，外围的购物中心、办公、酒店、会议、商务、公寓共同形成“绿色、低碳中央商务区”。“绿色中央商务区”将成为商丘的新名片。

神垕钧都

项目地点：河南 禹州
发 展 商：禹州新天地建设开发有限公司
设计单位：上海桑叶建筑设计咨询有限公司
总占地面积：10 123平方米
总建筑面积：34 201平方米
容 积 率：1.49
建筑密度：40.4%
设计时间：2009年

该宗地位于神垕镇的传统老城区，与解放路、市场路共同构成神垕镇的商业主线，由西向东贯通市场路、北寨街、瓷厂街和后公路。解放路为镇区对外联系的要道，即将改造的建设路通过市场路与解放路连接起来。

设计思路

依托改造后的神垕老街旅游文化资源，以“钧瓷”这张城市名片为媒介，在该地块内建立独具地方特色的集钧瓷采购、展示、文化体验、休闲餐饮为一体的现代商业步行街，营造“中国钧瓷文化街”的文化氛围，提高商业街的品质。

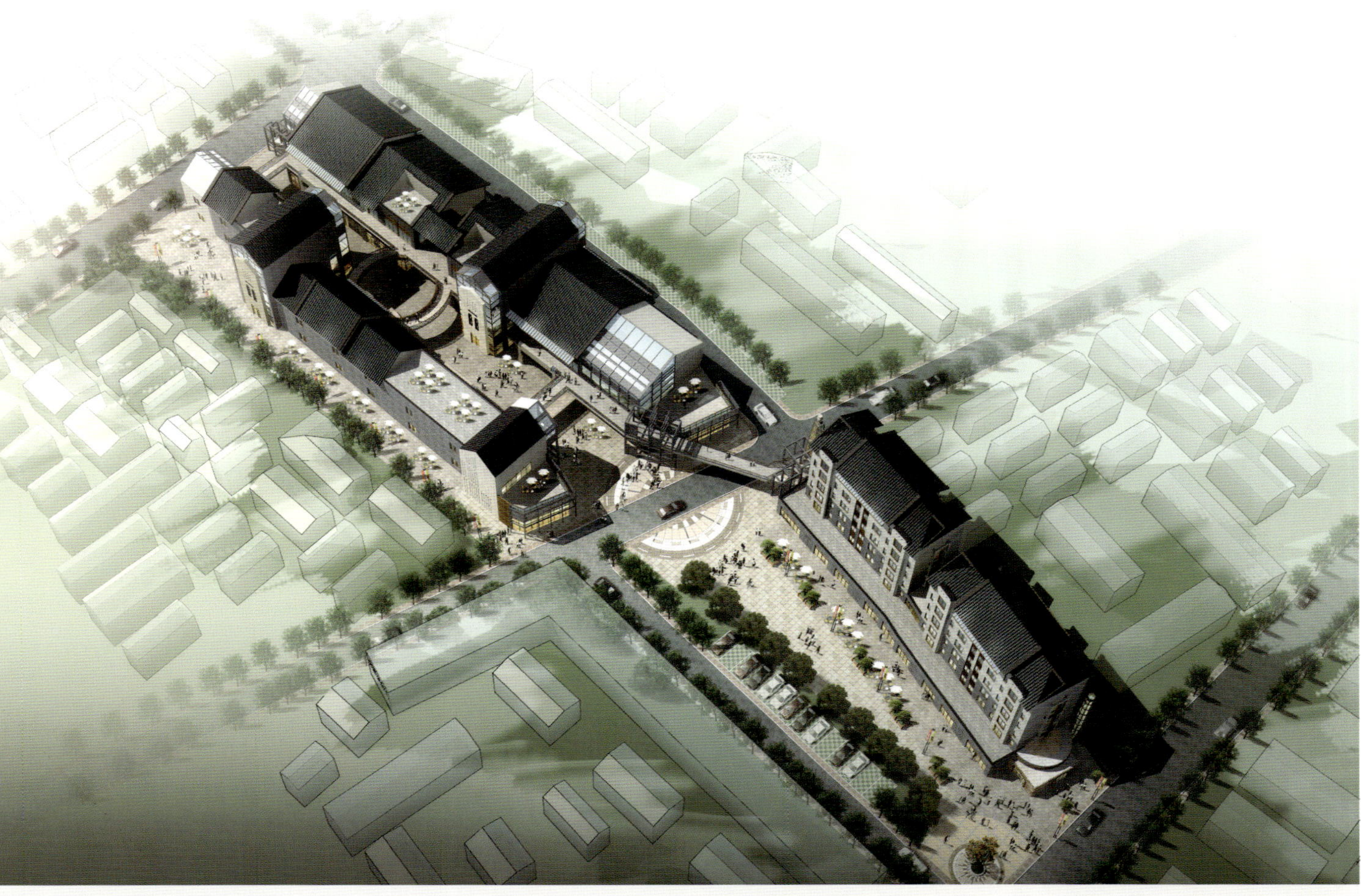

北京市建筑设计研究院
BEIJING INSTITUTE OF ARCHITECTURAL DESIGN

北京市建筑设计研究院（以下简称为BIAD），是与中华人民共和国同龄的大型国有建筑设计咨询机构。其业务范围包括城市规划、投资策划、大型公共建筑设计、民用建筑设计、室内装饰设计、园林景观设计、建筑智能化系统工程设计、工程概预算编制、工程监理、工程总承包等领域。

伴随着新中国的社会发展，BIAD在北京完成了许多重大项目的设计工作。如20世纪50年代象征新中国形象的人民大会堂、国家博物馆、民族文化宫、工人体育场；60、70年代体现我国自主科技实力的北京工人体育馆、首都体育馆和北京饭店工程；80年代体现改革开放的中国国际展览中心和第11届亚运会场馆；90年代完善国际大都市功能的北京首都机场2号航站楼、北京西站、北京植物园温室、首都图书馆、西单文化广场。进入21世纪，BIAD完成了包括2008年部分奥运工程、北京首都机场3号航站楼、中国石油大厦、全国人大机关办公楼等标志性建筑的设计工作。

近几年来，BIAD在全国范围内设计完成了一大批城市标志性建筑，如海南博鳌亚洲论坛、浙江中国美术学院、烟台市体育中心体育场、秦皇岛体育馆、重庆人民大厦、新疆体育中心、绵阳市政府办公楼、河南省体育中心体育场、西安国际金贸中心、三亚喜来登酒店等项目。

BIAD与许多国家的著名设计公司保持着良好的合作关系，合作设计了包括中国工商银行总行办公楼、北京顺义国际学校、深圳文化中心、国家大剧院、东方广场、LG大厦、中国电影博物馆、中国科技馆、凯晨世贸中心等工程。

改革开放以来，BIAD积极参与全国各地的经济建设，推动房地产业的健康发展，每年完成大量的民用建筑设计项目，累计完成设计面积超过1.5亿平方米，其中恒基中心、国际金融大厦、海淀剧院、投资广场、北京新闻文化中心、信远大厦等项目已成为北京城市新的地标建筑，而望京新城、北京现代城、万科星园、北京橘郡、万科西山庭院、深蓝华亭、颐源居、远洋山水、朗琴园、回龙观居住区均对房地产市场产生了较大影响，成为明星楼盘。

BIAD始终将设计精品工程和保持技术领先作为自己的责任，从1977年至2008年，设计作品获国家奖61项，获建设部奖180项，获北京市奖428项；科研成果获国家科技进步奖28项，获科技部进步奖75项，获北京市科技进步奖148项，获詹天佑土木工程奖10项。在多年的工作过程中，BIAD逐步形成了一个优秀的设计团队，并集中了一大批优秀的建筑师和各个专业的工程师，其中有工程院院士1名、国家设计大师9名、突出贡献专家10名、高级建筑师和高级工程师438人、建筑师和工程师361人、一级注册建筑师185人、一级注册结构工程师124人、注册造价工程师9人。

BIAD自成立以来的60多年中，始终致力于向社会提供高品质的设计服务，在行业中享有极高声誉，并逐渐形成了“建筑服务社会”的企业核心理念。面临新世纪设计行业的严峻挑战，全院1 800名员工深刻认识到，只有进一步强化市场意识，切实尊重业主需求，通过整合全院乃至国内外优势资源，逐步提升设计作品的完成度，才能体现国家大型设计机构的综合竞争实力。

开放的BIAD愿与社会各界进行广泛合作，为创造人类美好的生活环境而不懈努力。

The Beijing Institute of Architectural Design (BIAD) is a large-scale state-owned architectural design and consulting institute established in 1949, following the founding of the People's Republic of China. BIAD's business scope includes: city planning, investment planning, large-scale public building design, civil engineering, interior design, landscape design, design of intelligent systems for buildings, project estimating, project administration, and project contracting.

In the 1950s, BIAD designed the symbolic buildings that represented the image of the New China, including, the Great Hall of the People, the National Museum, the Cultural Palace of the Nationalities, and the Worker's Stadium. In the 60s and 70s, BIAD's projects reflected the independent scientific and technological strength of China, with projects such as the Beijing Worker's Gymnasium, the Capital Stadium and the Beijing Hotel. In the 80s, BIAD's designs displayed China's achievements in its reform and opening-up process, for example, the China International Exhibition Center and the gymnasiums and stadiums for the 11th Asian Games. In the 90s, BIAD's projects contributed to Beijing's function as an international metropolis with projects such as the Beijing Capital International Airport Terminal 2, the Beijing West Railway Station, the Conservatory of Beijing Botanical Garden, the Capital Library and Xidan Cultural Square. In the 21st century, BIAD has dedicated itself to some of the architectural projects for the 2008 Olympic Games and a number of landmark buildings in Beijing, including Beijing Capital International Airport Terminal 3, China National Petroleum Mansion and the office building for the Standing Committee of the National People's Congress. Other landmark commercial, office, and cultural buildings by BIAD included the Beijing Henderson Center, the International Financial Mansion, Haidian Theater, the China Academy of Art, Investment Plaza, Beijing Media and Cultural Center and Xinyuan Mansion. Some of BIAD's residential building complexes have exerted considerable influences on the real estate market, such as, Wangjing New City, Beijing SOHO New Town, Vanke Star Garden, Beijing Orange County, Vanke West Mountain Courtyard, Deep Blue Houses, Yiyuan Houses, Lar Valley, and Huilongguan Residential Area.

Since China's reform and opening-up, BIAD has been actively taking part in economic construction in other cities around China, making contributions to the healthy development of the real estate industry. BIAD undertakes many architectural design projects, and its total design area now amounts to over 150 million square meters.

BIAD has designed many landmark buildings nationwide, including Hainan Boao Forum for Asia, Yantai Sports Center Stadium, Qinhuangdao Stadium, Chongqing People's Mansion, Xinjiang Sports Center, the Mianyang Municipal Government Offices, Henan Sports Center Stadium, Xi'an International Finance and Trade Center, and the Sanya Sheraton Hotel.

BIAD has established good cooperative relations with many well-known foreign design offices. Some joint designs include the Head Office of Industrial and Commercial Bank of China, Beijing Shunyi International School, Shenzhen Cultural Center, China National Grand Theater, the Oriental Plaza, LG Mansion, China Movie Museum, Chemsunny Plaza, Beijing and the China Science and Technology Museum.

BIAD has always devoted itself to top-quality design and technological leadership. From 1977 to 2008, BIAD was awarded 61 national prizes for its designs, 180 awards from the Construction Ministry and 428 awards from the Beijing Municipal Government. Its scientific research has won 28 national scientific and technological awards, 75 from the Construction Ministry, and 148 from the Beijing Municipality. BIAD has been a five-time winner of the Zhan Tian You Construction Award. Over the years, BIAD has built up an outstanding design team with a large number of excellent architects and engineers specialized in various fields, including a member of Chinese Academy of Engineering, 9 national design masters, 10 experts with outstanding contributions, 438 senior architects and senior engineers, 361 architects and engineers, 185 Class 1 certified architects, 124 Class 1 certified structural engineers, and 9 certified cost engineers.

Over 62 years since its establishment, BIAD has devoted itself to providing good quality design services, thus enjoying a good reputation in the industry. BIAD has also gradually developed its key corporate concept "Architecture serves society". In face of the severe challenges to the design industry at this time, the 1800 staff members of BIAD have realized that it can fully display its comprehensive competitiveness as a large-scale design institute only if it: strengthens its market awareness, earnestly respects the needs of clients, and increases its design quality by integrating the resources of the whole institute and even those at home and abroad.

BIAD is willing to join hands with all circles of society, making sustained efforts to create a wonderful living environment for all human beings.

地址：北京市南礼士路62号
邮编：100045
电话：+86-10-88043999
传真：+86-10-68034041
网址：www.biad.com.cn

Add: No. 62 of South Lishi Rd., Beijing
P.C.: 100045
Tel: +86-10-88043999
Fax: +86-10-68034041
Http: // www.biad.com.cn

国家大剧院

建设地点：北京
建筑面积：219 400平方米
占地面积：120 000平方米
设计时间：2000—2002年
竣工时间：2007年
合作单位：法国安德鲁

National Grand Theater

Construction Location: Beijing
Building Area: 219,400 m^2
Site Area: 120.000 m^2
Design Time: 2000—2002
Completion Time: 2007
Cooperation: Andrew, France

首都机场T3航站楼

建设地点：北京
建筑面积：902 000平方米
占地面积：13 260 000平方米
设计时间：2003—2005年
竣工时间：2007年

Capital International Airport Terminal T3

Construction Location: Beijing
Building Area: 902,000 m^2
Site Area: 13,260,000 m^2
Design Time: 2003—2005
Completion Time: 2007

国际金融大厦

建设地点：北京
建筑面积：100 000平方米
占地面积：18 000平方米
设计时间：1996年
竣工时间：1998年

International Financial Building

Construction Location: Beijing
Building Area: 100,000 m^2
Site Area: 18,000 m^2
Design Time: 1996
Completion Time: 1998

北京植物园展览温室

建设地点：北京
建筑面积：7 250平方米
占地面积：55 000平方米
设计时间：1999年
竣工时间：2000年

Exhibition Greenhouse of Beijing Botanical Gardens

Construction Location: Beijing
Building Area: 7,250 m^2
Site Area: 55,000 m^2
Design Time: 1999
Completion Time: 2000

国家奥林匹克体育中心

建设地点：北京
建筑面积：110 000平方米
占地面积：660 000平方米
设计时间：1986年
竣工时间：1989年

National Olympic Sports Center

Construction Location: Beijing
Building Area: 110,000 m^2
Site Area: 660,000 m^2
Design Time: 1986
Completion Time: 1989

中国国际展览中心2–5号馆

建设地点：北京
建筑面积：75 000平方米
占地面积：150 000平方米
竣工时间：1985年

China International Exhibition Center Hall No. 2–5

Construction Location: Beijing
Building Area: 75,000 m^2
Site Area: 150,000 m^2
Completion Time: 1985

万科第五园

建设地点：广东 深圳
建筑面积：125 000平方米
占地面积：130 000平方米
设计时间：2004–2005年
竣工时间：2005年

Vanke Park V

Construction Location: Shenzhen, Guangdong
Building Area: 125,000 m^2
Site Area: 130,000 m^2
Design time: 2004–2005
Completion time: 2005

人民大会堂

建设地点：北京
建筑面积：171 800平方米
占地面积：150 000平方米
设计时间：1958年
竣工时间：1959年

Great Hall of the People

Construction Location: Beijing
Building Area: 171,800 m^2
Site Area: 150,000 m^2
Design Time: 1958
Completion Time: 1959

中国历史博物馆和中国革命博物馆

建设地点：北京
建筑面积：65 152平方米
设计时间：1958年
竣工时间：1959年

National Museum of Chinese History & Museum of the Chinese Revolution

Construction Location: Beijing
Building Area: 65,152 m^2
Design Time: 1958
Completion Time: 1959

毛主席纪念堂

建设地点：北京
建筑面积：28 225平方米
占地面积：57 200平方米
设计时间：1976年
竣工时间：1977年

Chairman Mao Memorial Hall

Construction Location: Beijing
Building Area: 28,225 m^2
Site Area: 57,200 m^2
Design Time: 1976
Completion Time: 1977

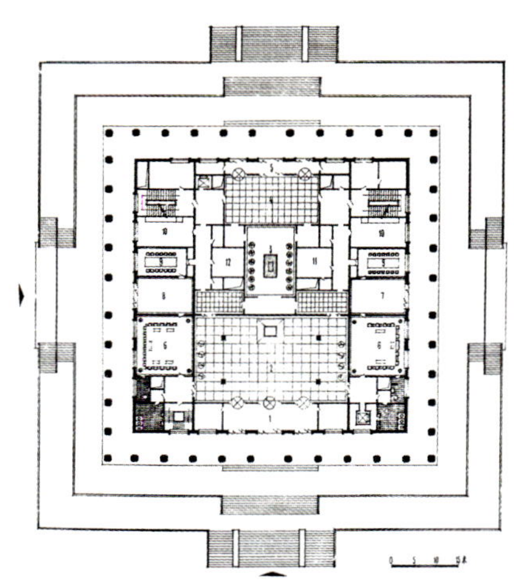

Bailin Architectural Design and Consultant Corporation Ltd.

北京白林建筑设计咨询有限公司

思想是作品诞生的基础

地址：北京市西城区西外大街1号西环广场T2 19层C1
电话：+86-10-58301336/7
传真：+86-10-58301336/7-606
邮箱：bailin.bailin@263.net
网址：www.bailindesign.com

Add: West Central Plaza T2 19-C1,
West Avenue, Xicheng District, Beijing
Tel: +86-10-58301336/7
Fax: +86-10-58301336/7-606
E-mail: bailin.bailin@263.net
Http://www.bailindesign.com

公司简介

Company Profile

基本原则

先做人，后做事，再做建筑。

办所理念

研究型，追求卓越，培育人才。

设计理念

追求"思想性，艺术性，精神性，前瞻性"。

设计方法论

"建筑是思想的容器"。
"以人为本，以环境为源，以科技为手段"。

学术目标

探索中国的建筑现代化与中国文化的融合。
探索中国人的价值观与建筑的结合点。
为中国的建筑学术事业作出贡献。

服务态度

超越自我，超越甲方，谋求甲方最根本的利益。

Principle

Firstly conduct ourselves, secondly conduct our business, and then the architecture.

Aim

Research, Pursuit of Excellence, Cultivation of Talents.

Design Concept

Pursuit of In-Depth Thinking, Artistic Spirit, Spiritual and Inspiring Prospect.

Design Methodology

"Architecture is the container of the thinking." "People as the base, environment as the source, science and technology as the means".

Academic Goals

To explore the intergration of the modernization of architecture in China and the Chinese culture.
To explore the joint of the Chinese values and architecture.
To make contribution to the architectural academic career of China.

Attitude for Service

Surpass ourselves, surpass Party A, and seek for the fundamental interests of Party A.

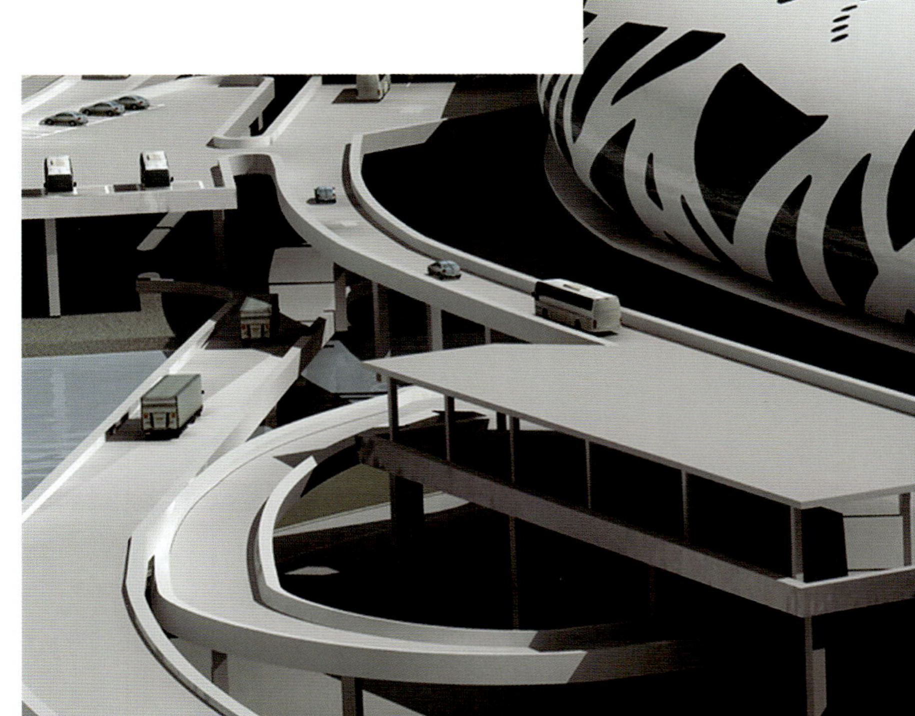

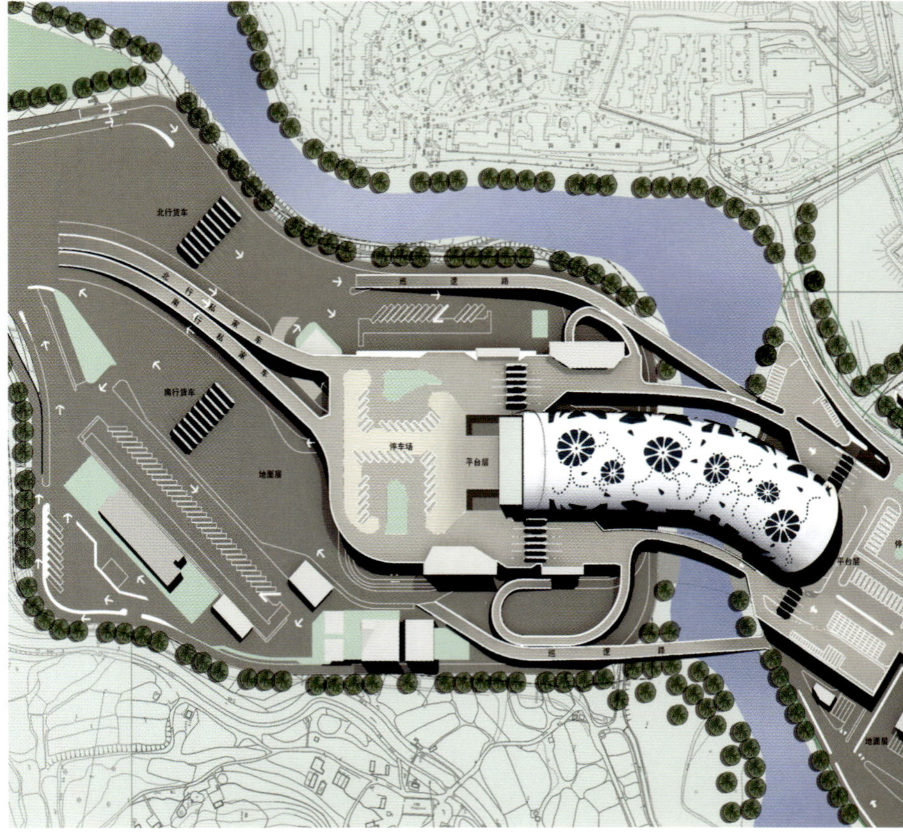

（深圳／香港）莲塘／香园围口岸联检大楼概念设计（国际竞赛）

建筑面积：46 900平方米
地　　点：深圳、香港
时　　间：2011年

设计理念

自然、简单、和谐、顺畅，合二为一。建筑是思想的容器。它应该表达人与人、人与自然、城市与城市相互依存、和谐共处的思想。

Liantang/Heung Yuen Wai Boundary Control Point Passenger Terminal Building (Shenzhen/Hongkong)

Conceptual Design (bid letter)

Building Area: 46,900 m^2
Location: Shenzhen; Hong Kong
Time: 2011

Design Concepts

It expresses the natural, simple, harmonious, smooth and combined design elements. Construction should be the smelter of different thoughts, but all in all it expresses the harmony and interdependence between people and people, people and nature, and city and city.

对我们来说什么是竞标？

竞标就是创造性地提出新思想、新观点、新思路。通过具体的地块条件，用建筑或规划的语言表达我们对时代精神和社会问题的思考和认识。

What does the bidding mean for us?

Bidding means the creative new ideas, new perspectives, new outlets, so we could use certain block and construction plan to express the world view and values.

总体鸟瞰图

山东潍坊滨海经济开发区中央商务区城市设计（邀标）

建筑面积：5 400 000平方米
地　　点：山东 潍坊
时　　间：2010年

Urban Design of Central Business District in the Binhai Economic Development Zone, Weifang, Shandong (bid letter)

Construction Area: 5,400,000 m^2
Location: Weifang, Shandong
Time: 2010

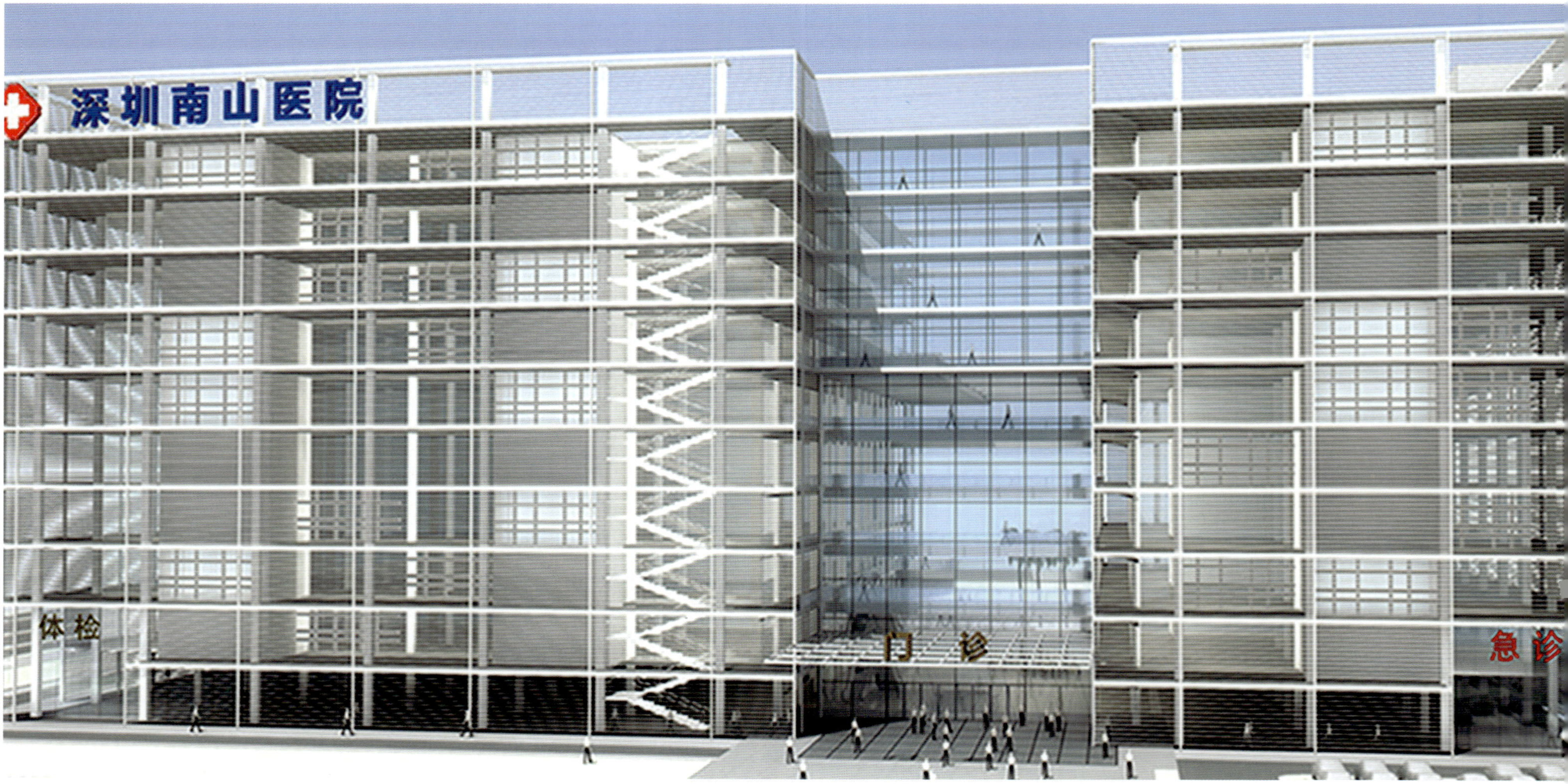

深圳南山医院改扩建工程方案设计
(资格预审方案)

用地面积：54 000平方米
建筑面积：250 000平方米
地　　点：广东 深圳
时　　间：2010年

Nanshan Hospital Expansion Project in Shenzhen City Conceptual Design (Prequalification Programme)

Land Area: 54,000 m^2
Construction Area: 250,000 m^2
Location: Shenzhen, Guangdong
Time: 2010

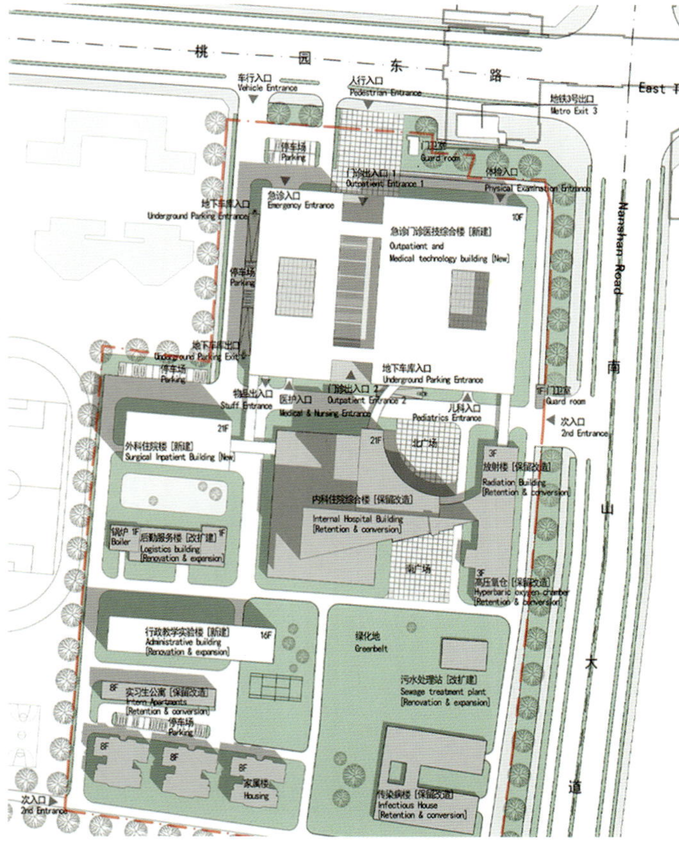

实景

太湖湾旅游度假区北大门方案及施工设计（议标）

Construction Design of the Northern Gate of the Tai Lake Bay Resort program (bid letter)

大　门：长101米，宽16.6米，高33米
地　点：江苏 常州
时　间：2006年

Door: Length 101 m, Width 16.6 m, Height 33 m
Location: Changzhou, Jiangsu
Time: 2006

设计理念
对建筑师来说最难做的设计就是大门。

Design Conception
It is more difficult for architects to design the door.

太湖湾旅游度假区西大门方案及施工设计

大　门：长32米，宽12米，高16.7米
地　点：江苏 常州
时　间：2007年

West Gate of Tai Lake Bay Resort Program and Construction Design

Door: Length 32 m, Width 12 m, Height 16.7 m
Location: Changzhou, Jiangsu
Time: 2007

01. 深港同城一体，共建世界都市
02. 理顺周边关系，描绘发展策略
03. 布局框架结构，构建整体形象
04. 紧凑城市空间，合理路网结构
05. 系统组织交通，绿色健康出行
06. 商业商务交通，相互支持体系
07. 运用山水资源，打造滨海景观
08. 地下空中结合，构筑立体城市
09. 倡导低碳行动，思考落实行动
10. 活用一国两制，推进开发策略

01. Integrate Shenzhen and Hong Kong; build a global city.
02. Harmonize peripheral relations; portray future blueprint.
03. Deploy a frame pattern; build up an overall image.
04. Refine the urban space; rationalize road network structure.
05. Systemize traffic organization; encourage green travel.
06. Combine commerce; business and common transport; establish a mutually supporting system.
07. Exploit natural resources; build a coastal landscape.
08. Link with underground; create a vertical city.
09. Promote low-carbon operation; set up an eco city.
10. Utilize "One Country, Two Systems" flexibly; push forward development strategy.

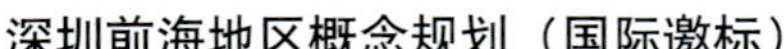

深圳前海地区概念规划（国际邀标）

建筑面积：18 040 000平方米
地　　点：广东 深圳
时　　间：2010年

International Bidding of the Conceptual Planning in Shenzhen Sea Region (bid letter)

Construction Area: 18,040,000 m^2
Location: Shenzhen, Guangdong
Time: 2010

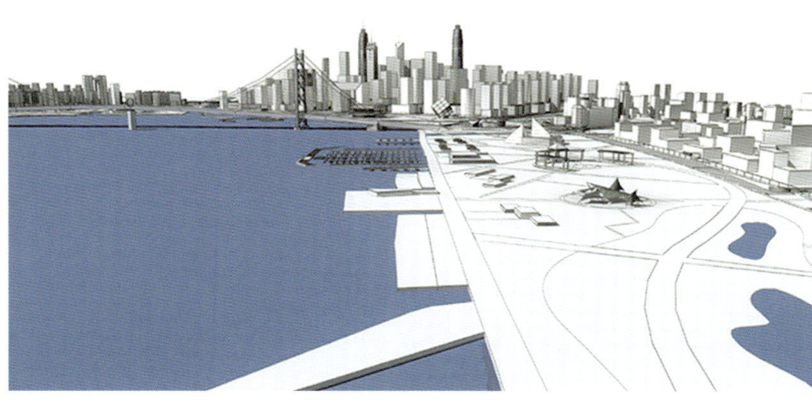

Institute of Architectural Design and Research Chinese Academy of Sciences Co., Ltd.

中科院建筑设计研究院有限公司

中科院建筑设计研究院有限公司前身为中国科学院北京建筑设计研究院，成立于1951年。公司拥有建筑行业建筑工程甲级资质、市政公用行业（热力）甲级资质、城乡规划乙级资质。目前公司有员工共409人，其中工程技术人员336人（高级以上职称75人，中级技术人员76人，一级注册建筑师26人、一级注册结构工程师20人、水暖电热力一级注册工程师28人）。除总部外，公司在广东东莞、浙江杭州、辽宁沈阳、四川成都、河南郑州、上海、陕西西安、江苏南京设有分公司。

公司致力于提供建筑行业全专业设计总承包业务，配备有各专项设计团队，包括热力所、综合一所、综合二所、综合三所、惠中设计所、环艺设计所、环境工程所、景观设计所、照明设计所、公共艺术中心、智能化设计所等专业设计机构。能够实现的全专业设计服务包括小区规划、建筑设计、市政设计、园林景观设计、照明设计、公共艺术（雕塑及VI等）、室内装饰装修设计（含配饰及家具选型）、智能化系统设计等。

公司能够完成大型文化、办公、商业、医疗、体育类建筑设计，城市规划设计，居住区规划和各类住宅设计，历史街区保护规划及设计，环境景观及室内装修设计、结构改造加固等，尤其擅长科研教育类建筑规划及设计，包括科研基地规划、科研办公建筑设计、学校规划及设计等。同时能够进行项目策划、项目管理、工程造价咨询等工作。

公司创作设计了一批具有社会影响力的建筑精品，如中科院文献情报中心、国家开发银行、中国人民银行重点库、中国驻埃塞俄比亚大使馆、中国驻贝宁大使馆、中科院研究生院教学楼、中科院苏州纳米技术与纳米仿生研究所、中科院生态中心实验楼、中科院青岛生物能源与过程研究所、中科院烟台海岸带可持续发展研究所、中科院动物所、中科院电子所总装备楼、中科院计算所、北京正负电子对撞机工程、国家天文台、石油部物探局计算中心及亿次银河机机房工程、北京基督教丰台堂和朝阳堂工程、上海万科城市花园、北京万科星园（二、三期）、天津万科假日风景、北京阜内大街历史文化保护区的保护规划等。曾获得国家级奖励10项、省部级奖励60余项。

1998年到2006年间（2007年未举办首都汇报展），在第5届到第13届首都建筑设计汇报展中，公司11个项目获得14个奖项，其中6个为方案设计最高奖。公司设计的“中国科学院文献情报中心”荣获全国优秀工程设计最高奖项“金奖”。2006年公司被地产界评为“北京地产十佳建筑设计机构”，获得万科集团最佳设计合作伙伴奖，并成为沿海集团的策略联盟——指定设计合作伙伴。

2009年，公司负责设计的中国科学院文献情报中心（中国国家科学图书馆）、九寨沟国际大酒店分别获“中国建筑学会建国60周年建筑创作大奖”；国家动物博物馆及中科院动物研究所科研实验楼、标本楼获北京市第十四届优秀工程设计公共建筑设计一等奖；北京万科紫台家园项目获北京市第十四届优秀工程设计住宅建筑设计三等奖；北京万科四季花城获2009年詹天佑大奖优秀住宅小区金奖；此外，为表彰公司在北京奥运会残奥会环境建设工作中作出的突出贡献，北京市“2008”奥运环境建设指挥部为其颁发了荣誉证书。公司先后与欧美、日本、澳大利亚、中国香港、中国台湾等国家和地区的有关专家机构，进行了广泛的学术交流或合作设计，与法国安东尼·贝叙建筑设计公司、美国Perkins Eastman建筑设计事务所、意大利罗马大学签订了长期合作协议，具有丰富的国际合作经验，并赢得了良好的声誉。公司设计专家团队相信物理环境对生活、工作及学习质量有重大的影响，而设计良好的空间则会影响个人对环境的感受与使用。公司专业化的团队同时密切关注所擅长专业领域的发展趋势，保持在创新中的领先地位。企业秉承精心设计、诚信守约、追求精品、锐意创新的经营理念，充分利用实力和优势为业主提供无边界的服务。

Institute of Architectural Design & Research, Chinese Academy of Sciences (CAS) grew out of Beijing Institute of Architectural Design & Research (BIAD), CAS. Founded in 1951, this company ranks Class-A in construction works and in heating power of municipal public utilities, Class-B in Urban-rural Planning. At present, the staff number of this company is 409 altogether, 336 of whom are engineering and technical personnel. Among them 75 people are qualified with senior professional title, 76 are intermediate technicians, 26 get PRC Class-1 Registered Architect Qualification, 20 are PRC Class-1 Registered Structure Engineers, and 28 are PRC Class-1 Registered Water and Electric Heating Engineers. Besides the headquarter in Beijing, it has also established branches in Dongguan of Guangdong province, Hangzhou of Zhejiang province, Shenyang of Liaoning province, Chengdu of Sichuan province, Zhengzhou of Henan province, Xi'an of Shaanxi province, Nanjing of Jiangsu province.

The company is dedicated to providing comprehensive design service and supports professional design institutions, including Heating Power Studio, Comprehensive Institutes No.1, No.2 and No.3, Huizhong Institute, Environment and Art Institute, Environmental Engineering Institute, Landscape Studio, Lighting Studio, Public Arts Studio, Intelligent Design Studio and so on. It has the ability to complete turn-key project such as residence planning, architectural design, municipal design, landscape design, lighting design, public arts (contain sculptures, VI etc.), interior design (contain decoration and furniture types), intelligent design and so on.

The business scope of this company covers architectural design of large-scale cultural buildings, office premises, hospital and medical buildings and sports venues. This company also works on urban studies, residential district planning, and all kinds of residential designs, protection planning and programming of historic district, landscaping and inner decoration designs, and reform and reinforcement of construction. This company boasts the excellent architectural drafting and designing of research and education buildings such as planning of science facilities, design of scientific research buildings, and planning and design of school buildings. Meanwhile, project planning, project management and cost consultation of construction project are involved in the business scope.

This company has designed and completed an array of high-quality construction projects which have immense social influence. Those brilliant construction projects include Library of Chinese Academy of Sciences, China Development Bank, key locations of the People's Bank of China, Chinese Embassy in Ethiopia, Chinese Embassy in Benin, the academic building of Graduate School, CAS, Suzhou Institute of Nano-tech and Nano-bionics, CAS, laboratory block of Biotechnology Center, CAS, Qingdao Institute of Biomass Energy and Bioprocess Technology, CAS, Yantai Institute of Coastal Zone Research for Sustainable Development, CAS, Institute of Zoology, CAS, general equipment building of Institute of Electron, CAS, Institute of Computing Technology, CAS, Beijing electron-positron collider project, Nation Astronomical Observatory, computing center of Geophysical Prospecting Bureau, Petroleum Department, Fengtai Church and Chaoyang Church, Beijing Christian Council, Shanghai Vanke City Garden, Beijing Vanke Xingyuan (Phase 2 and Phase 3), Tianjin Vanke Holidays Scenery, Protection planning of Fuchengmen Inner Street conservation of historic sites and so on. This company has obtained 10 state-level awards and over 60 province-level encouragement awards.

From the 5th to 13th capital planning exhibitions on the architectural design (from 1998 to 2006), the 11 projects accomplished by this company had won 14 awards, 6 of which were the highest conceptual design awards. Buildings of Library of Chinese Academy of Sciences designed by this company won gold medal, the highest prize of National Excellent Projects. In 2006, it was named as one of "Beijing Top Ten Planning Architectural Design Organizations" by the real estate industry, was honored as"'Best Designing Partner of Vanke"and became the unique authorized designing partner of strategic alliance of Coastal Group.

In 2009, the designs of buildings of Library of Chinese Academy of Sciences and Jiuzhaigou Valley International Hotel won Architectural Creation Prize held by Architectural Society of China in celebration of the 60th Anniversary of the Founding of PRC. The designs of The National Zoological Museum of China, the research activities and laboratory building and the specimen building of Institute of Zoology, CAS were awarded Meritorious Winners of Beijing the 14th Excellence Engineering and Public Building Designs. the project of Beijing Vanke Zitaiyuan Home won the third prize of Beijing the 14th Excellence Engineering in Residence Architectural Design. The design of Beijing Vanke Four Seasons Flower City won gold prize of outstanding residential areas of 2009 Zhan Tianyou Awards. In addition, 2008 Beijing Olympic Environment Construction Command issued certificates to this company which has made outstanding contributions in environment development during the Paralympics in Beijing in 2008.

This company has successively conducted extensive academic exchange and designs cooperative with the relevant exports and professional institutions in Europe, the USA, Japan, Australia, Hong Kong China, Taiwan China and other countries and regions. It signed a large long-term cooperation with Anthony Architects in France, Perkins Eastman in USA and Univ Roma La Sapienza Italy. Through the cooperation, this company has gained extensive experience and well-deserved reputation. Specialist team in this company believes that physical environment tremendously influence people's working and studying life, so the beautified exterior space is very important. The team, in the meantime, pays close attention to the present situation and the trend in this field and attempts to continue to rank among the leading position through innovation. This company will stick to operation principles of supplying with well-designed and competitive products, valuing honesty and integrity, upholding the spirit of innovation to provide the best service.

科学园南里

科学园南里项目，位于北京奥运主场馆西侧，大屯路与安翔北路之间，东临北辰西路。基地南侧是盘古大厦，北侧、西侧是风林绿洲等住宅小区。建筑用地东西向最宽处近200米，南北向沿北辰西路长约800米，总地块面积约10公顷。

后奥运时代，体现绿色、科技、人文的北京奥运精神，实现低碳城市设计、可持续发展策略、生态模式，为北京建设世界城市的新面貌探索，倡导城市设计与旧区改造的结合。

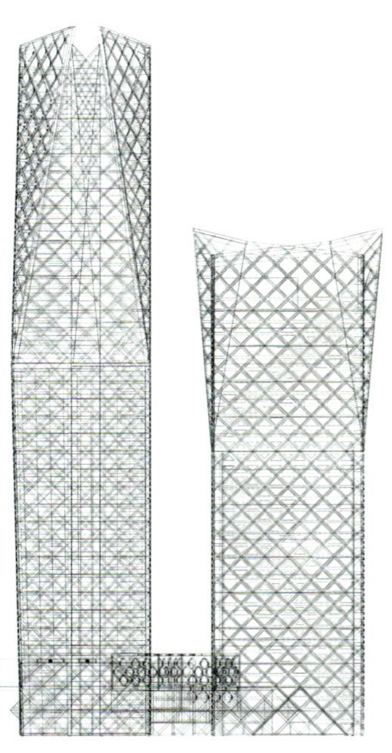

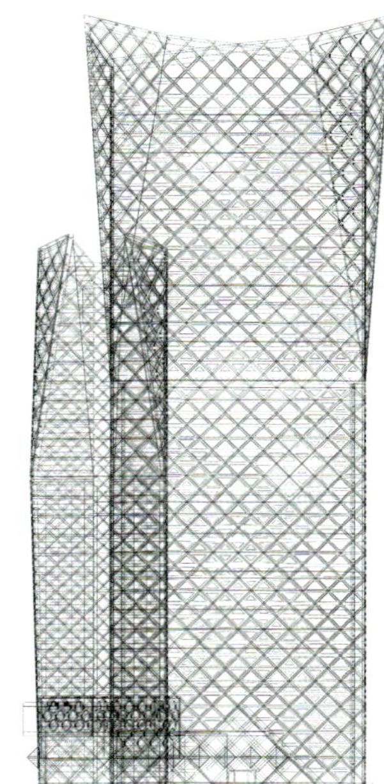

八里庄超高层办公楼

项目背景

项目定位为5A甲级写字楼，节能达到LEED银质认证标准。

区位分析

用地位于北京市朝阳区红领巾公园路20号、东四环红领巾桥东南角。用地东侧为中国音乐学院附中，南侧为二道沟河，西侧为东四环，北侧为朝阳北路。具体位置详见地形图。项目紧邻北京中央商务区东扩区，地理位置重要，区域位置标志性强。

创作理念

1. 结构美学
2. 自由开放空间
3. 形象——空中盛开的花

建筑群体在对外形象上形成了一组对拼的花瓣，双子座相互呼应，盛开在北京城市的上空，无论白昼都构成一幅完美的图画，寓意欣欣向荣，锦绣繁华的美好前景。

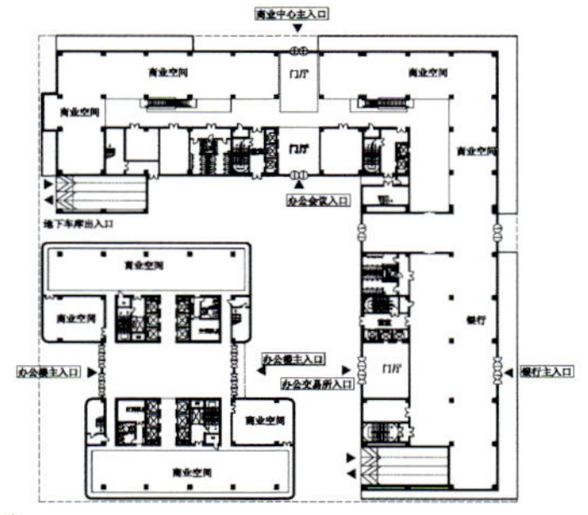
1

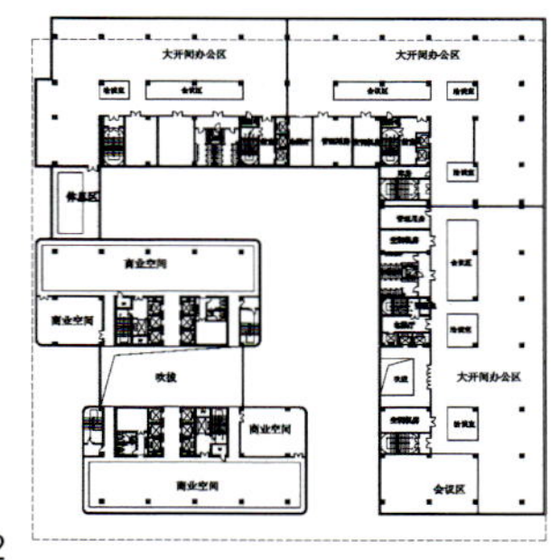
2

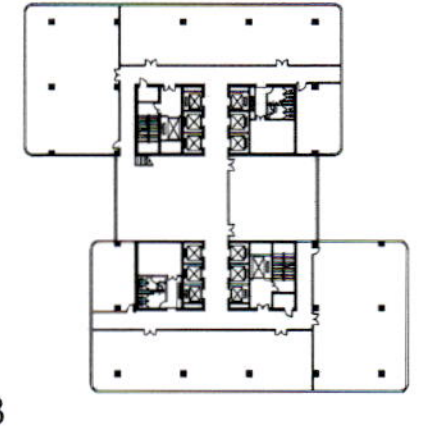
3

4

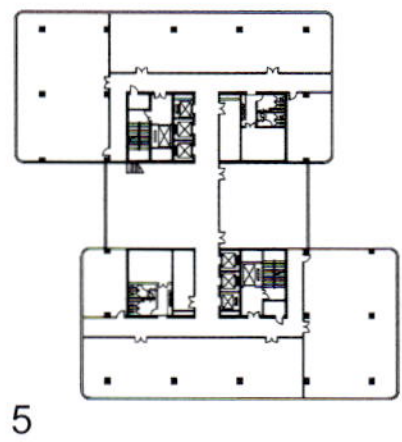
5

6

7

8

1 首层平面图
Plan of ground floor
2 3层平面图
Plan of 3rd floor
3 7至15层单数层平面图
Plan of 7th to 15th floor
(single-number floor)
4 8至14层双数层平面图
Plan of 8th to 14th floor
(double-number floor)
5 17、19层平面图
Plan of 17th and 19th floor
6 18、20层平面图
Plan of 18th and 20th floor
7 22、24层平面图
Plan of 22rd and 24th floor
8 23、25层平面图
Plan of 23rd and 25th floor

天津滨海新区于家堡金融区起步区03–20地块

用地面积：11 002.5平方米
建筑面积：77 017平方米
建筑高度：142.45米

天津滨海新区金融街首期启动项目以“集群设计”的面貌出现，旨在聚合各种能量，实现协同式的建筑实践。幸运的是我们的地块靠近海河，这便有一种可能使存在于自然与中央商务区之间的媒介体形成，并进而探索河边的城市建筑的敏感性，包括从场所周围吸收营养以滋养自身成长，同时将这种能量又回馈于环境。其一，体现在轻柔地触动海河以形成空透的界面；其二，中间喜悦而动人的城市空间把阳光、景观引入并贡献给城市。

鄂尔多斯 20+10

鄂尔多斯的财富是“天”的恩泽和“地”的造化，昔日的鄂尔多斯草原，今天成为最有发展潜力的城市，它的迅速崛起源于这片神奇的土地，脚下的“稀土、煤炭、天然气”足以让它富有百年。我们怎能不对地球所赐赞美，怎能不对远古的沉淀敬畏……我们的设计便源于这片独特的地脉文化。

在这个城市中，有一处别样的山谷和丘陵，我们编织着关于现代“游牧部落”的梦想：一个天地之间的神话；一个现在与过去的不期而遇；一个隐性自然与显性自然在此凝聚和风化中演绎……

建筑犹如“钻探”沉积的砂岩方体暴露于天地之间，任自然风吹雨打，雕蚀成形。建筑“记录”了流沙沧海，天光云影。

玉树藏族自治州游客服务中心

玉树藏族自治州游客服务中心，是应青海省建设厅、玉树藏族自治州政府的委托，由中国建筑学会牵头，针对玉树414地震灾后重建10大重点项目的设计援建工程。本项目位于玉树藏族自治州政府所在地，玉树结古镇，巴塘河与扎西科河交汇处半岛，海拔约3 600米，是具有独特服务属性与景观属性的重要城市节点，是典型的高原山地滨水建筑。

玉树藏族自治州位于青藏高原腹地，康巴藏区核心，自唐以来商贸发达，藏族为该地主体民族。建筑外观设计源于对玉树地区独特“帐房”建筑的研究与解析。营造了具有浓郁康巴风情，符合游客中心主题，汉藏通达的建筑意向。

广州亚运城岭南水乡民俗主体建筑

建设地点：广东 广州
建筑面积：1 893平方米
设计时间：2008年

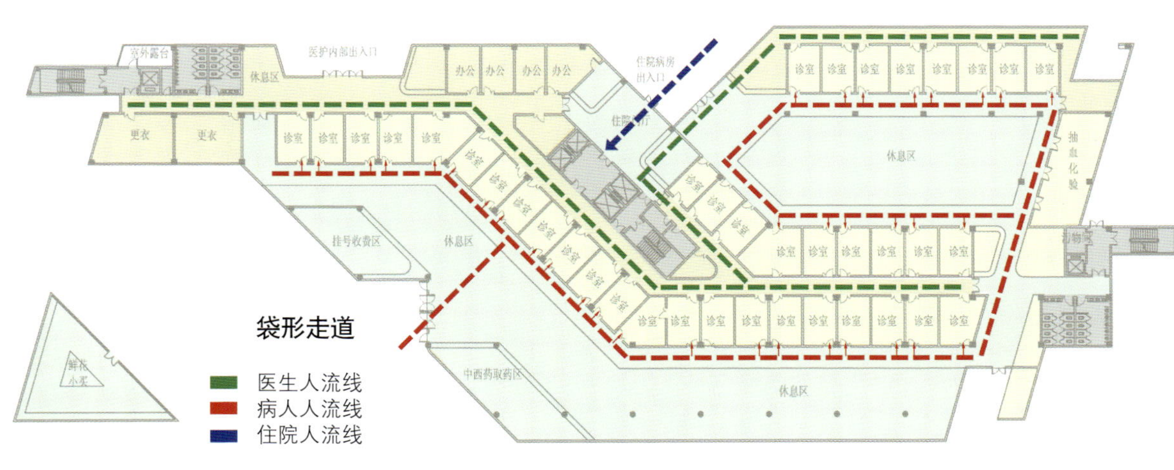

蒙中医院

建设地点：内蒙古 乌海
建筑用途：医疗
建筑面积：15 545平方米
建筑高度：50米
建筑层数：地上12层，地下1层
设计时间：2006年
竣工时间：2010年

本项目是集门诊、急诊、病房、手术、中心供应、办公等多功能为一体的地区医院，同时也是该地区唯一一家蒙医、中医相结合，具有浓郁民族特色的综合性医院。设计中提出了独特的袋形医疗空间构想，彻底解决了医患、洁污分流的设计难题。建筑体型采用相互咬合的双L形体形，在顺应平行四边形基地环境的同时，巧妙地隐喻了中医、蒙医的完美结合与强烈的衣钵传承，同时又迎合细胞DNA的链形关系。总体布局的分期建设表达出了强烈的医疗建筑细胞可持续发展理念。

北京清润国际建筑设计研究有限公司

Beijing Tsingrun International Architectural Design and Research Co., Ltd.

具备先进设计理念与杰出设计能力的股份制公司

设计骨干来自国内名校与海外，有浓郁的学院氛围提供策划、规划、建筑、景观、室内等全程服务高品质作品与坦诚交流赢得了广泛的认同与尊重众多知名公司、政府等组成了稳定恒久的客户群清润国际珍惜每一次设计机会

清润国际真诚呵护每一位员工

清润国际清新自然、润物育人

A joint-stock company with advanced design concepts and outstanding design capabilities.

With all key designers coming from the top universities or having international study background, the company is rich of academic atmosphere.

Providing full services covering scheming, planning, architecture, landscaping, interior design and so on.

High-quality works, open and honest exchange have won them wide acknowledgement and recognition.

Numerous well-known enterprise corporations and governmental departments have formed a stable and long-term customer base.

Tsingrun cherishes every design opportunity.

Tsingrun delivers considerate care for each employee.

Tsingrun, means purity and freshness, nourishing minds and cultivating people.

地址：北京经济技术开发区西环南路26号院11号楼
邮编：100176
电话：+86-10-67856060
传真：+86-10-67856060-104
邮箱：tsingrun2006@126.com

Add: 11th Building, 26th Yard, Xihuan South Road,
Beijing Economic-Technological Development Area
P.C. :100176
Tel: +86-10-67856060
Fax: +86-10-67856060-104
Email: tsingrun2006@126.com
Http://www.tsingrun.com.cn

康和盛世综合社区
工程位置：内蒙古鄂尔多斯
建筑面积：180 000平方米
设计时间：2009年
开工时间：2009年

领秀·尚城综合社区
工程位置：内蒙古 达拉特旗
建筑面积：210 000平方米
设计时间：2010年
开工时间：2011年

汇金国际综合社区
工程位置：内蒙古 达拉特旗
建筑面积：112 000平方米
设计时间：2011年

B15住宅小区配套会所
工程位置：内蒙古 鄂尔多斯
建筑面积：3 400平方米
开工时间：2011年

河套酒业（集团）股份有限公司办公楼
工程位置：内蒙古 陕坝
建筑面积：18 000平方米
设计时间：2010年
开工时间：2010年

公园会所
工程位置：内蒙古 鄂尔多斯
建筑面积：4 000平方米
设计时间：2010年
开工时间：2010年

创业大厦
工程位置：河北 廊坊
建筑面积：24 000平方米
开工时间：2009年
竣工时间：2011年

第二中学
工程位置：内蒙古 鄂尔多斯
建筑面积：38 000平方米
设计时间：2008—2010年
开工时间：2010年

第五中学
工程位置：内蒙古 鄂尔多斯
建筑面积：37 000平方米
设计时间：2010—2011年

第七小学
工程位置：内蒙古 鄂尔多斯
建筑面积：28 500平方米
设计时间：2010年

包钢集团职工医院综合病房楼
工程位置：内蒙古 包头
建筑面积：78 000平方米
开工时间：2008年
竣工时间：2009年

包钢集团第三职工医院综合门诊住院楼
工程位置：内蒙古 包头
建筑面积：43 000平方米
开工时间：2010年

包钢集团研发基地
工程位置：内蒙古 包头
建筑面积：120 000平方米
设计时间：2010—2011年
开工时间：2011年

包钢集团体育馆
工程位置：内蒙古 包头
建筑面积：13 560平方米
设计时间：2008年

内蒙古达拉特旗领秀 · 尚城　景观用地面积：7.3公顷　开工时间：2011年

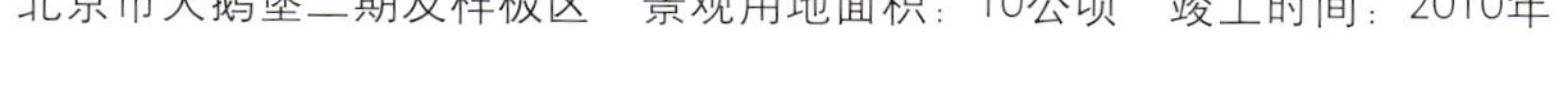

北京市天鹅堡二期及样板区　景观用地面积：10公顷　竣工时间：2010年

内蒙古鄂尔多斯市康巴什新区西公园　用地面积：15.98公顷　竣工时间：2011年

内蒙古鄂尔多斯市康巴什新区中公园　用地面积：19.73公顷　竣工时间：2011年

SYNarchitects

德国SYN建筑师事务所

&

北京华诚博远建筑设计有限公司
Beijing Huacheng Boyuan Architectural Designing Co., Ltd

OFFICE BERLIN
Tel: +49 30 322970400 Fax: +49 30 322970402 Add: Badensche Str.29
10715, Berlin, Germany
北京公司地址：北京西城区宣武门外大街10号庄胜广场中央办公楼北翼13A层
邮编：100052 电话：+86-10-57693518 传真：+86-10-57693520 邮箱：syn_lahc@163.com

北京华诚博远建筑设计有限公司 总部电话：+86-10-57693601 传真：+86-10-57693609 网址：www.lahc.com.cn

SYN建筑师事务所是放眼亚洲与欧洲，致力于建筑实践与建筑功能、空间学术研究的国际性事务所。

事务所2003年成立于德国柏林，并于2004年在北京设立工作室。通过与中国建筑设计师及设计院多方面多层面的合作，具备设计各种类型及深度设计任务的能力，并先后于不同领域探索及实现不同的作品。例如，各种类型公共建筑，商业建筑，博物馆、美术学院等文化设施，体育设施，住宅等。同时，通过与业主紧密的沟通与探讨，从整体规划开始，在完成建筑设计过程中，对景观设计、室内设计给予充分的重视。

SYN建筑师事务所坚持将全程设计服务的观念作为指导，从前期概念到实践过程，积极与业主配合，期望与业主一起，完成共同的建筑理想。

SYN Architects is an international architecture firm that keeps Asia and Europe in view. We are dedicated to architecture practice, architecture function and architecture space study.

SYN is founded in Germany in 2003 and set up a new branch in Beijing in 2004. Through the multifaceted and multi-layered cooperation with many Chinese architects and design institutes, we possess the ability to design for the various types and depth design mission. We researched on different fields of architecture and realized our design ideas in cultural facilities, sports facilities and housing. During that time, for example, we explored the various types of public buildings, offices, museums, public galleries and so on. Meantime, through work in close cooperation with our client, we provide our adequate attention to landscape design and interior design as well as from the overall planning to architecture design.

SYN Architects' main guiding principle is to provide the service of building whole-life design for the client. From conception to practice of design, we build a close cooperation relation with our client to make the common architecture ideal come true.

华诚博远（北京）建筑规划设计有限公司成立于1993年1月，具有建筑工程设计甲级资质、风景园林工程设计专项乙级资质、城市规划编制乙级资质。经北京市科委、北京科技咨询业协会评审，认定为“北京科技咨询信用单位”。2011年经北京勘察设计行业协会评为《北京地区工程勘察设计行业诚信单位》。

公司自成立以来，先后承接了文化体育、办公、居住、医疗、工业、室内、商业综合体等多领域工程设计及相关工程咨询服务，业务遍及全国。由资深设计师组成的高水平设计团队，以先进的设计理念、丰富的设计经验、高质量的设计作品、良好的设计服务信誉，赢得了业主肯定和业内广泛好评。发扬“诚信、创新、增长、高效”的企业精神，积极开拓市场。以先进的技术、周到的服务、高效的管理、最佳的效益为目标，不断实现技术与管理创新，努力成为国内知名企业。

Huacheng Boyuan (Beijing) Architectural Designing Co., Ltd. was founded in January 1993. The company has architectural engineering designing grade A, and was identified as "Beijing Science and Technology Consulting Credit Unit" by Beijing Science and Technology Industry Committee. Since its founding, the company has undertaken the cultural & sports, office, residence, medical, industrial and commercial buildings design and engineering consulting services. Business spreads all over the country. The high level design team of senior designers with advanced design concept, rich design experience high quality design works and good service prestige get clients" praise and affirmation. The company has "honesty, innovation, growth, high efficiency" as the spirit. And with"advanced technology, considerate service, efficient management, and the best benefit" as the goal, we work hard to realize technology and management innovation, hoping to become a domestic famous enterprise.

龙门博物馆

本项目是以龙门石窟1 500年的历史文化积淀为基础，并融合了广大佛教艺术的主题博物馆，是世界文化遗产之一“龙门石窟”1 500年来最重要的一项工程。本项目在5年的设计过程中，在干预下不断地修改，但是核心设计理念一直保留下来。该设计理念是以佛教哲学为中心，以“儒”、“释”、“道”三教哲学的互相影响与融合为背景，以中国人文艺术的“心性与直觉”为突破口，以龙门1 500年历史为基础的综合研究与分析的过程。总结如下：

①信仰的力量；
②空间是精神的载体；
③龙门——具有不可替代性的空间；
④可视世界与理想世界的过渡空间；
⑤龙门——反建筑的空间；
⑥佛教艺术的“空”与“圆”。

最终，龙门博物馆是以佛教艺术为背景的现代建筑空间，是“形而上”地展示龙门石窟精神而非表象的建筑。建筑设计的过程给予我们很多的启示，令我们更加认识到中国传统哲学中蕴涵的广阔的空间，成为我们进行设计时最重要的思想宝库，也奠定了我们深入挖掘中国传统文化、研究中国的现代建筑设计的方向，并将其定为职业生涯的主要任务。

宁波数码媒体产业园

宁波数码媒体产业园是宁波报业集团于宁波市南部商务区打造的包括“数字传媒基地”、“国家新媒体平台”、“文化产业中心”和其他配套的商务服务设施的以数字媒体为特色与主体的文化创意产业园。本项目建设用地约1.33公顷，建设面积6万平方米。其使用过程中除考虑大部分为宁波报业集团自用外，需同时具备便于管理与分区的相对独立的出租与出售的办公区。项目设计思路与理念首先是关注“数字”这一概念及其内在性质与外在的表现方式，以及体现在建筑上的影响。数字时代的来临，以超乎人类想象的速度与能量改变着我们的世界，并深刻地影响着我们的内心。而宁波报业集团不仅是数字媒体的先锋，更是文化产业的主力军，文化是深入人心的真正的力量所在，所以不能为了数字而数字。本方案将数字的力量、艺术的力量、时间的力量并在一起，方案理念分析时共同列举著名雕塑家亨利·摩尔的雕塑作品与饱历岁月的太湖石。从它们各自造型的共性与个性上分析，寻求建筑语言上的答案，求证出本项目的造型特点，并且通过周边项目与功能的分析，帮助解决本项目规划与建筑设计中的空间方案。通过中国传统文化中的“有无相生”的理念将建筑与场所结合，以建筑的“实有”表现视觉不易感到的数字的力量、艺术的气质与时间对建筑的长远影响。

从建筑表面上看为一幢建筑，实际上由两幢建筑组成，两幢建筑之间为高大的极富表现力的室内中庭，除空间质量优异外，还具备良好的室内外气候调节的作用，并有利于功能划分，各有侧重，符合业主使用的需求。

宁波宜家广场

本项目位于宁波市鄞州区潘火地区的下应镇，距市区中心、栎社国际机场、港口只有15分钟的车程；由国骅集团投资开发，随着宁波东部新城的建设，国骅宜家广场将地处极其重要的位置。本项目总占地面积约5.71公顷，南北方向东侧最宽处宽约235米、西侧最窄处约为142米，东西方向北侧最长处约为373米，南侧最窄处约为110米。

场地布置

根据场地现状、规划条件，用地被下应大道分为A、B两个地块，下应大道用地由南向北贴临B地块，并在北侧A地块处转为由西向东，故A、B两地块大小不等。A地块呈矩形，B地块近似正方形。

“一横、两纵”的规划结构把基地分为三个区，高层建筑按需分布。公寓位于地块的东侧，临近规划河道；酒店位于基地西侧，毗邻排水干河。两者都临水而布，拥有良好的景观。商业置于两者之间，二层以上设置商业环路，把商业合为一个整体。

为求合理利用地下空间，在满足技术指标、充分考虑了经济性的前提下。地下建设两层汽车库，A、B地块通过地下二层相互联通。地下二层局部战时作为人防工程使用。

竖向设计

场地依据用地现状在进行景观、道路设计的同时合理平衡土方量。因为项目所在地降水量较大，道路及广场雨水均采用有组织排水，设置雨水井和雨水管道，部分与市政雨水管相接或排至两侧河流，部分用于补充景观水池用水以达到节能环保要求。

交通设计

场地建筑及绿化覆盖率较大，场地内车行道路在满足消防与车行的前提下尽量少设计，场地地下整块为地下汽车库，满足了人车分流。

中央美术学院燕郊校区

中央美术学院燕郊校区项目是其继望京楼区后，另一大型校区，由中央美术学院附属中学及部分研究院及院系组成，建筑面积10万平方米左右，目前一期建筑5万平方米已完成。该项目难点在于一期建设部分为改扩建项目，将来的二期则相对自由。但是一期的规划应充分考虑到与将来二期建设的结合。所以本项目是在统一规划的基础上，对原来仅有框架结构的烂尾楼进行质的改变，并扩建一万多平方米的新建筑。校区规划重点在于解决场地中东西贯通的教学楼与其南北校区的关系，即如何令其成为整个校区的功能核心，而非南北校区交通、视线以及场地间的障碍。其次，就校区其余建筑与全新的交通流线及与二期建筑之间的组群关系借鉴中国园林的空间手法，通过新建的廊桥体系，以及相应产生的新的围合半围合的若干庭院，营造出丰富多变的主题庭院群，并在新建的建筑及扩建于老建筑周边的建筑部分采用与原建筑共同围合内庭院的手段，打造出通风良好、视线光线变化丰富的流动空间。

本项目建筑设计采用简约的现代风格，从自身功能的需求出发，特别是解决画室的采光与立面的关系问题。基于艺术院校自身的气质与审美特点，建筑设计中充分考虑的是如何预留其建筑空间的自我拓展与变化性空间，期望通过艺术家的自我完善与艺术作品的陈列与在建筑内部的创作来实现建设空间艺术的不善提升与多元化发展。目前，该项目已投入使用，实际使用效果良好，并得到艺术家与学生的一致好评。

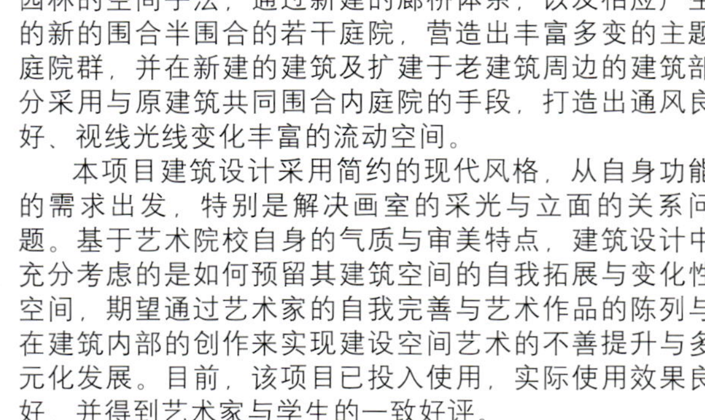

WESTERN

Beijing Western Architectural Design Co., Ltd.

北京威斯顿建筑设计有限公司

北京威斯顿建筑设计有限公司创立于1993年，经过近二十年的发展已成为国内颇具实力和极具影响力的甲级建筑设计公司。公司总部设在北京，在上海、武汉、石家庄、郑州和西安设有分公司。

威斯顿是一家在建筑规划、建筑设计、室内设计、建筑策划、工程总承包、项目管理、建设项目可行性研究以及技术咨询服务等领域均有建树的综合建筑设计公司。

威斯顿拥有国家一级注册建筑师15人，国家一级注册结构工程师18人，各专业技术人员100余人。威斯顿设计师具有极高的时代使命感和责任心，对建筑有全面的理解，对技术有细致的把握，对所承接的每项工程都给予最大的关注，并始终保持独立的思考和追求不断的创新。

Beijing Western Architectural Design Co., Ltd., founded in 1993, after nearly twenty years of development has become an influential architectural design company.
Western is headquartered in Beijing, with branch offices in Shanghai and Wuhan. Western now has 15 registered architects, and 18 registered structural engineers, professional technicians more than 100 people.
Our service ranges from urban planning, architectural design, interior design, construction planning, project management, construction projects, feasibility studies, to technical advisory services and other areas.
Western designers with a high sense of responsibility and mission, have a comprehensive understanding of the technology and detailed grasp of the undertaking of each project, and always maintain the independence of thought and pursuit of continuous innovation.

地址：北京市翠微路甲十号建筑大厦三层
电话：+86-10-68250274　68256311
传真：+86-10-68256313
邮编：100036
邮箱：westerndesign@sina.com

北京锋尚国际公寓

北京锋尚国际公寓是中国第一个"告别空调暖气时代"的"高舒适"、"低能耗"项目，这种没有传统空调和暖气片的高舒适度环保住宅，一年四季保持在20℃～26℃的人体舒适温度和湿度，置换式新风对人体健康极为有利。首次在中国实现了欧洲发达国家节能标准，引起社会的轰动，媒体对此进行了1 300多次报道，社会各界人士参观达5万人次，建设部有关部门发文件，要求全国各地建委组织到锋尚参观学习。

该工程施工中开发应用了天棚低温辐射采暖制冷系统、干挂饰面砖幕墙聚苯复合外墙外保温系统、健康新风系统、低辐射保温密闭外窗系统、垃圾处理系统、防噪音系统、水处理系统、屋面及地下系统等环保装饰施工成套技术。其中天棚低温辐射采暖制冷系统施工技术为核心技术，健康新风及干挂饰面砖幕墙聚苯复合外墙外保温施工技术为主要配套技术。

北京锋尚国际公寓已被国家住宅居住环境工程技术研究中心和北京市建筑与墙体革新办公室，分别评定为"高舒适度代低能耗住宅实验基地"和"北京市建筑节能试点小区"。

南京锋尚国际公寓

中国第一个零能耗住宅项目，低碳住宅代表。北京锋尚升级版，除具有北京锋尚的优点外，还具有房间温度、湿度和二氧化碳排放自动控制系统，住户可按照需要自行调节，居住生活更加舒适；同时，利用太阳能等可再生能源，使夏季制冷和生活热水零能耗，缓解高峰用电。零排放避免了一般的空调室外机排放的热量破坏环境。

南京锋尚国际公寓作为中国第一个"零能耗住宅项目"，2009年获得了联合国人居署颁发的"人居最佳范例奖"，2010年获得"精瑞住宅科学技术金奖"。国家"十一五国家科技支撑计划——可再生能源与建筑集成示范工程"。

恩施民族高中

建筑类型：教育　　建筑面积：140 000平方米
项目地点：湖北 恩施　　设计时间：2007年

上海西郊蟠龙源

设计时间：2005年
项目地点：上海
用地面积：32.8公顷

别墅的平面功能布置不仅采用了上海老洋房平面组织活泼、主要房间朝南等特征，又考虑了现代人的生活习惯。立面造型上吸取上海老洋房各种风格的精华之处，巧妙运用艺术处理手法，千姿百态，风格迥异，营造出美好的人文气息。

北京航丰园科技发展有限公司工业综合配套楼

设计时间：2011年
项目地点：北京
用地面积：3.38公顷
建筑类型：办公建筑
建筑面积：93 890平方米
建筑层数：17层

本项目的平面采用U形布局，自然围合形成园区内空间，三栋主楼巧妙呼应，给人以震撼的整体感，并展现工业办公建筑的简约、高效、典雅的国际化现代建筑特质。

造型处理注重空间、形体、色彩以及虚实关系的把握，采用大胆的黑、白石材对比，方窗、玻璃幕墙与石材墙面的虚实对比，板楼与塔楼的形体对比，以期和谐、简约，与环境互成风景，使得整栋建筑散发着自己独特的魅力。

北京中国国际新闻中心建筑设计

设计时间：2005年
项目地点：北京
用地面积：115 230平方米

昙花林艺术交流中心

建筑类型：文化建筑
建筑面积：4 887平方米
建筑规模：低层
项目地点：湖北 武汉

北京构易建筑设计有限公司
Beijing CO+E Architecture Design Co., Ltd.

地址：北京市海淀区清华大学科技园创业大厦906
电话：+86-10-62701980
传真：+86-10-62799030
邮箱：coe001@126.com

Add: 906, Chuangye Building, Tsinghua University Science and Technology Park, Haidian District, Beijing
Tel: +86-10-62701980
Fax: +86-10-62799030
E-mail: coe001@126.com

北京构易（CO+E）建筑设计有限公司，前身为中冶·欧伯麦尔设计咨询有限公司，成立于1987年1月，1993年获得建设部批准的工程设计甲级资质证书。公司原隶属于中华人民共和国冶金部。为了更好地面向市场，加强公司在建筑行业的竞争力，公司于2001年进行重组，在公司运行体制和人员结构等方面进行调整，以多名有经验的知名中青年建筑师为班底，实行了充满活力并具有较强市场竞争力的工作室运营模式，先后参与了全国多项大型工程的设计工作，并得到了社会广泛的认同及好评。

北京构易建筑设计有限公司主要合伙人均为国内著名建筑院校教授和具有丰富实践经验的一级注册建筑师。现有高级职称设计人员32人，中级职称68人，一级注册建筑师15人，一级注册结构工程师10人。主创团队普遍具有较高的学术素养，以清华大学、东南大学、荷兰Delft大学、意大利米兰工学院和挪威科技大学等著名建筑院校为依托，构易与国际建筑界展开了广泛密切的合作。

北京构易建筑设计有限公司集中精英力量，形成多元化的、独树一帜的设计风格，在城市规划、建筑设计、室内设计、景观设计、建筑策划、项目管理和技术咨询服务等领域均颇有建树。

Beijing CO+E Architecture Design Co., Ltd., formerly known as MCC - Oubo Maier Design Consulting Co., Ltd., established in January 1987, obtained the Grade A engineering design qualification certificate by the Ministry of Construction in 1993. The company originally under the PRC Ministry of Metallurgical Industry. In order to better market-oriented, strengthening the company's competitiveness in the construction industry, the company reorganized in 2001, including the company running the system and personnel structure, etc., the company is based on several well-known and experienced young architects, with studio business model dynamic and strong competitive market, has participated in the design of major construction projects all over the country, and won wide recognition and praise.

The partners of Beijing CO+E Architecture Design Co., Ltd. are all renowned professors of domestic architectural universities and registered architects with extensive practical experience. At present, there are 32 senior title designers, 68 intermediate grade designers, 15 registered architects and 10 registered structure engineers. Creative team generally has high academic quality, with the basis construction of Tsinghua University, Southeast University, the University of Delft, The Netherlands, Milan, Italy Institute of Technology and Norwegian University of Science and other prestigious institutions, CO+E starts to have a wide range of close cooperation with the international architecture industry.

Beijing CO+E Architecture Design Co., Ltd. focuses on elite forces, the formation of diverse, unique design, and has many achievements in urban planning, architectural design, interior design, landscape design, construction planning, project management and technology consulting services, etc.

北京电子城国际电子总部三号地项目

项目地点：北京
用地面积：142 023平方米
建筑面积：496 979平方米
施工时间：2010年
容 积 率：5.14

Beijing Electronics City International Electronic Headquarters No.3 Block Project

Project Location: Beijing
Land Area: 142,023 m^2
Building Area: 496,979 m^2
Construction Time: 2010
Floor Area Ratio: 5.14

有机的生命体

电子产业技术的视觉化表现——如何创造一个能代表高科技电子产业总部的建筑形象是我们设计的出发点：将高科技的含义壮观地视觉化、物质化。

电子线路板——我们最终选择了电路板来扮演承载物的角色，它是最被人熟知且深入生活的电子产物。它表面的线条有着繁复曲折，却思路清晰的纹理，纹理中又精密地安排了或方或圆大小不一的电路元件，这些元素分明组成了一幅理性精密的形态构成图，传达着智慧的气质。

"造城"——在我们的理想中，本方案应该有着整体统一的形态，统一中又有着无穷的变化。在震撼的第一观感过后，漫步其中，会惊讶地发现它又有着丰富的层次，新奇的发现层出不穷，仿佛一座有着丰富内涵的壮观城市。

Organic Forms of Life

Visual electronics industry technology performance – How to create a high-tech electronics industry to represent the image of the headquarters building is the starting point of our design: the meaning of the spectacularly high-tech visual materialistic.

Electronic circuit boards – We opted for the board to play the role of host material, it is the best known, and the depth of life electronic products. It has a complex surface of the line twists and turns, but lucid textures, texture and precise arrangements for the different sizes of round or square or circuit elements, these elements form a clear precise form constitute a rational map, to convey the wisdom of temperament.

"Making the city" – In our vision, the program should be consistent with the overall shape, unity and has endless changes in perception after the first shock, walk in them, you will be surprised to find that it has a rich level with novel discoveries emerging one after another, like a rich inner spectacular city.

信达东湾半岛A组团规划设计方案

项目地点： 吉林 长春
用地面积： 190 200平方米
建筑面积： 457 502平方米
容 积 率： 1.72
设计时间： 2011年

Xinda East Bay Peninsula Group A Concept Planning and Design

Project Location: Changchun, Jilin
Land Area: 190,200 m^2
Building Area: 457,502 m^2
Floor Area Ratio: 1.72
Design Time: 2011

百年商埠地，这里最长春

东湾半岛项目位于长春市中心城区，距火车站2.5公里，距人民广场约3公里，具有明显的地段优势。地段往往是大盘的短板，所以东湾半岛项目兼具大盘与中心城区的双重优势。

“地段”是决定价格的第一要素，大盘的出现使人们关注的焦点从“地段”转向“社区”。发挥大盘优势，形成社区文化和品牌效应，信达东湾半岛项目必将卓然闪耀于城市中心。

Century trading port, the most Changchun here

East Bay Peninsula project is located in Changchun city center, 2.5 km from the railway station, about 3 km from People's Square, the location has obvious advantages. Location is often the market's short board, so the East Bay Peninsula project has downtown market and city center the dual advantage.

"Lot" is the first element to determine the price, the broader market appeared to focus people's attention from the lot to community. With the market advantage, the formation of community culture and brand, Xinda East Bay Peninsula project will shine in the city center.

国家工商行政管理总局商标档案业务用房

项目地点： 北京 宣武区
用地面积： 16 094平方米
建筑面积： 48 086平方米
容 积 率： 3.0
施工时间： 2009年

State Administration for Industry Trademark Files Business Building

Project Location: Xuanwu District, Beijing
Land Area: 16,094 m^2
Building Area: 48,086 m^2
Floor Area Ratio: 3.0
Construction Time: 2009

可持续发展的绿色建筑

本项目贯彻生态优先的准则，尊重生态格局和自然生态过程，依自然设计，有限度、合理地利用自然生态环境，旨在提高环境品质，增强环境吸引力，力图创造人工生态和自然生态和谐共存的可持续发展的绿色建筑。

建筑物由上部主体和下部裙房两部分组成。裙房部分采取对称布局，南侧由三层高的柱廊形成空间界面，创造了庄重恢宏的建筑气质，体现了作为行政办公用房的稳重与大气。北侧二层屋顶为屋顶花园丰富了建筑空间，为其东西两侧提供自然通风、采光条件的同时，为内部人员提供安静的休息场所，改善了建筑办公环境，使人置身于绿色的自然环境中。

Sustainable Green Building

The criteria for project implementing ecological priority, respect for ecological patterns and natural ecological processes, by natural design, limited, reasonable use of the natural environment to improve environmental quality, and enhance the environment attractive, trying to create artificial ecology and ecological harmony sustainable green building.

The building is composed of two parts, main building the upper and lower skirt. The podium of symmetrical layout and the south portico formed by the three-storey space interface, create a magnificent stately building qualities, reflecting the stable atmosphere as an administrative office. North side of the second floor roof enriches roof garden, provides to its east and west sides with natural ventilation, lighting conditions, as well as a quiet internal place to rest, improving the building office environment in which exposure to the natural environment and green in.

阿拉善歌舞剧院

项目地点：内蒙古 阿拉善盟
用地面积：43 000平方米
建筑面积：6 094平方米
池　　座：876座
竣工时间：2010年

Alax Theater

Project Location: Alax, Inner Mongolia
Land Area: 43,000 m^2
Building Area: 6,094 m^2
Seats in the Orchestra: 876
Completion Time: 2010

"地景建筑"

——城市形象标志与文化标志的双重载体

作为一个城市大型的观演建筑群体，无论从规模还是建筑形象上都已经成为了一个新的城市标志，这种标志蕴涵着物质和文化两层意义。首先是观演建筑，完成日常和大型的城市观演活动和大型集会与政务会议的功能。其次是景观建筑，它的形体与布局应该更加紧密地与城市绿化和景观轴带相关联，形体的视觉冲击力足可以让它成为新的地标景观。

"Landscape Architecture"

-- The dual carrier of city's visual identity and cultural symbols

As a large building complex for performing arts in a city, in terms of image size and construction the theater has become a new city logo, which implies two layers of material and cultural significance. The first is performing buildings, complete daily and large performance and activities of large gatherings and meetings of government functions. Second, landscape architecture, its type and layout of the body should be more closely with the urban greening and landscaping associated with the shaft, the full visual impact of the body can make it become a new landmark in the landscape.

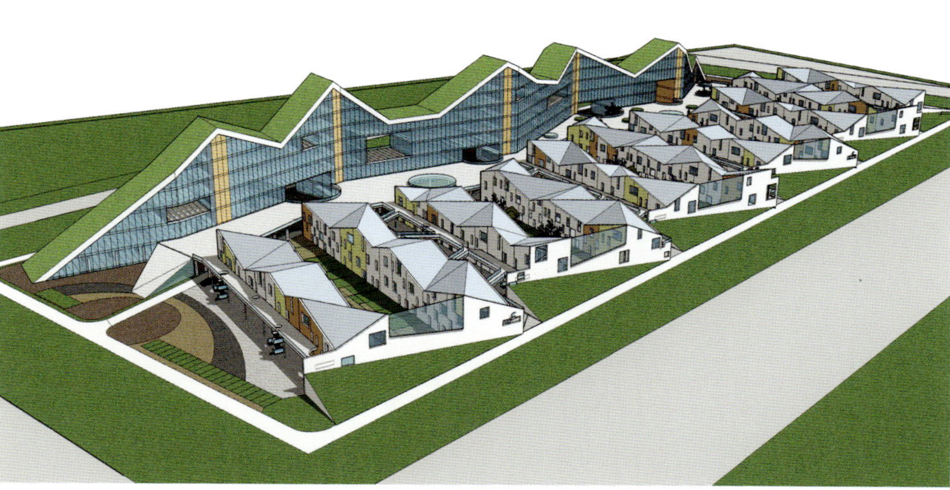

概念生成

噪音是影响本案的主要因素，如何解决噪音污染是本项目的关键所在。
本方案从几何学的角度入手，用声波曲线的拓扑图形作为主要的建筑语汇。

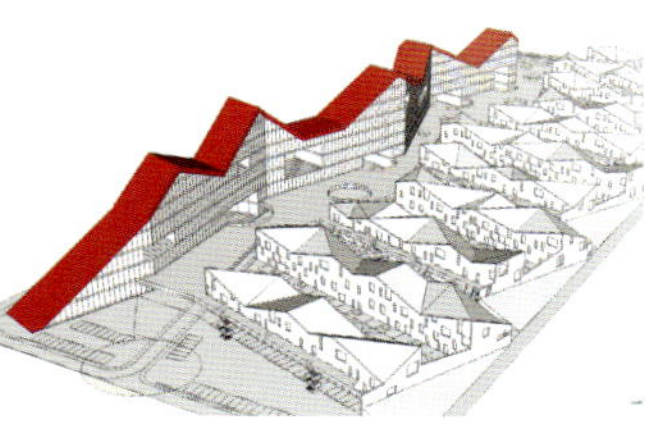

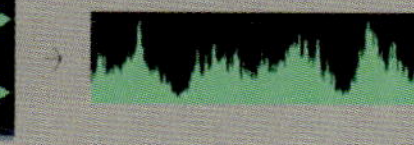

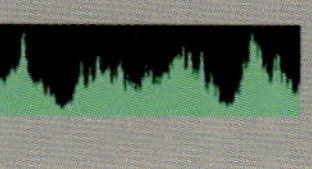

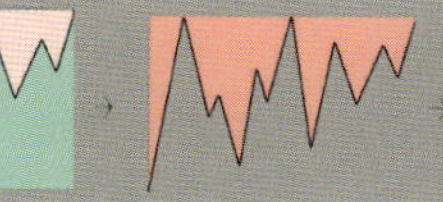

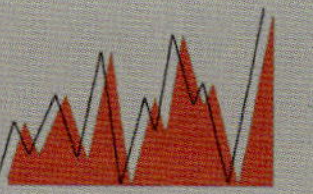

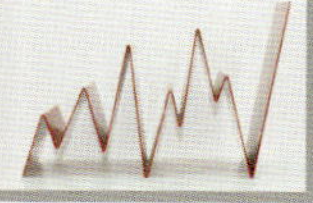

声音波形图　截取　拓扑　负图形　翻转　提取

李峤商务中心概念设计方案

项目地点：北京 顺义区
用地面积：43 489平方米
建筑面积：110 234平方米
容 积 率：1.5
设计时间：2009年

Li Qiao Business Center Conceptual Design

Project Location: Shunyi, Beijing
Land Area: 43,489 m^2
Building Area: 110,234 m^2
Floor Area Ratio: 1.5
Design Time: 2009

历史

朗东创立于 2000 年 10 月的深圳。
2004 年在北京设立分部。
2006 年总部迁至北京并设立深圳分部。

团队

专业包括建筑、结构、机电等，形成专业齐全配套的服务体系。

业务

朗东的业务范围涵盖了城市设计、居住区规划与居住建筑设计、公共建筑设计、园林景观设计、室内设计及结构、机电等工程设计。

理念：品德、品质、品位、品牌

朗东追求根植于本土文化，以国际先进的理念和技术实现的设计创新，强调设计质量与服务并重，为客户提升最优质的价值，与客户实现共赢。

History

In October of 2000, Landon was founded in Shenzhen.
In 2004, Landon Beijing branch was established.
In 2006, the headquarter moved to Beijing and the Shenzhen branch was set up.

Team

A complete system of building, structure, electromechanical services.

Business

Landon's business scope covers urban design, residential area planning and residence architecture design, public architecture design, landscape architecture design, interior design, engineering design of mechanical electronics and so on.

Philosophy: character, quality, taste, brand

Taking root in the local culture, realizing the design innovation through international forefront ideas and technology, stressing on the design and service quality, to provide the best value for the clients and achieve the win- win cooperation.

北京朗东建筑设计咨询有限公司
地址：北京市海淀区西三环北路 72 号世纪经贸大厦 B 座 2600
电话：+86–10–51798330
传真：+86–10–51798330-801

Beijing Langdong Architecture Design Consulting Co., Ltd.
Add: 2600, Block B, Millennium Plaza.No.72, Xisanhuan Bei Road, Haidian District, Beijing
Tel: +86–10–51798330
Fax: +86–10–51798330-801

深圳市朗东设计咨询有限公司
地址：深圳市福田区景田路擎天华庭擎天阁 32A
电话：+86–755–83544377
传真：+86–755–83285144

Shenzhen Langdong Design Consulting Co., Ltd.
Add: 32A Qingtiange, Qingtian Huating, Jingtian Road, Futian Disctrict, Shenzhen, Guangdong
Tel: +86–755–83544377
Fax: +86–755–83285144

1

2

3

4

5

1　大连友服传媒办公楼
项目地点：辽宁 大连
建设规模：5 万平方米
设计时间：2011 年

2–5　盘锦金远宝兴・狮城
项目地点：辽宁 盘锦
建设规模：203 万平方米
设计时间：2011 年
竣工时间：在建

6　通州文化活动中心
项目地点：北京
建设规模：3 万平方米
设计时间：2010 年
竣工时间：在建

7–8　林西博物馆
项目地点：内蒙古 赤峰
建设规模：8000 平方米
设计时间：2011 年
竣工时间：在建

9　迁西地质公园博物馆
项目地点：河北 迁西
建设规模：3 800 平方米
设计时间：2009 年
竣工时间：在建

1 Youfu Media Office Building
Project Site: Dalian, Liaoning
Floor Area: 50,000 m^2
Design Time: 2011

2-5 Jinyuan Baoxing – Lion City
Project Site: Panjin, Liaoning
Floor Area: 2,030,000 m^2
Design Time: 2011
Completion Time: in progress

6 Culture Center
Project Site: Beijing
Floor Area: 30,000 m^2
Design Time: 2010
Completion Time: in progress

7-8 Linxi Museum
Project Site: Chifeng, Inner Mongolia
Floor Area: 8,000 m^2
Design Time: 2011
Completion Time: in progress

9 Qianxi Geographic Park
Project Site: Qianxi, Hebei
Floor Area: 3,800 m^2
Design Time: 2009
Completion Time: in progress

6

7

8

9

10　固安办公楼
项目地点：河北 固安
建设规模：12000 平方米
设计时间：2009 年
竣工时间：在建

10 Gu'an Office Building
Project Site: Gu'an, Hebei
Floor Area: 12,000 m^2
Design Time: 2009
Completion Time: in progress

11　林西青少年活动中心
项目地点：内蒙古 赤峰
设计时间：2011 年
竣工时间：在建

Linxi Youth Center
Project Site: Chifeng, Inner Mongolia
Design Time: 2011
Completion Time: in progress

12　华业・荷叶山项目
项目地点：湖北 武汉
建设规模：48 万平方米
设计时间：2010 年
竣工时间：在建

12 Homyear – Lotus Hill Project
Project Site: Wuhan, Hubei
Floor Area: 480,000 m^2
Design Time: 2010
Completion Time: in progress

10

11

12

中国建筑科学研究院建筑设计院，隶属中国建筑科学研究院，是国家住房和城市建设部系统甲级设计单位，主要从事各种类型的大中型民用与工业建筑设计。多年来设计院在建筑设计领域做了大量的工作，积累了丰富的设计工作经验，拥有一支由高素质人才组成的设计队伍。设计队伍包括院士、国家级设计大师、教授级建筑师、注册建筑师、注册结构师等，人员结构配备合理，其中博士、硕士占1/3，且有相当一部分设计人员拥有国外工作和学习的经历。

设计院依靠中国建筑科学研究院这一国家级综合性科学研究实体，在设计领域具有雄厚的技术基础。中国建筑科学研究院设有结构、地基、抗震、防火、装修、建材、建筑物理、机械化施工以及研发中心等十多个研究所，故能充分、迅速地把国内外建筑领域的新技术、新产品应用到设计院主持的建筑设计中去；对技术复杂、难度大的工程具有综合设计优势；有正确运用、深入理解规范与规程，完成高难度设计的能力。这对确保工程设计质量、节约材料、降低造价有着独特的优势。凭借这一优势，我们赢得了与SOHO中国、华润置地、华远地产、中粮地产等国内知名开发公司长期合作的机会，使设计院得以持续快速的发展。

设计院在国际合作领域方面，具有独特的优势和丰富的经验。依托设计院综合技术实力，我们长期与贝聿铭等世界级建筑大师以及其他如美国、德国、法国、日本、瑞士、新加坡等国家和中国香港地区的国际著名建筑设计事务所密切合作。在合作过程中，设计院员工迅速、及时地掌握了当前国际最先进的技术和设计标准，熟悉了国际建筑设计、施工承包的操作模式，从而提高了设计院的学术水平，促进了设计技术的快速发展。

设计院在公共建筑设计领域方面，凭借高超的设计技术和精益求精的工作态度，承接的具有广泛社会影响力的项目有：中国国家博物馆、中国疾病预防控制中心、国家信访局办公楼、国务院国有资产监督管理委员会、中华世纪坛、中国银行总部大厦、北京燕莎中心、人民大会堂改造等不同功能的知名建筑。

设计院一贯以重信誉、保质量、高效率作为工作宗旨，是住房和城乡建设部系统甲级综合设计院中第一家通过ISO9001质量认证的单位。以顾客为关注焦点的市场服务意识，始终是设计院及全体职工的工作标准。这为设计院在设计领域中保持技术领先提供了可靠的质量保证，赢得了较好的社会声誉和客户的信任，并占领了相当广泛的市场。设计院与设计人员已经形成同时注重单位品牌与个人品牌，单位与个人是互相支持和依托的整体这样一个共识。愿以单位的信誉和高精技术，向客户提供优质的产品和服务。

China Academy of Building Research Architectural Design Institute that belongs to Chinese Architecture and Science Research Institute, is A level design institute in the system of Ministry of Housing and Urban-Rural Development. It mainly deals with various kinds of designs of large and middle sized civil construction and industrial construction. For years, the Architectural Design Institute has made a lot of projects in the architectural design field and accumulated much working experience of designing. It has a design team which is made of well educated talents and it includes academician, national senior designer, professional architect, registered architect and registered structure engineer etc, of whom the masters and the doctors account for 1/3, and many designers have foreign working and studying experiences.

Because the institute depends on Chinese Architecture and Science Research Institute, it has strong technical base in the design field. China Academy of Building Research Architectural Design Institute has more than ten research institutes of structure, building foundation, shock resistance, fireproofing, decoration, building materials, building physics, construction mechanization etc. Therefore, it can apply the new technique and products from home and abroad into the projects which are held by our institute quickly and efficiently. We have comprehensive design advantage for the complex and difficult project; we also have ability of finishing the difficult design by using and understanding the norms and rules correctly. These advantages are helpful for us to ensure quality, save materials, sell our product in a low price. Thanks to their advantages, we have won the opportunities of making long term cooperation with SOHO China, Huarun Land, Huayuan Real Estate, Zhongliang Real Estate and other famous development companies, so our institute can develop rapidly and continuously.

The Architectural Design Institute has unique advantages and rich experience in the field of international cooperation. Supported by the comprehensive technique strength, we can make long term cooperation with Pei Cobb Freed and other famous international architects as well as some famous international architect offices which are located in America, Germany, France, Japan, Switzerland, HongKong of China, Singapore etc. During the process of cooperation, we learn the most advanced international techniques and design standards of modern time rapidly and we are familiar with the operation mode of international architectural design and contracted construction. Consequently, we improve the level of scholarship and promote the development of the design technique rapidly.

By using the high technique and excelsior working attitude, the Architecture Design Institute have finished many projects with wide social influence in the field of public construction design, such as National Museum of China, Chinese Center for Disease Control and Prevention, office building of State Bureau for Letters and Calls, State-owned Assets Supervision and Administration Commission of the Sate Council, China Millennium Monument, Bank of China Headquarters, Beijing Lufthansa Center, the reconstruction of Great Hall of the People etc.

The institute always follows the working principle of credit standing, ensuring quality and high efficient. It is the first unit which passed the ISO9001 Quality Certificate in the A level comprehensive design institutes of national ministry of construction system. The customer service orientation is the working standard for all the staffs forever, it can provide reliable quality assurance for us to keep advanced techniques in the field of design, and win good social reputation and customer credit. The institute and the designers put emphasis on the institute brand and personal brand, and the staffs are aware that the unit and the individual support with each other and benefit with each other. They hope provide excellent service and products with all the customers by good credit and high techniques.

中国国家博物馆
National Museum of China

中国国家博物馆前身是中国历史博物馆和中国革命博物馆。原有馆舍建成于1959年8月，是新中国成立十周年的"十大建筑"之一，建筑面积65 000平方米。改扩建后的中国国家博物馆，是一座以历史与艺术为主、系统展示中华民族悠久历史文化、具有国际先进水平的综合性国家级博物馆，建筑面积将增加到191 900平方米。

National Museum of China origins from Museum of Chinese History and Museum of Chinese Revolution. The original museum was built in August 1959, which is one of "Top Ten Building" of the 10th Anniversary of PRC Creation. Its floor area is 65,000 m^2. Based on history and arts, the reconstructed and expanded National Museum of China is an international-level comprehensive national museum, which systematically shows the long and rich culture and history of the Chinese nation. The floor area will be increased to 191,900 m^2.

安徽合肥新地中心
Anhui Hefei Xindi Center

该项目位于合肥政务文化新区潜山路与祁门路交汇处，是政务文化新区最核心位置。项目总建筑面积为60万平方米，总投资额30多亿元，是合肥政务文化新区投资额最多、建设规模最大、档次最高的城市综合体。该城市综合体包括：一幢210米超高层办公楼、两幢130～150米超高层办公楼、五幢150米超高层住宅及一幢五层的商业中心。项目建成后，必将成为政务区新地标，大幅提升政务区商业、文化氛围，成为合肥政务文化新区最核心的现代化商业中心。

The project is located at the interchange of Qianshan Road and Qimen Road (the most central location) in Administrative and Cultural New Area of Hefei. The total floor area is 600,000 m^2, and the total investment is more than RMB3 billion. With the biggest investment, it becomes the largest-scale and highest-level urban complex of Administrative and Cultural New Area of Hefei. It includes: a 210 m ultra high-rise office building, two 130-150 m ultra high-rise office buildings, five 150 m ultra high-rise residential building and one five-storey business center. After completion of the project, it will definitely become the new landmark of Administrative and Cultural New Area, will greatly promote the commercial and cultural atmosphere of this area and become the core of modern business center in Administrative and Cultural New Area of Hefei.

英特宜家北京购物中心
Inter Ikea Shopping Center (Beijing)

该项目位于北京市大兴区，为一大型商业综合体建筑，其西侧为宜家家居，总建筑面积约50万平方米，其中地上4层，建筑面积约25万平方米，地下3层，建筑面积约25万平方米，为超大型汽车库，可停放5 000辆汽车。项目建筑设计上沿用了英特宜家集团在欧洲建造的大型商业项目的设计理念，以多条室内购物街组合成超大规模的多层商业综合体建筑。

The project is located in Daxing District of Beijing. It is a large-scale commercial complex, with Ikea on its west side. The total floor area is 500,000 m^2, including 4 aboveground floors and 3 underground floors, the area of which is about 250,000 m^2. The underground area is built as a large-scale garage, with a capacity of 5,000 cars. Project architectural design follows the design idea of Intel IKEA Group used in large commercial projects in Europe, which is to build an ultra-large-scale multi-storey commercial complex by a number of indoor shopping streets.

珠海市博物馆和城市规划展览馆
Zhuhai Museum and Zhuhai Urban Planning Exhibition Hall

该项目位于广东省珠海市香洲区情侣路与梅华路东南角，东临情侣路和香炉湾，正对海面，与野狸岛隔海湾相望。项目规划总占地面积50 336平方米，总建筑面积56 000平方米。“两馆”项目是珠海市政府批准设立的地方性综合馆，是珠海市重要的公益性公共文化设施，将成为珠海标志性建筑和城市的新名片。

The project is located at the southeast corner of Lover Road and Meihua Road in Hong Kong Area of Zhuhai, Guangdong, facing Lover Road and Incense Burner Bay on the east. It faces against the sea, with Wild Fox Island on the other side. The total land area is 50,336 m^2, and the total floor area is 56,000 m^2. The construction of “Two Halls” is a local comprehensive museum project approved by Zhuhai Municipal Government. It is an important public cultural facility with public interest, and will become a landmark and new business card of Zhuhai.

金隅·万科城
Jinyu ▪ Vanke City

该项目位于北京昌平区南环路与创新路交口，是万科金隅项目的一个重要组成部分，同时也是昌平区第一个集购物、餐饮、娱乐和休闲于一体的综合购物中心。项目占地23 000平方米，总建筑面积约17.7万平方米。商业部分地上总建筑面积约为82 000平方米，包括大百货、电影院、餐饮、超市和各类娱乐休闲配套。另外，还包括一栋总建筑面积约21 000平方米的住宅建筑。

The project is located at the intersection of South-ring Road and Innovation Road in Changping District of Beijing. It is an important component of Vanke Jinyu Poject, which is also the first comprehensive shopping center of Changping integrated with shopping, dining, entertainment and leisure. Its total land area is 23,000 m^2, and the total floor area is 177,000 m^2. Total floor area of commercial part on the ground is about 82,000 m^2, including large department stores, cinemas, restaurants, supermarkets and all kinds of supporting recreational facilities. In addition, it also includes a residential building, the total floor area of which is about 21,000 m^2.

清华大学人文社科图书馆
Humanities and Social Science Library of Tsinghua University

该项目位于清华大学主校区，西临校园内部南北主干道。用地面积：9 297.9平方米，建筑面积：约20 000平方米。与瑞士马里奥·博塔建筑事务所合作设计。该设计充分考虑场地形状，布置为北侧的长方形与南侧的倒圆台体两个体量穿插的形式。建筑的西侧结合环境地形布置为高差4米的斜坡绿地，东南侧与规划建筑共同围合而成中心绿地。建筑外装饰采用红色石材与玻璃幕墙相结合的形式，风格现代。内部空间处理变化丰富，充分体现现代化的图书馆空间和开放的建筑设计理念。

The project is located in the main campus of Tsinghua University, neighboring with the north-south trunk road of campus on the west. The land area is 9,297.9 m^2, and the floor area is about 20,000 m^2. It is designed by cooperating with Switzerland Mario Botta Architecture Office. The design takes the space shape into fully consideration. The place is designed into an interspersed form with rectangular building on the north and inverted rounding building on the south. Integrating with surrounding landform, the west side of the building is a green slope with an elevation of 4 meters. Its southeast part and planning buildings enclose the central green land. On the outside, the building is decorated with red stones and glass curtain wall. Its modern style and various treatment of interior space fully show the modern library space and the open idea of architecture design.

北京W酒店
Beijing W Hotel

该项目位于建国门立交桥东南角，建国门南大街2号，建设基地面积约6 980平方米。总建筑面积63 799平方米。改造后成为喜达屋酒店管理集团旗下W顶级酒店品牌，拥有350间客房。

The project is located at the southeast corner of Jianguomen Overpass, No.2 Jianguomen South Avenue. The land area is about 6,980 m^2. The total floor area is 63,799 m^2. After reconstruction, with 350 rooms, it becomes the W top hotel brand under the management of Sheraton Hotel Management Group.

三亚凯莱度假酒店改扩建项目
Reconstruction and Expansion of Sanya Gloria Hotel & Resort

该项目位于三亚市亚龙湾度假区，南临大海，北临滨海大道，环境优美，交通便利，由现有的三亚凯莱酒店及其未来扩建两部分组成。新建建筑结合现有建筑进行整体设计，融为一体，共同形成一栋新的酒店建筑。项目规划用地面积106 666.59平方米，总建筑面积107 666平方米。其中新建建筑面积68 961.39平方米，改建建筑面积38 704.61平方米。

The project is located in Yalong Bay Resort Area of Sanya, with sea on its south and Binhai Road on its north. With beautiful environment and convenient transportation, it includes existing Sanya Gloria Hotel & Resort and future expansion part. The design of newly-constructed buildings follows the style of existing building. The old and new integrate with each other and form a new hotel. Planned land area is 106,666.59 m^2. The total floor area is 107,666 m^2, in which the floor area of newly-constructed building is 68,961.39 m^2, and the area of reconstructed building is 38,704.61 m^2.

汶川青少年活动中心
Wenchuan Youth and Children Center

该项目位于汶川县映秀镇旋口中学遗址南侧，岷江和鱼子溪交汇口西北侧，是共青团中央特殊团费援建项目，是共青团中央设在灾区的国家青少年活动基地。建筑总面积约4 600平方米，地上3层。

On the northwest side of the junction of Minjiang River and Yuzi Stream, the project is located in Yingxiu county, Wenchuan on the south of the Site of Xuankou Middle School. It is a supporting project funded by central special league membership dues of Communist Youth League, and a Youth and Children Center established in the affected areas by Communist Youth League. The total floor area is about 4,600 m^2, with three floors on the ground.

中粮地产北京后沙峪项目
Houshayu Project of COFCO Property, Beijing

该项目位于北京市顺义区后沙峪镇吉祥庄村京承高速以东，顺义空港工业区西侧，包括A－10、C－03、C－06三个地块，总建设用地290 360平方米，总建筑面积（地上）520 822平方米，其中住宅建筑面积323 097平方米，两限房建筑面积48 464.5平方米。商业金融（含配套商业和文体活动站）建筑面积194 265平方米，配套服务设施（不含配套商业和文体活动站）建筑面积3 460平方米。

The project is located in Jixiang Village of Houshayu Town, Shunyi District of Beijing, which is on the east side of Beijing-Chengde Expressway and the west side of Shunyi Airport Industrial Park. It includes three land lots: A-10, C-03 and C-06. The total land area is 290,360 m^2. The total floor area (over-ground) is 520,822 m^2, in which the floor area of residential building is 323,097 m^2. Floor area of price-control and model-control residential building is 48,464.5 m^2. Floor area of commercial finance building (including supporting commercial and cultural activities station) is 194,265 m^2. Floor area of supporting service facilities (excluding supporting commercial and cultural activities station) is 3,460 m^2.

Insight Architects Ltd. (Beijing)
英思特（北京）建筑设计咨询有限公司

Insight Architects (Beijing) is composed of a group of passionate architects who would like to pursue high-standard career lives. The firm is set up by the architects who have accumulated over 20 years of professional experience and won numerous awards in national and international architectural competitions. Their multicultural backgrounds and well-balanced experience contribute valuable visions and choices to their clients. The firm works in an interactive manner with all the consultants but acts as a leading role in the process of design to deli- ver integrate works. Architects are also involved in the control of construction to guarantee the quality of finished buildings.

英思特（北京）建筑设计咨询有限公司为加拿大外资建筑专业设计公司，公司核心建筑师在建筑规划和设计领域积累了20余年的职业经验，设计作品多次在国内外建筑设计大赛中获奖，多元的文化背景及中外结合的职业经验使得他们能为建设方提供更多的视角和可供借鉴的因素。所有设计师均为建筑学专业毕业，富有激情、热爱建筑并有着追求完美的职业精神。英思特不仅强调要为建设方提供先进的设计理念，而且十分注重缜密的思考和细节的完善，并在项目实施过程中把控和协调各相关环节以保障设计作品的完整性以及设计效果的充分实现。

电话: +86-10-5869 4181　　传真: +86-10-5869 4185
邮箱: insight@insightbj.com　　网址: www.idschina.com
地址: 北京市朝阳区东三环中路39号建外SOHO东区B座2206室　　邮编: 100022

Flying Colours Of Dubai
第十一届蒂森·克虏伯国际注册建筑师设计大赛“迪拜标志性构筑物”公开设计竞赛三等奖

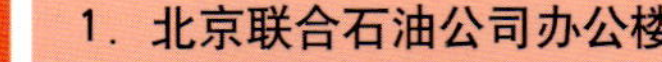

1. 北京联合石油公司办公楼
2. 胶州科润广场商业
3. 上海荣胜综合体
4. 北京亿时酒店

1

2

3

4

1. 长春净月潭森林高尔夫会所
2. 宜昌龙盘湖高尔夫练习场
3. 北京翡翠湖高尔夫会所
4. 天津滨海湖高尔夫会所

2

1

3

4

1. 南通玖珑城
2. 天津泰达城–万通上游国际三期
3. 胶州科润广场

1. 北京CBD国际高尔夫会员公寓
2. 天津滨海湖高尔夫别墅
3. 天津滨海湖高尔夫别墅
4. 北京CBD国际高尔夫会员别墅
5. 天津泰达城–万通上游国际三期别墅

3

1

4

2

5

1. 北京翡翠湖高尔夫会所
2. 天津滨海湖高尔夫会所
3. 宜昌龙盘湖高尔夫练习场
4. 北京翡翠湖高尔夫会所
5. 长春净月潭瓦萨滑雪博物馆

TOPACE 拓尚

Beijing TOPACE Architecture Design Co., Ltd.

北京拓尚建筑设计有限公司

名称：北京拓尚建筑设计有限公司
地址：北京市朝阳区霞光里15号霄云中心1单元708室
电话：+86-10-83282836
传真：+86-10-83289036
邮箱：public@a-topace.com
网址：www.a-topace.com

Name: Beijing TOPACE Architecture Design Co.,Ltd.
Add: Room 708, Unit 1, No. 15, Xiaoyun Center, Xiaguangli Chaoyang District, Beijing
Tel : +86-10-83282836
Fax: +86-10-83289036
E-mail: public@a-topace.com
Http: //www.a-topace.com

北京拓尚建筑设计有限公司（TOPACE）是英国伦敦TOPACE Design LTD.在北京成立的专业酒店及住宅设计公司。业务范围主要包括酒店、度假村及别墅、高档住宅等，服务内容涵盖了规划设计、建筑设计及室内设计三方面。拓尚（TOPACE）的主创设计师均来自国内外著名设计公司，具有丰富的设计经验、创新意识和敬业精神，作品遍布全国许多城市及地区。

As a branch of London TOPACE Design Ltd., Beijing TOPACE Architecture Design Ltd. provides expertise in hotel and residence design. Combining planning, architecture and interior design, we offer a range of design services including hotels, holiday resorts and villas, luxury apartments, etc. All senior designers in Topace Design are from famous design companies around the world, with high practical experience, innovation and professionalism. Their projects spread across various cities and districts in China.

蓬莱中国湾大酒店

蓬莱中国湾大酒店为集会议、娱乐功能于一体的酒店建筑群。总建筑面积65 156.96平方米，酒店包括高档客房、不同特色的餐厅、完备的宴会和会议设施、康体健身中心等。

设计理念

本案中，中式传统建筑的精华与现代主义建筑的简约精致相结合，打造出现代中式风格。以此体现现代度假酒店的时代性和雅致、高洁的风格，并与周边的自然环境相契合。在文化传承、呼应自然、地域特色三方面相互交融，做出深度尝试。建筑群由四栋建筑构成，酒店主楼和两翼客房楼相连通，客房楼入口由大堂进入，会议中心则独立布置。现代中式风格是对如何传承传统建筑文化的探索。设计中我们借鉴了传统建筑与自然的关系，力图通过自然与建筑的相互衬托，利用丰富多样的灰空间处理，精确地表达出传统空间之美。建筑语言方面则从传统文化入手，将其抽象为具有现代美学特征的建筑语汇，以一种东方式的构成方式予以组织，从而创造出具有中式意境的空间形式和建筑形象。

阆中山水城酒店

阆中山水城酒店位于中国四大古城之一阆中古城西北侧，东临有“天下第一江山”之誉的嘉陵江，西接盘龙山余脉，景观资源极佳。酒店地下1层、地上9层，总高度为35.05米，包括酒店精美宏伟的公共区、307间豪华客房、设施完备的会议中心、健身中心等功能。

设计理念

设计以创造体现时代精神和阆中文化地域特点的新式建筑为最终目标。

在建筑外观与空间设计中汲取中国传统建筑中步移景异、空间层次多样的理念，同时大胆采用现代简约的手法予以表现，在现代美学与技术的范畴内营造出“庭园深深深几许”式的东方意境。

在功能方面以满足人们对现代度假休闲环境所要求的舒适性、健康性、生态性和经济性为出发点，力求打造一个设计布局合理、功能齐备、交通便捷、环境优美的现代度假天堂。

青岛时代锦江大酒店

青岛时代锦江大酒店位于山东青岛胶州市新城区，北京路与福州路交叉口东北角，是以酒店为主，集商业、娱乐于一体的综合体建筑，总建筑面积68 956平方米。酒店部分包括314间豪华客房、不同特色的餐厅、完备的宴会和会议设施、康体健身中心等；商业部分包括底层的沿街商铺和可自由划分的主力店；娱乐部分位于建筑裙房的3、4层。

设计理念

时代国际中心位于胶州市蓬勃发展的新区，因而在建筑设计中也充分体现了在中国快速发展的时代背景下建筑特点的精髓。以现代的设计语言与构造技术在东方传统文化的基础上表达超前的东方现代主义设计理念。设计从建筑外观到室内设计都把这一理念贯穿始终，外部设计简约现代，通过自下而上一气呵成的竖向石材装饰板，表现建筑锐意向上的动势，与所处区域特征吻合，而内部却现代而不失优雅豪华。

葫芦岛富都东戴河旅游度假村

建设地点：辽宁 葫芦岛
占地面积：30.5公顷
建筑面积：354 140平方米
容 积 率：1.24
设计时间：2011年

设计理念

项目南临渤海湾，北邻“五点一线”滨海公路，东临公共绿化带（军事管辖区），西侧为规划商住区。

总体规划

“两横两纵”的轴线式布局，纵向为统领，横向为联系，将整个项目划分为四个部分：用地南侧为超五星级大型滨海度假村，北侧为度假公寓区，东侧为度假村的高端接待度假别墅，以及向南延伸的海上游乐区。西侧地块强调山与海的呼应，打开一条视线通廊，从而形成一条以水为主题的中央景观轴线，建筑围绕水系展开布局。居住产品形态多样，别墅、洋房、高层公寓分区布置，有节奏地起伏，形成丰富的城市轮廓，酒店独特的轮廓将这一韵律推向高潮。东侧以保留的泄洪沟为轴线，将其改造为一条景观峡谷，架起三座桥与西侧的地块发生联系。区域内部为轻松自由的组团式布局，院落型度假别墅成组围合布置，形成园中园的格局。海上游乐区为度假村的延伸区域，酒店正前方为海滨浴场，海岸转折的地方为游艇码头和直升机停机坪。

芭东和园·国际养生健康中心

建设地点：山东 青岛
占地面积：65 500平方米
建筑面积：65 422平方米
容 积 率：1.0
设计时间：2011年

设计理念

芭东和园·国际养生健康中心项目位于青岛即墨市温泉镇，大田路以南，镇政府以东，毗邻芭东小镇项目。分别规划建设28 893.8平方米的养老中心和36 528平方米的康复中心。面对2011年以后的30年里中国人口老龄化将呈现加速发展态势，本项目定位为养生、养老、度假新模式复合地产，带着崇高的社会使命感，推崇一种养生养老概念的生活观。

总体规划

本项目规划用地由两个区块组成：东部为养老中心区，西部为康复中心区。养老中心区体现典型的德式风格特色，整体为高效的集中式布局，南面面向开阔的园林和生态景观湖；主体平面呈十字形，将基地划分为四个各具特色的庭院；而康复中心区的规划则是力图营造传统中式园林的内涵和品位，设计以蜿蜒曲折的水系为中心，结合地形的落差，形成相对分散、自由的布局。在大田路设置三个出入口，分别服务于养老中心、康复中心及超市、康复中心地库。两个区块具有各自独立的到达流线，互不干扰，同时，又具有相互贯通的内部景观线路与中心景观连通。

建筑风格

为了呼应德式养老理念和服务体系，本项目建筑设计采用了体现“简洁、精致、朴素、自然”特色的现代德国风格。无论是养老中心还是康复中心，都追求简洁明确的体量关系，从形态上体现现代简约的美感。同时注重细节的刻画，强调窗口等部分的线条，提升建筑在细节部分的品质感；同时注重人与材料接触时的心理感受，在阳台等部位使用具有温暖感的木材饰面。

海信天玺

建设地点：山东 青岛
建筑面积：102 079平方米
占地面积：37 049平方米
容 积 率：2.16
设计时间：2009年
竣工时间：2012年

设计理念

海信天玺设计概念来源于装饰艺术风格，强调建筑物的高耸、挺拔，给人以拔地而起、傲然屹立的非凡气势，体现出工业革命技术革新所带来的、不断克服地心引力而达到的新的高度，表达出不断超越的人文精神和力量。

设计整体上汲取了装饰艺术风格的构图手法。面对辽阔的海面，建筑整体形象采用对称的三段式布局，底部采用深色石材，主体以列柱组成立面的框架，强化建筑的垂直线条，结合顶部层层退进的造型，形成了建筑整体向上，直入云霄的趋势，强烈的视觉感受，以及独特的韵律感，引领古典奢华与简约时尚相结合的建筑潮流；材料上以浅色石材为主，突出石材、金属、玻璃等材料之间的配合，追求高贵典雅、庄重大气的效果。

建筑细部的处理上，讲究线条的丰富细腻，使建筑在转折、进退、高低错落中增强了动感和光影的和谐变化。壁柱通过石材的凹凸强调了竖向线条的微妙层次，精美的灯具既丰富了建筑的细节，又使壁柱在垂直方向上产生了节奏。壁柱底部收分较大的细部处理，让壁柱具有了十足的力量感和工艺感，使建筑在近人尺度上具有很强的可读性。为了充分表达装饰艺术因结合工业文化所产生的机械美学的文化内涵，立面上采用大量颜色丰富、对比强烈、线条细腻的金属几何图案作为装饰。这些图案既增强了立面的层次感，又形成了鲜明的建筑标志，彰显了天玺独特的建筑品位及艺术情调。

胶州湾财富中心

建设地点：山东 青岛
占地面积：9 786.88平方米
建筑面积：63 111.29平方米
容 积 率：6.4

设计理念

本工程为大型城市商业综合体，位于两条主干道交叉口处，区位优势明显。对于这样一个重要的城市节点，我们提出以下原则：

从城市的角度考量，建筑主体布局围合形成最大化前广场，强化主干道交叉节点的城市空间。

充分考虑本期商业与城市干道及一期商业密切的空间关系，形成良好的城市商业氛围。

建筑主体将公寓、SOHO办公和酒店分为两个体量，既便于管理，又有利于分期建设。

公寓、SOHO办公结构上满足灵活分割的需求，同时，将公寓设置在顶部以获得良好的日照和景观。

为了将更多的地面面积用于商业，酒店大堂置于裙楼的第6层，并利用屋顶花园将景观与大堂及咖啡厅大堂酒廊融为一体，营造了与众不同的都市桃源特色。

商业内部考虑中庭空间，并与自动扶梯结合，创造充满活力的商业气氛。

总体建筑平面形式及布局以8.4米x8.4米的正交柱网络为结构基准，充分考虑了办公、公寓、酒店、商场及地下车库的合理性和经济性。

建筑外立面要很好地呼应二期高层住宅建筑的新古典风格，同时，通过现代材料技术和设计形式来凸显自身作为商业、办公、酒店综合体的公共建筑的特性，树立了新的城市地标。

建筑风格

建筑立面采用反映当代高层建筑美学特征的装饰艺术风格，突出高耸挺拔的特点，重点表现城市重要公共建筑的现代感和高档感，并能与迅速发展的城市环境相适应。

裙房部分运用强烈的虚实对比来突出商业的主入口，同时充分考虑了商业的广告展示诉求；公寓和SOHO办公L形高层通过简洁的体量、干净利落的竖线条、明确的虚实对比体现作为城市地标卓尔不群的独特格调；而酒店则既兼顾了整体性和协调性，也考虑到作为酒店的需要，强调了差异性。酒店顶端设计了一个明亮的发光体，夜晚将成为酒店醒目的标志。

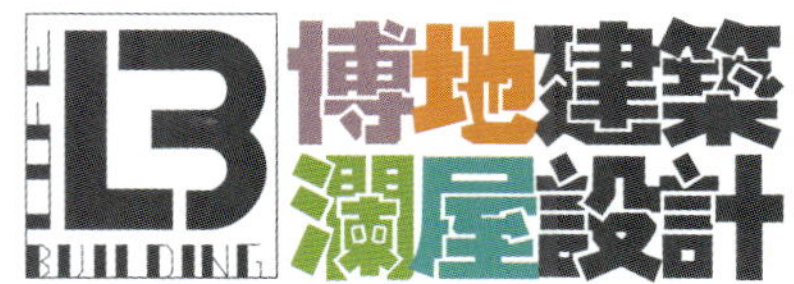

北京博地澜屋建筑设计咨询有限公司

Beijing Building Life Architectural Design & Consulting Co.,Ltd.

博地澜屋作为一家专业设计机构，拥有规划、建筑、景观及室内四大专业，其核心设计师荣获多项国家级、省部级优秀设计大奖。四大主导专业相互融合是博地澜屋的独特优势，提升项目价值是博地澜屋的设计目标，卓著的设计水平和优质的服务质量是博地澜屋的经营理念。博地澜屋是为您提供全专业、全程设计的服务平台。

Building Life is a specialized design agency. We provide unique services resulting from interaction amongst planning, architecture, landscape and interior with the goal to maximize value for our clients. Top design and high-quality service are what we strive for. The core design team is represented by winners of important prizes set up by the state or provincial construction administration, who aspire to furnish clients with professional one-stop service powered by excellent project consulting, ingenious conception as well as abundant engineering experience.

北京市海淀区中关村南大街31号神舟大厦8层(100081)
Tel: +86-10-68118690 Fax: +86-10-68118691
mail@buildinglife.com.cn www.buildinglife.com.cn

1–2 承德双滦松树沟区域规划

3–4 承德松树湾温泉度假村规划

5 承德滨水新区城市设计

1–5 威海南海香水海四期——米兰尚都

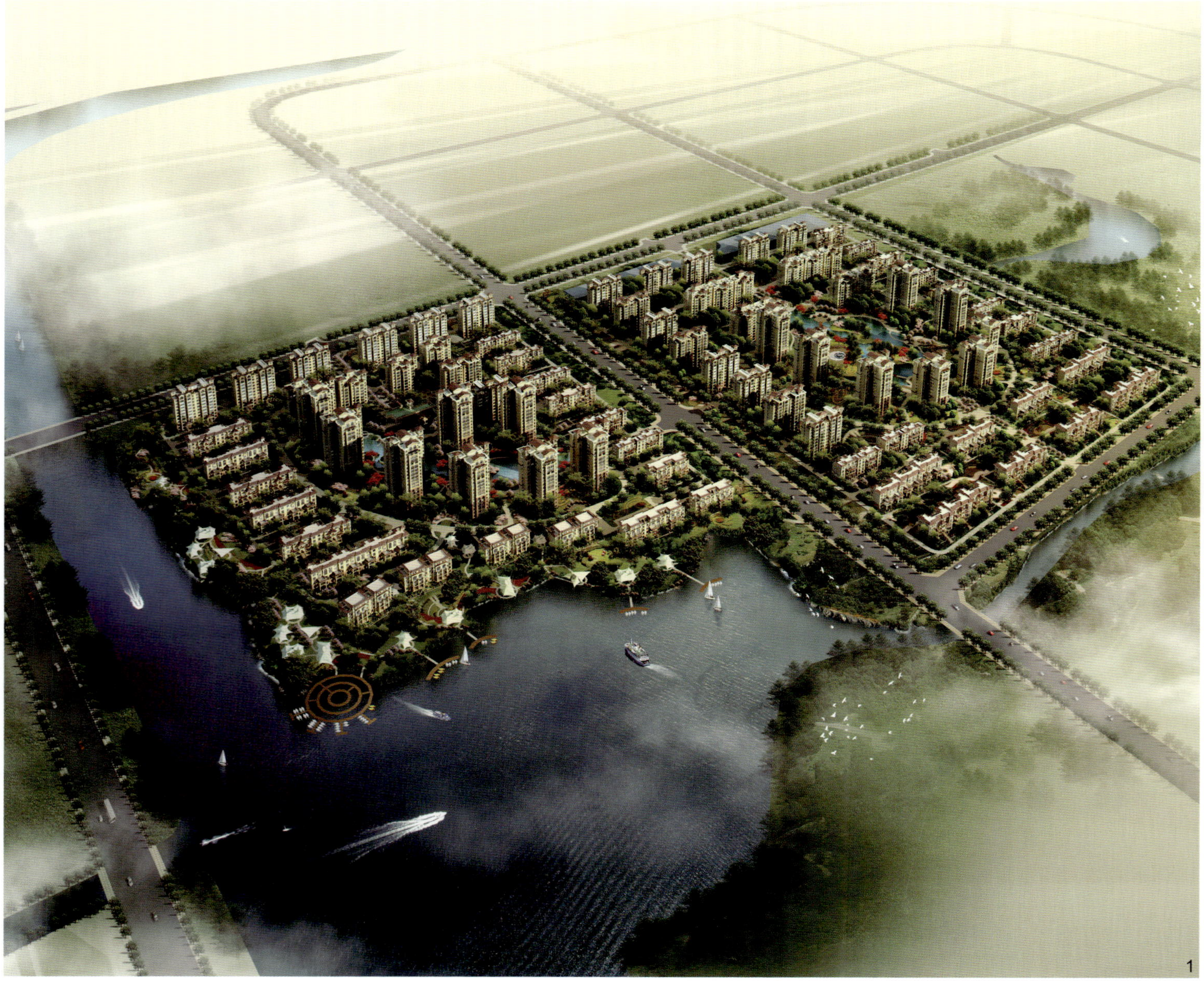

建设地点：山东 威海
占地面积：33公顷
建筑面积：42万平方米
景观面积：14.5万平方米

Project Location: Weihai, Shandong
Site Area: 33 ha
Floor Area: 420,000 m^2
Landscape Area: 145,000 m^2

1–3 保定阳光海洋花园D区

1

2

3

4

5

6

4–5 山东泰钢嬴牟社区

7–11 承德塞上江南住宅小区

7

8

9

10

11

1–6北京东亿国际传媒产业园二期项目

1

2

3

4

5

6

建设地点：北京 朝阳区
占地面积：6.7公顷
建筑面积：14.2万平方米
景观面积：2.8万平方米

Project Location: Chaoyang District, Beijing
Site Area: 6.7 ha
Floor Area: 142,000 m^2
Landscape Area: 28,000 m^2

1 承德环保局综合办公楼

2 天津泰达海邻售楼处

3 新疆神宇集团创业基地

4–5 银川经济开发区中小企业创业大厦

1–2 保定毛主席视察纪念馆

3–5 承德博物馆

1–2 承德御道风情商业街

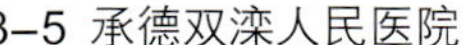
3–5 承德双滦人民医院

1–2 莱芜文化广场

3–4 保定东方红城

5–7 中国移动大厦景观设计

Architectural Design and Research Institute of Tsinghua University Co., Ltd.

清华大学建筑设计研究院有限公司

清华大学建筑设计研究院有限公司成立于1958年，是国家甲级建筑设计院，拥有建筑工程设计、古建筑与文物保护等甲级设计资质、工程咨询甲级资质、施工图设计文件审查资质和水利行业（水库枢纽）及公路行业（公路、交通）乙级设计资质。

主要承担各类公共与民用建筑工程设计、城市设计、文物与古建筑保护、景观及室内设计等工程设计与咨询服务。设计并建成了一批如北京菊儿胡同居住区、清华大学图书馆、北京天桥剧场、中国美术馆改造装修工程、清华大学医学院等精品之作，独立完成2008年北京奥运会射击馆等多项奥运工程项目。

Architectural Design and Research Institute of Tsinghua University Co., Ltd. was established in 1958, is a national architectural design institute with grade A qualification, has the Grade A qualification of architecture engineering design, ancient buildings and heritage conservation, the Grade A qualification of engineering consulting, the qualification of construction drawing design document review and the Grade B qualification of water industry (reservoir hub) and road sectors (roads, transport).

It mainly undertakes architecture design and consulting services, such as public and civil architectural project design, urban design, the protection of cultural relics and ancient buildings, and scenic and indoor design. Several selected projects have been designed and constructed, such as Beijing Ju'er Hutong Residential Area, Tsinghua University Library, Beijing's Tianqiao Theater, China Art Gallery Rebuilding Project, and School of Medicine of Tsinghua University. THAD independently has also undertaken many 2008 Beijing Olympic Games projects.

地 址：北京市清华大学建筑设计中心楼
邮 编：100084
电 话：+86–10–62789999
传 真：+86–10–62784727
网 址：www.thad.com.cn
邮 箱：jzsjy@tsinghua.edu.cn
法定代表人：庄惟敏

Add: Architectural Design Center of Tsinghua University, Beijing
P.C.: 100084
Tel: +86-10-62789999
Fax: +86-10-62784727
Http: //www.thad.com.cn
E-mail: jzsjy@tsinghua.edu.cn
Legal Representative: Zhuang Weimin

北京奥林匹克公园中心区下沉花园2号院

建设地点：北京 朝阳区
总用地面积：4 480平方米
总建筑面积：590平方米
设计起止时间：2007—2008年
竣工时间：2008年
获　　奖：
荣获2009年度教育部优秀勘察设计奖园林一等奖；
荣获2009年度全国优秀工程勘察设计行业奖一等奖

奥林匹克中心区下沉花园2号院取名"瓦院"。通过诠释一组北京四合院建筑空间片断，以具有代表性的中国传统建筑材料——瓦作为主要表达手段，结合树木、室外水景等手段，展现北京的地域文化特色，形成有地域文化背景的室外休憩场所。建筑设计在现实与传统之间建立联系纽带的归属感，传统文化中积淀的、受大众文化包容认可的社会审美价值和人文意境是设计中寻求与传统结合的法则。院落的主题是表达传统文化背景下市民生活的意境。

Sunk Garden Yard No.2 in Beijing Central Olympic Area

Construction Site: Chaoyang District, Beijing
Total Land Area: 4,480 m^2
Total Floor Area: 590 m^2
Design Duration: 2007–2008
Completion Time: 2008
Honors:
Garden First Prize of Excellent Survey & Design by Ministry of Education, 2003;
First Prize of National Excellent Project Survey & Design Award, 2009

The Sunk Garden Yard No.2 in Olympic central area is named as "Tile Yard". By showing a series of construction spaces of Beijing Quadrangle House and combining trees and outdoor waterscape etc, "Tile Yard" is made of a traditional Chinese construction material – tile, which shows distinctive cultural characteristics of Beijing and becomes an outdoor relaxing place with regional and cultural backgrounds. By the principle of combining social aesthetic value and human culture accumulated in traditional culture and accepted by people with traditions, architectural design creates a sense of belonging between reality and tradition. The theme of Yard No.2 is to show citizen's living environment with traditional culture background.

乔波冰雪世界滑雪馆及配套会议中心

建设地点：北京
总建筑面积：52 728.6平方米
滑 雪 馆：28 691.5平方米
配套会议中心：24 037.1平方米
设计起止时间：（滑雪馆）2003—2004年、
（配套会议中心）2005—2006年
竣工时间：（滑雪馆）2005年、（会议中心）2006年
获 奖：荣获“2008年度全国优秀工程勘察设计奖银奖”

乔波冰雪世界滑雪馆及其配套会议中心项目位于北京市顺义区牛栏山镇。本项目东邻潮白河奥林匹克水上运动中心，滑雪馆部分在奥运会期间及其后提供了相应的体育休闲设施，会议中心部分的住宿、餐饮、会议等也提供了相应的配套服务，缓解了奥运会期间观看奥运项目游人汇聚的压力。奥运会后滑雪馆及会议中心成为丰富市民业余文化生活的重要体育休闲设施。

工程总用地面积48 048平方米，总建筑面积52 728.6平方米，其中滑雪馆28 691.5平方米、配套会议中心24 037.1平方米，最高点建筑高度54.36米。滑雪馆部分主体由滑雪大厅及服务区组成，滑雪厅位于会议中心和管理用房之上，滑道终点部分深入地下，滑道长261米，滑道坡度6°～18°，总落差49.5米，分初级与专业两个滑道区，滑道区平面呈梯形，起点处窄终点处宽，滑道顶部结构跨度为24米～40米，滑雪馆满负荷使用人数约1 000人。

配套会议中心建筑面积26 700平方米，设有176套客房及辅助用房。设计注重公共空间的开合变化及室内外空间的流通与视线的畅通，满足了其复杂而特殊的流线要求。该中心的造型与立面设计充分结合并尊重滑雪馆建筑的整体控制性体量，追求简洁平实、现代精致，使之与滑雪馆的既定风格协调并为之增色。

Ski Hall and Auxiliary Conference Center of Qiaobo Ice-snow World

Construction Site: Beijing
Total Lloor Area: 52,728.6 m^2
Ski Hall: 28691.5 m2
Auxiliary Conference Center: 24,037.1 m^2
Design Duration: (Ski hall)2003—2004;
(Auxiliary Conference Center)2005—2006
Completion Time:(Ski hall) 2005; (Conference Center) 2006
Honors: Silver Prize of National Excellent Project Survey & Design Award, 2008

Ski Hall and Auxiliary Conference Center of Qiaobo Ice-snow World is located in Niulanshan Town of Shunyi District, Beijing. Neighboring with Chaobai River Olympic Water Sports Center in the east, ski hall is provided with appropriate sports and leisure facilities during and after the Olympic Games, while the conference center provides proper supporting services of accommodation, catering, conferences, which release people-stream pressure of Olympic visitors during the Olympic Games. After the event, ski hall and conference center have become important sports and leisure facilities to enrich citizens' cultural life in their spare time.

The total land area is 48,048 m², and the total floor area is 52,728.6 m², including 28,691.5 m² area of ski hall and 24,037.1 m² area of auxiliary conference center. Height of the tallest building is 54.36 m. Ski Hall part is mainly composed of ski lobby and service area. The ski lobby is located above the conference center and the management area. Going deep into the ground, the slide-way is 261 m long with the slide slope of 6°–18° and a total drop length of 49.5 m. It is divided into two slide-ways: junior and professional. The slide-way is in the shape of trapezoid, which is narrow in the starting and broad in the finishing point. The span of its top structure is between 24 and 40 meters. Full capacity of the ski hall is about 1,000 people.

Floor area of the auxiliary conference center is 26,700 m^2, with 176 hotel rooms and ancillary rooms. Focusing on opening and closing changes of public space, the flow of indoor and outdoor space and clear sight, the design meets its complex and specific requirements of flow lines. The shape and facade design of conference center is fully integrated with the overall controlling body mass of the ski hall upon consideration. Pursuing simplicity and plainness, modern and delicacy, the conference center stands coordinately with ski hall and has added luster to its established style.

徐州水下兵马俑博物馆及汉文化艺术馆

建设地点：江苏 徐州
总用地面积：64 660平方米（含水面）
总建筑面积：4 150平方米
设计起止时间：2005年
竣工时间：2006年
获　　奖：荣获2007年度教育部优秀勘察设计奖一等奖；荣获2008年度全国优秀工程勘察设计行业奖三等奖。

徐州水下兵马俑博物馆与汉文化艺术馆是徐州汉文化景区里的两个主要博物馆建筑。建筑设计从所处景区悠久的历史文化背景与山水交融的优美自然环境出发，将历史文化要素作为一种建筑设计中的审美原则，采用写意的建筑表达手法，通过空间、体形、材质、色彩等方面的建筑语言，营造浓厚的传统文化意境，实现与历史文化背景及山水自然环境的和谐统一，探索性地尝试历史文化地段的建筑创作方法。

水下兵马俑博物馆通过较小的建筑体量、简洁的建筑形态、内敛的建筑性格追求写意的表现效果；汉文化艺术馆则通过传统院落空间的组织方法，营造出静谧、含蓄的文化氛围。二者的建筑性格张弛搭配、收放结合，相得益彰。

Xuzhou Underwater Terra-cotta Warriors and Horses Museum / Art & Culture Center of Han Dynasty

Construction Site: Xuzhou, Jiangsu
Total Land Area: 64,660 m^2 (incl. water surface)
Total Floor Area: 4,150 m^2
Design Duration: 2005
Completion Time: 2006
Honors: First Prize of Excellent Survey & Design by Ministry of Education, 2007; Third Prize of National Excellent Project Survey & Design Award, 2008

Xuzhou Underwater Terra-cotta Warriors and Horses Museum / Art & Culture Center of Han Dynasty are the two main museum buildings of Xuzhou Art & Culture Center of Han Dynasty Scenic Spot. In consideration of the long historical and cultural background, and the beautiful natural environment of water and mountain landscapes, taking historical and cultural factors as the aesthetic principle of architectural design, the design creates a strong sense of traditional culture, by using freehand methods of construction expression through architectural languages of space, shape, texture and color etc. The design realizes the harmony between historical and cultural backgrounds and the natural environment of water and mountain landscapes, and has tried the architectural creation method in historical and cultural areas.

By a small body mass, simple architectural form and an introverted architectural character, the Underwater Terra-cotta Warriors and Horses Museum pursues a freehand method of appearance expression; by using traditional organizational method of courtyard space, the Art & Culture Center of Han Dynasty creates a quiet and subtle cultural environment. Characters of the two buildings match and complement each other, bringing out the best in both of them.

北京市北外附属外国语学校

建设地点：北京
总用地面积：82 748平方米
总建筑面积：68 846平方米
设计起止时间：2003—2005年
竣工时间：2006年
获　　奖：荣获2007年度教育部优秀勘察设计奖二等奖；
荣获2008年度全国优秀工程勘察设计行业奖三等奖。

北外附属外国语学校位于北京海淀区中关村高科技园区内，是以寄宿制为主的、具有教学特色和完善设施的高标准现代化学校。它不仅完善了中关村高新科技园区的服务功能，而且是政府建设海淀区基础教育的标杆学校。

学校总建筑面积6.8万平方米，包括小学教学及宿舍楼、中学教学及宿舍楼、游泳馆及体育活动中心、综合服务楼、食堂及锅炉房等建筑。总体设计在功能构成、校园布局及建筑设计构思方面提出了“多义空间”等新的方法，并在园林化校园、人文化校园等方面进行了探索。

建筑采用咖啡色墙面及白色构件点缀的外观色彩，表现出脱俗的气质与文雅的风格。与此形成对比的是大楼内部的走廊和活动空间的墙面被涂上各种纯色，表现出孩子们活泼好奇的天性。

Beijing Foreign Language School, BFSU

Construction Site: Beijing
Total Land Area: 82,748 m^2
Total Floor Area: 68,846 m^2
Design Duration: 2003—2005
Completion Time: 2006
Honors:
Second Prize of Excellent Survey & Design by Ministry of Education, 2007;
Third Prize of National Excellent Project Survey & Design Award, 2008

Located in Zhongguancun High-tech Park of Beijing Haidian District, Beijing Foreign Languages School, BFSU is a boarding-based high-standard modern school with teaching features and complete facilities. Not only has it improved the service function of Zhongguancun High-tech Park, but also it has become the yardstick school of basic education in Haidian District for the Government.

Its total floor area is 68,000 m^2, including primary school building and dormitory building, middle school building and dormitory building, swimming pool and sports center, integrated service building, canteen, boiler room and other buildings. The overall design brings “multi-vocal space” and other new methods in function formation, campus layout and idea of architectural design, and has explored in the campus of garden, the humanistic campus etc.

The brown-color wall and white-color components decorated on it show a refined temperament and an elegant style. In contrary, walls of the corridor and activity space are painted colorfully, showing the lively and curious nature of children.

Beijing ArtPureLux Building Advisory Co., Ltd.

北京亚特朴华建筑咨询有限公司

北京亚特朴华建筑咨询有限公司于2010年5月注册成立。主要发起人有着丰富的设计、施工、管理和房地产从业经验，本着为社会和设计领域作出应有的一份贡献的愿望，联合有志之士，组建本公司。

“朴华”的来源：“艺术的风格有两大类。一种是华，一种是朴；华近乎雕琢，朴近乎自然，华朴相错是为妙品。人们对艺术的欣赏是华久则思朴，朴久则思华，两种风格轮流交替，互补互济，以求得某种平衡。”

设计人员主要以经验丰富的设计师为主导，以敢于创新和能够创新的青年设计师为主体；以项目的设计创作为载体，实现艺术和技术的有机结合；为项目的业主和社会实现最大化的经济及社会效应，为企业本身实现合理化的效益。

设计项目领域：大型规划、大型商业综合体、住宅、别墅、写字楼、酒店、及各类公建项目。主要以中高端项目的规划设计为主。

公司理念：

以市场为导向，以技术、经验为依托，追求实用、适用和设计完美。以人性为前提，以时间、效益为目标，追求团结、协作和共同提高。

This Company was incorporated in May 2010. Its major founder, who is experienced in design, construction, management and real estate professions, creates this organization in cooperation with a group of talents, with a vision to contribute to the society and the design field.

The origin of company name “ArtPureLux”: There are two styles of art (“Art”), luxury (“Lux”) and pure (“Pure”). Luxury is delicate, Pure is natural, while the mixture of the two is perfect. People immersed in luxury art will long for pure art, vice versa. So the two styles alternate and complement each other, coming to a balance.

The team of designers is mainly led by experienced designers, and innovative young designers are the main body. They integrate art and technique by means of design creation of projects; they maximize the economic and social benefits of the project owners and the society, and realize reasonable corporate benefits.

Design Projects: Large-sized planning, large-sized commercial complex, residential, villa, office buildings, hotel, and other public buildings. Major in planning and design of medium- and high-end projects.

Corporate Philosophy:

Market-oriented, backed by technology and experience, we pursue practical, applicable and perfect design. Human-based, targeted to time and effect, we pursue teamwork, collaboration and common promotion.

地址：北京市朝阳区广渠路21号（百子湾西里403）金海商富中心B座1907
电话：+86–10–57102853
传真：+86–10–59574997
邮箱：bjytph@163.com

Office Address: Room 1907, Block B, Gimhae Fortune Center, 21 Guangqu Road (403 Baiziwan Xili), Chaoyang District, Beijing
Tel.: +86–10–57102853
Fax: +86–10–59574997
Email: bjytph@163.com

潍坊农村信用社办公楼

总用地面积：0.8公顷
总建筑面积：3万平方米（地上约2万平方米，地下约1万平方米）
建筑层数：9层（地上最高7层，地下2层）
建筑高度：30米
设计时间：2010年

Weifang Rural Credit Cooperative Office Building

Total Land Area: 0.8 ha
Total Floor Area: 30,000 m^2 (Aboveground: 20,000 m^2; underground: 10,000 m^2)
Building Storey: 9 (Aboveground: 7; underground: 2)
Total Height: 30 m
Design Time: 2010

潍坊农民画博物馆

总用地面积：2公顷
总建筑面积：10万平方米
（地上约8万平方米，地下约2万平方米。其中主体博物馆约5万平方米，配套建筑约3万平方米）
建筑层数：7层（地上最高5层，地下2层）
建筑高度：24米
设计时间：2010年

Weifang Farmers Painting Museum

Total Land Area: 2 ha
Total Floor Area: 100,000 m^2
(Aboveground: 80,000 m^2, underground: 20,000 m^2. Incl.: museum main body: 50,000 m^2, supporting building: 30,000 m^2)
Building Storey:7 (Aboveground: 5; underground: 2)
Total Height: 24 m
Design Time: 2010

潍坊杨家埠沿街商业中心

总用地面积：5公顷
总建筑面积：9万平方米（地上约7万平方米，地下约2万平方米）
建筑层数：7层（地上最高5层，地下2层）
建筑高度：24米
设计时间：2010年

Weifang Yangjiabu Commercial Street

Total Land Area: 5 ha
Total Floor Area: 90,000 m^2 (Aboveground: 70,000 m^2, underground: 20,000m^2)
Building Storey: 7 (Aboveground: 5; underground: 2)
Total Height: 24 m
Design Time: 2010

威海荣威新天地名人广场

总用地面积：6公顷
总建筑面积：25万平方米
（地上约19万平方米，地下约6万平方米。其中：住宅约13万平方米，商业及配套公建约6万平方米）
建筑层数：33层（地上最高31层，地下2层）
建筑高度：98米
设计时间：2011年

Weihai Rongcheng Xintiandi Fame Plaza

Total Land Area: 6 ha
Total Floor Area: 250,000 m^2
(Aboveground: 190,000 m^2, underground: 60,000 m^2. Incl.: residential building: 130,000 m^2, commercial and supporting public building: 60,000 m^2)
Building Storey: 33 (Aboveground: 31; underground: 2)
Total Height: 98 m
Design Time: 2011

烟台中拓

总用地面积：6公顷
总建筑面积：18万平方米(地上约13万平方米，地下约5万平方米)
建筑层数：20层（地上最高18层，地下2层）
建筑高度：54米
设计时间：2010年

Yantai Zhongtuo Project

Total Land Area: 6 ha
Total Floor Area: 180,000 m^2 (Aboveground: 130,000 m^2, underground: 50,000 m^2)
Building Storey: 20 (Aboveground: 18; underground: 2)
Total Height: 54 m
Design Time: 2010

总平面图 1:600

烟台昆嵛山旅游项目

总用地面积：135公顷
总建筑面积：80万平方米（地上约65万平方米，地下约15万平方米）
建筑层数：7层（地上最高5层，地下2层）
建筑高度：24米
设计时间：2010年

Yantai Mt. Kunyu Tourism Project

Total Land Area: 135 ha
Total Floor Area 800,000 m^2 (Aboveground circa 650,000 m^2, underground circa 150,000 m^2)
Building Storey: 7 (Aboveground at most 5 storeys, underground 2 storeys)
Total Height: 24 m
Design Time: 2010

济宁火炬大厦

总用地面积：3.5公顷
总建筑面积：18万平方米
（地上约15万平方米，地下约3万平方米。其中：写字楼、酒店约9万平方米，公寓约2.5万平方米，商业约3.5万平方米）
建筑层数：43层
（地上约40层，地下3层）
建筑高度：160米
设计时间：2010年

Jining Torch Building

Total Land Area: 3.5 ha
Total Floor Area: 180,000 m^2
(Aboveground circa 150,000 m^2, underground circa 30,000 m^2. Incl.: office building and hotel circa 90,000 m^2, apartment circa 25,000 m^2, commercial building circa 35,000 m^2)
Building Storey: 43
(Aboveground circa 40 storeys, underground 3 storeys)
Total Height: 160 m
Design Time: 2010

烟台蓁山屯低密度

总用地面积：6公顷
总建筑面积：7.5万平方米
（地上约6万平方米，地下约1.5万平方米）
建筑层数：5层（地上最高4层，地下1层）
建筑高度：12米
设计时间：2010年

Yantai Zhenshantun Low-density Project

Total Land Area: 6 ha
Total Floor Area: 75,000 m^2
(Aboveground: 60,000 m^2, underground: 15,000 m^2)
Building Storey: 5 (Aboveground: 4; underground: 1)
Total Height: 12 m
Design Time: 2010

烟台邵家疃

总用地面积：3公顷
总建筑面积：6万平方米(地上约4.5万平方米，地下约1.5万平方米)
建筑层数：7层（地上最高6层，地下1层）
建筑高度：18米
设计时间：2011年

Yantai Shaojiajiang Project

Total Land Area: 3 ha
Total Floor Area: 60,000 m^2 (Aboveground: 45,000 m^2, underground: 15,000 m^2)
Building Storey: 7 (Aboveground: 6; underground: 1)
Total Height: 18 m
Design Time: 2011

烟台上曲家

总用地面积：35公顷
总建筑面积：83万平方米
（地上约65万平方米，地下约18万平方米。其中住宅约40万平方米，写字楼、酒店及配套公建约25万平方米）
建筑层数：26层（地上最高24层，地下2层）
建筑高度：98米
设计时间：2011年

Yantai Shangqujia Project

Total Land Area: 35 ha
Total Floor Area: 830,000 m^2
(Aboveground: 650,000 m^2, underground: 180,000 m^2. Incl.: residential building: 400,000 m^2, office building, hotel and supporting public building: 250,000 m^2)
Building Storey: 26 (Aboveground: 24; underground: 2)
Total Height: 98 m
Design Time: 2011

北京中联环建文建筑设计有限公司于2000年成立，至今已有十年历史，是首都地区拥有建筑工程设计甲级、规划设计乙级和风景园林设计乙级资质的股份制设计公司。特有的股份制结构，让企业焕发生机。十年来，公司经济高速发展的同时规模也在不断地扩大，现已发展为以北京为中心、覆盖全国的大型品牌建筑设计机构。公司业务现已拓展至全国30个省、市、自治区，分别在上海、深圳、天津和济南等地设有分支机构和办事处。在建筑市场异军突起，市场占有率和品牌影响力居于行业前列。

公司拥有国家一级注册建筑师、国家一级注册结构工程师、国家注册城市规划师、博士、硕士和高级建筑师（含正教授级）、高级工程师（含正教授级）等众多专业人才。在项目前期可行性研究与策划、城市规划与城市设计、居住区规划与居住建筑设计、大型公共建筑设计、大学校园规划与学校建筑设计、风景园林与环境艺术设计、城市导向系统及夜景照明设计、室内装饰装修设计以及工程咨询与项目管理等领域都有着骄人的业绩。

公司十分重视加强设计研发力量，提高自主创新能力。建筑方案设计是品牌的灵魂，在方案创新方面公司曾多次在众多国内外知名设计机构参与的国际竞标中脱颖而出，拔得头筹。公司设计的众多明星项目现已成为地标性建筑，很多作品曾分别荣获国家建设部、北京市、中国建筑艺术网、全国工商联房地产商会、《中国不动产》、《北京晚报》以及《新地产》等机构和传媒颁发的各种大奖。

中联环曾和法国享誉全球的著名设计集团公司ADPI合作，并与澳大利亚、德国等国家的著名设计公司成立中外建筑师设计联盟，大幅度开拓国际市场，向品牌全球化的目标逐步迈进。

United Architects & Engineers Co., Ltd. (UA Design), founded in 2000, having ten years' development history, is vigorous with its special shareholding structure. UA Design has obtained Grade-A qualification for engineering design, Grade-B qualification for planning design and Grade-B qualification for landscape design. In the past ten years, the Beijing-based company has witnessed a fast growth and has developed itself into a famous architectural design organization in Beijing. At present, it has spread its business to 30 provinces and has established branches and offices in Shanghai, Shenzhen, Tianjin and Ji'nan. It has occupied the front position on the aspect of market share and brand influence.

UA Design possesses a powerful design force, including national Class-1 registered architect, national Class-1 registered structural engineer, national registered urban planner, doctor, master and senior architect (including full professor) and senior engineers (including full professor). The company is pretty good at feasibility study and planning, urban planning and urban design, residential community planning and design, large public facilities design, campus planning and building design, landscape and environment design, urban guide system, lighting design, interior decoration and planning, engineering consulting and project management.

The company attaches great importance to the improvement of design ability and creativity. Project design is the key to building brand. The company has competed against many famous design agencies and has won important awards. Some buildings the company designed have become landmarks in some cities, and its works have won many famous awards granted by Ministry of Construction, Beijing Municipality, China Architecture and Art Net, China Real Estate Chamber of Commerce, *China Real Estate*, *Beijing Evening Newspaper*, *New Real Estate*, etc.

UA Design has established a Sino-France joint venture with ADPI, which an internal prestigious design group, design association of Chinese and foreign architects with famous design companies in Australia and German, to stride towards the foreign market and to create an international brand.

刘光亚
清华大学建筑学学士
清华大学建筑学硕士
正教授级高级建筑师
国家一级注册建筑师
国家注册城市规划师

Liu Guangya
Bachelor of Architecture, Tsinghua University
Master of Architecture, Tsinghua University
Senior Architect at professor level
First Class Registered Architect of China
Registered Urban Planner of China

孟伟康
清华大学建筑学学士
清华大学建筑学硕士
高级建筑师
国家一级注册建筑师

Meng Kangwei
Bachelor of Architecture, Tsinghua University
Master of Architecture, Tsinghua University
Senior Architect
First Class Registered Architect of China

王泉
大连理工大学建筑学学士
东南大学建筑研究所研究生
比利时鲁汶大学建筑学硕士

Wang Quan
Bachelor of Architecture, Dalian University of Technology
Postgraduate of Architecture Research Institute, Southeast University
Master of Architecture, Catholic University of Louvain, Belgium

张兵
天津大学建筑学学士
清华大学建筑学硕士
高级建筑师
国家一级注册建筑师

Zhang Bing
Bachelor of Architecture, Tianjin University
Master of Architecture, Tsinghua University
Senior Architect
First Class Registered Architect of China

刘常青
清华大学建筑学学士
清华大学建筑学硕士
国家一级注册建筑师

Liu Changqing
Bachelor of Architecture, Tsinghua University
Master of Architecture, Tsinghua University
First Class registered Architect of China

李晓光
东南大学建筑学学士
国家一级注册建筑师

Li Xiaoguang
Bachelor of Architecture, Southeast University
First Class Registered Architect of China

方楠
清华大学建筑学学士
荷兰戴尔夫特工业大学（TU Delft）建筑学博士

Fang Nan
Bachelor of Architecture, Tsinghua University
PHD of Architecture, Delft University of Technology, the Netherlands

赵成豪
华中科技大学工学学士

Zhao Chenghao
Bachelor of Engineering, Huazhong University of Science and Technology

战立光
山东建筑大学建筑学学士
天津大学工程硕士
国家一级注册建筑师

Zhan Liguang
Bachelor of Architecture, Shandong University of Architecture
Master of Engineering
First Class Registered Architect of China

杨杰
北京工业大学工学学士
高级结构工程师

Yang Jie
Bachelor of Engineering, Beijing University of Technology
Senior Structure Engineer

曾宪丁
天津大学建筑学院城市规划学士
北京大学环境学院景观设计硕士

Zeng Xianding
Bachelor degree of urban planning, Tianjin University Architecture College
Master degree of landscape design, Beijing University Environment College

郑悦琦
西安建筑科技大学建筑学学士
国家一级注册建筑师

Zheng Yueqi
Bachelor of Architecture, Xi'an University of Architecture and Technology
First Class Registered Architect of China

宋兵
天津工业大学工学学士

Song Bing
Bachelor of Engineering, Tianjin University of Technology

内蒙古阿拉善经济开发区金融中心

（国内中标、第三届百年建筑优秀作品奖•商业类建筑风格优秀奖）

设计时间：2009年

建设地点：内蒙古 阿拉善盟经济开发区

规　　模：50 000平方米

沙丘是大自然的伟大雕塑艺术品，是风和岁月留下的痕迹。该项目选取当地环境中最美、最富有诗意的沙漠为方案原型，从中抽取出两个要素——锐利连绵的曲线和清晰细密的纹理。选取能展示当地风格的方案作为类比，抽取出我们想要的元素。该案中选择了连贯、绵延的建筑形式来展现当地的大气之风。

为了满足金融中心内部使用功能的需求，抽取出"独立"这个元素，即在一个连续的统一体内划分出独立的功能块。在统一的表皮的覆盖下，内部划分为若干个独立的功能块，每个单元有独立对外的出入口，满足对功能的需求。单体之间围合出的空间，形成了有活力的商业休闲、街、中庭等空间。细密的沙痕如阿拉善的文化般丰富，源远流长……将它与简洁、现代的风格融合在一起，即形成了建筑身上流畅的纹理。建筑形态仿佛低缓、连绵的沙丘，在基地蔓延……它是风造就的形态，在风的吹拂下，涌动成一首美丽的诗。韵律连绵起伏，一直推向最高潮……诗歌高潮部分形成的高层位于整个基地的东南角，成为整个基地的新地标，它可以俯览广场及文化中心，亦可在城市中心接受他人瞻仰。

内蒙古阿拉善经济开发区文化艺术中心
（国内中标、第三届百年建筑优秀作品奖•公建类建筑风格优秀奖）
设计时间：2009年
建设地点：内蒙古 阿拉善盟经济开发区
规　　模：50 000平方米

内蒙古阿拉善经济开发区文化艺术中心，位于内蒙古阿拉善盟经济开发区内，项目用地东西长约112米，南北长约288米。用地北侧为吉兰大街，东侧为开发区商业文化艺术中心广场和商业中心，它们共同组成阿拉善开发区经济文化中心。该中心位于中央公园和那达慕公园之间，是城市主要景观带的一部分，是开发区最重要的城市文化艺术中心和居民活动场所。

文化艺术中心作为经济区社会事业的重要组成部分，是最能够体现阿拉善城市文化的物质载体。

太原市和泰龙城项目规划及建筑设计
设计时间：2010年
建设地点：山西 太原
规　　模：113 672平方米

北京像素
（与日本SAKO建筑设计工社合作）
设计时间：2008年
建设地点：北京 朝阳区
规　　模：800 000平方米

自在香山

（《时代楼盘》杂志社第五届金盘奖•年度综合楼盘）
设计时间：2005年
建设地点：北京 海淀区
规　　模：238 990平方米

西南民族大学新校区
（建设部"中国人居环境范例奖"）
设计时间：2001年
建设地点：四川 成都
新校区占地面积：68公顷
新校区建筑面积：500 000平方米

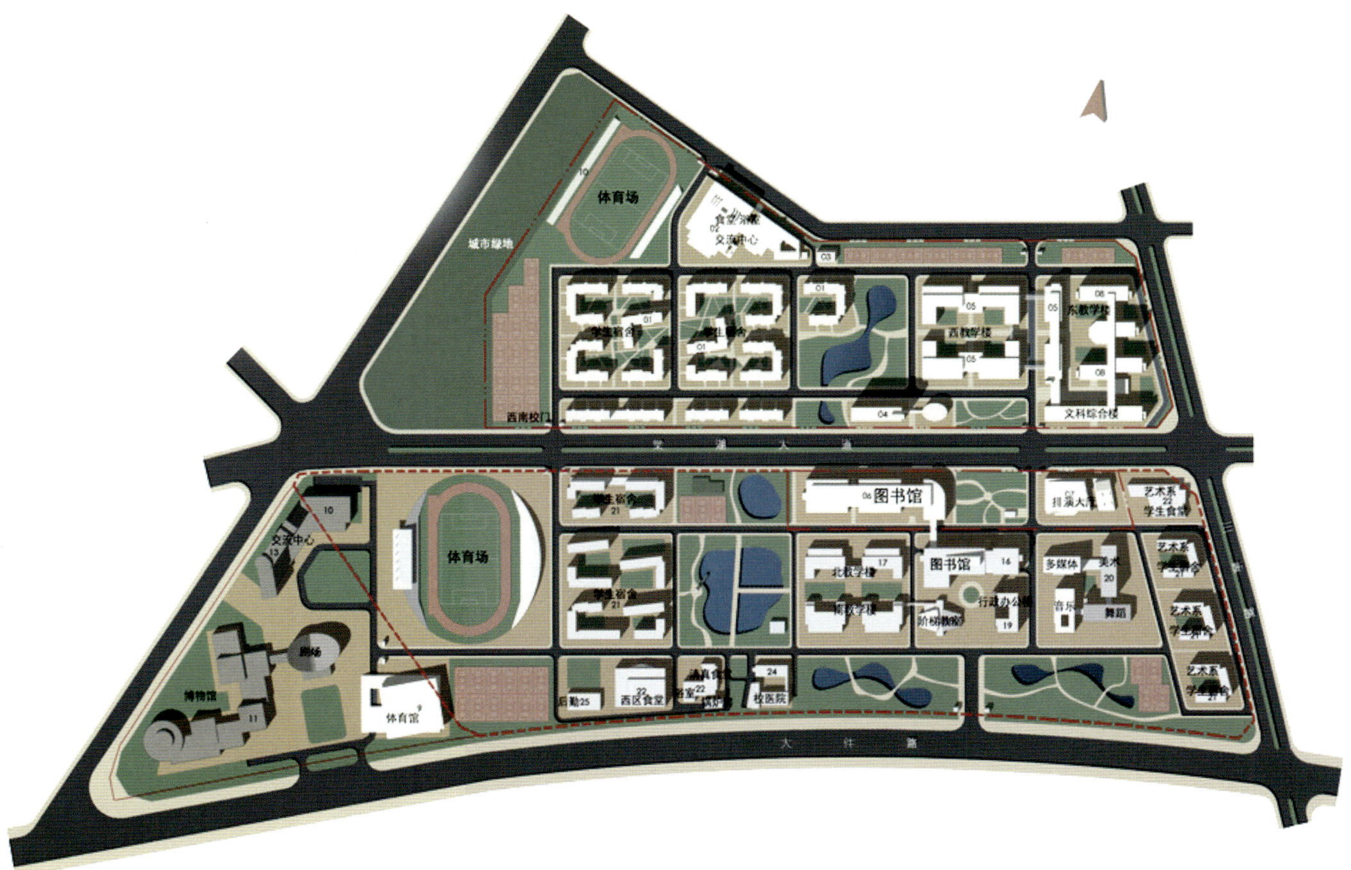

1 **天狮健康产业园产品展示中心**
设计时间：2006年
建设地点：天津
规　　模：6 000平方米

2 **天狮国际健康产业园会议中心**
设计时间：2007年
建设地点：天津
规　　模：32 000平方米

3 **天狮健康产业园生命医学研究中心**
设计时间：2006年
建设地点：天津
规　　模：27 000平方米

4 **天狮产业园全球研发质检中心**
设计时间：2007年
建设地点：天津
规　　模：8 732平方米

5 **天狮国际健康培训基地**
设计时间：2007年
建设地点：天津
规　　模：70 000平方米

产业园位于武清开发区内，占地54公顷，它的北面有静静的龙凤河紧贴着流过，不远处为京津塘高速路，西面还有一条通往华东和上海的高速路。

该产业园，建筑面积有40多万平方米，包括：国际生命研究中心6.5万平方米，教育培训中心3.2万平方米，高级研究人员宿舍4.9万平方米，员工大食堂2.8万平方米。研发一、二所两幢共3.6万平方米，生命医学研究中心2.7万平方米，研发化检中心0.75万平方米，产品展示中心0.3万平方米，厂房仓储10万平方米，员工公寓三幢共计1.5万平方米，研发三所1.7万平方米，锅炉房0.5万平方米，后勤配套用房1.4万平方米；它还有：四季植物园0.3万平方米，办公接待、会议三幢共计1万平方米。

天狮健康产业园办公楼
设计时间：2006年
建设地点：天津
面　　积：70 000平方米

天狮集团的总部办公楼建筑呈弧形，总长270米。办公楼前粗壮浑圆的柱子，显得底气十足。办公楼幕墙外有一层会翻动的“鳞甲”，用做遮阳系统，能够给办公楼带来灵动和生机的感觉。

在办公楼前的大平台下面“藏着”一个5 000平方米的博物馆，在它的中央还有一个直径36.6米的透光水池。

曲阜游客服务（集散）中心

（2009年度山东省优秀城市规划设计奖二等奖）

设计时间：2008年

建设地点：山东 曲阜

规　　模：16 311.67平方米

涿州市博物馆
（国内中标）
设计时间：2009年
建设地点：河北 涿州
规　　模：20 000平方米

根据博物馆功能设置，出入口包括观众出入口（一般观众出入口和专业观众出入口）、工作人员出入口、藏品出入口。固定陈列厅的一般观众出入口位于博物馆二层，观众可通过自动扶梯进入展区；专业观众出入口与办公人员出入口合设于入口广场东南角，南临华阳路；藏品出入口则位于用地北侧，通过坡道进入博物馆库区。另外，临时展厅设单独出入口，直接面对入口广场；行政办公出入口位于东北角，并设置了绿化设施；其他功能展示空间出入口安排在固定陈列区北侧临观音堂街，自成一区。

鄂尔多斯蒙古族学校
（中联环设计与瑞士EM2N联合设计）
设计时间：2008年
建设地点：内蒙古 鄂尔多斯
规　　模：120 000平方米

该项目位于鄂尔多斯市康巴什新区核心地带，学校总用地约21公顷，周围有城市道路环绕。该校按寄宿制设计。

该项目旨在建立一所世界一流的、现代化的、新型教学模式的民族学校，响应党的民族政策，积极推进民族教育，并借鉴现代西方教育模式，将该校建成向国内外展示自治区及我国少数民族教育和学校建设探索的窗口。

该项目采用瑞士EM2N公司的设计方案，以蒙古族传统元素——部落为设计理念，对蒙古族传统建筑进行现代诠释。在设计手法上把当地最有代表性的天空、草地、云、蒙古包等自然和人文景观进行提炼、简化，运用到建筑造型和装饰上来。重现群居于部落的人们与自然亲密接触的生活方式，为学生们创造第二个家，一个融合学习、生活的小型城市。

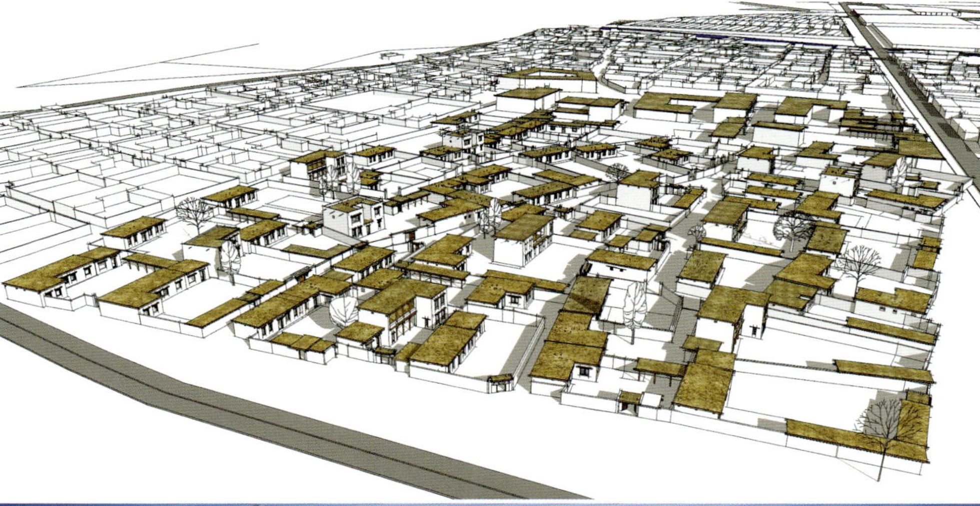

玉树新寨民居灾后重建

（北京市规划委员会"北京工程勘察设计测绘行业援建工作贡献奖"）
建设地点：青海 玉树藏族自治州
总建筑面积：88 000平方米
占地面积：175.86公顷
设计时间：2010—2011年
容 积 率：0.3

在充分调研的基础上，以四川灾后重建的经验为指导，通过对玉树地区藏式建筑认真细致的研究工作，总结出结古镇新寨村藏式民居的建筑特点和建造方式，用于指导灾后重建的民居建设。

朱雀门

（获中国建筑艺术网“2004人文建筑奖”）

设计时间：2004年

建设地点：北京 宣武区

规　　模：231 000平方米

北京太古三里屯Village商业街
（与英国Oval等公司合作）
设计时间：2007年
建设地点：北京 朝阳区
规　　模：170 000平方米

A&S International Design Design The Future of Chinese Buildings

北京京澳凯芬斯设计有限公司
BEIJING JING-AO KANN FINCH DESIGN GROUP LTD.

翰时国际建筑设计
INTERNATIONAL DESIGN

翰时国际简介

翰时（A&S）国际建筑设计咨询有限公司是由国内外建筑师共同创立的建筑设计公司。公司创立于美国亚特兰大，2002年在中国北京正式注册。公司的目标是把国际上先进的建筑设计与规划理念和技术介绍到中国，并按照国际标准，为业主提供高质量的设计服务，为新世纪中国城市与建筑的发展作出贡献。翰时（A&S）国际可在建筑设计、城市设计、室内设计、景观设计等各领域，为业主提供全方位的咨询服务。

A&S Introduction

A&S international Design is an international architectural design consulting firm which was jointly established by the architects both at home and abroad. It was established in Atlanta, USA, and then registered in Beijing, China in 2002. A&S bends itself to creating modern and functional construction with international standard, by integrating the international and advanced concepts and techniques with an architectural philosophy, in order to provide its worldwide clients with high standard design and consulting services. A&S is positioned to provide services in architectural design, urban planning, interior design, landscape design and etc.

翰时国际设计理念

翰时国际采用国际化的理念、专业化的设计、地域化的服务，强调对客户的理解与尊重。翰时国际的敬业精神、专业知识、实践经验、市场活力已经成为有建设任务的业主们强有力的帮手和顾问。

A&S Design Philosophy

A&S provides an international design perspective and professional services with regional design fees. It excels at providing a high standard of service and respect for its clients. A&S becomes its clients' powerful consultant and assistant with passion, vision, professional knowledge, advanced project management experience and marketing energies.

A&S International Design Design The Future of Chinese Buildings

KFDG 北京京澳凯芬斯设计有限公司
BEIJING JING-AO KANN FINCH DESIGN GROUP LTD.

建筑设计项目类型
- 住宅建筑
- 办公建筑
- 医疗建筑
- 实验室建筑
- 商业建筑
- 教育建筑

公司主要服务内容
- 建筑规划
- 总体策划
- 总体规划
- 景观设计
- 工程及建筑投资估算分析

建筑设计
- 基地分析
- 方案设计
- 初步设计
- 施工图设计
- 室内设计
- 施工配合

Project Types

Residential Facility
Office Facility
Hospital Facility
Lab Facility
Commercial Facility
Educational Facility

Main Services

Architectural Design & Planning
Integrate Programming
Master Plan
Landscape Design
Investment Analysis

Architectural Design

Site Analysis
Scheme Design
Design Development
Construction Document
Interior Design
Construction Consulting

医疗实验室工艺流程设计
- 工艺流程设计
- 设备规划

工程设计
- 暖通空调工程
- 电气工程
- 给排水工程
- 消防工程

Medical /Lab Process Design

Process Design
Equipment Planning

Engineering Design

HVAC
Electrical Engineering
Water Supply & Sewerage Engineering
Fire-protection Engineering

A&S International Design Design The Future of Chinese Buildings

北京京澳凯芬斯设计有限公司
BEIJING JING-AO KANN FINCH DESIGN GROUP LTD.

中国北京
北京市宣武区宣武门外大街10号，
庄胜广场中央办公楼北翼1301室
邮编：100052
电话：+86-10-63109869
传真：+86-10-63109870
邮箱：office@as-arch.com
网址：www.as-arch.com

美国佐治亚州
美国佐治亚州德卢斯市
Wild Dunes大街1054号
邮编：30097
电话：+1(770)4971645
传真：+1(770)4979148
邮箱：office@as-arch.com
网址：www.as-arch.com

Beijing, China
No.1301, Central Tower North Wing,
Junefield Plaza,
No.10, Xuanwumenwai Street,
Beijing, 100052 China
Tel: +86-10-63109869
Fax: +86-10-63109870
E-mail: office@as-arch.com
http:// www.as-arch.com

Georgia, USA
1054 Wild Dunes Way,
Duluth, GA 30097
Tel: +1(770)4971645
Fax: +1(770)4979148
E-mail: office@as-arch.com
http:// www.as-arch.com

公司主要负责人

余　立　董事长、总设计师
　　　　建筑学博士
陈娟娟　董事、总裁助理
张　兵　董事、副总设计师
张广亮　董事、副总设计师
林载舞　董事、副总建筑师

公司成员

90人

Directors

Yu Li Ph.D.　Director / Chief Designer
Chen Juanjuan　Director / Office Director
Zhang Bing　Associate / Vice Chief Designer
Zhang Guangliang　Associate / Vice Chief Designer
Lin Zaiwu　Director / Vice Chief Architect

Number of Employees

90 persons

A&S International Design Design The Future of Chinese Buildings

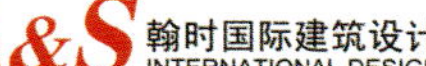

翰时-凯芬斯简介

北京京澳凯芬斯设计有限公司由澳大利亚北京投资股份有限公司、北京中铁工建筑工程设计院、北京翰时国际建筑设计咨询有限公司共同出资组成。京澳凯芬斯设董事会，由三方人员组成，董事会决定公司的战略规划、经营计划和投资方案等。北京京澳凯芬斯设计有限公司组织机构如下：凯芬斯总部、凯芬斯一所、深圳分公司、四川分公司、杭州工作室。

A&S-KFDG Introduction

Beijing Jing-Ao Kann Finch Design Group Ltd. (KFDG) is formed by Australia-Beijing Investment Pty. Ltd., China Railway Engineering Design Institute and A&S International Design Consulting Co. Ltd. KFDG has a board of directors, composed of the three parties, and the Board decided the company's strategic planning, business plans and investment programs, etc. Orgnizations are as follows: Headquarter KFDG, First Branch, KFDG Shenzhen Branch, KFDG Sichuan Branch, KFDG Hangzhou Studio.

公司理念

京澳凯芬斯采用国际化的理念、专业化的设计、地域化的服务，强调对客户的理解与尊重。公司对国内建筑设计市场存在的问题有着清醒的认识，为自身的发展确定了明确的目标，并在努力为业主提供最卓越的服务的同时，不断探索适合中国文化的建筑设计工作模式，以此为中国建筑师职业化的发展作出贡献。

Company Philosophy

KFDG absorbs international vision, professional design, and regional services. It emphasizes the comprehension and deference of the clients. The company has a clear understanding of the problems in current architectural design market. Clear objectives had been made for its development. Based on supplying excellent services to the clients, KFDG explores an architectural design style which matches with the Chinese culture, and devotes itself to the development of architect professionalism.

地址：北京市海淀区杏石口路65号
电话：+86-10-62856463
传真：+86-10-62856490
邮编：100195

Add: No.65 Xingshikou Road, Haidian District, Beijing
Tel: +86-10-62856463
Fax: +86-10-62856490
P.C.: 100195

北京新松建筑设计研究院有限公司脱胎于开发企业10年的发展历程，并以产品研发及创新能力著称于业内，拥有一支高素质的专业团队和成熟的业务发展平台，为房地产投资与开发提供项目策划、产品研发、建筑设计、技术服务等一条龙专业服务。

10年实战经验，20个城市和地区实战案例

120位专业人士，平均年龄32岁

高级职称及资深专业人员45人

一级注册建筑师、结构工程师10人

3大业务平台、3大业务延伸

超过30家合作伙伴

展望未来，新松将一如既往地秉承创新理念，以求真务实的行事作风，凭借富有经验而充满激情的专业团队，矢志成为卓越的房地产投资与开发专业服务机构。

服务于客户，让客户满意。

Beijing Xinsong Architectural Design Institute Co., Ltd. is born out of development history of development company for 10 years, and is well known for the product R & D and innovation in the industry, with a high-quality professional team and a mature business development platform, providing one-stop professional services of project planning, product development, architectural design, technical service for real estate investment and development.

10 years of practice experience, 20 cities and regions in real case

120 professionals, average age of 32

45 people with senior and senior-level professional titles

10 are the First Class Registered Architects and Structural Engineers

3 major business platforms, an extension of 3 major business

More than 30 partners

Looking ahead, Xinsong will continue to uphold the innovative concept, a realistic acting style, with rich experience and passionate team, committed to being an excellent professional service organization for real estate investment and development.

Serve the customers, make them satisfied.

哈尔滨长江路规划

建设地点：黑龙江 哈尔滨
占地面积：170公顷
建筑面积：480万平方米
容 积 率：2.8
设计时间：2010年
项目主创：张 杨

北京上东8号

建设地点：北京
占地面积：1.5公顷
建筑面积：10万平方米
容 积 率：4.5
设计时间：2010年
联合主创：张 杨

▲ 西安中新浐灞半岛A1地块九年制学校设计说明

——创造立体开放、多重构成的学校交流空间

项目主创：陈志刚

设计中突出学校国际化、高品质的氛围，功能齐全，使用舒适。方案从校园空间的特色出发，借鉴中国传统建筑以“庭院”虚空间为中心的布局模式，结合周边的自然资源，力求创造出立体开放、多重构成的学校交流空间。

一、因地制宜、巧妙合理地安排功能布局。利用平台、廊道组织人流集散，生成核心空间；

二、平台廊道既是校园的交通核心，也是全天候的校园生活中心场地；

三、全天候、多向开放、全方位、多层交融的交流空间。

总之，设计的重点放在教室内外的交流空间上，注重学生个体之间、个体与群体之间的交流与交往。创建一个活泼、开朗、自然而富有时代感和浓郁文化内涵的校园环境。

▲ 西安浐灞半岛A1地块五星级酒店（2009年）

项目主创：陈志刚

项目用地位于西安浐灞半岛A1地块酒店，会展区的收官之作，总建筑面积5万平方米，建筑高度150米，建筑设计既具有鲜明的标志性、时代性、又体现了西安悠久的历史文化传统。

立面形象从西安的大雁塔出发，采用平面旋转的现代手法，形成独具特色的流线型旋转造型，象征“塔”在新时代下的崭新形象，使建筑具有较强的标志性，进而成为西安浐灞新区内的新地标。

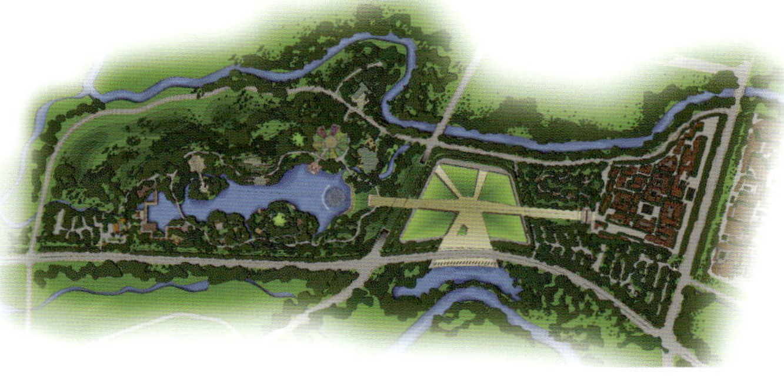

▼ 成都非物质文化遗产国家公园

项目主创：杨映国

在“198”政策指引下，依循统一规划、分步实施、建管并重、持续发展的思路，以非遗文化为主题、非遗产业为支撑、非遗节会为品牌，将成都非物质文化遗产国家公园打造成为文化定位有主题、产业发展有支撑、运作机制有活力、片区发展可持续的大型文化主题会展旅游复合型项目。

◀ 北京青年汇小区三期综合商业楼（2009年）

项目主创：陈志刚

方案设计中运用城市设计的现代手法，为城市创造一座收放有序、功能与形象统一的建筑。立面设计中高度重视城市界面和空间的设计，通过建筑材料的虚实间隔和透明及半透明材料的变化，避免了板式建筑容易造成的单一城市界面，对城市空间做出了良好的呼应。

▲ 徐州枫林天下 项目主创：杨映国

▲ 天津老城厢11号地 项目主创：杨映国

▲ 西山美墅馆　项目主创：冯 慈

▲ 江苏徐州中新森林海　项目主创：冯 慈

▼ 北京青年工社　项目主创：冯 慈

▼ 天津老城厢15号地　项目主创：冯 慈

▼ 西安浐灞A11 联合主创：张燕宏

▲ 西安浐灞A3 联合主创：张燕宏

▲ 玉树小学 项目主创：张燕宏

▶ 玉树幼儿园 项目主创：张燕宏

易 肯 设 计

易肯设计机构（E–TOWN International Corp.）在全球致力于城市发展过程中的综合顾问服务，业务涉及战略咨询、城市规划与设计、项目开发管理与投融资等服务。机构拥有一个在行业内具备领先声誉的专家团队，并在中国的业务发展过程中更加注重与中国本土的专业资源紧密结合，专注于解决城市发展中的关键问题。

北京易肯建筑规划设计有限公司作为美国易肯设计机构在中国发展的分支机构，于2010年正式在北京成立。公司以"创新(Innovation)、合作(Cooperation)、责任(Responsibility)、执行(Implementation)"作为企业文化之本以及竞争力核心，倡导6E城市发展理念[经济的(Economic)、活力的(Energetic)、快乐的(Enjoyable)、舒适的(Easeful)、高效的(Efficient)、生态的(Eco–friendly)]，向政府及企业客户提供包括区域发展战略咨询、城市总体规划、城市规划与设计、景观规划设计、环境评价与可持续发展规划、旅游规划及酒店设计、项目开发咨询与投融资顾问等方面的咨询服务，旨在以国际化的专业能力及经验、结合本土的项目实施需求，提升客户的资源价值效应和项目竞争能力。

E-TOWN International Corp., dedicated to the comprehensive consultancy service in the process of urban development, involved in strategic consulting, urban planning and design, project development and management and investment and financing service. The agency has a reputation in the industry with leading a team of experts, and pays more attention to get close to local professional resources of business development in China, focusing on solving the key issues of urban development.

Beijing E-TOWN Architecture Planning and Design Corp. as the branch of E-Town International Corp. International Corp., founded in 2010 in China. Taking "innovation, cooperation, responsibility, implementation" as corporate culture and core competitiveness, We promote the concept of 6E urban development (Economic, Energetic, Enjoyable, Easeful, Efficient and Eco-friendly), provide the consulting services of regional development, strategic consulting, urban master planning, urban planning and design, landscape planning and design, environmental assessment and sustainable development planning, tourism planning and hotel design, project development consulting and consultancy services in investment and financing for government and corporate clients. We combine the international expertise and experience to the needs of local implementation, aiming to enhance our customer's resource value and project competitiveness.

地址：北京市朝阳区东四环双新桥朝阳公园东五门内
朝阳规划艺术馆南侧办公楼4层
邮编：100125
电话：+86–10–65006766
传真：+86–10–65001806
联系人：武宁
邮箱：etowndesign@qq.com
网址：www.etowndesign.com.cn

Add: The 4th Floor of the Office in the South of Chaoyang Planning Art Gallery,
the 5th East Gateway of Chaoyang Park, Shuangxin Bridge, East Fourth Ring Road, Chaoyang District, Beijing
P.C: 100125
Tel: +86–10–65006766
Fax: +86–10–65001806
Contact: Wu Ning
E-mail: etowndesign@qq.com
Http//: www.etowndesign.com.cn

成都“东村”文博艺术产业核心区城市设计

业主：成都市政府、华熙集团
内容：开发策划、规划调整、详细城市设计
规模：400公顷

成都市政府和文化产业投资商将共同在成都“东村”文化创意新城核心区打造世界级的中华文博艺术中心、国内首家专业化文化艺术交易中心、西南文化艺术消费旅游中心，本项目探讨了如何创造连续、开放、有趣和人性化的空间，以吸引更多的艺术家、游客和居住者。将从全国各地收集来的100个传统民居院落作为传统文化的载体，以视觉焦点和景观中心的形式重新进行城市设计，形成成都东村区域开发建设的一个文化脉络和区域特色，为区域建设注入鲜活的生命力。

Urban Design for Core Area of Chengdu “Dongcun” Creative Industry

Owners: Chengdu Government, Huaxi Group
Content: Development Planning, Planning Adjustment, Detailed Urban Design
Size: 400 hectares

Chengdu government and the cultural industry investors will co-invest in Chengdu heart area of East Village cultural and creative new town to build a world-class Chinese culture arts center, the first domestic professional culture and art exchange center, southwest culture and art consumer tourism center, the project explores how to create a continuous, open, interesting and user-friendly space to attract more artists, tourists and residents. Taking 100 traditional residential compounds collected from all over the country as the traditional carrier of culture, the city was redesigned with the visual focus and form of landscape center, forming a cultural context and regional characteristics for development and construction of Chengdu East Village area, and bringing the region fresh vitality.

成都"东村"文化创意新城 华熙启动项目

业主：华熙集团
内容：开发策划、概念规划设计方案
规模：60公顷

华熙集团作为文化产业投资和开发商将参与成都"东村"文化创意新城的开发建设，本项目是其启动项目，也有望成为成都"东村"的启动项目。规划设计方案围绕"时尚活力、书香动力、文化魅力"三大主题，利用有文化传承意义的中国民居老院子，打造多样化的立体景观、商业、休闲空间，实现艺术与商业的联姻、文化与生活的交融、自然与都市的重叠，创造能够带动区域发展的、充满活力的、动态可持续发展的、以人为本的新城市功能区——一个人人乐在、欢聚一堂的家。

Chengdu East Village Cultural and Creative New City Huaxi Group Start-up Project

Owners: Huaxi Group
Content: Development Planning, Conceptual Planning and Design
Size: 60 hectares

Huaxi Group as cultural industry investment developers will participate in Chengdu "East Village" cultural and creative development of new city construction. It is the start of the project, and also is expected to be Chengdu East Village start-up project. The planning and design programs are around "fashion vitality, scholarly power, culture charm" the three main themes, and use cultural heritage significance of the Chinese old courtyard houses to create a variety of three-dimensional landscape of business leisure space, to achieve the integration of art and commerce, culture and life, nature and city, to promote regional development and create vibrant, dynamic sustainable development, people-oriented features of the new urban area – a place where everybody happy in and the family gathered.

哈尔滨兴隆水库区域水景小镇定位策划与规划设计

业主：伟东集团、祥阁房地产
内容：开发策划、概念规划
规模：400公顷

兴隆水库是一座天然的活水景观水库，周边千顷良田，有着哈南工业新城强大的产业支撑。本项目围绕“自然原野水岸人家、生态低碳绿色节能、商务会议休闲娱乐”三大主题，借鉴北欧可持续发展社区的先进经验，充分保护并利用水岸湿地生态景观，规划“运动、居住、旅游、酒店、商业、农业”六大功能板块，将水库周边打造成兴隆小镇水景湖城——哈南工业新城现代服务业商住配套综合区、哈尔滨旅游休闲度假生活品质小镇、黑龙江新农村建设有机农业示范区。

Harbin Xinglong Reservoir Area Waterscape Town Positioning and Planning Design

Owners: Weidong Group, Xiangge Real Estate
Content: Development Planning, Concept Planning
Size: 400 hectares

Xinglong Reservoir is a natural landscape of the reservoir, around one thousand hectares fertile, with a strong industry support of Ha'nan Industrial Park. This project, around the three main themes "natural wilderness and waterfront people, low-carbon ecology and green energy saving, leisure and entertainment and business meeting", learns advanced experience of the Nordic community sustainable development, protects and uses adequately waterfront wetland ecological landscape planning "movement, housing, tourism, hotel, commercial, agricultural", the six functional blocks. The town is booming around the reservoir caused water features to play Lake City – modern service commercial and residential district of Ha'nan Industrial Park, tourist holiday resort of Harbin, organic agriculture demonstration area for the construction of Heilongjiang new countryside.

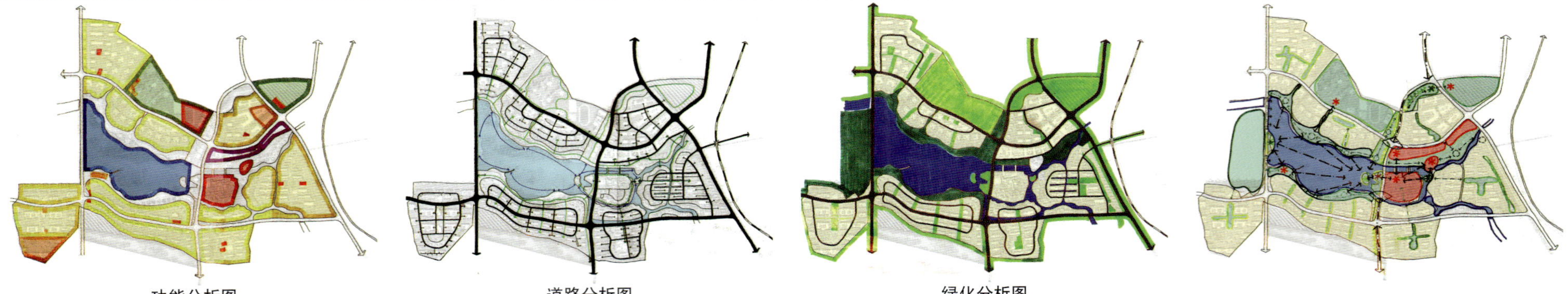

功能分析图　　道路分析图　　绿化分析图

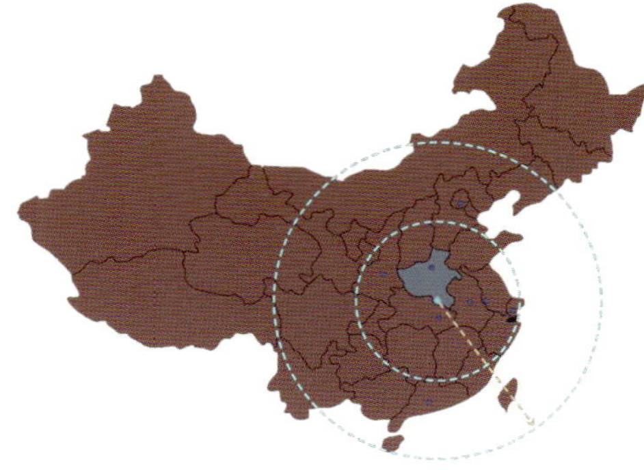

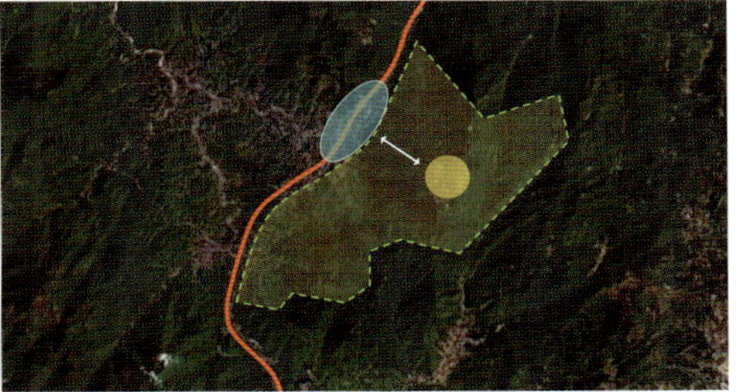

河南信阳鸡公山万国文化小镇总体规划及景观设计

Henan Xinyang Jigong Mountain International Culture Town Master Planning and Landscape Design

业主：信阳市政府
内容：总体规划、开发策划、景观设计
规模：8.6平方千米

Owners: Xinyang Government
Content: Overall Planning, Development Planning, Landscape Design
Size: 8.6 square kilometers

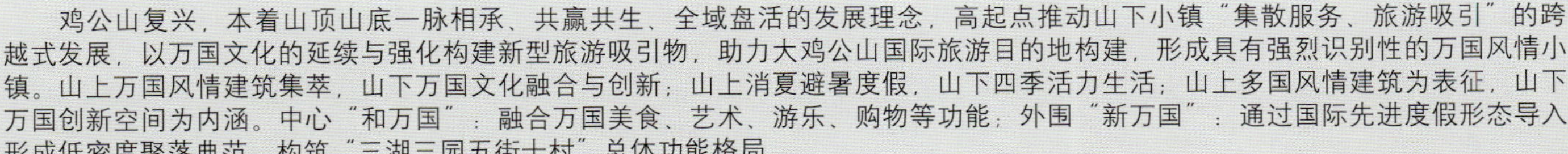

鸡公山复兴，本着山顶山底一脉相承、共赢共生、全域盘活的发展理念，高起点推动山下小镇“集散服务、旅游吸引”的跨越式发展，以万国文化的延续与强化构建新型旅游吸引物，助力大鸡公山国际旅游目的地构建，形成具有强烈识别性的万国风情小镇。山上万国风情建筑集萃，山下万国文化融合与创新；山上消夏避暑度假，山下四季活力生活；山上多国风情建筑为表征，山下万国创新空间为内涵。中心“和万国”：融合万国美食、艺术、游乐、购物等功能；外围“新万国”：通过国际先进度假形态导入形成低密度聚落典范。构筑“三湖三园五街十村”总体功能格局。

Jigong Mountain revives with the development philosophy of the top and bottom of mountain at the same strain, win-win symbiosis and a global operation, and highly promotes the great-leap-forward development of the mountain town with "distribution services, tourist attraction"; the new tourist attraction will be built through the extension and enhancement of the international culture, helping the construction of large Jigong Mountain international tourist destination, and forming a strong recognition of the international flavor town. In the mountain, flavor architectures gather, down the mountain, cultures of different are integrated and innovative; summer vacation up, seasons vitality life down; flavor architectures as the surface feature up the mountain, international innovation space as the content. The center "harmonious world": integrates the international foods, art, recreation, shopping and other functions; the peripheral "new world": through taking the international advanced holiday style into the formation of low-density settlement pattern model, builds a "three lakes, three parks, five streets and ten villages" pattern of overall function.

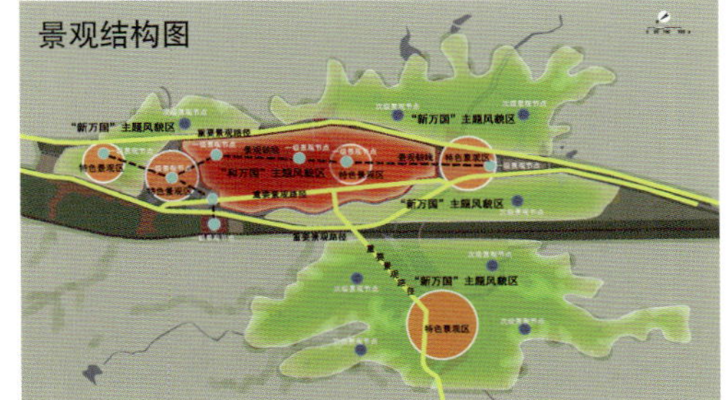

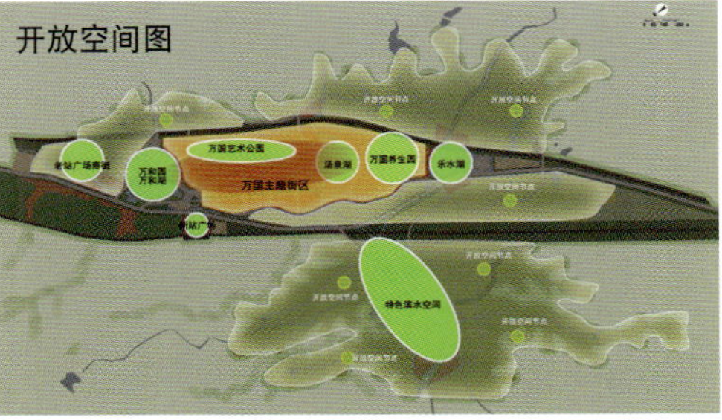

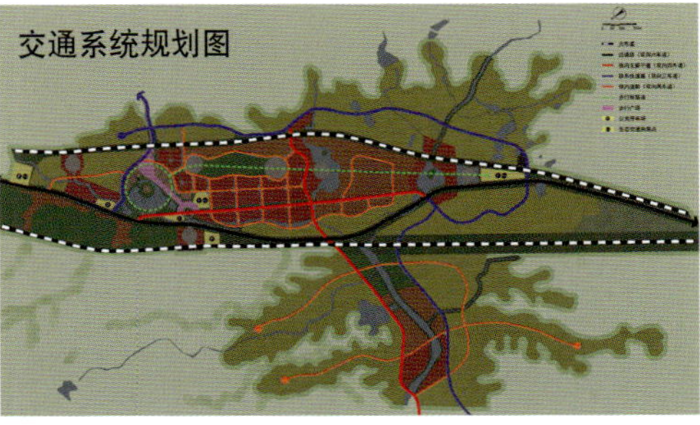

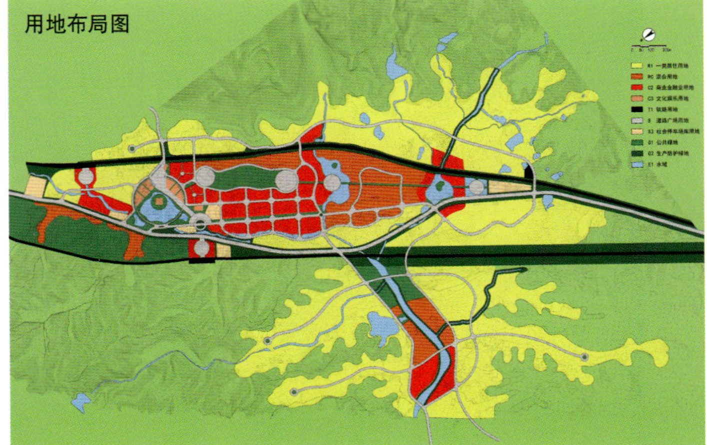

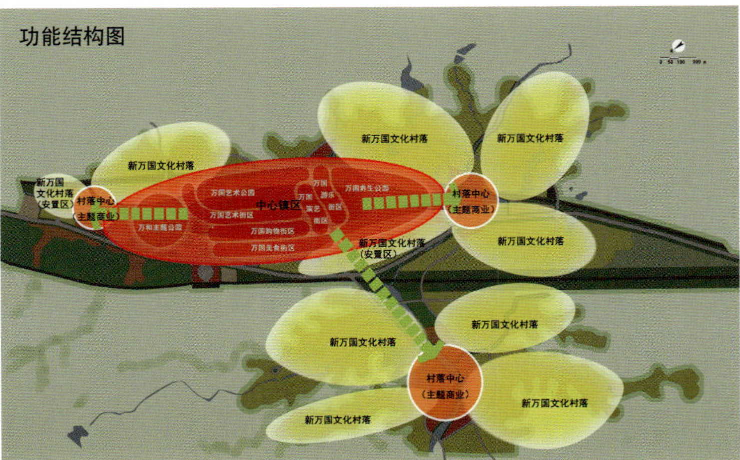

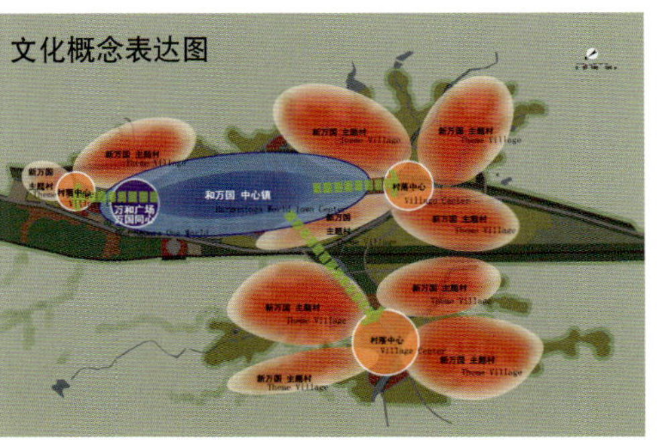

四川彭山彭祖新城策划与规划设计

业主：彭山市规划局、彭山控股
内容：开发策划、概念规划、城市设计
规模：11平方千米

一水相依，一山相望，代表养生城市模型的彭祖新城，与代表养生文化的彭祖山脉，构成养生文化新旧二元支点，将引领彭山未来的跨越式崛起。本规划基于对彭山优势发展条件研判，根植本土文化，立足全球城市类型研究，率先提出一种新的城市类型——"养生城市"，以"文化养神、产业养身、生活养心、空间养形"构建山、水、城、圣四体合一的新城市结构，成就一幅复兴养生传统、传播养生文化、引领自在生活的"中华首席养生城市"愿景蓝图。

Sichuan Pengshan Pengzu New City Scheme, Planning and Design

Owners: Pengshan City Planning Bureau, Pengshan Holdings
Content: Development Planning, Concept Planning, Urban Design
Size: 11 square kilometers

The landscape dependent Pengzu New City on behalf of health city models and Pengzu mountains on behalf of health culture make up of the new and old support of health culture, leading Pengshan rise leaps and bounds in the future. The planning based on research and judgment of advantages of Pengshan conditions for development, as well as the type research of global cities, first proposes a new urban type — the "health city" through native culture; while build the new city structure of mountains, water, city, saint four-body in one with "cultural repose, industry self-cultivation, the heart of life support, space for shape", creating a "China's Chief Health City" vision blueprint that revives the tradition of health, spreads health culture and leads comfortable life.

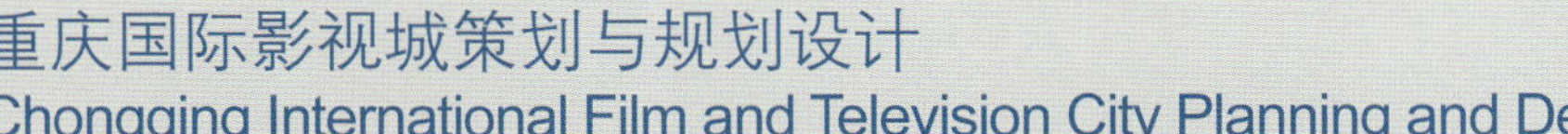

重庆国际影视城策划与规划设计
Chongqing International Film and Television City Planning and Design

业主：重庆两江新区管委会
内容：开发策划、概念规划城市设计
规模：5平方千米

Owner: Two Rivers Area Administrative Committee of Chongqing
Content: Development Planning, Concept Planning, Urban Design
Size: 5 square kilometers

落户重庆两江的中国国际影视城，集成高端影视业的最新成就，是具有重庆特色、中国文化主题、世界顶级品牌的超级影视文化产业园。它将独享"影视文化、影视产业、影视游乐、城市休闲"复合发展模式，成为影视文化旅游的超级城市综合体。围绕"山上主题园，水边活力岛，城中不夜城，重庆新都心"的规划思想，形成"一体三翼四区，主副双轴纵横"的功能结构，它将是一座文化圣殿、影视之都和游乐王国，成为重庆的新文化地标和走向世界的名片。

China International Film and Television City, located in two rivers in Chongqing and integrated the latest achievements of high-end film and television industry, is a super film and television industrial park with Chongqing features, Chinese cultural theme and the world's top brands. It will only use the complex development model of "film and television culture, industry, play, city leisure", and become a super city complex in film and television culture tourism. Around the planning ideas "mountain theme park, waterfront vitality Island, ever bright city and Chongqing new city center" and the functional structure of "three-wing one main and four districts • vertical and horizontal axis" , it will be a new culture landmark of Chongqing and the world business cards combining culture, film, and amusement.

淄博文昌湖策划与规划设计 Zibo Wenchang Lake Planning and Design

业主：中海投资
内容：概念规划、城市设计
规模：5平方千米

Owner: China Shipping Investment
Content: Concept Planning, Urban Design
Size: 5 square kilometers

文昌湖即萌山水库，紧邻滨博高速，处于淄博市半小时生活圈内。规划将建成以优异的生态环境、山水景观及地域文化为依托，以齐文化为主打品牌，以主题娱乐、节事庆典、休闲度假和商务会展为综合发展方向的主题型滨湖度假地。在结构上形成“十一个主体功能区、五个中心旅游小镇、五个次中心旅游点、一个核心环湖景观带”。其中五个中心旅游小镇分别为：滨湖风情小镇、精品购物休闲小镇、浪漫温泉小镇、水岸湖湾养生小镇、运动休闲小镇。

Wenchang Lake is Mengshan Reservoir, close to Binbo high-speed way, half an hour in the life circle of Zibo City. The planning will build theme lakeside resort with excellent ecological environment, geographical and cultural landscape, taking Qi culture as the main brand, and entertainment, festivals and celebrations, leisure and business exhibitions as the main direction of comprehensive development. The structure formation is "eleven main functional areas, five tourist town centers, five tourist attractions, with a central lake landscape." The tourist town centers are: lakeside flavor town, boutique shopping and leisure town, romantic spa town, waterfront health town, sports and leisure town.

河南项城市政和广场景观设计 Henan Xiangcheng Zhenghe Square Landscape Design

业主：河南项城市住建局
内容：景观方案设计及施工图
规模：15万平方米

Owners: Henan Xiangcheng City Building and Housing Bureau
Content: Landscape design Scheme and construction drawing
Size: 150,000 m^2

政和广场位于河南省项城市北部行政教育新区的中轴线上，北接新市政府党政办公楼，南至驸马沟河边，是新城绿地景观系统中重要的组成部分。景观设计中将带状广场由北向南分为细水长流、浑天立运和芳华流转三大部分，通过丰富的景观语汇完成从礼仪、人工、规整的政府前广场到休闲、自然、自由的市民滨水广场的良好过渡。项目建成后将是一条亮丽的城市景观轴线、一个老百姓茶余饭后乐于聚集的公园，同时也是一片展示项城文化的精品广场。

Zhenghe Square is located on the central axis of Executive Education New District, the north of Xiangcheng City, in Henan Province, close to the new government and party office buildings and south to the Fumagou River, it is an important part of the new city landscape system. In the landscape design, the strip square from north to south is composed of three parts: a steady flow, celestial globe up and fairview turn, through a rich landscape vocabulary to complete from the ritual, artificial, neat square in front of the government to leisure, natural, free public waterfront square, a good transition. It will be a beautiful urban landscape axis, a park people relax themselves, and a high quality square displaying culture of Xiang City.

北京市古代建筑设计研究所初创于1980年3月，是全国成立最早的、实力最雄厚的古建筑专业设计研究单位，具有文物古建筑保护修缮设计甲级资质和建筑设计乙级资质。

本所自成立以来已完成建筑设计项目600余项，其中境内代表作有：中共中央党校汇名园、中国紫檀博物馆、北京钓鱼台国宾馆养源斋、中央电视台无锡外景基地、北京万佛华侨陵园、山东龙口南山寺、天津大悲院、辽阳广佑寺建筑设计、武汉归元寺保护扩建规划及设计、麦积山风景名胜区保护及扩建设计、北京大学国际数学研究中心规划及设计；境外代表作品有：华盛顿中国城牌楼、莫斯科北京饭店室内装修、扎伊尔金沙萨恩塞莱总统庄园中国园林、英国曼彻斯特中国餐馆、瑞典北欧中国聚龙城方案设计等。文物古建筑保护项目代表作品有：北京历代帝王庙保护修缮设计、北京古观象台修缮、山海关六国饭店保护修缮等。

本所始终坚持古建筑传统技术及理论研究，由本所专家撰写的专业学术技术著作《中国古建筑木作营造技术》、《中国古建筑瓦石营法》、《中国清代官式建筑彩画技术》、《中国建筑彩画选》和行业标准《古建筑修建工程质量检验评定标准》、《北京四合院建筑要素图》，是这些成果的代表之作。

《古建园林技术》杂志是本所主办的学术期刊，自1983年创刊至今，在宣传普及古建园林专业知识、交流古建文物保护经验、弘扬中华传统建筑文化、促进中外文化交流等方面发挥了重要作用，成为古建园林界不可缺少的重要学术技术期刊，是从事古建园林工作者的必读之书。

本所拥有国内一流的古建筑技术专家，他们在古建筑研究、设计、教学、办刊、编制规范、培养人才等方面成绩显著、硕果累累，他们的学术观点和主张在行业内有重要影响，是古建园林行业的学术带头人。

Beijng Traditional Chinese Architectural Design and Research Institute (TCAD) was founded in 1980 and specializes in the design and research of ancient (traditional) and traditional style architecture and gardens.

Since the founding of TCAD more than 600 design projects have been completed, including The Huiming Garden of the Central Party School, The China Sandalwood Museum, Yangyuanzhai in the Diaoyutai State Guest House Beijing, The Outdoor Setting of Wuxi for CCTV, The Overseas Wanfo Cemetery in Beijing, The Nanshan Temple in Longkuo Shandong, and the Great Mersey Monastery in Tianjin the architectural design of the Tianyou Temple in LiaoYang, the Protection expansion planning and design of the Guiyuan Temple in Wuhan, the protection and expansion design of the Maijishan scenic spot, the planning and design of the international math research center of the Peking University; TCAD has also completed various overseas design projects, such as The Classical Chinese Archway Tower in Washington, The Interior Design for the Beijing Hotel in Moscow, and The Chinese Garden in the Manor of the President of Zaire. The chinese restaurant of Manchester in England, The design scheme of Sweden Northern European China Julong city and so on. TCAD is China's oldest institute that researches ancient architecture and includes the study of traditional wood working techniques, roof tiling and decorative painting of the Ming and Qing Dynasties.

The following publications have all been edited by our institute specialists: *Chinese Ancient Architectural Wood Work Techniques, Chinese Architectural Tile and Stone Construction Techniques, Selected Works of Chinese Architectural Paintings and Ancient Architecture Renovation of Quality Examining and Evaluating Standard*. These works have played an important role in furthering and disseminating design and construction concepts of Chinese ancient (traditional) architecture and gardens.

TCAD publishes the journal *Traditional Chinese Architecture and Gardens*. Our magazine, was published in 1983 due to the increasing development of the Chinese architectural industry and heritage protection movement.

A first class organization, TCAD is at the forefront of its field and continues to a source of authority in Traditional Chinese Architectural Design and Research.

地址：北京市东城区安德里北街甲20号
紫萱园写字楼三层
邮编：100011
电话：+86-10-84126198
传真：+86-10-84126698
邮箱：gj_cad.@163.com
网址：www.bjgjsj.com

Add: 3rd Floor, Office Building of Zixuan Garden,
No.20, North Andeli Street, Dongcheng District, Beijing
P.C.: 100011
Tel: +86-10-84126198
Fax: +86-10-84126698
E-mail: gj_cad.@163.com
Http: //www.bjgjsj.com

澳门渔人码头中国城方案设计

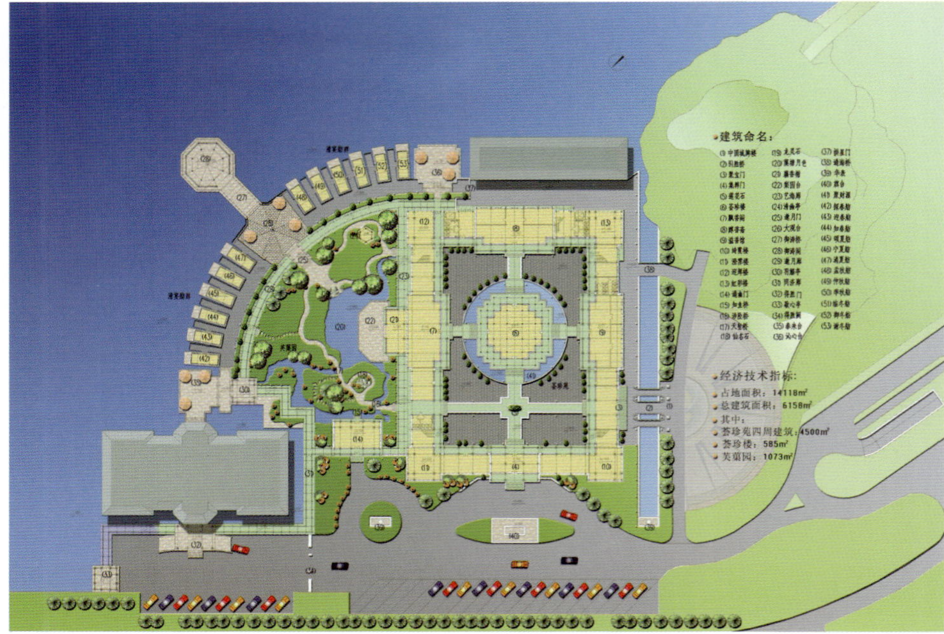

澳门渔人码头项目是"澳门特别行政区政府成立以来投资最大的旅游项目"，"是一个综合中西饮食文化，古典与现代、粗犷与细致于一体的大型旅游景观"。其中中国城"拟建成中国明清古典风格的园林建筑，功能有中国文化特色的饮食、民间工艺、杂耍表演、观光旅游等"（引自《澳门渔人码头方案设计任务书》）。

中国城主体建筑群设计构思主要有以下两方面的考虑：首先是对"城"概念的理解和体现，本方案按甲方给定的用地范围，将功能建筑围合成一个城的形式，并在"城"的中央圆形水池中设立一座三重檐十字脊歇山顶抱厦复合式楼阁建筑，这座建筑每层屋檐都有12个翼角向外挑出，共有36个翼角、36个凹角、102条屋脊，设计复杂精巧，充分展现中国木结构古建筑的精华，具有很强的观赏效果，是中国优秀传统建筑文化在澳门的经典再现。

中国城的建筑风格是明清皇家宫廷建筑的风格，红墙黄瓦，玉阶丹楹，飞檐翼角，雕梁画栋，体现出浓重的中华民族建筑文化色彩。

主要经济技术指标

占地面积：14 118平方米
建筑面积：6 158平方米
园林绿化率：30%
容 积 率：0.44
设计时间：2001年
主要设计人：马炳坚、张 越
建筑设计：张 越、相炳哲、肖 东
结构设计：易 斌
设 备：单成福、周启玲
电 气：王培杰

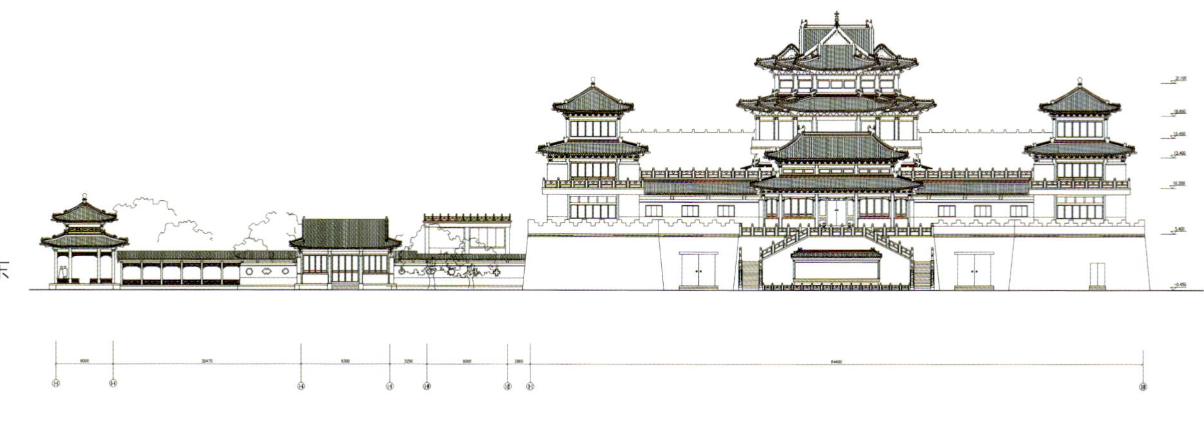

BEIJING TRADITIONAL CHINESE ARCHITECTURAL DESIGN AND RESEARCH INSTITUTE
北京市古代建築設計研究所

浙江文成县安福寺修建性详细规划

安福寺位于浙江省西坑畲族镇始建于唐宪宗元和三年，距今已有近1 200年的历史。历史上，多次重建重修，现该寺已破烂不堪，仅存半厅四方殿和石碑等遗物。

安福寺修建遵循四项原则，即：“继承与创新”原则；“与周边环境和发展趋势相协调”的原则；“以人为本”的原则；“可持续发展”的生态原则。

安福寺的布局主要规划为五个部分：

1. 寺庙区

建筑分为三路：

东路建筑为：钟楼、客堂、僧僚、方丈室、斋堂。

中路建筑为：山门、大雄宝殿、藏经阁。

西路建筑为：鼓楼、云水僚、僧僚、上客堂。

2. 养生堂

主要建筑：药师殿、忏悔堂、健身房、休闲松林。

3. 禅味茶堂

主要建筑：寂静禅堂、闲趣茶室、禅风茶室、茶室演示厅、茶文化展厅、画廊。

4. 游客中心

主要建筑：接待室、咖啡厅、小卖部、法物流通处。

5. 禅堂、安心僚

建筑风格：

安福寺始建于唐代，因此将建筑外观定位为唐代风格。建筑形式除庑殿式外，分别采用歇山式、悬山式，以及非常朴素的农舍式形式。

道路系统：

由寺院外道路、主环路、园路步道共同构成完整的游览路线。

园林规划：

以维护本地区生态平衡为宗旨，保护自然环境的自然景观，突出绿化主题和个性，达到植物景观与人文景观相协调；树种选择既考虑组织风景、丰富林相、又本着因地制宜、适地适树的原则，优先选用当地适生树种。

主要设计人：张丽鲜

项目参数

总用地面积：41 504平方米

建筑占地面积：9 027平方米

总建筑面积：16 437平方米

绿化面积：20 918平方米

其中地上建筑面积：15 111平方米

地下建筑面积：1 326平方米

道路及广场面积：11 559平方米

容 积 率：0.396

绿 化 率：0.504

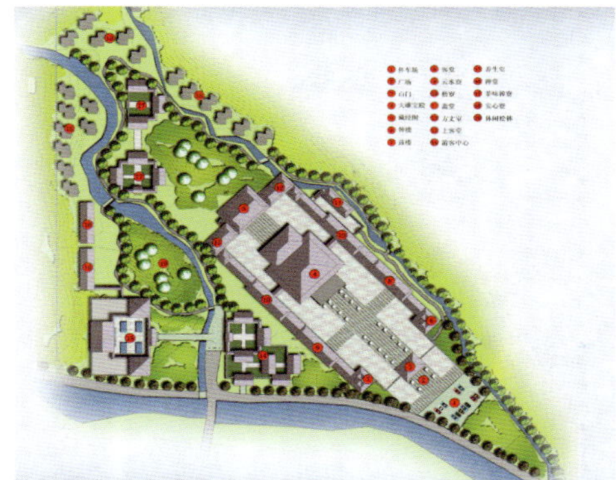

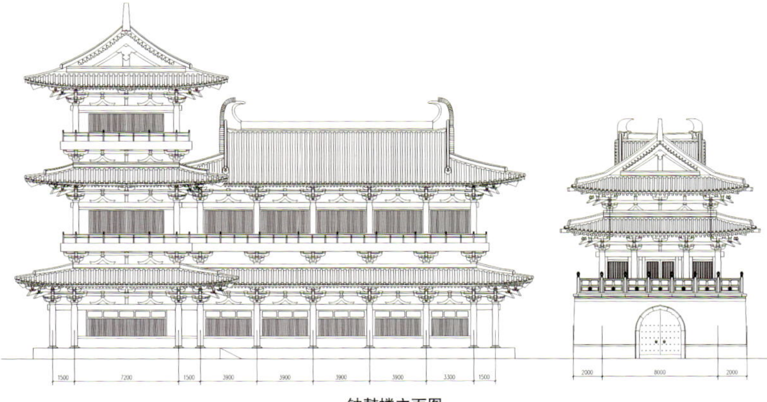

钟鼓楼立面图

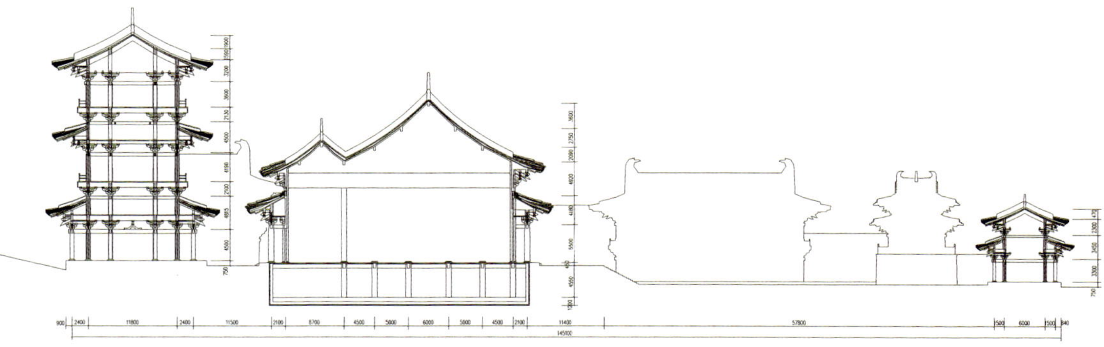

寺庙区纵剖图

山东桓台王渔洋故居保护及相关景区建设规划

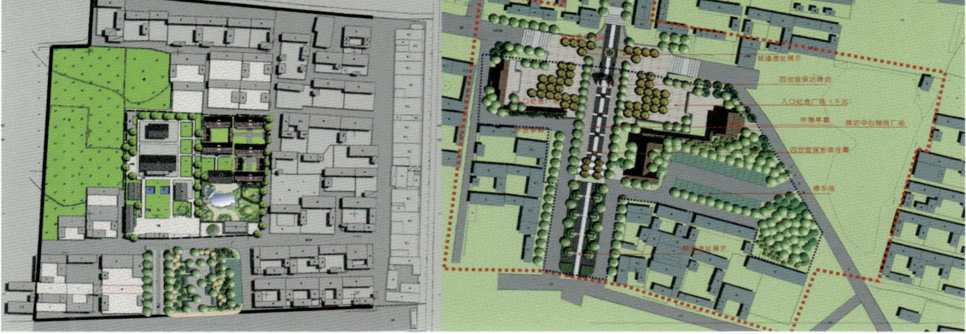

山东桓台县历史悠久，距今已有8 000年的发展历史，现存于县城内的王渔洋故居、王渔洋祠堂、忠勤祠、四世宫保坊等被定为省级文物保护单位。以王渔洋为代表的王氏家族在历史上声名显赫，王渔洋曾任清代刑部尚书，一生恪守“清、慎、勤”的为官原则，还是历史上著名的诗人、文学家、诗词理论家、教育家，著述很多，在文学界享有很高的声誉。

王渔洋故居位于古县城西南侧，现有保留建筑3 749平方米，但因年代久远，迭经改造，大部分已非原来面目。对王渔洋故居的保护修缮是在对现存建筑进行充分调查研究的基础上进行的。保护修缮指导思想恪守三条原则，即：不改变文物原状的原则、充分尊重当地建筑特色的原则和安全第一的原则。对现有建筑的保护修缮方案经专家充分论证，文物主管部门批准，对已倾圮的建筑按原有风格进行恢复。

该项目是一个集保护和规划建设为一体的综合项目，第一期工程还包括忠勤祠、四世宫保坊、沙帽树街等三个景区的规划和建设。

主要设计人
项目负责人：张　越
规划及建筑设计：张　越、王思盟、李小龙、李　琳、任一帆、谢成涛
结构设计：姜　津、王大欣
设备电气设计：单成福、李　栋、周启玲
项目顾问：马炳坚

BEIJING TRADITIONAL CHINESE ARCHITECTURAL DESIGN AND RESEARCH INSTITUTE
北京市古代建築設計研究所

河北邯郸邺城博物馆

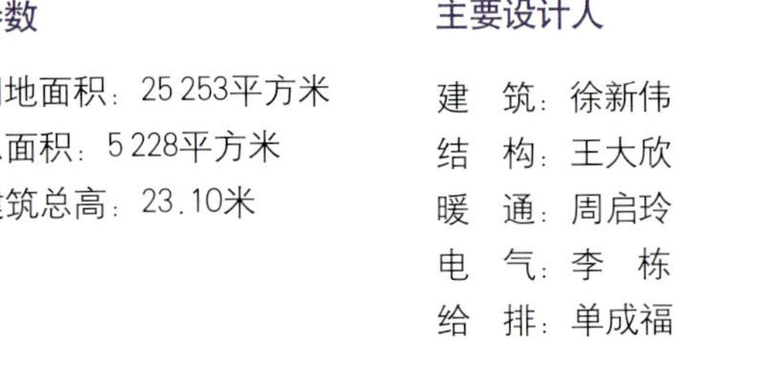

邺城博物馆位于河北临漳县王村以北，为综合性公共建筑，具有介绍古邺城历史与文化，收藏整理、展示邺城出土文物等功能，同时也是邺城考古工作站。

博物馆为主体二层、局部三层汉代风格城楼式建筑，钢筋混凝土框架结构。

项目参数

建筑用地面积：25 253平方米

建筑总面积：5 228平方米

主楼建筑总高：23.10米

主要设计人

建　筑：徐新伟

结　构：王大欣

暖　通：周启玲

电　气：李　栋

给　排：单成福

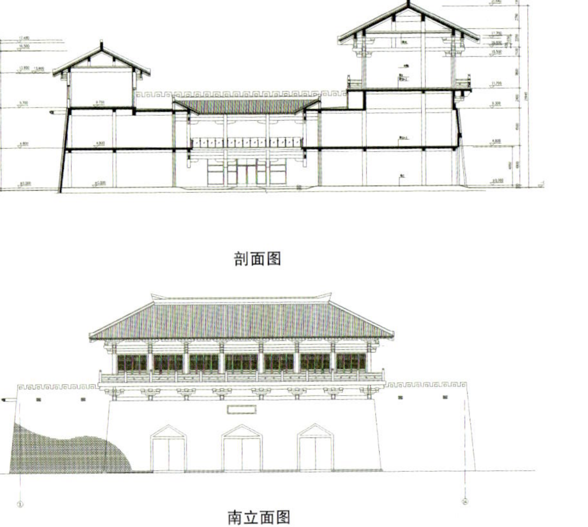

剖面图

南立面图

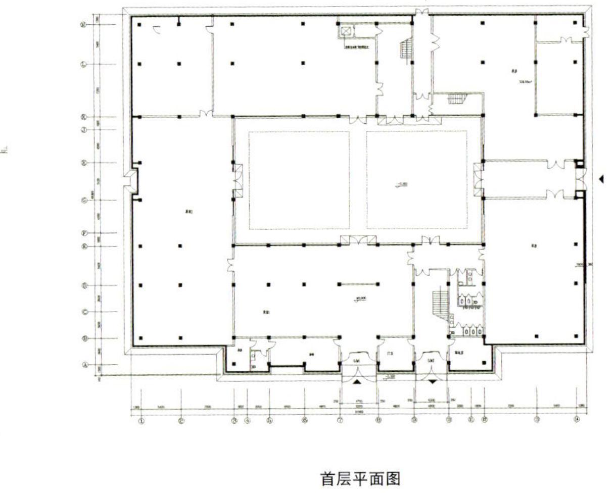

首层平面图

昂塞迪赛
On-Site Design Group

On–Site Design Group总部位于加拿大渥太华市。是一家有着丰富的城市规划设计、建筑设计和景观设计经验的国际化创新型设计公司。On–Site Design Group于2010年将业务范围扩展至中国，与中国的业主及建筑师共同探讨城市的未来，了解中国的文化。运用专业技能和丰富经验为客户提供设计方案，在设计中充分体现城市与自然、人与自然之间的平衡及共存。昂塞迪赛提供包括城市规划设计、建筑设计、景观设计等在内的全套解决方案。设计作品在展现自身特色的同时充分体现了客户的品牌价值。

昂塞迪赛（北京）建筑设计有限公司是由早年在德国留学并在海外知名设计机构担任首席设计师多年、具有丰富城市规划和景观设计经验的刘亮博士领衔，并汇聚了一批在国内著名设计院履职多年的高级建筑师和青年海归规划师、景观设计师。

昂塞迪赛积累了一批优秀的设计作品，并致力于创作出更多更好的、代表现代精神的设计精品。相关作品得到了世茂、金地、鲁能，以及南京、南昌、海南、哈尔滨和桂林等业内甲方和政府部门的肯定。

The headquarters of On-Site Design Group is located in Ottawa, Canada. It is an international and innovative design company with rich experience in urban design, architectural design and landscape design. On-Site Design Group has extended its business scope to China in 2010 and it will explore the future of cities, understand Chinese culture with the owners and architects of China together. We use professional skills and rich experience to provide design scheme. In the design we fully embody in coexistence between the city and nature, human and nature. On-Site Design Group provides customers with complete set of solutions which contain urban planning and design, architectural design and landscape design. Our design not only reflects characteristics of ourselves but also shows the brand value of clients.

Doctor Liu Liang, the leader of On-Site Design Group studied in Germany in the early days and worked in famous design institute overseas as chief designer for many years, he has rich experience in urban planning and landscape design. It has been gathering a batch of senior architects who have worked in domestic famous design institute for several years and young planners who studied in overseas before and landscape architects.

On-Site Design Group accumulates a number of outstanding design works, and is committed to creating more and better design that represent the spirit of the modern and high-quality goods. All the relevant works got affirmation from party A such as Shimao, Gemale, Luneng, and government departments of Nanjing, Nanchang, Hainan, Harbin, Guilin.

Http:// www.on-site.com.cn

刘亮（加拿大籍），首席设计师

黄非　管理总监

刘海伦　副总规划师

Andreas Luka（德国籍）
副总景观设计师

成都青城 · 堰道

Chendu Qingcheng Mountain • Yandao

南昌 · 伟梦清水湾

Nanchang · The Great Dream Clear Water Bay

西安醴泉湖
Xi'an Li Quan Lake

海南顺泽福湾庭院景观设计

Hainan Shunze Fuwan Courtyard Landscape Design

平谷峪口镇商务休闲区
Pinggu Yukou Town
Business Leisure Areas

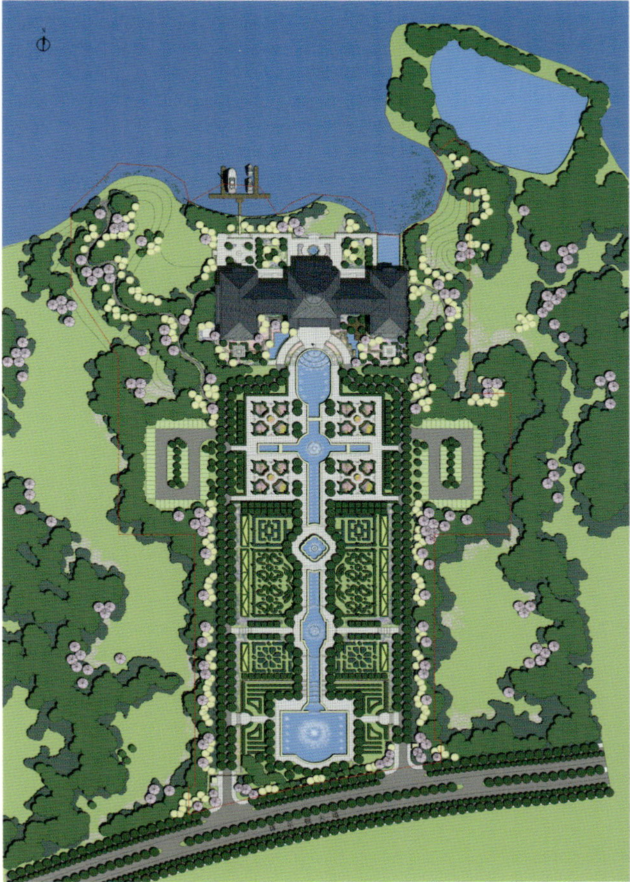

武汉世茂嘉年华游艇码头

Wuhan Shimao Group Carnival Yacht Dock

世茂福州鼓岭国际会议中心

Shimao Group Fuzhou Guling International Conference Center

哈尔滨市202国道城市设计
Harbin 202 National Highway Urban Design

哈尔滨月亮湾湿地酒店及湿地公园

Harbin Moon Bay Wetland Hotel and Wetland Park

世茂成都未来城 >

Shimao Group Chengdu Future City

< 沈阳金地东郡

Shenyang Gemdale Dongjun

北京宝鸿建设项目管理有限公司

地址：北京市丰台区海鹰路1号院2号楼4F-08　邮编：100070
总机：+86-10-51269622　电话：+86-10-63796601 / 63796602 / 63796603
传真：+86-10-51269622-8026　邮箱：baohongjianshe@126.com

北京宝鸿建设项目管理有限公司成立于1995年，是为响应国资委和建设部“代建制”而成立的较早的项目管理公司之一。发展至今已成为注册资金5 100万元，集项目管理、建筑设计及工程勘察等业务为主的国内知名公司之一。宝鸿建设拥有资深管理和建筑设计人员70余人，业务遍及全国。作为“代建制”推广的先驱，宝鸿建设通过十多年的实践，拥有一套较为完整、成熟、科学的项目管理模式，不仅能为项目业主在项目招投标、勘察、设计、施工、采购和监理等全过程和中间过程提供专业性服务，同时，宝鸿建设改变传统经营模式，拓展服务空间，为业主提供项目管理、项目组织、项目目标、项目招标、项目合同、项目目标控制等阶段的策划服务。

宝鸿建设坚持与时俱进、不断创新，愿与业界同行以广阔的视野和积极的态度真诚合作，参与到祖国的经济建设中，并为取得更加辉煌的成绩而努力奋斗！

Beiing Baohong Construction Project Management Co., Ltd. was established as one of the earliest project managers in response to the “agent construction system” of the SASAC and MOHURD in 1995. Now it has developed into a famous company with registered capital RMB51,000,000 integrating project management, architectural design and project survey etc. Baohong has over 70 senior managers and architectural designers, operating nationwide. As a pioneer of “agent construction system”, Baohong has a complete, mature, and scientific mode of project management through more than 10 years' practice, which can not only provide the owners with professional services in the full process and intermediate process of bidding, survey, design, construction, procurement and supervision, but also change its traditional management mode, expand service space and offer the owners planning service in project management, project organization, project target, project tendering, project contract, project target control and other phases.

Baohong insists on progressing with the time and constant innovation, is willing to cooperate with competitors in a broader vision and an active attitude, participate in the economic construction of China, and strive for a more glorious performance!

1

2

3

1 2 3 保定市农业生态科技教育中心
总用地面积：4.48公顷　总建筑面积：1.55万平方米　容积率：0.35

4 5 保定北国先天下广场
总用地面积：1.22公顷　总建筑面积：6.68万平方米　容积率：3.35

建筑设计：李　志、易碧华、谭　越、齐瑞生
结构设计：李淑珍、陈宏斌
机电设计：冯　伟
效果图绘制：牟未秋、苗　昆

4

5

1

2

3

4

5

6

7

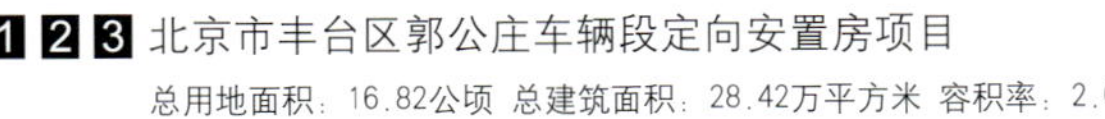

1 2 3 北京市丰台区郭公庄车辆段定向安置房项目

总用地面积：16.82公顷 总建筑面积：28.42万平方米 容积率：2.68

4 5 6 河北易县普华新区规划方案

总用地面积：19.20公顷 总建筑面积：48.28万平方米 容积率：2.99

7 8 湖南常德中原 · 都会春天设计规划方案

总用地面积：24.21公顷 总建筑面积：63.99万平方米 容积率：2.0

建筑设计：李宇星、李　志、易碧华、谭　越
结构设计：李淑珍、陈宏斌
机电设计：冯　伟
效果图绘制：牟未秋、苗　昆

8

苏州太湖国际高尔夫别墅

项目地点：江苏 苏州

建筑面积：20 535平方米

占地面积：28.4公顷

项目性质：高尔夫别墅

林秀江南——英格堡小镇

项目地点：浙江
占地面积：21.52公顷
建筑面积：90 382平方米
项目性质：酒店、别墅、商业、休闲

archland ARCHITECTS AND ENGINEERS

中星红庐

项目地点：上海
用地面积：35.4公顷
项目性质：高档别墅区

海南什进村乡村旅游区

项目地点：海南 三道湾

用地面积：39.11公顷

建筑面积：35 051平方米

项目性质：旅游度假区

archland

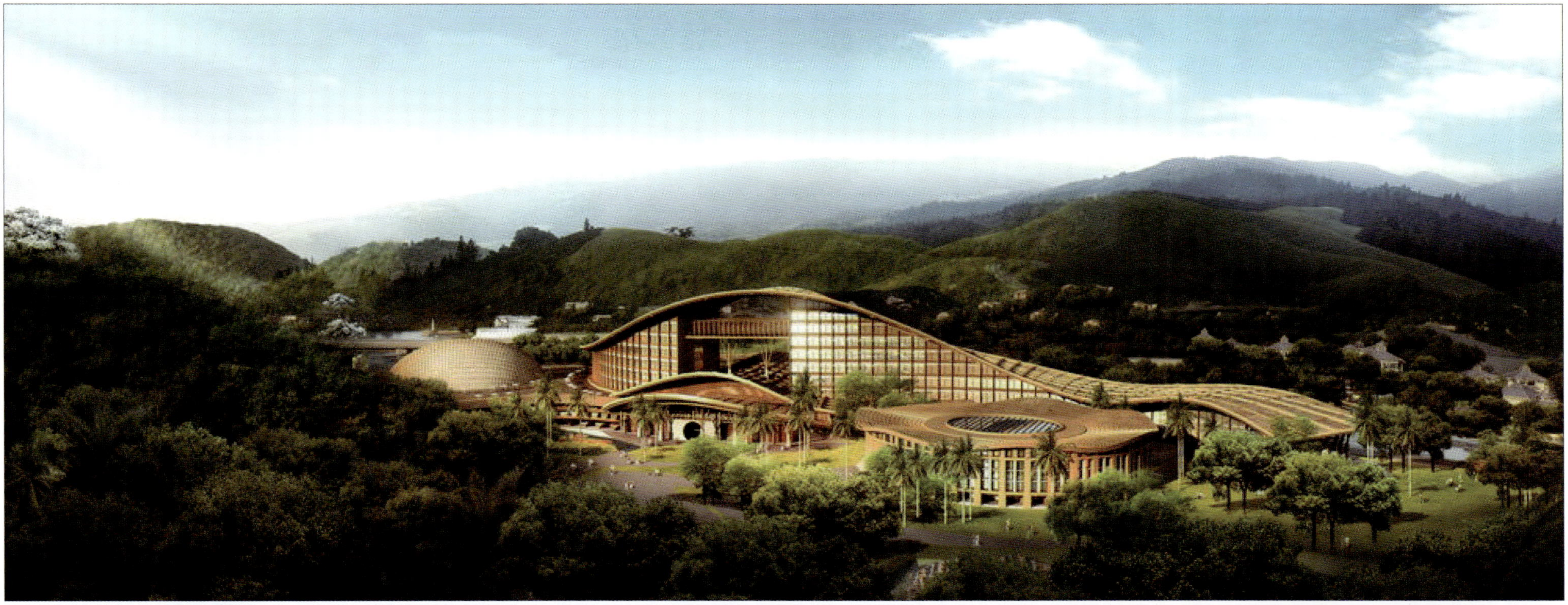

桂湖温泉城市设计国际咨询

项目地点：福建　　占地面积：624公顷

建筑面积：63.3万平方米　　项目性质：综合旅游度假区

archland ARCHITECTS AND ENGINEERS

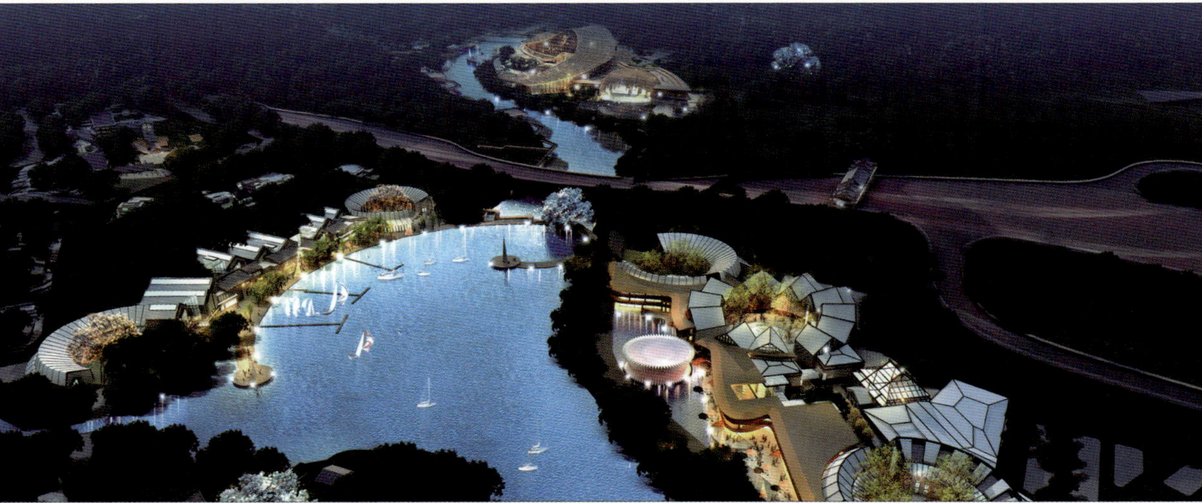

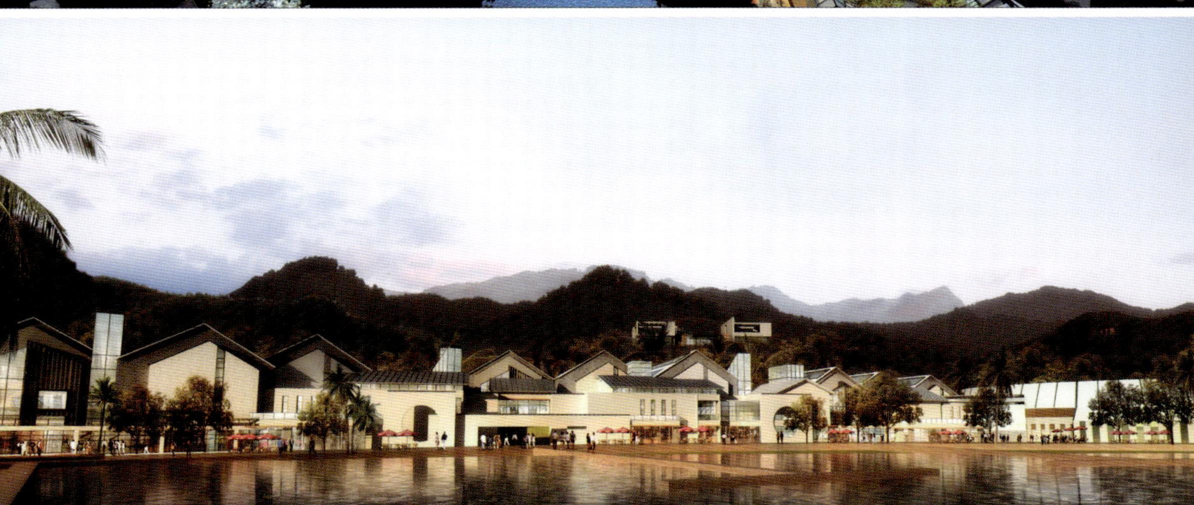

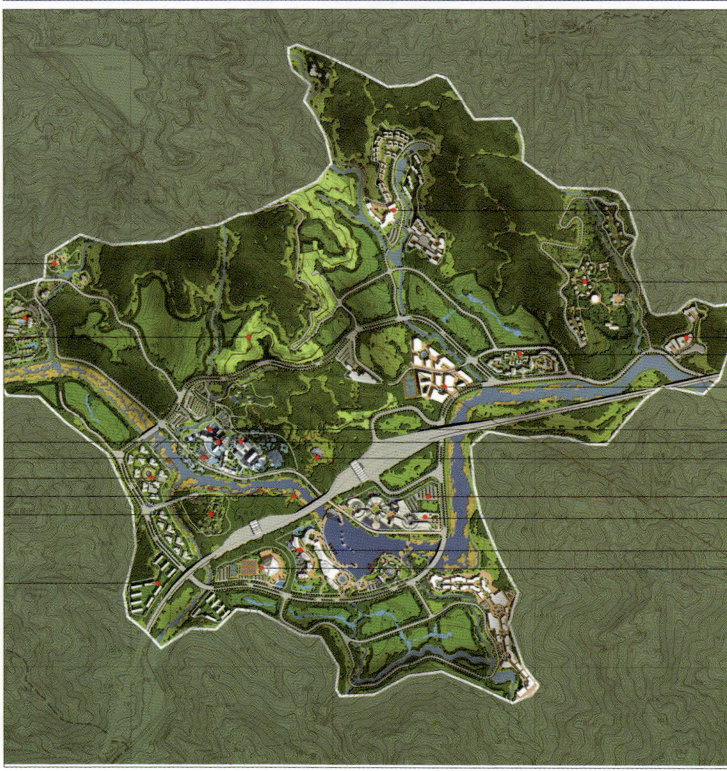

重庆缙云山展示中心

项目地点：重庆
用地面积：2.33公顷
建筑面积：2 461平方米
项目性质：艺术家园区

archland

亚特兰蒂斯——黄金时代

项目地点：四川 成都
占地面积：26.67公顷
建筑面积：190 620平方米
项目性质：综合性旅游度假区

archland

贵阳中天高尔夫概念规划设计

项目地点：贵州 贵阳　　占地面积：492公顷

建筑面积：488万平方米　　项目性质：综合旅游度假区

遂宁河东新区五星级酒店

项目地点：四川 遂宁

占地面积：3公顷

建筑面积：355 001平方米

设计类型：五星级酒店

archland

Zone Architecture & Engineering Design Co., Ltd.

中鸿建筑工程设计有限公司

中鸿建筑工程设计有限公司，拥有建设部颁发的建筑工程设计、总承包及项目管理甲级资质。总部位于北京，下设昆明、成都、海外事业部等分支机构；业务范围涵盖规划、建筑、结构、机电、景观、室内概预算及项目管理。业务以北京为中心，辐射全国十几个省市。公司以清华大学等知名学府培养的专业人才为核心，拥有各类设计人员约150人。近15年来已设计完成约260个项目，其中大型工程约160个，竣工项目总建筑面积超过800万平方米。

作为植根于中国、秉承国际化的设计企业，中鸿坚持携手伙伴、探索梦想，以勤奋和创新，成就建筑与环境的永恒价值。

Zone Architecture & Engineering Design Co, Ltd. is licensed by the MOHURD for construction design, general contract and project management Grade A. It is headquartered in Beijing, with branches in Kunming, Chengdu, and overseas offices; its business covers planning, construction, structure, mechanical & electrical, landscaping, interior estimate & budget and project management. Its business is based in Beijing, radiating to over 10 provinces/municipalities in China. The Company has about 150 designers of all kinds, centered by special talents educated by Tsinghua University among other well-known educational institutions. Over last 15 years, it has designed or completed about 260 projects, including about 160 major projects, with completed floor area exceeding 8,000,000 m^2.

As an international design firm rooted in China, Zone will always work with partners, to explore dreams, and realize the perpetual value of architecture and environment, with diligence and innovation.

公司理念

倾听梦想
探索使命
和你像伙伴一样
用创造性的解决方案
成就建筑和环境恒久价值

MISSION

Listen to the Dreams
Explore Vision
With You as Partners
Create Solutions
Everlasting Value of Built Environment

地址：北京市海淀区北四环西路68号左岸大厦1418
邮编：100080
电话：+86-10-82676060
传真：+86-10-82676166
邮箱：office@zonearch.com
网址：www. zonearch.com

Add: Left Bank Building, 1418 North Fourth Ring Road 68, Haidian District, Beijing
P.C.: 100080
Tel: +86-10-82676060
Fax: +86-10-82676166
E-mail: office@zonearch.com
Http: // www. zonearch.com

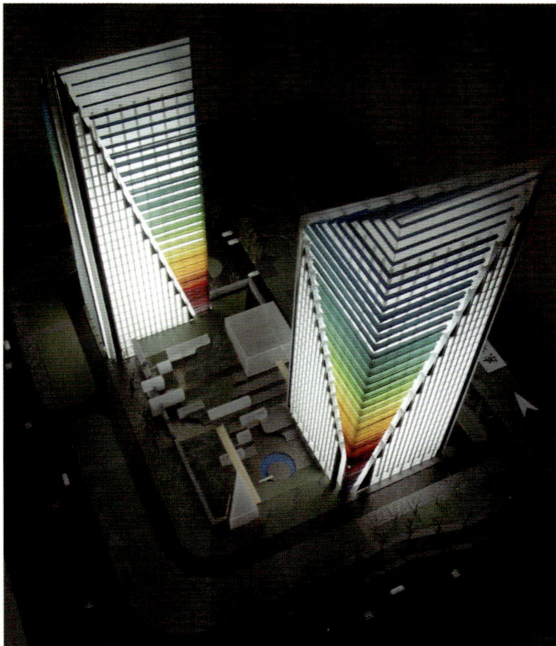

北京中央商务区乐成国际中心

项目地点：北京
项目规模：9.9万平方米
工程性质：公建
设计时间：2003年
建成时间：2005年

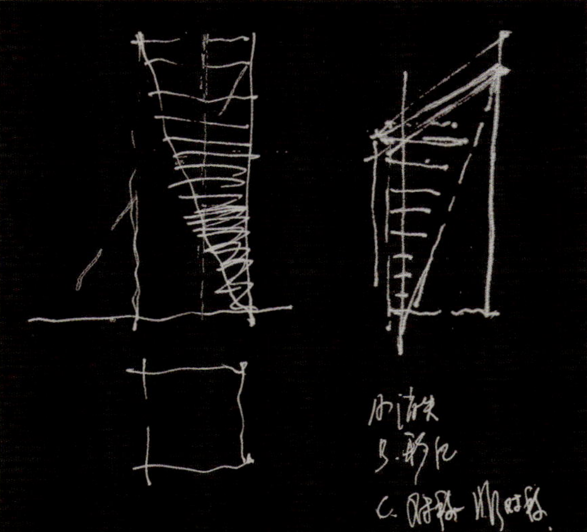

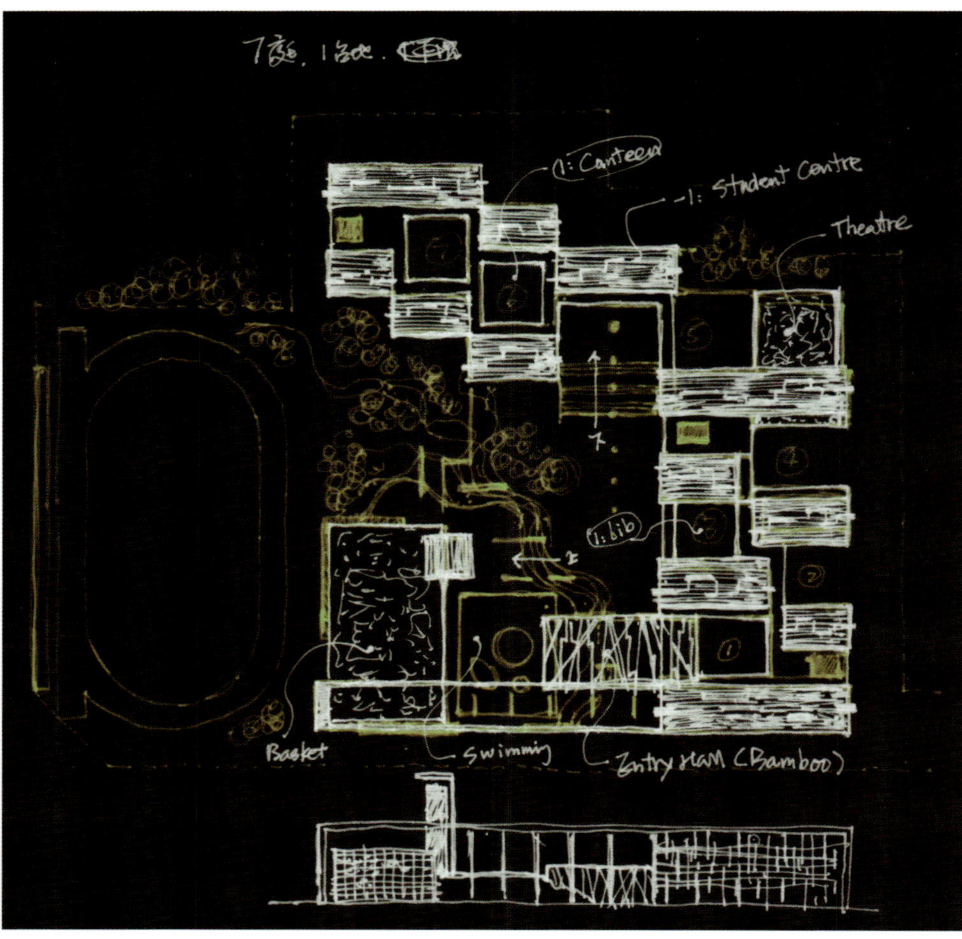

乐成国际学校

建设地点：北京
项目规模：3.8万平方米，5层
工程性质：公建及学校
设计时间：2004年
建成时间：2005年

空间

基于方格构图的错落式布局。4万平方米的庞大建筑，被化解成了一个小村庄的模样：曲折的小溪，散落的房子，缓缓起伏的丘陵，拾阶而下的广场，高地上的钟塔。

校园有一条隐隐约约的主线，连廊蜿蜒迂回，串起若干相对独立的院落；它们被赋予不同的主题，自然地聚集在一块不大的用地中，朝向空旷的中心。校园完全符合中国园林的构架模式。

建筑本身的数学逻辑单纯清晰，尺寸由9米x9米教室和双侧3米走廊（阳台）的轴线网确定。两个图书馆都在首层，有几百平方米的两层高空间；屋顶上还是庭院。这高高低低，大大小小的7个院子，是不同年级的室外领地。庭院深深，步移景异，也正是中国传统建筑的精神所在。

教室群的西南方，在起伏的丘陵中半掩着两座体育馆，它们烘托起一方钟塔；屋顶上的室外球场和雕塑展览台地，是校园的制高点。教室群的东北角，则是一个下沉式广场，那里连接着学校的剧场。这一高一低之间是开阔的中心空地，隔着从南大门向操场曲折流去的校园小溪相望。这些场地是全校师生都可以方便到达的。场地高低相宜，空间层层渗透，或开或阖、或幽或敞，各遵其法。

表皮

简单的立面框架，强烈的材质对比。在规整的结构梁柱间，放置木本色遮光百叶和窗下木墙板；仿砂岩的白色外墙；内装石子或花盆的钢丝护网；暖色金属穿孔栏板和窗框。木材、石子、铝板、仿真涂料——一切材质力求天然。简单纯粹，是理性主义的必然结果，也是现代化建筑技术的要求。质朴天然，是都市对自然的向往，“天人合一”也正是中国建筑的内在精神。

朝来绿色家园

项目地点：北京
项目规模：91.71万平方米
工程性质：公建及住宅
设计时间：2000—2009年
建成时间：一期2006年
二期2008年
三期2011年

朝来绿色家园，是北京市来广营乡绿化隔离地区项目，含有农民安置房、商品房、教育配套、商业配套、社区服务设施等。是综合性的大型居住区片。自2000年开始规划至今，主要地块已基本建成。

金盏乡长店组团

一、项目概况

本项目为北京市朝阳区金盏乡长店组团。用于北京金盏乡绿化隔离地区农民安置和乡镇企业发展。总体上居住区片性质明确，规划限制极强。 规划依据北京市规委2006版控规，容积率1.8～2.0，控高18～30米。中小学各一所，托幼二所、卫生所一所，公交站场一处。可规划用地75.43公顷，建设总规模133.08万平方米，安置户数9 910户，约2.77万人（户均人口2.8）。

二、项目特色

地处城市边缘，周边休闲商业与创意文化产业初步聚集，大区带整体品质高，发展趋势好，自然生态条件上佳。 本方案突出商业设施与周边类似项目的衔接，组团内公共设施力求成为大区片休闲产业链中的一个节点，在服务本区的同时，丰富本区公共生活，提高本区品质，促进本区经济活力。

三、规划结构

一轴双中心，十一组团。调整路网：5条道路走向进行调整，有利于整体结构塑造，有利于组团布置。

四、绿化系统

一轴多心串联系统，共分三个层次。步行系统借助绿化系统的串联而贯通，公共绿地与公建设施相匹配，特别强调本规划与西部小龙河绿带的衔接，本规划设想在此处设置文化广场等开放公共空间，构建滨河城市公园，形成整体的开放感。

五、民俗遗迹重构

选择性进行点滴民俗痕迹的保留与整合（并非严格意义的文物保护），例如一块碑、一棵树、一段路、一片墙、一口井…… 或直接保留，或部分异地摆放。 同时特色商街在风格上也民俗化、民居化。 这些遗迹散布于居住组团及公共区域，相互有一定串联关系，重构出本区特色之一，旨在唤起对以往的某些记忆。力求使城市规划后的新区仍然能延续而不是完全割裂长店村过去的历史。

六、建筑形态

含蓄而不失时尚，塑造多姿多彩的形态感受。规划避免简单行列布局，风格上尊重多样化市场需求。 本方案提供简约式、中式、自然式及经典欧式四类风格取向的多样化产品，相互对比衬托，其中A2-1、B2、B3-1、C3四个区设局部坡顶；以建筑设计手法确保高品质视觉感受。

Beijing Yanhuang United International Engineering Design Co., Ltd.

北京炎黄联合国际工程设计有限公司

北京炎黄联合国际工程设计有限公司是一家股份制设计公司，拥有建筑工程设计甲级资质及城市规划乙级资质。总部设于北京中关村南大街甲18号北京国际大厦B座9～12层，设计人员近700人。公司为适应国家城市化进程和与日俱增的设计任务的需要，正在稳步发展壮大中，公司在天津、南京、合肥、厦门等地还设有分公司。

公司由我国资深建筑师刘永梁先生于20世纪80年代末创建，至今已有二十余年的历史。2003年改成股份制，改称北京炎黄联合建筑设计有限公司。2010年3月，与中外建建筑设计有限公司第一分公司合并重组，更名为北京炎黄联合国际工程设计有限公司（简称炎黄国际）。从此，这家创立20余年的设计企业开始了全新的历史时期，将以前所未有的活力和生命力，去迎接新的挑战，为实现“建设全国一流的设计公司”的宏伟目标而努力奋斗。

炎黄国际的业务范围包括各种建筑工程设计、室内装修设计、园林景观设计、城市规划和风景区旅游规划等多个领域，尤其在高尔夫社区、酒店及商业建筑设计上拥有自己的专业队伍。公司专门设有高尔夫社区规划院和商业设计院，完成和正在设计的项目从住宅、别墅、公寓、学校、医院、商场、购物中心、体育馆到写字楼、宾馆酒店、度假中心、高尔夫社区，几乎包括所有民用建筑领域，项目所在地点除北京外，从最北端的伊春、呼伦贝尔、哈尔滨到最南面的昆明、贵阳、三亚，遍及全国各地。公司还设有海外设计部，中东、北非的一些设计项目也正在运行中。

炎黄国际技术力量雄厚，专业配备齐全，拥有一支业务精湛、经验丰富、骨干稳定、充满活力的设计团队。公司的宗旨是：用先进的设计理念和技术，用精益求精、一丝不苟的工作态度，为客户提供国际化、专业化、品牌化、本地化的设计服务，并为国家的经济建设和人民工作、生活环境的改善作出自己应有的贡献。

地址：北京市海淀区中关村南大街甲18号北京国际B座9-12层
邮编：100081
法定代表人：冈钢
电话：+86-10-62117746
传真：+86-10-62122704
邮箱：yanhuangguoji@vip.163.com
网址：www.yuie.com.cn

Add: Floor 9-12, Block B Beijing International, No.18 Zhongguancun South Street, Haidian District, Beijing
P.C.: 100081
Legal Representative: Gang Gang
Tel: +86-10-62117746
Fax: +86-10-62122704
E-mail: yanhuangguoji@vip.163.com
Http:// www.yuie.com.cn

Beijing Yanhuang United International Engineering Design Co., Ltd. is a joint-stock design company, with Class-A License of Architectural Engineering Design and Class-B License of Urban Planning. The Company is headquartered in 9-12F Beijing International Building B, No. A-18 Zhongguancun South Avenue, Beijing. It has nearly 700 designers. To meet urbanization process of China and increasing design tasks, it is steadily growing and expanding. Company has branch offices in Tianjin, Nanjing, Hefei, Xiamen etc.

The Company was founed by a senior architect of China — Mr. Liu Yongliang in late 1980s, with a history of twenty years till now. It became joint-stock company in 2003 and was renamed as Beijing Yanhuang Union Architectural Design Co., Ltd. In March 2010, it was merged and reorganized with CCI Architectural Design Co., Ltd (First Branch), and was renamed as Beijing Yanhuang United International Engineering Design Co., Ltd. (Yanhuang International). Since then, this twenty-year old design company began a new historical period. It uses its unprecedented vigor and vitality to meet new challenges and work for the grand goal of "building a national first-class design company".

Business fields of Yanhuang International includes: all kinds of architectural engineering design, interior decoration design, landscape design, urban planning and travelling planning of tourist sites etc. Especially, it has its own professional design team for golf community, hotels and commercial buildings. The company has a Golf Community Planning and Commercial Design Institute, which has completed or working on projects from residential community, villas, apartments, schools, hospitals, shopping malls, shopping centers and stadiums to office buildings, hotels, resorts and golf communities, including almost all areas of civil construction. Besides Beijing, locations of projects include Yichun, Hulun Buir and Harbin in the north and Kunming, Guiyang and Sanya in the south, all over China. The company also has overseas design department. Some projects in the Middle East and North Africa are also running.

Yanhuang International has strong technical force and complete professional equippments. It has a skillful, experienced and active design team with stable backbone of designer. Purpose of company is: using advanced design idea and technology, with excellence and meticulous working attitude, to provide customers with international, professional, branded and localized design services and make contribution to economic construction of China and improvement of people's working and living environment.

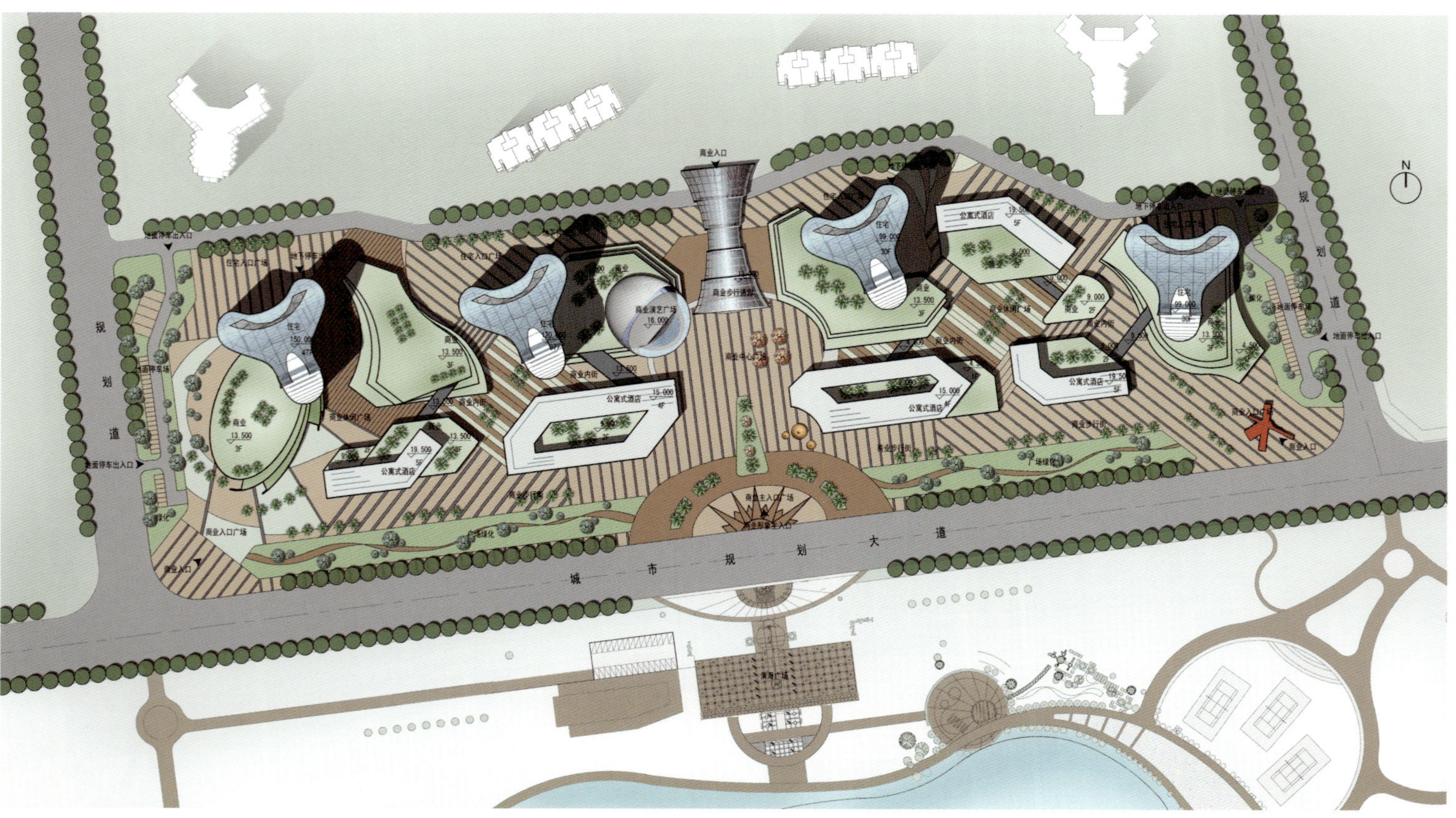

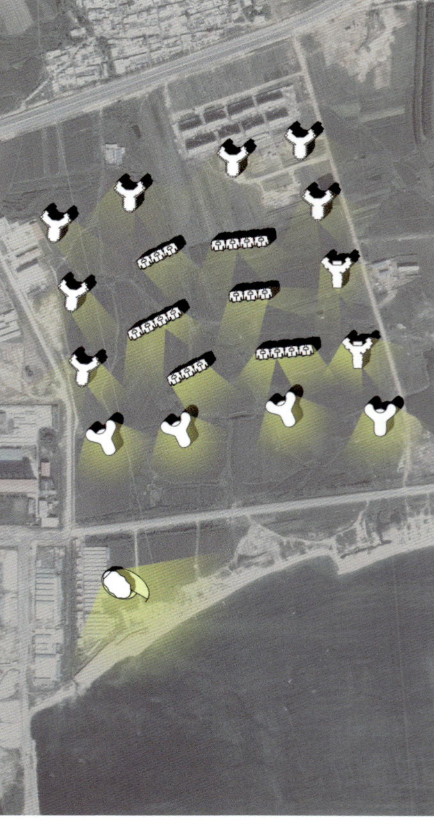

恒泰地产・辽宁绥中滨海综合性商业建筑群落概念性规划设计方案

商业景观空间——商业公园

商业的游线组织与景观的设计布局紧密联系。在整体规划中，将建筑与景观融合，沿街面提供给人们足够的流通空间和商业外延空间，在中心广场设有表演的剧场空间，这是艺术家们和商家们展示的天堂和乐园。将内外大小空间有机地与商业相互的渗透，打造一个现代新型的休闲商业公园。

现代风情街景——体验消费

拥有一条东西向的现代沿海度假式风情商业街，良好的采光和通风效果，在每个角落都可以嗅到阳光的气息，“坐在街旁的咖啡店门口，浅尝一口在休闲桌上的咖啡，满满的都是阳光的味道”，这就是休闲生活的缩影，通过景观营造生活感受，逛街不仅仅是购物，更是品味生活。

地域景观要素——海洋主题

设计结合海边风情，以海洋的肌理为设计理念，商业建筑就好比漂浮在海洋之上的石头，与景观交织在一起。在一些公共空间中增加一些海洋主题的元素，如地面铺砖和雕塑小品等；演艺广场则直接面海，让观众在观赏表演的同时，更能欣赏优美的海景，享受极致的视觉冲击。

贵阳生态旅游体育基地及配套设施项目修建性详细规划

项目位于贵阳市乌当区航天路北侧，绕城高速东侧，周边有新天二小、新天四小、红华超市等生活配套设施。

基地为自然山地，北侧坡地向北可以眺望湖水，南侧毗邻航天路，东部有良好的生态资源，西侧为绕城高速。现今基底周围地块的成熟度日益升高，地理位置在未来的城市发展中属于较好的区域。

该方案以尊重自然、再造自然、融入城市空间环境这一理念为核心，总体布局以人为本，通过一系列与环境相融的人性化设计，创造出尊贵、自然、优雅的人居社区，使居民真正感受到亲和、随意、自由的生活方式。

在“经济、实用、美观”的基本前提下，以“环保、节能、安全、舒适”为设计原则，为居住者提供人性化、生态化的设计产品。在处理好总体功能布局、环境景观、交通组织以及整体建筑风格几大要素的同时，也要满足规划、消防、安全和经济合理等各方面的要求。努力打造一个理想的人居境地，使该项目成为一个极具个性的城市住宅小区，力争将经济、社会和环境效益三者统一起来。

大庆让胡路中心区规划概念设计

西宾路为让胡路区最为重要的大街，西宾路与昆仑大街交汇处西南侧地块则为让胡路区最为重要的商业区 。该商业区内最重要的商业建筑为三层的大庆商厦、商厦美食城及电信家电城。从整体商业环境上分析该区域商业环境存在以下缺陷。

1. 缺乏统一长远的商业规划与合理的商业功能分布，基本上属于随机建设，过分追求短期效益。
2. 缺乏完整、鲜明、有特色的商业建筑形象和标志性建筑。
3. 缺乏合理的商业流线规划，人流与车流相互混杂，不能形成安全舒适的商业环境。
4. 商业餐饮不成规模，层次较为单一，没有吸引力。

功能较为单一，只有普通百货商店与餐饮，没有高档的影城、溜冰场、娱乐城等目的性消费场所，不能吸引多个年龄层的消费群。由于这些缺陷的存在，造成该区的商业结构与氛围远不能符合让湖路区核心商业中心的地位，该区现状阻碍了该区向更高一级商业层次的发展。

一、佲金新都会规划原则

大庆让胡路中心区规划概念设计应体现六个结合：与大庆市城市总体规划相结合；与大庆光辉的石油文化相结合；与著名的铁人精神相结合；与特有的湖泽湿地地貌相结合；与温泉文化相结合；与新型生态低碳发展趋势相结合。

二、佲金新都会规划理念

1.加强城市轴线，形成双通道复合步行商业景观轴线

佲金新都会路网的走向延续了大庆城市肌理走向，形成了“两纵、两横”的方格状路网格局，沿西宾路购物中心南侧设置城市和步行商业街，形成商业轴，沿昆仑大街设置绿化轴，突出绿色生态城市理念，将城市中心绿轴与中央商务区结合在一起，为大庆人民提供了系统的、相对独立的商业、居住、商务等多功能公共城市空间。

2.塑造美好社区生活，创建大庆绿色天际线

建立由“绿轴、绿心、水面”组成的景观居住空间结构。规划购物、商务、生活与休憩相互平衡的未来复合型新城市中心，现代的商业及办公建筑与绿轴和生态城市广场相呼应，新古典主义的居住形象与绿色水景及自然形态的景观环境相融合，一起创建湖在城中、城在绿中的美好生活。

3.彰显大庆中心区CBD形象 ，营造城市客厅

作为中心区的焦点，在佲金新都会的西北角，以现代气息浓厚的130米的标志性五星级酒店、高级办公公寓群、城市级大型购物中心，共同围合成大庆让胡路中心区极具城市张力的圆形城市广场——佲金中心广场，形成高层林立的CBD中央商务区，体现大庆团结奋进的铁人精神，营造大庆城市客厅。

三、商业规划

佲金新都会将按照一流的商业模式对该区的商业进行全新的规划 。

根据规划将沿西宾路建设一座规模为20万平方米的一流购物中心，内部功能包括主力店，次主力店 、精品店 、百货公司 、国际品牌超市 、国际美食城 、国际影城 、IMAX影院 、室内四季溜冰场和娱乐城等等，各项功能一应俱全，面向大众，吸引不同年龄层及不同购物取向的消费群体 。

除沿西宾路建设集中的购物中心，还在购物中心南侧设置商业内街，形成纵横交错的商业街区，商业内街分别平行于西宾路和昆仑大街设置，交汇于佲金中心广场，使商业环境纵深化、立体化 。

建设大规模的地下停车库，实现人车分流，创造安全舒适的购物环境 。与此同时，在西宾路与昆仑大街西南角建设一座超高层的五星级酒店，成为未来商业中心的地标性建筑，酒店裙房内将设国际一线奢华品牌精品店，面向大庆不断壮大的高端消费群体 。

四、城市设计

沿西宾路购物中心北退建筑红线20至30米，形成宽阔的商业街，便于吸引和集散人流，形成与城市主干道之间的缓冲和过渡。在用地的东北角规划一个约4 000平方米的城市公共广场，在这里设置水池、喷泉、绿地、休闲座椅，创造一个绿色宜人的公共休闲环境。

广场通过一条斜向的道路与内街的中心广场连接，形成对景，将沿城市干道的人流引入商业内街，使商业向地块内部延伸，充分挖掘地块的商业价值，激活内街的商业氛围，提升内街的商业价值。

昆仑大街是让胡路区重要的城市南北干道，承担着体现让胡路区绿色生态理念的角色，因此沿昆仑大街将不设置任何商业，把沿街的空间完全留给城市绿化，设置20米宽的城市绿化带，形成城市的绿轴。

五、建筑形象设计

沿西宾路和昆仑大街规划现代化的面向未来的高层建筑群，130米高的五星级酒店构成建筑群的制高点，水晶体般的建筑造型高耸挺拔，简洁有力，与高低错落的办公公寓建筑一同勾勒出富于变化的城市天际线。

购物中心的立面强调活力与灵动，突出商业气氛，通透的玻璃把商场内部中庭的商业活动表现出来，配合外墙广告及夜间照明设计，打造溢彩流光的立面效果，创造充满吸引力的购物环境。商场屋面进行屋顶绿化，形成建筑的第五立面，注重周边高层建筑俯瞰购物中心的效果。

居住区的立面采用新古典主义立面风格，营造出温馨舒适的居住环境，与现代风格的办公楼和酒店形成鲜明的对比，力求打破整个地块采用一种建筑风格的单调感。

随着这些规划的逐步实现，佲金新都会将成为让胡路区功能最齐全 、最具商业活力和吸引力的核心商业中心，成为大庆西部的核心商务区，成为大庆的城市名片，其商业辐射力最终将超出大庆远及整个黑龙江西部地区。佲金新都会将成为：一个商务商业活动的枢纽；一个实现节能、能源高效利用、生活契合自然的区域；一个体现绿色工作和生活新概念的地方；一个融合石油文化和大庆精神的活力区域；一个可持续发展，展示大庆发展之雄心的区域。

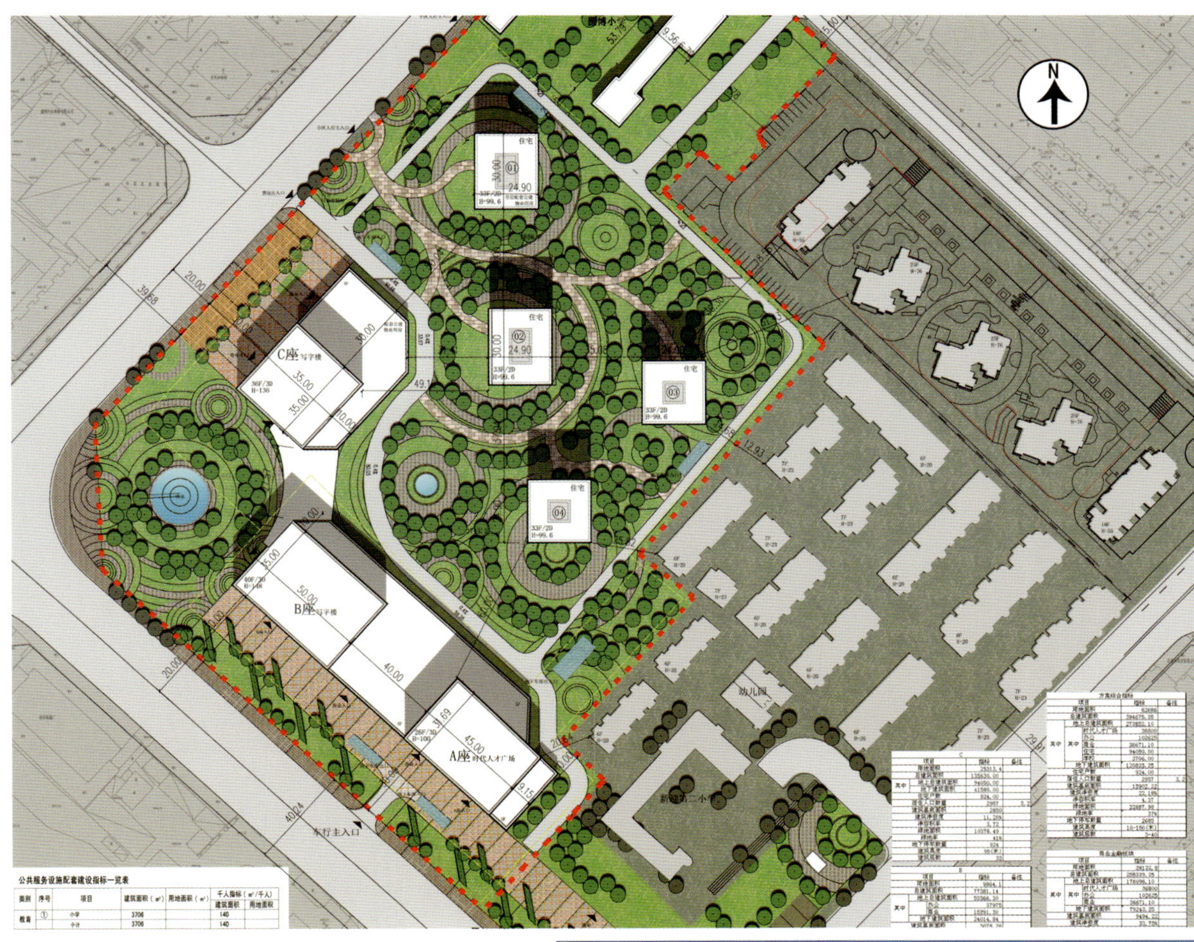

昆明时代人才广场规划设计方案

本地块位于中心城区二环路内，属迎新片区，北靠穿金路，西南临白云路，东南接新兴路。公共交通较为方便，有多条公交专线经过，并且离正在修建的地铁二号线步行只有不到十分钟的路程，距二环快速交通系统不超过3公里的距离。周边医院、幼儿园、小学、中学等生活配套设施非常完善，属于城市成熟地块。项目规划总用地面积一期为20 945平方米，二期为32 522平方米。

本项目定位为构建面向国际的现代化人才交流平台，营造新昆明城市人才管理、培养的城市功能空间。依托人才交流中心平台文化背景，提升该片区城市风貌，完善城市配套，改造提升城市面貌，充分融合休闲娱乐、商务办公和城市广场的功能。

结合城市景观，与区域内绿化带相结合：将西侧城市广场东、南侧城市道路绿化带纳入到整体规划中，美化区域环境。北侧为大面积中心绿地，结合水系布置绿化景观。区内南北形成一条主要绿化景观带，并将水系引入小区，极大地优化了小区景观，大大提升了小区品质。

合理的区内外交通与停车规划：通过合理的道路布局、竖向设计和宅间空间规划，提供了地面、地下等多种停车解决方式。

以小区外环路为主要交通：二期小区外环路将机动车与小区内部环境隔离，满足停车的条件下做到了人车分流和小区景观的最佳化。停车方式分地面停车及地下停车。

与城市干道相连接的小区入口区域：小区西北设置一个小区主入口，与小区南北景观结合成完美轴线。

北京华英设计顾问有限公司

INTEGRAL DESIGN SERVICE LTD.

通信地址：北京市朝阳区望京中环南路9号望京大厦A座18F 100102
联系电话：+86-10-64717408 / 64718525 / 64718508
传 真：+86-10-64727409 E-mail: hoybi@public.bta.net.cn

公司名称

北京华英设计顾问有限公司，是一家外商独资经营的公司，主要提供建筑规划及设计的咨询服务，包括景观处理、室内设计、媒体形象及工程策划的咨询。

"华英"，顾名思义，是希望能聚合一批"中华精英"，为投入祖国重建大业而尽心尽力。英文取名为"Integral Design Service"，提供完整无缺、全方位的设计服务是公司一贯的追求目标及经营理念。

Integral Design Service Ltd., Beijing is a foreign-invested company, which provides comprehensive land planning, architecture design, interior design, landscape design, media image, and construction management services.

"Hua Ying"means that the company intends to collect the most outstanding professionals together in pursuit of design innovation in China."Integral Design Service" means to provide full professional services towards fulfilling the company's goals and objectives.

公司规模

北京华英设计顾问有限公司目前全部人员18名，其中包括美国、瑞士注册的资深建筑师2名；持有硕士学位的建筑师3名；持有学士学位的人员13名。

IDS is organized as an architecture firm in the USA and offers all employees the best benefits and a good working environment. There is a team of 18 members including 2 senior architects registered in the USA and Switzerland, 3 architects with Master's Degree and 13 members with Bachelor's Degree.

经营理念

1. 全员参与，永续经营；
2. 坚守"事业培养事业"原则；
3. 确立品质，兼顾效率，稳步发展；
4. 坚守专业原则，提供业主全面、整体、最佳的服务。

The policy of IDS management is as follows:
1. All employees are invested in the long term business.
2. IDS believes a great business in the future starts with good business today.
3. IDS constantly develops greater efficiency and the highest quality.
4. IDS will offer all its clients the best integrated service available.

专业特长

根据国家制度和规范要求，除负责项目的前期规划、方案的初步、扩初设计及节点大样设计之外，根据项目需要，与国内设计院配合，并督导其完成结构、设备及施工图的设计，以达到整体完善的专业服务。服务范围涉及整体规划、建筑设计、室内设计等领域。

Due to Chinese laws today, IDS's professional involvement ranges from predesign, schematic design, design development, and quality control for structural, MEP and field services. Depending on the project, IDS usually subcontracts to local architectural engineering institutes to complete structural, MEP as well as working drawings and field services.

工作业绩

由主持建筑师谢国权先生负责规划设计的"北京大西洋新城"多次荣获《北京晚报》、《北京精品购物指南》主办的楼盘评选大奖15个以上。

此外，完成的其他项目有：

年份	项目	指标	数值
2001年	大庆义耕时代丽景	总楼地板面积	70 000 m^2
2001年	吉林远东石油大厦	总建筑面积	26 624 m^2
2001年	天津南开大学数学研究所	总建筑面积	20 952 m^2
2001年	大庆奥维馨苑	总楼地板面积	84 000 m^2
2001年	大庆时代广场商业中心	总建筑面积	30 892 m^2
2002年	义耕工区厂房	总楼地板面积	56 000 m^2
2003年	大连优豪斯花园	总楼地板面积	72 000 m^2
2005年	北京大湖山庄	总用地面积	300 000 m^2
		总楼地板面积	80 000 m^2
		别墅155单元	
2005年	太利住宅小区改建	55户	
2006年	海南三亚天际大厦改建	建筑面积	12 000 m^2
2006年	大西洋新城E2区精装修设计	总楼地板面积	18 000 m^2
2006–2007年	大庆奥龙小区规划	总楼地板面积	455 000 m^2
2006–2007年	世界客属总商会郑州住宅小区规划	总楼地板面积	307 000 m^2
2008年	海南保亭神仙湾温泉大酒店规划	总用地面积	123 000 m^2
		建筑面积	32 000 m^2
2008年	海南阳光绿苑小区规划	总楼地板面积	161 000 m^2
2008年	海南什岭乡村度假中心	建筑面积	20 000 m^2
2008年	牡丹江第二医院商住楼建筑设计	建筑面积	75 000 m^2
2008–2009年	大西洋新城F3区精装修设计	精装修面积	60 000 m^2
2008–2009年	大西洋新城G区建筑及精装修设计	楼地板面积	90 000 m^2
2008–2009年	大庆湿地富苑住宅小区规划	总楼地板面积	515 600 m^2
2008–2009年	海南保亭荔苑温泉山庄改扩建项目	改扩建面积	8 200 m^2
2008–2009年	海南新东方九所开发项目	总用地面积	26 400 m^2
		建筑面积	32 000 m^2
2010年	青岛李沧区十梅庵住宅区及老虎山规划设计	总用地面积	2 500 000 m^2
2010年	青岛李沧区东南渠住宅区规划设计	总用地面积	430 000 m^2
2011年	齐齐哈尔扎龙湿地温泉旅游度假区概念性规划	总用地面积	2 385 000 m^2
2011年	天津台商总部基地暨台湾名品城规划及建筑设计	总楼地板面积	150 000 m^2

规划设计

Interior Architecture Planning

目的

- 追求最合理的土地分区和使用；
- 实现政府、开发商及项目参与者之使命和期望；
- 以"人为本，家为中心"，塑造与市中心截然不同、达到居民的认同与归属及永远住下去的愿望的小区；
- 创造整体最高的价值。

方法

- 针对开发商的意向、居住者的需求、规范限制、市场需求及基地环境等条件，做深入的分析；
- 以道路、景观塑造与市中心截然不同的住宅环境：
  - 最大的安全感
  - 最美最温馨的形象
  - 机能造型于一体
  - 人为与自然、理性与感性结合

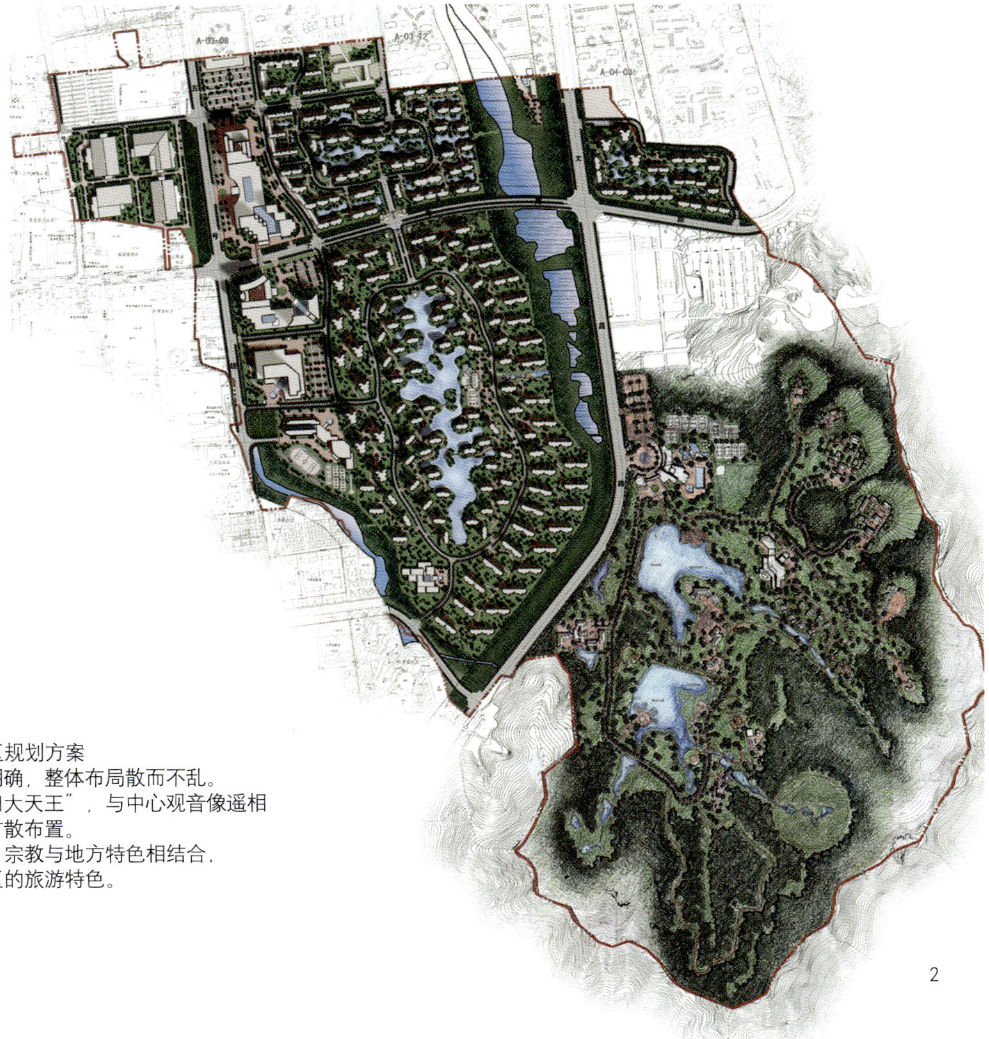

1

青岛十梅庵小区及老虎山景区规划方案

1. 佛教特点明显，动静分区明确，整体布局散而不乱。
2. 充分利用地形高程布置"四大天王"，与中心观音像遥相呼应，景点从中心向四周扩散布置。
3. 景区功能齐备，交通便利。宗教与地方特色相结合，最大限度地挖掘十梅庵景区的旅游特色。

2

3

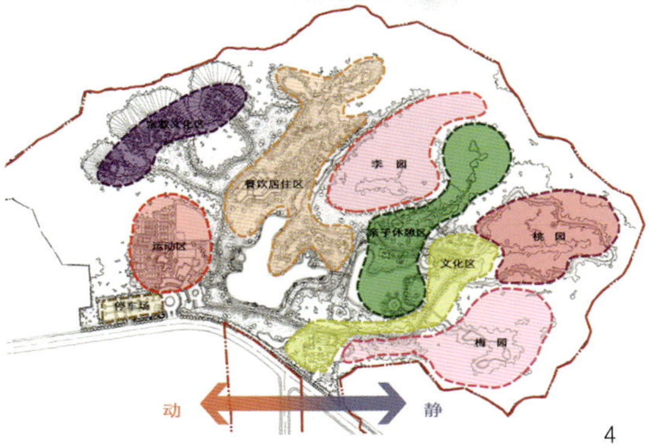

动 静

功能分区分析图

4

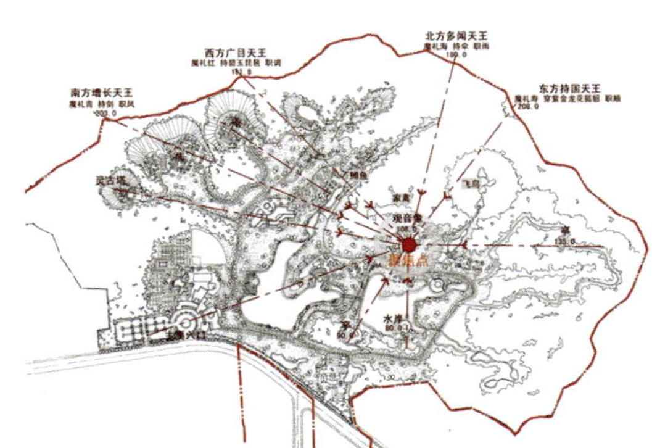

景点高程分析图

5

交通分析图

6

1 青岛李沧区东南渠旧村改造规划设计方案
2 青岛李沧区十梅庵小区及老虎山景区规划设计方案
3 青岛李沧区老虎山景区规划设计方案
4–6 青岛李沧区老虎山景区规划方案设计分析图

规划设计

Interior Architecture Planning

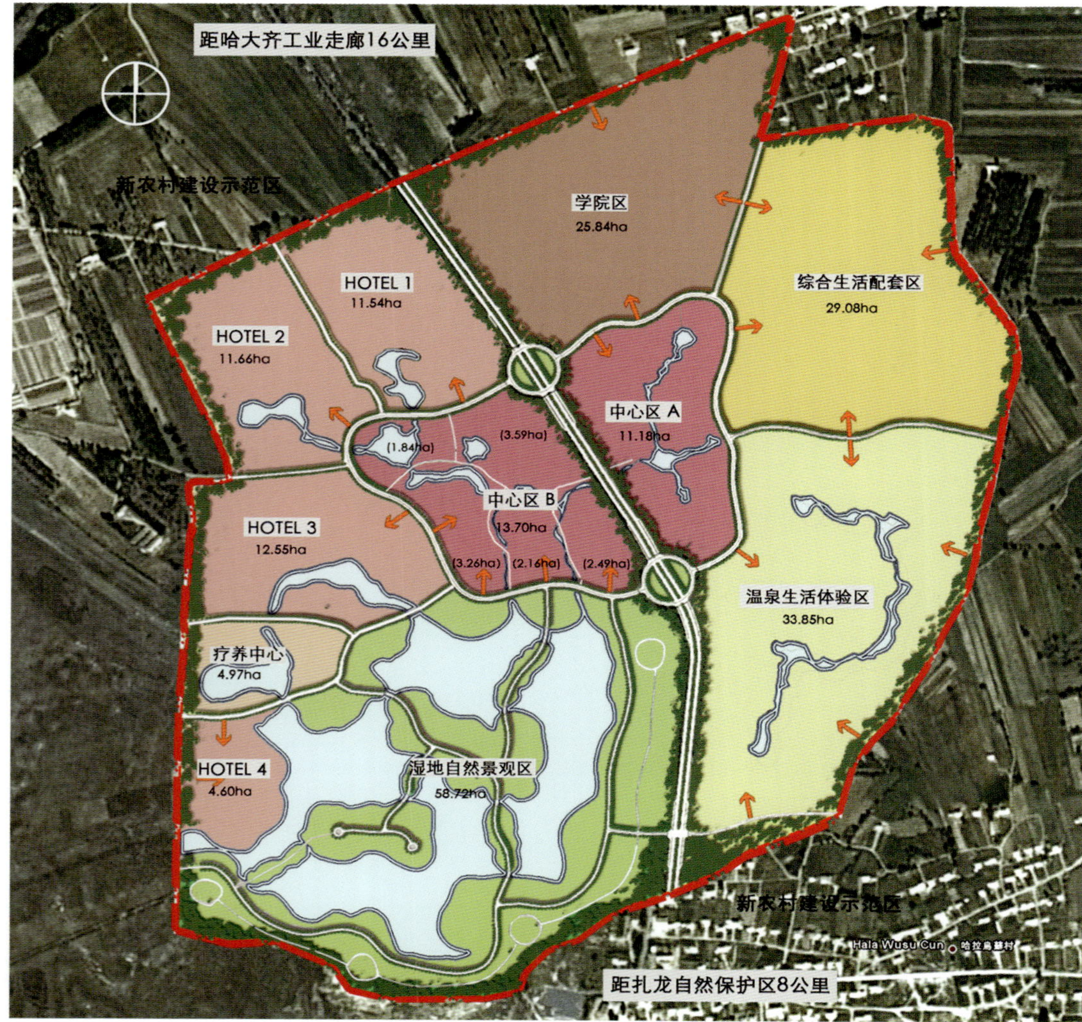

1–5 齐齐哈尔扎龙湿地温泉旅游度假区概念性整体规划方案及分析图

水系统

道路系统

用地系统

2 基地地形

1 系统空间分析图

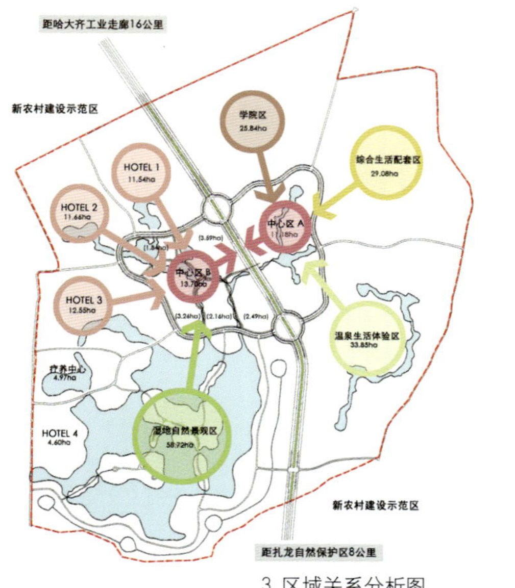

3 区域关系分析图

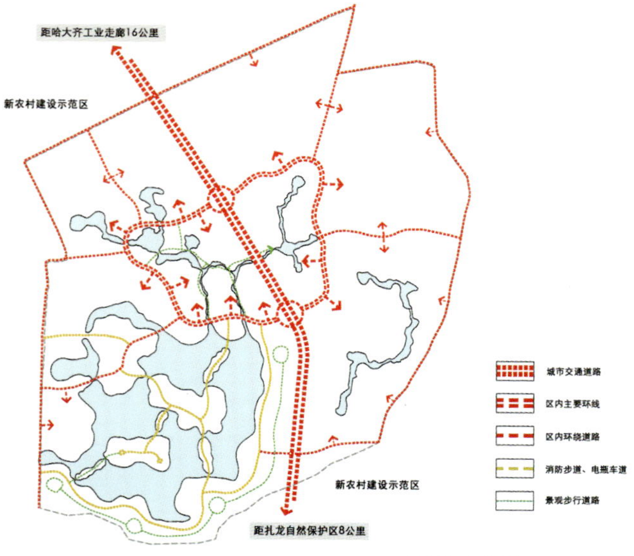

4 交通分析图

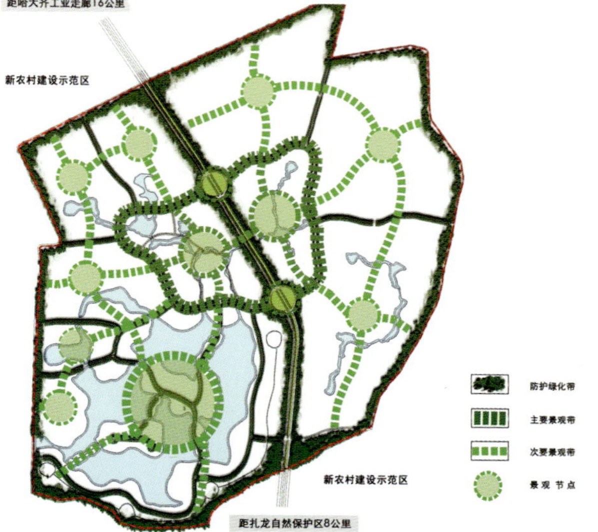

5 绿化分析图

建筑设计

Planning Interior Architecture

目的

- 追求"以人为本"，达到专业、经济地使用空间的目的；
- 符合时尚、市场的需要；
- 创造整体性最高的价值。

方法

- 针对建筑所在地点、环境、违规及本身条件做深入分析；
- 追求功能、造型、经济性及安全性于一体；
- 提供开发商、设计专业人员、技术营造单位密切配合。

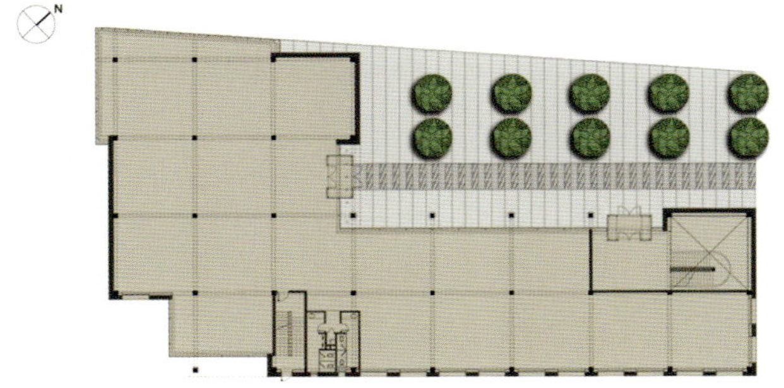
1

2

3

4

1 大庆湿地福苑售楼处首层平面图
2–4 大庆湿地福苑售楼处实景照片
5 澳洲别墅总平面图
6–9 澳洲别墅模型图片

7

6

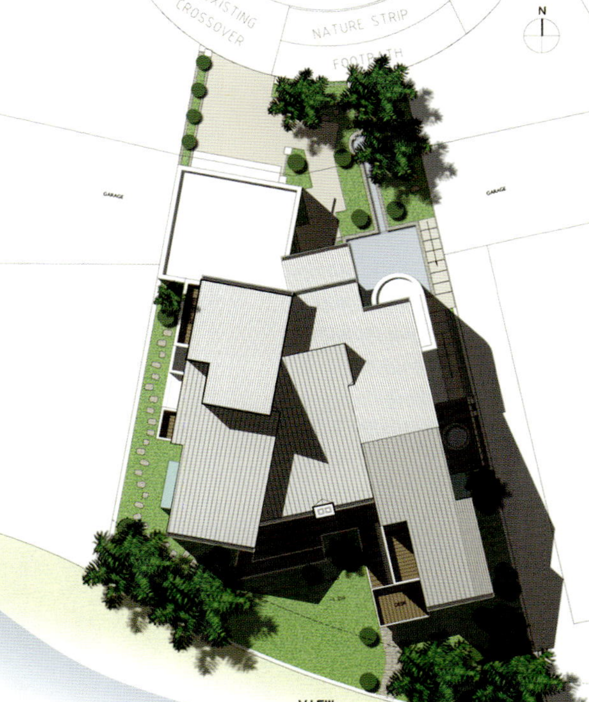

5

9

8

建筑改建项目

Planning Architecture Interior

目的

- 赋予原有建筑新的生命；
- 创造销售和长期使用整体性最高的价值。

方法

- 针对建筑所在地点、环境、规范及本身条件做深入分析；
- 追求功能、造型、经济性及安全性于一体；
- 提供开发商、设计专业人员、营造单位密切配合。

1

2

7

4

5

3

8

6

1 海南荔苑温泉山庄别墅原貌实景照片
2 海南荔苑温泉山庄别墅改造方案外景照片
3–8 海南荔苑温泉山庄别墅改造方案庭院内景照片

室内设计

Planning Architecture Interior

目的

- 追求“以人为本”，达到专业、经济地使用空间的目的；
- 符合时尚、市场的需要；
- 创造销售和长期使用整体性最高的价值。

方法

- 针对使用的需要和空间现况，做深入的探讨和分析；
- 追求功能、造型、经济性及安全性于一体；
- 提供开发商、设计专业人员、技术营造单位密切配合。

3

2

4

5

6

7

1

8

9

1 海南荔苑温泉山庄大堂室内实景照片
2 海南荔苑温泉山庄主要交通空间室内实景照片
3 海南荔苑温泉山庄多功能厅休息室室内实景照片
4–5 海南荔苑温泉山庄普通标准间室内实景照片
6–9 海南荔苑温泉山庄豪华大床房室内实景照片

地址：北京市海淀区增光路45号综合楼601
电话：+86-10-88560230
传真：+86-10-88560310
网址：www.aadoffice.com

Add: 601 General Building, 45 Zengguang Road, Haidian District, Beijing
Tel : +86-10-88560230
Fax : +86-10-88560310
Http: //www.aadoffice.com

AA · 都市建筑设计

北新桥西南口

天坛站内部

刘家窑站出入口

雍和宫站出入口

老城区出入口

天坛站西北口

Archi And design office

AA · 都市建筑设计由刘弘（日本注册建筑师，加拿大安大略省建筑师协会会员）于2003年成立于北京，为外资设计事务所。

从成立初期承接的北京总部基地规划设计到近期的地铁系列空间设计，AA · 都市建筑设计都把独特的设计理念和有效的设计手法相结合，通过融合规划、建筑、室内、景观的一体化设计，为人们提供具有时代精神、高品质的空间场所。

Archi And Design Office was founded in Beijing in 2003 by Liu Hong who is a registered architect in Japan and the member of Canada Ontario Association of Architects.

From the early days of undertaking Beijing Advanced Business Park Master Planning to the recent series of subway space design, Archi And Design Office combined the unique concept and effective design method together through the integrated design of planning, architecture, interior and landscape design, providing a modern and high-quality space for people.

西土城站出入口

农展馆站出入口

金台路出入口外部

国贸站西北口

北京地铁5、6、10号线出入口系列

老城区出入口

通用站出入口

通用站出入口

通州新城出入口

牡丹园站细部

模型

金台夕照站出入口细部

模型

Beijing BEDA Institute of Architecture Urban Planning & Design

北京北达建筑规划设计院

北京北达建筑规划设计院（Beijing BEDA Institute of Architecture Urban Planning & Design）地处人文荟萃、思想交融的中关村科技园核心区，东临清华大学，南靠北京大学，正北俯视圆明园文化古迹，具有与生俱来的浓厚的文化底蕴。

北达系由北京大学、清华大学、加州大学伯克利分校等世界一流大学的教授创建而组成的精英团队，是一家集城市规划设计、发展战略规划、国民经济和社会发展规划、产业规划、区域规划、旅游规划、市政基础设施规划、园林景观规划设计、遥感与地理信息系统开发、规划支持系统开发、项目策划、项目可行性论证以及投融资规划等业务于一体的综合性规划设计院。属于北京市高新技术企业，拥有城市规划甲级、旅游规划乙级资质，已于2009年8月13日通过GB/T19001–2008 idt ISO9001：2008质量管理体系认证。与北京大学中国公共规划研究院、林肯设计集团（中国）实现整合运作。

Beijing BEDA Institute of Architecture Urban Planning and Design is located in the core area of Zhongguancun Science Park, east of Tsinghua University, with Peking University in the south, overlooking the Old Summer Palace and cultural heritage in the north, born with a strong cultural heritage.

BEDA is an elite team, created by the professors coming from Peking University, Tsinghua University, the University of California at Berkeley and other world-class universities, is an integrated planning and Design Institute with a collection of urban planning and design, strategic planning, economic and social development planning, industrial planning, regional planning , tourism planning, municipal infrastructure planning, landscape planning and design, remote sensing and geographic information systems development, planning support system development, project planning, project feasibility and business planning, investment and financing. It is a high-tech enterprise in Beijing, with Grade A urban planning, and Grade B tourism planning qualification, has passed GB/T19001–2008 idt ISO9001: 2008 quality management system on August 13, 2009. It has cooperated with Peking University's China Institute of Public Planning, Lincoln Design Group (China) to achieve integrated operations.

地址： 北京市海淀区中关村北二条13号（中科科仪）1号楼4层
编码： 100190
总机： +86–10–62571096, 62571097, 62571098, 62571099
传真： 总机转8200
邮箱： beidaguihua@superplanner.com.cn
网址： www.superplanner.com.cn

Add: Fourth Floor, No.1 Building (Zhongke Keyi), No.13 North Second Street, Zhongguancun Area, Haidian District, Beijing
P.C.: 100190
Tel: +86–10–62571096, 62571097, 62571098, 62571099
Fax: +86–10–62571096–8200, 62571097–8200, 62571098–8200, 62571099–8200
E-mail: beidaguihua@superplanner.com.cn
http//: www.superplanner.com.cn

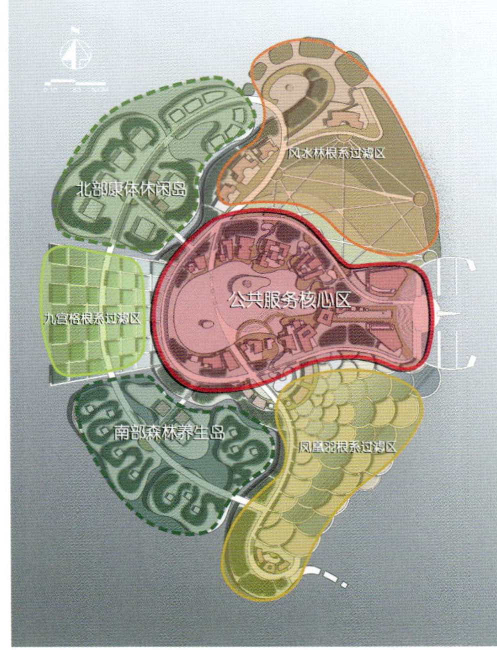

唐山湾三岛旅游区总体规划

该规划最早由乐亭县人民政府托我院编制，是我院承接的三岛系统规划（战略规划、总体规划、岛域控制性详细规划、保护与利用规划、开发利用方案设计）核心工作之一。规划提出把唐山湾国际旅游岛建设成为以假日休闲、文化休闲和商务休闲为先导，以滨海生态观光、海洋休闲渔业、分时度假酒店等为支撑，以海湾生态基础设施和旅游基础设施建设为基础，世界一流的海洋度假旅游目的地、国际经济合作和文化交流的重要平台、全国生态文明建设示范区和唐山市旅游业改革创新的试验区。在规划中，我院项目组坚持以系统规划的理念统领全部规划工作，与战略规划和下层次的控制详细规划进行了无缝对接，无论从理念到方案均得到了甲方各级领导的高度认可，规划工作得到唐山市、河北省的高度重视，使该规划从一个县级主导的项目，逐级上升为省级项目。由于我院在该规划中的出色表现，唐山市三岛开发建设指挥部于2010年授予我院“优秀规划设计单位”奖杯，以表彰我院在三岛开发建设中作出的巨大贡献。

广东佛山“智慧新城”城市设计

通过全球信息化背景下的城市发展要素分析，探讨城市发展优势的要素组合方法，对佛山的区位、交通、生态、产业、土地等要素和总部基地、现代服务中心、创意研发中心、科技孵化中心等创新要素进行组合，提出智慧型总部经济综合体、智慧型云计算综合体、智慧型创意研发综合体、智慧型现代服务综合体、智慧型运动休闲综合体等智慧城市综合体构想。

以智慧型总部创意为目标，智慧型客机服务为驱动，智慧型数字休闲为提升，智慧型生态空间为依托，提出本项目发展的“多元互动”联动机制和祥云智慧新城含义，打造珠三角智慧产业总部基地与智慧城市示范基地的城市发展目标和主题定位。

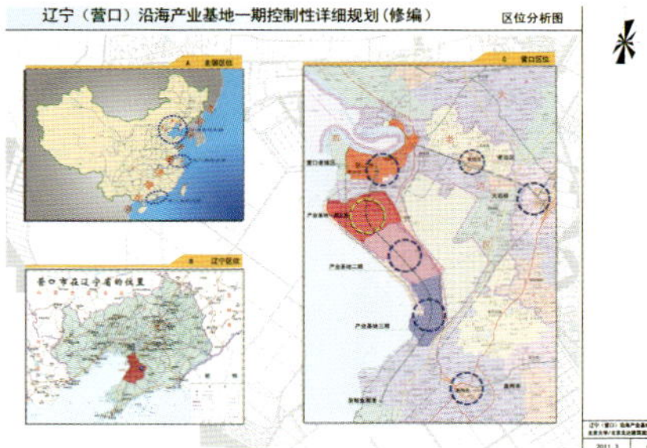

辽宁（营口）沿海产业基地规划国际竞赛

项目地点：辽宁 营口
项目规模：1 600平方千米

为了更好地整合营口滨海地区资源，科学引导辽宁（营口）沿海产业基地的开发建设，营口市政府组织实施了“辽宁（营口）沿海产业基地暨营口滨海经济区发展战略规划国际竞赛活动”。规划范围分为1 600平方千米的滨海经济区、120平方千米的产业基地和15平方千米的两湖地三个空间层次。此次竞赛吸引了国内外44家规划设计单位参与，经过评审，6家单位进入最终竞赛，除了我设计院外，其他的分别为中国城市规划设计研究院、南京大学规划设计院、日本州设计有限公司、美国MAD建筑与城市设计公司和哈工大设计研究院联合体、香港都市规划顾问公司。最终我院的方案获得了一等奖。

通过实施本战略规划提出的以沿海产业基地为“卫星”，以人（口）、港（口）、产（业）、城（市）、生（态）五大要素及其相互关系为“燃料”，以滨海经济区、辽中南城市连绵带，包括东北和环渤海地区在内的东北亚为三级“推进器”的“星箭战略模型”，力争打造“新港口、新产业、新营口”的城市形象，实现新的历史时期营口城市的跨越式发展，将辽宁（营口）沿海产业基地和营口滨海经济区建设成为具有国际竞争力的滨海生态文化科技新城。

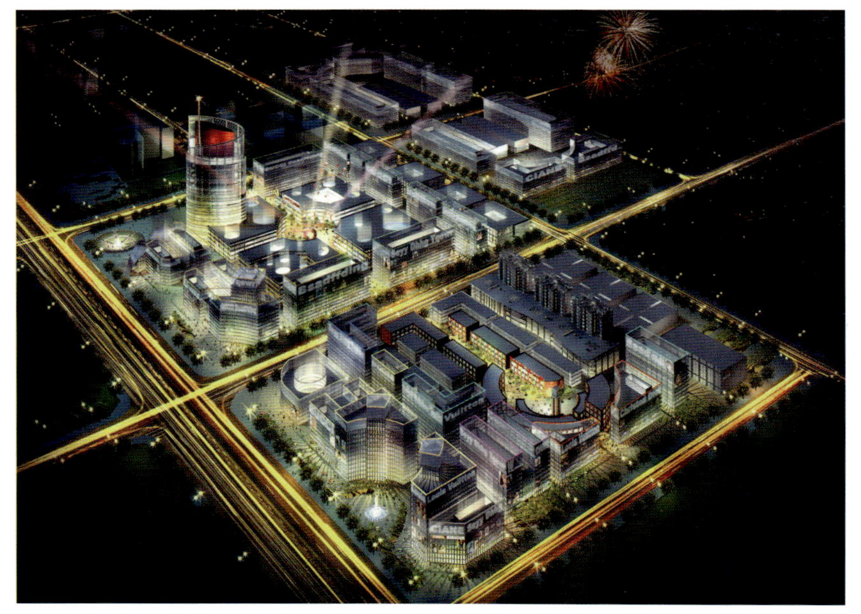

营口开发区圣水河地块修建性详细规划

在系统规划思想的指导下，我院系统编制了鲅鱼圈区从宏观的战略到微观的控详规等规划项目，圣水河地区修规是我院在鲅鱼圈战略规划、总体规划和片区控规的指引下，以系统发展的思路谋划的一个项目，规划以"圣水人家，水墨江南"为主题，设计主旨上坚持弘扬城市精神，推动城市运营，在方案设计上突破对传统意义上城市空间的理解，创造城市新的生活方式，着意于文化的创新与进步，江南情怀的北国演绎，赋予空间以文化灵魂，创造新的生活方式。

山东省烟台市牟平区西部控制性详细规划

规划区位于烟台市牟平区与莱山区的交汇处，牟平区是烟台六大发展组团中重要的城市功能区之一，位于烟台市的东部沿海。规划在规划理念上针对规划区的特色，从理念创新、理论创新、模式创新三个方面指导方案的规划与设计。以“景观都市主义”理论为基础，坚持“上天入地”的系统控制详细规划原则，提升城市策划手段，注重城市文化建设，结合投融资规划，有效地促进了项目的实施。

三亚凤凰国际机场陵水城市候机楼

项目概况

根据海南省国际旅游岛建设计划，未来将高标准建设一批精品度假区如海棠湾、香水湾、清水湾等，随着高端旅游设施的完善和相关产业的聚集，这些区域势必成为三亚高端旅客的集中地。

三亚凤凰国际机场陵水城市候机楼的建设既缓解机场现场保障压力，又为旅客提供了无缝式乘机服务和购物休闲场所。为海南省的旅游事业和相关产业的发展带来了生机。

设计理念

在海南省政府，三亚市政府的大力支持和海航集团的正确领导下，三亚凤凰机场进入了高速发展阶段。为了满足机场的快速发展需求，机场迫切需要建设城市候机楼为旅客提供优质快速的服务和减轻分流压力，于是陵水城市候机楼项目在这种背景下诞生了。

由于海南旅游岛这个特性和陵水县整体经济发展状况，本项目的定位不能和大部分城市综合体类似。应借助整个海南岛和三亚市的旅游优势，在满足候机楼和仓储物流需求的基础上，大力发展旅游度假型产业，例如度假酒店、免税购物、文化消费、休闲娱乐等。在建筑方案上既要体现地域文化特点又要符合国际潮流，在景观塑造上要充分展示海南岛亚热带的景观特点。并且举办一些具有民族风情的特色活动来吸引海内外的游客，促进当地经济的发展。

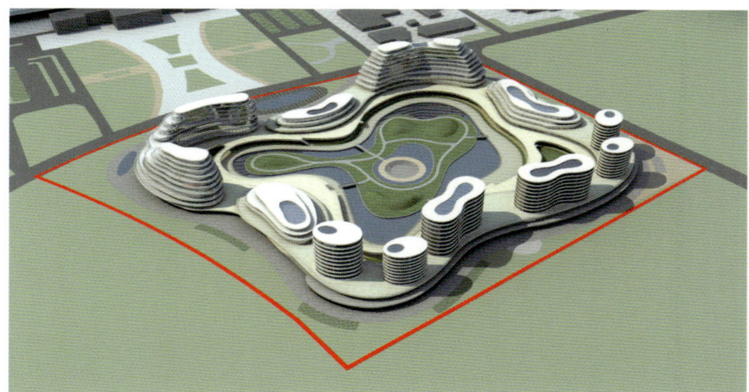

河南省许昌市老城区修建性详细规划

规划坚持“系统规划、提升价值、保护与发展并重、延续利用相结合、可操作性、合理强性控制”等原则。明确以汉魏古都、曹魏风情为定位，并通过借用、更新、再现等规划手法，本着“尊重历史地段传统风貌，整治历史文化环境，提升城市形象品质，合理利用历史遗存，注入地区活力”的规划指导思想，强化古城原有历史文化特色及风貌特征，营建古城特有吸引力，使古城重现昔日辉煌。

针对汉魏古都和曹魏风情主题，规划设计强化了城市设计和文化艺术主题专题设计等，以汉魏风情文化广场为核心，以魏都古城为历史背景，以地方民俗文化为纽带，力求全面展示出许昌丰厚的历史文化底蕴和当代文化建设的时代风采。形成一个符合现代传承和弘扬古城文化的规划作品。

北京建院约翰马丁国际建筑设计有限公司
BIAD & JAMA CO., LTD.

法定代表人：李海南
地址：北京市南礼士路66号建威大厦302–308室
邮编：100045
电话：+86–10–68086516
传真：+86–10–68060528

Legal Representative: Li Hainan
Add: Suite 302-208 Canway Building, No.66 Nan Lishi Road, Beijing
P.C.: 100045
Tel: +86–10–68086516
Fax: +86–10–68060528

吉安文化艺术中心

设计单位：北京建院约翰马丁国际建筑设计有限公司
主持设计师：邹雪红、王 鹏、朱 琳
参与设计师：沈 桢、韩 涛、葛亚萍、曾 劲
项目面积：24 000平方米
主要材料：混凝土、钢、玻璃
项目所在地：江西 吉安
工程造价：3亿元
完工时间：2011年
设计风格：现代

吉安市文化艺术中心以一个近1 300座大剧院为主体，综合了群艺馆、书法创作展览馆、文化休闲服务产业区、文化局机关办公等多种功能，是集观演、会议、展示、休闲、办公于一体的大型综合文化建筑。

建筑设计以整体塑造城南中心区的城市空间为出发点，将用地东侧的市政府前广场引入文化艺术中心建筑内部，形成开放姿态的内广场，营造出连续丰富的城市公共开放空间。

建筑各功能块南北分散布局，具有分散人流、简化流线、降低运营成本的优势。建筑整体造型清雅、灵动，以“弦动庐陵”为主题，用细密钢柱承托出檐深远的大屋顶。细柱的排布形成富有韵律的节奏变化，似被拨动的琴弦，令观者产生有关于音乐、艺术的联想，使人在建筑场所中亲身感受到“看得见的声音”。

新天津站至秦皇岛铁路客运专线秦皇岛站设计方案

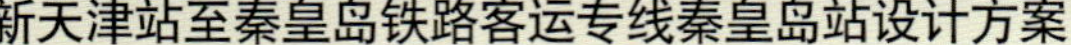

设计单位：北京建院约翰马丁国际建筑设计有限公司
主持设计师：李 琳
参与设计师：杨 凡（建筑设计）、赵 琦（室内设计）
项目面积：（车站建筑总面积）79 800平方米、（站房）19 880平方米、（站台雨棚）55 000平方米
项目所在地：河北 秦皇岛
设计风格：现代

本设计的重要设计理念之一就是以旅客为中心，“以人为本，以流线为主”，建设高效换乘、舒适便捷的现代化综合交通枢纽及良好的城市景观。设计遵循“立体开发、高效换乘，综合布局、配套齐全，环境优先、生态自给”的理念。

中国电能成套设备有限公司教育培训基地

设计阶段：方案设计
设计主持人：李 琳
参与设计师：马 凯
总占地面积：1.3公顷

本项目利用现代建筑技术与现代建筑材料，具有鲜明中国特征（包括中国各地地域特征）与鲜明时代特征、能够表达当代中国文化现代化特征的当代中国建筑，抽象模仿传统建筑特征为主的"形之'象'"的现代建筑与脱离象形的、表达中国文化精神的时代建筑 。结合中国文化精神特别是中国传统美学精神，提取其中重要并适合现代建筑表达的范畴——"崇尚自然"、"中和"、"虚静"、"阴阳"、"模糊"、"气韵生动"、"混沌"等相应的哲学与美学思想，在现代建筑中表达相应的精神或美学特征。

希尔顿烟台大酒店规划设计方案

设计阶段：方案设计
设计单位：北京建院约翰马丁国际建筑设计有限公司
合作单位：美国瑞正国际建筑设计
设计主持人：张彤梅
参与设计师：杨 林、张 扬
建筑规模：400 000平方米

本工程由地面上一栋150米高五星级酒店主楼、一栋150米高5A级办公大楼、六栋100米高公寓塔楼组成，在理念上，建筑与海洋有着密切的联系，赋予曲线美的形态充分利用了本地块靠山面海，得天独厚的位置，将其打造成地标性的建筑。拥有曲线美的外墙采用玻璃材质，强调了流动感的概念，使得所有建筑的外形上更加和谐，侧面矩形外墙材质为石材和玻璃相结合，与富有曲线美的玻璃幕墙形成对比。玻璃材质与曲线造型的玻璃幕墙品质相同。为了强调与玻璃表面的对比关系，石材的颜色为白色，略带点灰。不同材质颜色的建筑与海水浪花相互反射、交相辉映，在滨海形成一条壮美亮丽风景线。

ZhengDong International Architectural & Engineering Design (Beijing) Co., Ltd.

北京正东国际建筑工程设计有限公司

北京正东国际建筑工程设计有限公司成立于1998年，是经中华人民共和国建设部批准成立的建筑工程设计企业，拥有建筑行业（建筑工程）甲级和城市规划乙级资质（资质证书编号：A111010311；规划设计乙级【京】城规编第082044）。一直从事建筑工程规划、设计与咨询、城市及园林景观设计等业务，可从事资质证书许可范围内相应的建设工程总承包业务以及项目管理以及相关的技术与管理服务，并可承担相应范围相应等级的建筑装饰工程设计、建筑幕墙工程设计、轻型钢结构工程设计、建筑智能化系统设计、照明工程设计和消防设施工程设计等专项工程设计业务。

公司自成立以来，以积极的姿态活跃于设计领域，已完成600多项不同类型的建筑设计任务。从整体设计的原则出发，在对社会、城市、技术及公众需求的综合考虑上，在对地方性大众文化的充分了解上，公司力求引导与提升市场对建筑艺术的理解，创作了不少优秀作品。公司能敏锐感知客户的需求，协助客户追求卓越的市场业绩，公司鼓励创新，致力于专业领域的探索，不断提升面对复杂问题寻找多种解决方案的能力，力求让综合实力和先进技术产生最现实的社会效益。公司提供从策划、方案、初设到施工图的整体建筑设计服务，且一直在为客户创造超越客户期望的产品和服务而努力。

ZhengDong International Architectural & Engineering Design (Beijing)Co.,Ltd. which was established in 1998, is approved by the PRC Ministry of Construction. The construction engineering design enterprise has the Grade A construction industry (Engineering) and Grade B urban planning qualification (Certificate No. A111010311; Grade B planning design [Beijing], No. 082044). Has been engaged in construction planning, design and consulting, urban and landscape design business, it can engage in the corresponding qualification certificates to the extent permitted construction general contracting and project management services and related technical and management services, and undertake the appropriate level of the corresponding range of architecture decoration engineering design, architectural curtain wall engineering design, light steel structure engineering design, building intelligent systems design, lighting design and fire engineering design and other specialized facilities engineering services.

Since its foundation, with a positive attitude active in design field, the company has completed over 600 different types of architectural design tasks. With the principle of the overall design, taking the community, city, technology and public demand into account, fully understanding the popular culture of the local, the company sought to guide and enhance the understanding of the market for architectural art, created a lot of good works. The company keenly compreneds customer needs, assists our customers with market performance excellence. The company encourages innovation, dedicates to the professional field of exploration and improves the ability of facing a variety of complex problems to find solutions, strives for producing the real social benefits through the overall strength the advanced technology. The company provides from the planning, programming, preliminary design to construction drawings of the overall architectural design services, and has been making efforts for our clients exceed customer expectations and service.

地址：北京市西城区北三环中路乙6号伦洋大厦11层
电话：+86-10-62018703
传真：+86-10-62017605
邮箱：bjzdiad@zdiad.com
网址：www.zdiad.com.cn

Add: the 11th floor, Lunyang Building, No.6 North 3rd Ring Middle Road, Xicheng District, Beijing
Tel: +86-10-62018703
Fax: +86-10-62017605
E-mail: bjzdiad@zdiad.com
Http:// www.zdiad.com.cn

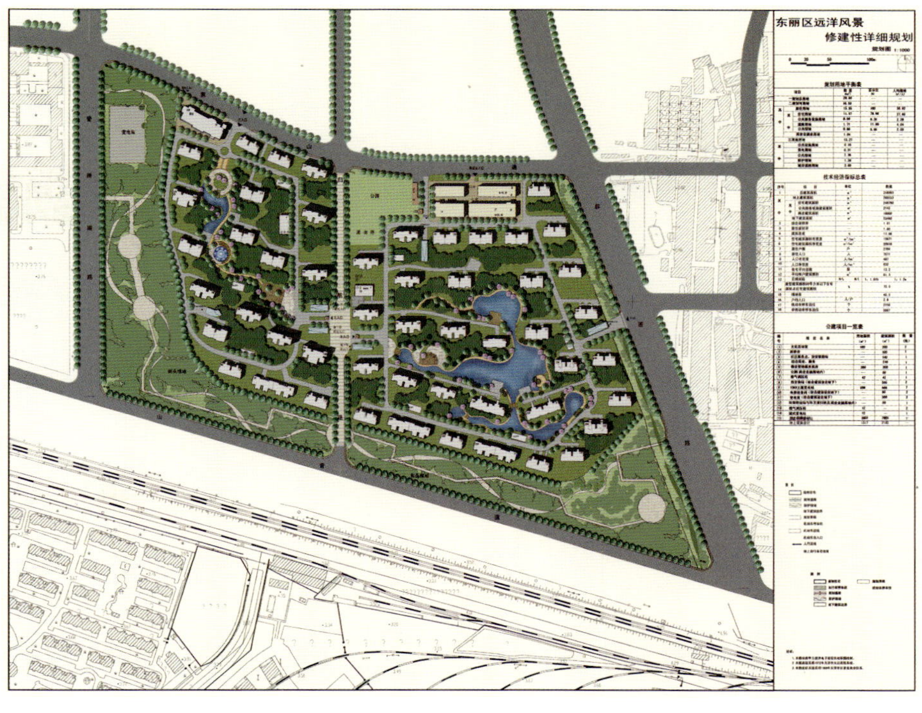

天津远洋风景

项目地点：天津
用地面积：240 000平方米
建筑面积：300 000平方米
设计时间：2010年

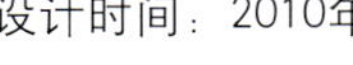

石家庄星湖国际

项目地点：河北 石家庄
用地面积：250 000平方米
建筑面积：900 000平方米
设计时间：2010年

鄂尔多斯鼎誉龙城

项目地点：内蒙古 鄂尔多斯
用地面积：34 000平方米
建筑面积：140 000平方米
设计时间：2011年

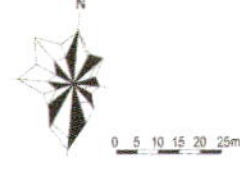

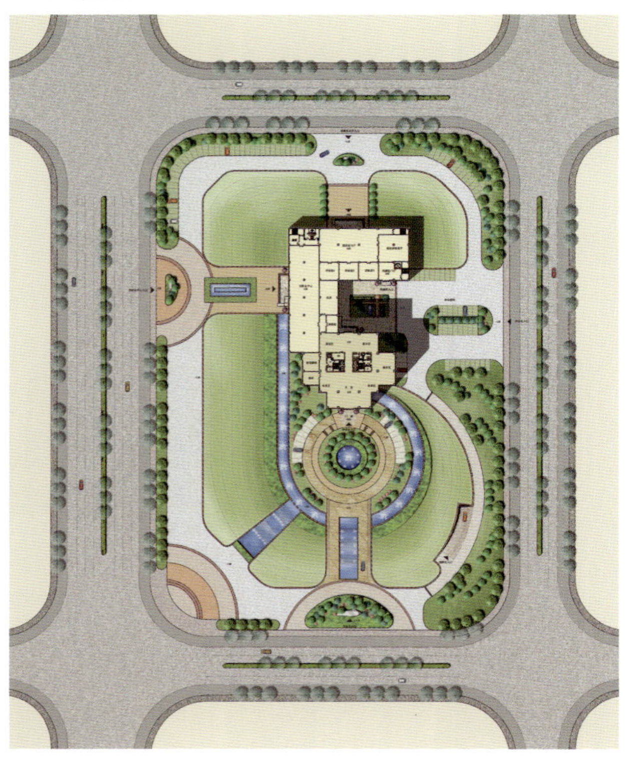

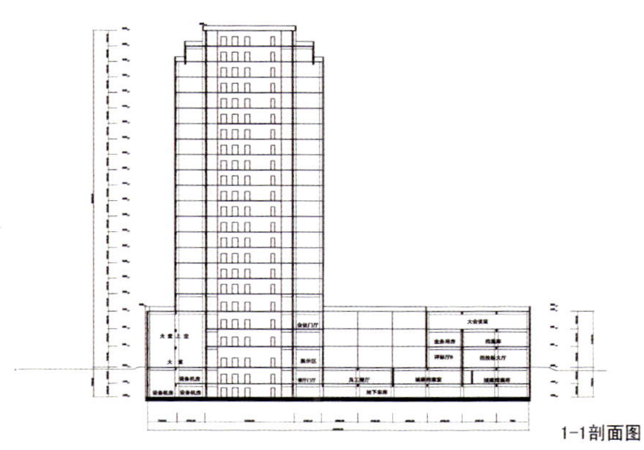

1-1剖面图

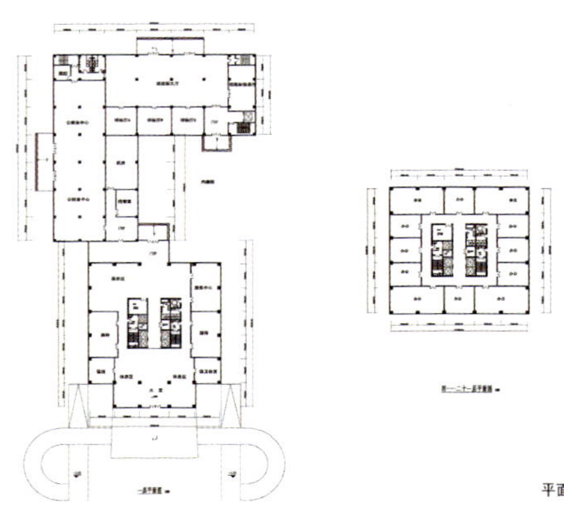

平面图

竖向功能分区示意图

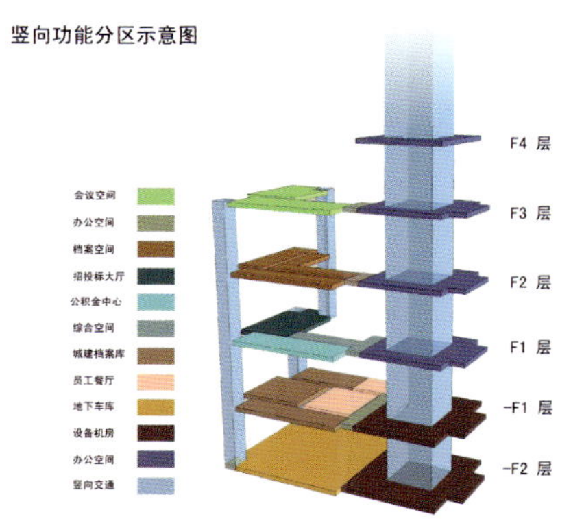

点式楼的布局形式

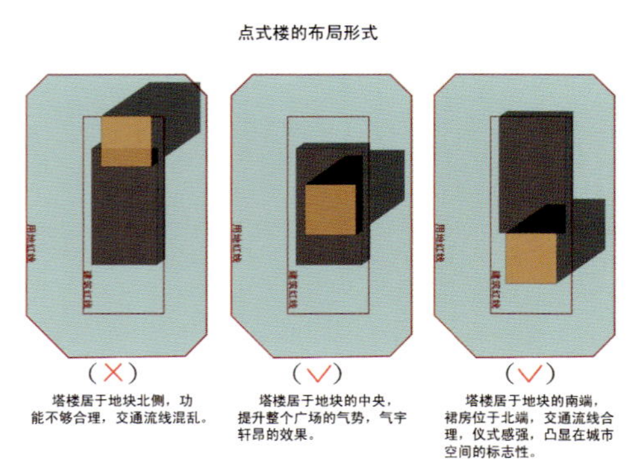

塔楼居于地块北侧，功能不够合理，交通流线混乱。

塔楼居于地块的中央，提升整个广场的气势，气宇轩昂的效果。

塔楼居于地块的南端，裙房位于北端，交通流线合理，仪式感强，凸显在城市空间的标志性。

通辽市住房和城乡建设服务中心

项目地点：内蒙古 通辽
用地面积：25 000平方米
建筑面积：45 000平方米
设计时间：2010年

3H ARCHITECTS PARTNERSHIP BEIJING

北京三和创新建筑师事务所

城市设计　建筑设计　室内设计　景观设计

北京市海淀区阜成路42号中裕商务花园33A二层
电话: +86-10-88154421
传真: +86-10-88154431
邮箱: hjhm@vip.sina.com

Add: 2nd Floor, 33A, Zhongyu Business Garden,
No.42 Fucheng Road, Haidian District, Beijing
Tel: +86-10-88154421
Fax:+86-10-88154431
E-mail: hjhm@vip.sina.com

北京三和创新建筑师事务所（3H）是由胡绍学、何玉如、黄星元三位国家设计大师共同发起并于2007年3月成立的建筑设计专项事务所，致力于全球环境中城市及建筑设计的创新研究与实践，追求人与建筑、城市及自然的和谐，追求设计与历史、文化及社会的共鸣。3H已经在城市概念规划及空间形态设计、公共建筑设计、居住社区规划和室内设计等领域，开始展现其理论创新的勇气、综合把握的能力及不懈探索的精神。

3H Architects Partnership, Beijing was sponsored by Hu Shaoxue, He Yuru and Huang Xingyuan and founded in March, 2007. We devote ourselves to creative research and practice of city and architecture design in global context, working for the harmony between human and architecture, city & nature, and for the resonance between design and history, culture & society. 3H has begun to show its courage of theory innovation, its integrated abilities and its exploring spirit.

中国武钢博物馆

合作单位：清华大学建筑设计研究院
建设地点：湖北 武汉
建筑规模：12 000平方米
设计时间：2008年
项目状况：已建成

台州城市规划展览馆

合作单位：清华大学建筑设计研究院
建设地点：浙江 台州
建筑规模：10 000平方米
设计时间：2010年
项目状况：建设中

杭州阮家桥项目

合作单位：北京城建设计研究总院
建设地点：浙江 杭州
建筑规模：200 000平方米
设计时间：2009—2010年
项目状况：建设中

烟台福山宾馆

合作单位：烟台市建筑设计研究股份有限公司
建设地点：山东 烟台
建筑规模：27 000平方米
设计时间：2008–2010年
项目状况：建设中

长岛长园宾馆

合作单位：中房集团建筑设计有限公司
建设地点：山东 烟台
建筑规模：10 000平方米
设计时间：2008–2010年
项目状况：已建成

苏丹总统府室内设计

合作单位：浙江省建工建筑设计院
建设地点：苏丹
建筑规模：18 000平方米
设计时间：2009–2011年
项目状况：建设中

都市意匠城镇规划设计(北京)中心

Urban Design and Technology Limited

都市意匠城镇规划设计（北京）中心创建于2007年，专业团队目前由资深城市规划师、景观规划师、建筑师、经济规划师及生态规划师构成，服务项目范围覆盖：发展战略规划、总体概念性规划、城市设计、旅游度假区规划、景观规划、修建性详细规划、平面设计等。

都市意匠有一群对规划和设计怀着至诚热情的专业工作者，国际化的视野、创新思维以及对中国文化的理解，使得我们对城市规划的决策、开发、保护，以及提倡可持续发展、生态敏感和社会完整性的城市规划和设计提出具有前瞻性和可操作性的方案。

回顾2010年，我们对于沟域经济发展、山地旅游区，以及旅游小镇的开发深有体会。都市意匠的团队以“发现问题、解决问题”的工作方法，创造具有当地独特价值的设计，以强调生态、经济发展和空间的个性体验。

网址：www.urban-dtek.com
电话：+86–10–65525201
传真：+86–10–65525203

长城脚下的旅游小镇——天津下营镇

项目位置：天津 蓟县
用地规模：205.21平方千米

福灵旺岛山地生态主题游乐园

项目位置：天津 蓟县
用地规模：237.21公顷

Beijing Power Era Architectural Design Co., Ltd.

北京大国时代建筑设计有限公司

北京大国时代建筑设计有限公司成立于2002年6月，前身为北京大国时代建筑设计事务所，是一家专业从事工业与民用建筑设计的建筑设计公司。公司业务主要涵盖各类建筑及景观的规划设计、方案设计、施工图设计以及投资咨询、可行性研究报告等方面。

多年来，公司始终坚持"技术先进、质量第一、管理科学、服务周到"的方针，积极在建筑行业开拓创新，始终以优秀的质量、优良的服务、优惠的价格服务于新老客户，我们一直认为：设计一个负责任的作品既是我们的理想，更是每个建筑师的使命，能够获得您的满意是我们最大的骄傲。我们期待您的青睐，与您携手共创美好未来！

Beijing Power Era Architectural Design Co., Ltd. was established in June 2002, formerly known as the Beijing Power Era Architecture Design Firm, and now it is a professional architectural design company engaged in industrial and civil architecture design. Its business mainly involves various types of planning and design for construction and landscape, program, architectural drawing, investment consulting, as well as feasibility studies, etc.

Over the years, the company adheres to the principle of "technologically advanced, first class quality, scientific management, considerate service", positively innovating in the construction industry by virtue of excellent quality, excellent service, and preferential prices to serve customers both new and old equally. We believe all the time: to design a responsible work is not only our aspiration but also the mission of each architect; meanwhile, to won your satisfaction is our biggest pride. We look forward to your favored, and wish to create a better future hand in hand with you.

地址：北京市海淀区蓝靛厂南路55号金威大厦
邮编：100029
电话：+86-10-88859729 88859799
邮箱：Daguoshidai@163.com
网址：www.daguoshidai.com

Add: North Anwar Building, 2nd Floor, Cherry Street West, No. 8, Chaoyang District, Beijing
P.C.：100029
Tel: +86-10-88859729 88859799
E-mail: Daguoshidai@163.com
Http: //www.daguoshidai.com

胜利小区

规划用地面积：353 000平方米
总用地面积：267 800平方米
总建筑面积：309 969平方米
住宅建筑面积：267 572平方米
商业建筑面积：26 738平方米
公建面积：4 895平方米
车库面积：10 764平方米
容 积 率：1.2
绿 化 率：35%

巨泰房地产华典佳苑

规划用地面积：144 970.97平方米
建设用地面积：105 325.24平方米
总建筑面积：162 891.80平方米
商业建筑面积：69 708.12平方米
公寓建筑面积：31 358.62平方米
酒店建筑面积：22 803.8平方米
容 积 率：2.496
绿 地 率：31.4
建筑密度：29.8%

杭锦旗恒利国际锦河苑

用地面积：218 231.00平方米
总建筑面积：532 818.8平方米
商业面积：107 456.9平方米
住宅面积：289 253.7平方米
容 积 率：1.94
绿 化 率：35%
建筑密度：27.1%

东海晶都华府住宅小区

用地面积：68 007.00平方米
总建筑面积：199 123.586平方米
商业建筑面积：26 614.882平方米
住宅建筑面积：133 573.391平方米
酒店建筑面积：21 275.804
容 积 率：2.704
绿 化 率：32.8%

准格尔旗沙圪堵镇锦丽苑小区

规划用地面积：110 996.279平方米
总建筑面积：248 217.280平方米
商业建筑面积：41 432.58平方米
住宅建筑面积：131 779.36平方米
酒店建筑面积：19 934.4平方米
容 积 率：1.89
绿 化 率：32.8%
建筑密度：25.2%

河北赞皇住宅小区

用地面积：181 414.736平方米
总建筑面积：261 070.49平方米
商业面积：70 714.22平方米
住宅面积：190 356.27平方米
容 积 率：1.43
绿 化 率：35%
建筑密度：23.9%

易兴住宅小区

规划用地面积：122 527.4平方米
建设用地面积： 101 362.7平方米
总建筑面积：279 045.92平方米
地上建筑面积：252 233.18平方米
地下建筑面积：26 812.74平方米
容 积 率：2.49
绿 化 率：33%
建筑密度：24.5%

CSA 中天伟业
做建筑设计专家

中天伟业（北京）建筑设计集团(甲级)

CHINA SKY (BEIJING) ARCHITECTURE GROUP

中天伟业（北京）建筑设计集团成立于1995年，是经由建设部部属国有企业、社团企业改制重组而成的股份制企业。具有建设部颁发的建筑工程设计甲级资质以及工程监理资质。集团注册资金1 000万元。中天伟业现拥有150余名工程设计人员，其中一级注册建筑师、注册规划师、一级注册结构工程师、注册设备工程师及注册电气工程师数十名，并已通过中国质量认证中心ISO 9001质量体系认证。

中天伟业专业齐全，人才济济，不仅具有原国有企业雄厚的技术实力和高度的社会信誉，同时还拥有国内外优秀建筑设计企业所具有的核心竞争力，以及独特的客户意识和优良的运营管理模式。

中天伟业坚持“质量第一、信誉第一、服务第一”的原则和“为客户创造价值”的理念，广泛涉及建筑创作领域，为客户提供优质服务。从区域规划到城市设计，从办公楼宇、星级酒店到商业地产，从文教建筑、医疗建筑到体育建筑，从高档别墅到居住建筑，从旧城改造到仿古建筑，再从室内装饰设计到环境景观园林设计，中天伟业无不创作出优秀的设计精品，在业内和市场上建立了良好的口碑。中天伟业于2009年荣获中国银行北京分行年度（2009—2012）设计单位。

中天伟业凭借合理的组织结构、科学的管理方法、雄厚的技术实力以及先进的服务模式，为客户提供包括区域规划与城市设计、建筑与工程设计、室内装饰设计、景观园林设计、结构设计、给排水、采暖空调、电气照明、楼宇自控、弱电智能设计、房地产前期策划、工程项目的可行性研究、咨询，项目管理、工程监理等全方位的服务。

中天伟业以北京为基础，业务辐射全国，同时为拓展业务需要，在乌鲁木齐、兰州已设立分公司，并陆续将在上海、天津、重庆、沈阳、山东等地设立办事处或分公司。中天伟业的目标是立足北京，面向全国，走向世界，同时，中天伟业亦与世界著名的美国雅马萨奇设计公司开展紧密和广泛的合作，努力设计出中国最好的建筑。

作为低碳绿色建筑设计的先行者，中天伟业长期致力于绿色建筑的设计和探索，成绩卓越，2010年被《中国企业报》评为“中国建筑设计绿先锋”。2010年被中国建筑文化中心评为“中国建筑设计绿色百强设计机构”。2011年在北京市规划委员会组织的全球招标《北京市绿色建筑设计标准》大纲编制竞赛中，中天伟业荣获三等奖，最终成为《北京市绿色建筑设计标准》的编制单位。

China Sky (Beijing) Architecture Group, founded in 1995, is a corporate enterprise restructured and recombined from the state-owned enterprise and community enterprise affiliated to Ministry of Construction. It owns the Class A qualification of constructional engineering design issued by Ministry of Construction as well as qualification of construction supervision. The registered capital of the group is 10 million Yuan. There are more than 150 engineering design staffs, among which tens of Class 1 registered architects, registered planners, Class 1 registered structural engineers, registered facility engineers and registered electrical engineers. It has passed quality system certification of China Quality Certification Center ISO9001.

China Sky has got complete majors and plenty of talents. It possesses not only solid technical strength and high social reputation of the former state-owned enterprise, but also the core competence of the excellent constructional design enterprises home and abroad, as well as unique client consciousness and eminent operation management mode.

China Sky sticks to the principle of “Quality first, reputation first and service first” as well as the concept of “create value for courtesy”. It involves widely the domain of architectural creation and provides the clients with high-quality service. It has created excellent design works from regional planning to urban design, from office buildings and starred hotels to commercial properties, from cultural and academic constructions, medical constructions to sports constructions, from top-grade villas to residential architectures, from old town reformation to pseudo-classic architectures, from indoor decorative design to environmental landscape design and gained great public praises in the industry and in the market. In 2009 it has won the award of Annual (2009-2012) Design Unit of Beijing Branch of China Bank.

By virtue of reasonable organizing structure, scientific managing method, solid technical strength and advanced service mode, China Sky provides the clients with all-around services as the following: regional planning and urban design, constructional and engineering design, indoor decorative design, landscape design, structural design, water supply and drainage, heating and air conditioning, electric lightening, building automation, weak current intelligent design, real estate pre-job planning, feasibility research and consultation of engineering projects, project management and construction supervision etc.

Based in Beijing, China Sky exposes its business all around the nation. For the purpose of extending business, it has established branches in Urumqi and Lanzhou, and will successively establish agencies or branches in Shanghai, Tianjin, Chongqing, Shenyang and Shandong etc. It has got the goal of keeping a foothold in Beijing, facing the whole nation, and stepping into the world. Meanwhile it has carried out intimate and wide cooperation with world-famous Yamasaki Design Company of USA, endeavoring to design the best architectures in China.

As a pioneer of low carbon green constructional design, China Sky devotes itself to long-term design and exploration of green architecture and has got outstanding achievements. In 2010 it has been rated by *China Enterprise News* as “Green Pioneer of China Constructional Design”. In 2010, it was entitled China Green Building Design The Top 100 Design Agencies by China Architectural Culture Center. In 2011, Beijing Municipal Planning Commission's Global tender *Beijing Green Building Design Standards* outline the preparation of the race, transit Albert won the third prize, eventually became *Beijing Green Building Design Standards*, the establishment of units.

地址：北京市海淀区车公庄西路乙19号华通大厦B座9层
邮编：100048
电话：+86-10-88018011/8012/8569
传真：+86-10-88018011/8012/8569转202
邮箱：csabj@163.com
网址：www.csabj.com

Add: 9th Floor of Huatong Mansion Building B, Yi No.19 of Chegongzhuang West Road, Haidian District, Beijing
P.C.: 100048
Tel: +86-10-88018011/8012/8569
Fax: +86-10-88018011/8012/8569 transferring to 202
E-mail: csabj@163.com
Http: //www.csabj.com

城市规划

成都双流县彭镇中心区改造

昆明泛亚国际冷链物流基地

湖北黄石汪仁镇低碳经济示范区

居住建筑

北京田村雪芳园

北京孙河未来家园

北京中堂五期工程

昆明永和花园

新疆喀什重庆城

辽宁葫芦岛宏运奥园

新疆伊宁国电家园

河北宣化中宣嘉城项目

新疆喀什唐城国际

唐山乐亭水悦华庭项目

昆明绿辰商业中心项目

昆明俊发商业建筑项目

北京马驹桥镇安置楼

绿色建筑

北京通用时代国际中心办公楼

神华可再生能源展示中心

新疆唐布拉旅游度假项目

新疆阿克苏旅游度假村皇家五号

北京汇江大厦

北京曙光计算机科研楼

北大资源生命科技园

山西梵王寺煤矿办公区建筑群

青海格尔木民族文化中心

北京现代艺术馆

中冶时代北京办公中心

公安部北京边防局基地

郑州兴亚建国酒店

首师大附中西校区

北京奥北产业园

天津中企仓储公司办公楼

室内设计

北京五环大酒店

北京五环大酒店

北京珠宝城

国家电网保定电力学校综合教学楼

北京兴隆公寓

北京兴隆公寓

北京珠宝城

中国精密机械进出口总公司办公楼

中国精密机械进出口总公司办公楼

昆明宝善大酒店改造

昆明宝善大酒店改造

北京曙光计算机科研楼

北京曙光计算机科研楼

公安部北京边防局基地

公安部北京边防局基地

4–5 西安荣华 北经城

4-5 Ronghua North Town, Xi'an

6 青白江 经典上城

6 Classic Uptown, Qingbaijiang District, Chengdu

7

9

7–8 四川成都 西南物流中心二期

7-8 Southwest Logistic Center Phase II, Chengdu, Sichuan

9–10 四川首飞 首航欣城

9-10 Shoufei Shouhang Xincheng, Sichuan

青岛原创工程设计有限公司是建设部批准的具有建筑工程甲级资质的综合性设计公司，并已通过IS09001：2000版质量管理体系认证，青岛原创工程设计有限公司拥有建筑、规划、结构、给排水、暖通空调、电气、装饰等各类专业大批具有设计实力和创新能力的优秀设计师，并拥有先进的现代化专业设施和设备。具有较强的方案创作能力、施工图设计能力、装饰景观施工能力。在长期以来的工作中一向注重质量与诚信的结合，享有良好的社会声誉，是一支在市场经济中诞生、具有现代经营管理理念和超前意识的生力军。

近年来，公司秉承一贯的精品设计理念，创作设计了大量的居住、教育、医疗、公共建筑及花园式工业厂区作品，尤其擅长总体设计协调，在力求规划合理，交通组织、空间组织流畅，富有韵律和景观特色的前提下，竭尽所能为建设单位增加和提升开发的社会价值和经济价值，并在工程设计中善于总结，积累大量的设计经验。许多项目以其优良的设计品质、完善的后期服务，赢得了广大客户的信赖。

我们公司秉承创新的设计理念，在功能、经济、美观的基础上，奉献具有长久生命力和社会价值的设计作品。

我们的经营理念是：“诚信,互利,双赢”
我们的管理理念是：“以人文本,科学管理”
我们的企业精神是：“团结、创新、开拓、进取”

青岛原创工程设计有限公司愿同各界人士广泛合作，共求发展，同创辉煌！

Qingdao Yuanchuang Design Co., Ltd. is a comprehensive design company with class-A construction qualification granted by the Ministry of Construction, and has adopted the IS09001:2000 Quality Management System. It has a group of excellent designers with design and creation capabilities in the fields of building, planning, structure, water supply and drainage, heating, ventilation and air-conditioning, electrical, decoration and others, with advanced modern specialized facilities and devices. It has strong conception creativity, construction drawing design capability, decorative and landscaping construction capabilities. For a long time, it is paying great attentions to the combination of quality and credit, with good social reputation, it is a new force of modern operation and management ideas and avantgarde senses, born in market economy.

Over recent years, we follow the exquisite design idea, to create and design a large number of residential, teaching, medical, public, building and industrial park works, especially proficient in general plan coordination; in addition to reasonable planning, smooth traffic and space organization, rich rhythm and landscape features, we try the best to increase and promote socioeconomic values for the developers; we are ready to conclude from project designs, and accumulate a great many design experiences. Many projects with good design quality and complete after services have won vast trust from customers.

We follow the creative design idea, to contribute long-living design works with social value on the basis of functionality, cost-effectiveness, beauty.

Our operating philosophy is “credit, mutual benefit, win-win”.
Our management philosophy is “human-based, scientific management”.
Our entrepreneurship spirit is “union, creation, expansion, progress”.

Qingdao Yuanchuang Design Co., Ltd. is willing to cooperate with all social communities, in seek for common development, and common victory!

地　　址：青岛市崂山区山东头路58号盛和大厦1号楼1-3层
联系电话：+86-532-83950136、83950137、83950138、83950139
传　　真：+86-532-83950117
邮　　箱：qingdaoyc@vip.sina.com
网　　站：http://www.qdyc-arch.com

Add: Floor 1-3, Building 1, Shenghe Building, 58 Shandongtou Road, Laoshan District, Qingdao
Tel: +86-532-83950136, 83950137, 83950138, 83950139
Fax: +86-532-83950117
E-mail: qingdaoyc@vip.sina.com
Http://www.qdyc-arch.com

1—3. 烟台滨海休闲娱乐中心

4—5. 肥城君悦国际花园

6. 某实验楼

7. 青岛畜牧局

青岛原创
QINGDAOYUANCHUANG

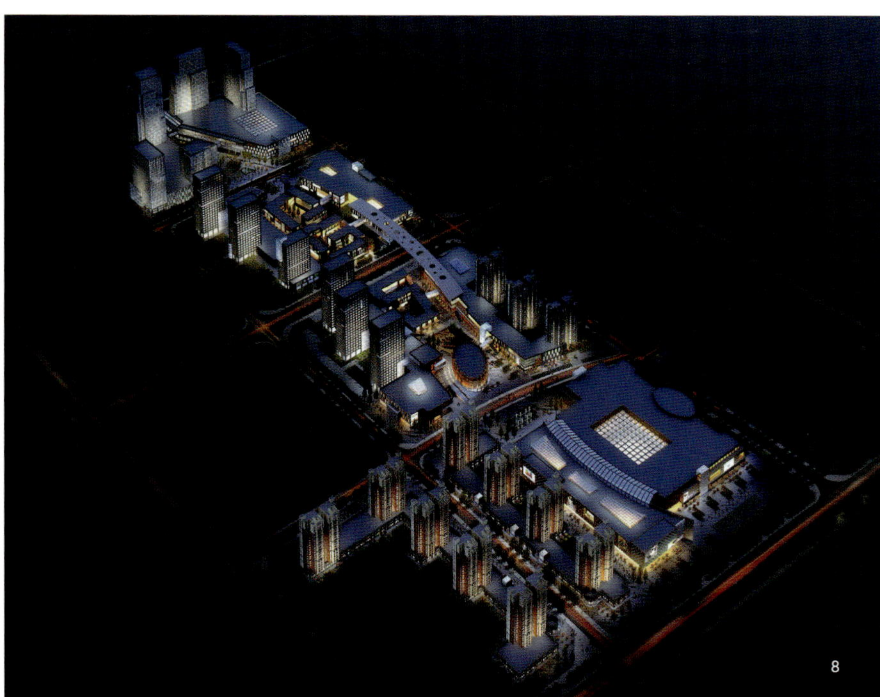
8

10

9

11

8–9. 济宁某购物中心

10–11. 肥城市某文化中心

青岛原创
QINGDAOYUANCHUANG

12. 某中学

13. 青岛沙子口小学

14. 某电视演播中心

15. 某职业学校

16. 肥城某培训中心

青岛原创 QINGDAOYUANCHUANG

17—18.三尚社区　19—20.临沭沭中名苑　21—22.青岛金沙滩别墅　23.万科即墨南杨头社区　24.泰安某温泉别墅

青岛原创
QINGDAOYUANCHUANG

25. 青岛某综合楼

26. 泰安环山路综合社区

27. 泰安环山路改造局部

28. 肥城中超大厦

29. 某城市综合体

30. 泰安某酒店

精心设计 热情服务
持续改进 客户满意

建筑工程甲级资质 A137012305
城市规划乙级资质 [鲁]城规编第（102B06）号
风景园林乙级资质 A237012302
ISO9001质量管理体系认证

青岛北洋建筑设计有限公司

初创于1995年，多年来我们与山东半岛的经济建设共同成长，从小到大，从弱变强；在此山东半岛贯彻国家蓝色海洋经济战略奋进转型之际，我们将抓住机遇，努力创新求变，立足市场、服务客户、创造精品、奉献社会。

Qingdao Beiyang Architecture Design Co., Ltd.

Founded in 1995, with the Shandong Peninsula over the years we build a common economic growth, from small to large, from weak to strong. At the time for economic restructuring of the Blue Sea implementation in Shandong Peninsula, we will seize the opportunity to innovate change, base on the market, customer service, create high quality, and contribute to society.

地址：中国山东青岛市香港东路23号
中国海洋大学浮山校区图书馆A区
邮编：266071
电话：+86-532-86669156
传真：+86-532-86669155
网址：www.qdbysj.com
邮箱：qdbysj@yahoo.cn

Add: No. 23 Eastern Hong Kong Road, Qingdao, P.R.China, 266071
Area A of Library, Fushan Campus, Ocean University of China
Tel: +86-532-86669156
Fax: +86-532-86669155
Http: //www.qdbysj.com
E-mail: qdbysj@yahoo.cn

1

2

3

1. **文化项目**/莱西市民活动中心
 · 竞标第一名 （山东 莱西）
2. **文化项目**/中国水文科技馆
 · 竞标第一名 （山东 青岛）
3. **文化项目**/胶南四馆
 · 竞标第一名 （山东 青岛）

2

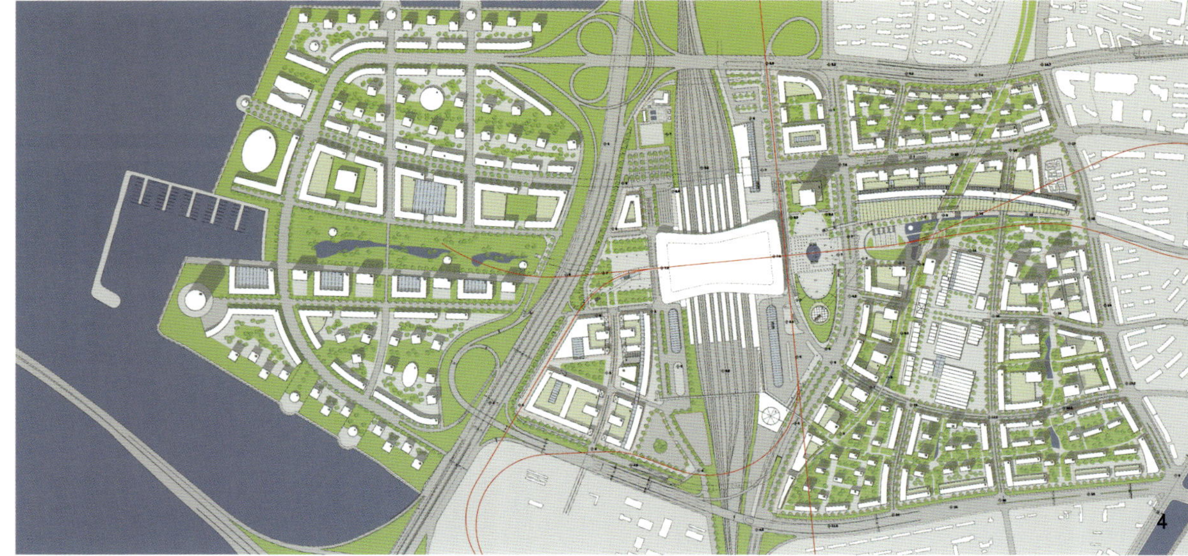
4

6

8

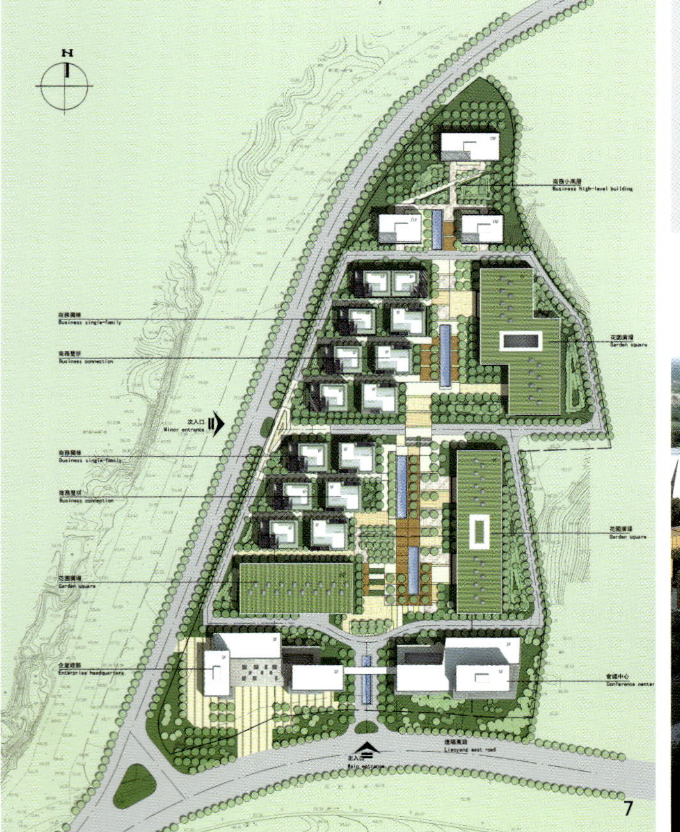
7

4. **城市设计**/青岛交通商务核心区 （山东 青岛）
• 中标实施方案
5. **概念规划**/国信康庭嘉苑 （山东 青岛）
6. **概念规划**/福日沙子口项目 （山东 青岛）
7. **概念规划**/福日高尔夫项目 （山东 青岛）
8. **园区规划**/青岛国家大学科技园 （山东 青岛）
9. **园区规划**/高密高新产业园区中心 （山东 潍坊）

9

10

10. **办公项目**/董家口港区行政办公楼 （山东 青岛）
11. **办公项目**/青岛新华锦集团总部 （山东 青岛）
12. **办公项目**/电业局漳州路办公楼 （山东 青岛）
13. **办公项目**/石油大厦 （山东 青岛）
  - 青岛市优秀工程勘察设计一等奖
  - 国家优质工程银质奖
  - 山东省优秀工程勘察设计二等奖
  - 山东省第二届城市设计精品工程三等奖
14. **酒店项目**/海上海大酒店 （山东 青岛）
15. **校园项目**/烟台二中高新区分校 （山东 烟台）
  - 竞标首选方案

11

12

15

13

14

15

16

17

18

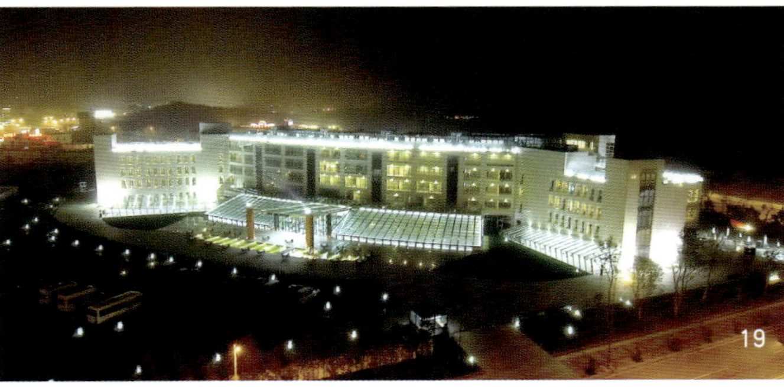
19

16. **校园项目**/青岛二中分校　（山东　青岛）
17. **校园项目**/高密银鹰文昌中学　（山东　潍坊）
- 青岛市2010年度优秀工程勘察设计二等奖

18. **别墅项目**/中信旅游璞玉岛　（山东　青岛）
19. **综合项目**/弄海园二期　（山东　青岛）
- 青岛市优秀工程设计一等奖
- 山东省优秀工程设计三等奖

20. **公寓项目**/青岛体育中心运动员公寓　（山东　青岛）
- 山东省优秀建筑设计一等奖
- 青岛市优秀工程勘察设计一等奖
- 山东省优秀工程勘察设计二等奖
- 蓝星杯威海国际建筑设计大奖赛优秀奖

20

20

21

22

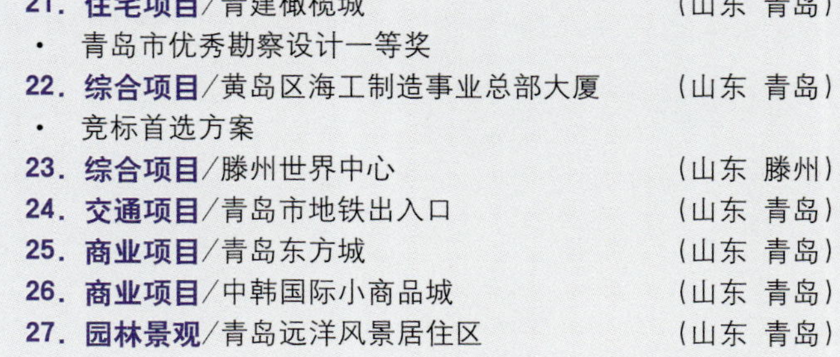
21. **住宅项目**/青建橄榄城 （山东 青岛）
- 青岛市优秀勘察设计一等奖

22. **综合项目**/黄岛区海工制造事业总部大厦 （山东 青岛）
- 竞标首选方案

23. **综合项目**/滕州世界中心 （山东 滕州）
24. **交通项目**/青岛市地铁出入口 （山东 青岛）
25. **商业项目**/青岛东方城 （山东 青岛）
26. **商业项目**/中韩国际小商品城 （山东 青岛）
27. **园林景观**/青岛远洋风景居住区 （山东 青岛）

23

24

25

26

27

山东建大 建筑规划设计研究院

SHANDONG JIANZHU UNIVERSITY ARCHITECTURE & URBAN PLANNING DESIGN INSTITUTE

山东建大建筑规划设计研究院成立于1960年，原名为山东建筑学院设计室，1985年、1992年分别改为设计研究所、设计研究院。1997年国家建设部审定批准为建筑工程设计甲级资质。目前设计研究院已具有建筑工程设计、城市规划、工程咨询三种甲级资质以及施工图设计文件审查机构认定书（建筑工程一类——钢结构），同时具有风景园林、市政工程设计两种乙级设计资质。

设计研究院下辖办公室、生产经营部、技术部、财务室四个行政管理部门，以及设计一分院、设计二分院、城市规划分院、建筑方案创作研究所、咨询审图中心等5个生产部门。主要从事民用与工业建筑设计、城市规划、风景园林以及相关专业的工程设计。全院共有员工140余人，其中系统内人员80余人。现有国家一级注册建筑师、注册结构工程师22人；国家注册规划师、注册公用设备工程师、注册电气工程师、注册咨询工程师、注册造价工程师26人。香港注册建筑师、结构工程师学会会员5人。设计人员中具有中高级技术职称人员约占60%以上。经过50余年几代人的努力，我院已经成为一所集设计、教学与科研相结合，技术实力雄厚，管理先进，在省内外有良好影响和较高知名度的设计研究单位。

20世纪90年代以来，该院进入较快发展时期，一支颇具实力的中青年技术骨干队伍已经形成，依托山东建筑大学强大的师资优势，充分发挥自身的勤奋和智慧，踏踏实实、锐意创新，先后承担了数百项较大工程项目的建筑设计、规划设计，以及工程咨询项目。涌现出一大批在省内外有一定影响的优秀作品。如山东建筑大学新校区、山东建筑工程学院新校区图书信息中心、济南军区燕子山庄、淄博港澳广场等50余个单项工程的建筑设计。同时，还承担了临沂市中心广场、聊城师范大学新校区、山东建筑工程学院新校区、曲阜师范大学日照新校区、山东农业大学科教园区、山东大学南外环新校区等20多个大型综合群体项目的规划设计和全部主体建筑设计；完成50余个项目的可行性研究报告。

设计研究院通过科学的管理、热情周到的设计服务、先进系统的质量体系，赢得了良好的经济效益与社会效益，2001年通过了ISO9001国际质量管理体系认证。近十年间，先后获国家建设部优秀勘察设计奖1项、建设部优秀规划设计二等奖3项、建设部优秀勘察设计三等奖3项、获中国建筑学会第四届建筑创作设计佳作奖1项；同时，获山东省优秀勘察设计一等奖、山东省优秀规划设计一等奖20余项；获山东省优秀工程咨询、装修设计一等奖15项。多项科研成果获国家、省部级奖励。

特别是近几年，设计研究院强化设计队伍建设，在参加全国性投标中，先后有十几项大工程中标，如占地81公顷的“第十一届全运会省建比赛场馆与配套设施”项目。这些大工程的竞标获胜，标志着该院在迈向21世纪的进程中，已经具有较强的竞争力。

Founded in 1960, Shandong Jianzhu University Architecture & Urban Planning Design Institute (preceded by Design Studio of Shandong Architecture Institute) was renamed as Design School in 1985 and Design Institute in 1992. In 1997, it received the Grade-A License of Architectural Engineering Design from Ministry of Construction. At present, SJUAUDI has Grade-A License of Architectural Engineering Design, Grade-A License of Urban Planning, Grade-A License of Engineering Consulting, Inspection Organization Confirmation Letter of Design Documents of Construction Drawing (Architectural Engineering – Steel Structure), as well as Grade-B License of Landscape Design and Grade-B License of Municipal Engineering Design.

SJUAUDI includes four administrative management departments: office, production and management department, technology department and finance office, as well as five production departments: the first branch of design institute, the second branch of design institute, branch of urban planning, research institute of construction scheme creation and consulting drawing-approval center. It is mainly engaged in the design of civil and industrial buildings, urban planning, landscape architecture and related engineering design. It has 140 employees in total, 80 employees of which work within the system. The number of existing Class-1 registered architect and registered structural engineers is 22; the number of registered planner, registered public facility engineer, registered electrical engineer, registered consulting engineer and registered cost engineer is 26. The number of Hong Kong-registered architect and members of Structural Engineer Society is 5. Designers with senior or medium technical titles account for about 60%. Integrated with design, teaching and scientific research, combining technical strength and advanced managing methods, our institute has become a well known design research organization with good reputation inside and outside Shandong province after 50 years of development by several generations.

Since 1990s, our institute has been developing rapidly, while a competitive team of young technical experts has been formed. Relying on strong teaching advantages of Shandong Jianzhu University, the team is stable and innovative, and gives full play to its diligence and wisdom. It has undertaken hundreds of larger-scale projects of architectural design, planning and design, as well as engineering consulting projects, so that a large number of excellent works with certain influences inside or outside Shandong province are born, such as Shandong Jianzhu University New Campus, Library Information Center of Shandong Institute of Architecture and Engineering New Campus, Swallow Mountain Villa of Jinan Military Region, Hong Kong and Macao Plaza of Zibo and more than 50 individual projects of architectural design. At the same time, it has undertaken the planning design and overall design of main building of Linyi City Center Square, Liaocheng Teachers' College New Campus, Shandong Institute of Architecture and Engineering New Campus, Rizhao Campus of Qufu Normal University, Science and Education Park of Shandong Agricultural University, South Outer-ring Campus of Shandong University and more than 20 large-scale integrated-group projects; it has completed feasibility research reports of over 50 projects.

Through scientific management, considerate design services, advanced quality system, SJUAUDI gains good economic and social benefits. It received the ISO9001 International Quality Management System Certification in 2001. In the past ten years, SJUAUDI has won one Award for Excellent Exploration and Design of the Ministry of Construction, three Second Award for Excellent Planning and Design of the Ministry of Construction, three Third Award for Excellent Survey and Design of the Ministry of Construction, one Masterpiece Award in the Fourth Architectural Creation and Design issued by China Society of Architecture; at the same time, it has won the First Prize for Excellent Survey and Design of Shandong, the First Prize for Excellent Planning and Design of Shandong and more than 20 prizes; it has won fifteen first prizes for Outstanding Engineering Consulting and Decoration Design. Many scientific achievements have won national, provincial and ministerial awards.

Especially in recent years, SJUAUDI strengthens the construction of design team. During national biddings, it has won the biddings of over ten large-scale projects, such as the project of "Stadiums and Supporting Facilities of the Eleventh National Games" (810,000 m^2). Winning biddings of these big projects shows that SJUAUDI has become strong and competitive marching into the 21st century.

地址：山东省济南市历山路96号
法定代表人：赵学义
电话：+86-531-86367169
传真：+86-531-86956156
邮箱：SDJDSJY@163.com

Add: No. 96 Lishan Road, Jinan, Shandong Province
Legal Representative: Zhao Xueyi
Tel: +86-531-86367169
Fax: +86-531-86956156
E-mail: SDJDSJY@163.com

第十一届全运会自行车比赛馆

自行车比赛馆是2009年第十一届全运会的主要场馆之一，建筑的屋盖结构系统采用在国内同类工程中实践较少的管桁架结构形式，长轴约为134.6米，短轴约为101米，最高点标高约为36.1米。根据用地地形特点，在设计采用椭圆形主体与东侧起伏的山地相结合的平面布局。36米高的头盔状建筑主体设于用地西南侧，形成良好的视觉景观。主体的头盔状帽檐由南向北延伸，对用地北面进场的观众人流起到很强的引导作用。巨大的椭圆形场馆主体为自行车赛车手头盔的抽象表现，给人以气势恢弘的感觉，充分表现出体育建筑特有的性格特点和结构美感。

山东大学高速公路研发中心

山东大学高速公路研发中心是一座以专业工程实验为主要功能的实验楼，建筑面积9 992.5平方米，独立基础，框架结构，地上四层，建筑高度19.2米。主体建筑呈"U"形布局，一方面满足内部使用空间的采光和通风要求，另一方面与周围道路、广场及建筑物连廊组成高效的交通网络。建筑物中央围合半开敞的庭院作为交通、交流的场所，灵活地改善了单调的平面肌理，结合时代的特征和信息化、数字化教学研究的要求，创造一种开放式、参与式、交流式的全新的空间模式，并综合了类似用途以及同类型空间，使共同利用及面积分配的变更更加容易，确保将来建筑可进行增建或改善的空间。

山东大学南校区综合实验楼

山东大学南校区综合实验楼总建筑面积41 061.18平方米，其中地上12层，地下1层，建筑主体高度49.05米。主体建筑呈“L”形布局，对中心广场形成围合的趋势，与已建成的综合教学楼和拟建图书馆共同构成完整的建筑群体及外部空间序列。多层部分有效利用了建筑用地的地面高差，满足了有特殊层高要求的功能用房，兼顾了设施的开放性。两翼的高层实验楼和面向中心广场的多层部分在形体构成上力求清晰、明确、简洁和富有节奏感，同时实墙面与穿插其间的大片玻璃幕墙形成明快的虚实对比，充分体现设计中对校园建筑提出的“凝重、高雅、深厚、兼容”的风格要求。

济宁贸易职工大学图书馆

本图书馆位于济宁贸易职工大学新校区东门南侧，是校区主入口广场的主要建筑。其总体布局强调统一协调，建筑东部形体采用与道路平行的外边缘，与道路取得一定的协调关系。图书馆分为图书藏阅区、办公区和公共资源区等3个部分，在功能分区上力求简洁合理，并且通过内外部的空间设计，力求使图书馆在采光通风及节能方面取得很好的效果，创造通透的内外部空间环境。立面设计简洁而协调，建筑色彩根据体块的变化分为灰、白两色，立面的开窗强调韵律，让图书馆能够很好地融入整个校区当中。

山东省公安厅综合训练基地综合楼

本工程建设场地为山地环境，室内外高差较大。办公楼与学员宿舍楼的设计依山顺势，协调自然，顺应基地高差的变化，形成了整体统一的效果。其立面设计简洁而协调，且不失庄重。功能布局清晰合理，综合楼一层入口大方明朗，三层的共享中庭丰富了建筑空间，又呈现了建筑的气势。综合楼的内庭丰富了建筑的空间，又丰富了景观的变化，内庭的使用使室内外空间紧密地联系在一起，形成了一个流动空间，为综合楼提供了一个良好的景观点。外部空间设计除了绿化以外，又设计了多处水面，既可以起到改善建筑局部小环境的作用，也可以收集本地区比较多的雨水，起到节水作用。

章丘市福泰小学

教学实验综合楼围绕校园主入口三边布置，主体地上5层，局部4层；建筑控制高度20.20米，总建筑面积21 770.94平方米。立面造型以红、白两色为主基调，体现了小学的色彩格调所形成的人文精神。竖向格栅、走廊，带有儿童色彩的片墙设计，体现小学的生机与活力。建筑的水平线条与竖向框架，在平实的形象上勾勒出高低错落、变化有致的轮廓，从而烘托其个性与可识别性，为立面创造出强烈的虚实对比。设计通过不同空间的尺度大小、开合方式、个性表达等多方面的协调对比，塑造空间序列的不同层次，为各空间之间相互渗透、对比和自由流动创造了可能，从而形成多样化的空间品质。

第十一届全运会省建比赛场馆指挥中心及附属建筑（山东体育学院教学区）

第十一届全运会省建比赛场馆基地位于济南市东部新区，处于世纪大道和泉港路交叉口，总用地面积62公顷，建设用地面积62公顷，总建筑面积为281 000平方米（含地下）。组团总体设计本着“以人为本、服务第一”的原则、优先服务于第十一届全运会，并考虑赛后能充分利用，其建筑布局、造型、高度、层数、色彩等均应符合规划的要求，并与周边环境协调；应充分展示21世纪的时代风貌。力争建设设计新颖、美观而又实用的现代化建筑，使其成为具有浓郁人文特色的文化窗口和亮点之一。

• 公司简介／Company Profile

山东大卫国际建筑设计有限公司，成立于1998年10月，是经国家建设部批准成立的甲级勘察设计股份制公司。公司主要从事民用建筑设计，尤其对房地产开发、工程前期咨询、策划及规划设计有较深入的研究。

公司由建筑师申作伟先生作为主要创办人。申作伟先生从事建筑、规划设计26余年来，设计作品获全国性优秀设计奖15项，获山东省优秀设计奖50余项，申先生历任临沂市规划设计研究院院长、山东省规划设计研究院院长、中国居住区规划专业委员会委员等职，是首批国家一级注册建筑师、国家注册规划师、山东省工程设计大师、山东省优秀建筑师、山东省十佳注册师。2003年申先生被评为中国新本土居住设计名家，2007年，获"中国城市建设卓越人物"称号，2008年3月，被授予"山东省十大杰出青年勘察设计师"，同年8月，被授予"中国建筑设计行业优秀管理人才成就奖"，2009年，被评为"人居十年 2009中国建筑规划十大影响力设计大师"，同年，荣获"2009中国建筑业十大创新建筑设计师"称号，与此同时荣获2009年度"法国金牌建筑设计师"称号。2010年4月被山东省住房和城乡建设厅授予"山东省勘察设计行业管理优秀勘察设计院长"称号。申作伟先生现已成为建筑设计界的学术带头人。

公司从成立之日起即以"设计精品，服务社会"的口号，不断提高设计水平，坚持走品牌设计之路，公司现有设计师150余人，是一支具有高度职业荣誉感和良好职业道德的设计队伍。公司注重建筑设计创新，立足于中国本土文化，融汇国际先进的设计理念和技术，精益求精，完成了一批享誉省内外的精品设计项目，得到了社会各界的一致好评。公司每年均有设计作品在各类评比中获奖。

公司于2003年通过了ISO9000质量体系认证，设计项目已遍及山东、北京、海南、沈阳、重庆、四川、山西、宁夏、云南、河南、河北、广西、吉林等地。公司以创建百年设计公司为奋斗目标，以质量和服务求生存促发展，本着"团结、责任、真诚、创新"的设计理念，在激烈的市场竞争中，赢得了较高的声誉。

我们真诚地期待和社会各界的交流与合作！

Shandong David International Architecture Design Co., Ltd. founded in October 1998, is a joint stock company and has the Grade A qualification of survey and design approved by the Ministry of Construction, P.R.C. The company is mainly engaged in civil construction design, has more depth research on real estate development, preliminary engineering consulting, planning, and planning and design.

Mr. Shen Zuowei is the main founder, he engaged in the construction, planning and design for more than 26 years, his design works won more than 15 national Excellent Design Award, more than 50 items of Shandong Province Excellent Design Award. Mr. Shen served as president of Linyi City Planning and Design Institute, president of Shandong Province, planning and design Research Institute, member of residential planning committee of China, The first batch of registered architect of China, registered planner of China, design master of Shandong province engineering, excellent architect of Shandong province, outstanding registered engineer. Mr. Shen was named Master of New Local Living Design in China, 2003, received the title "China Urban Construction Excellence" in 2007, was awarded Ten Outstanding Young Survey Designer, Shandong Province, in March 2008, and in August, was awarded outstanding management talent achievement Award for China Architectural Design Industry, was named Habitat Decade • 2009 Top Ten Most Influential Chinese architectural design master in 2009, the same year, won the title of 2009 China top ten innovative building construction designer, at the same time, won the title of 2009 French Gold Architect. He was awarded the title of excellent president for management of survey and design in Shandong Province Survey and Design industry by Shandong Department of Housing and Urban-Rural Development in April 2010.Mr. Shen Zuowei has become the academic leaders of architectural design industry.

From the date of its establishment, with the spirit of design quality and service society, the company constantly improves the design standards, and adheres to the way of design brand. There are more than 150 designers, it is a design team with highly professional pride and ethic. The company focuses on innovative architectural design, based on the Chinese local culture, integrates advanced design concepts and technology, excellence, completed a number of renowned boutique design projects outside the province, and has won the praise from the society. The design works get awards in various competitions each year.

The company passed the ISO9000 quality system certification in 2003, the design projects are located in Shandong, Beijing, Hainan, Shenyang, Chongqing, Sichuan, Shanxi, Ningxia, Yunnan, Henan, Hebei, Guangxi, Jilin and other places. With the goal of creating a century design company, survival and promote development of quality and service, the design concept of "solidarity, responsibility, sincerity, innovation", the company has gained a high reputation in the fierce market competition.

We respectfully await commercial exchanges and cooperation from all circles!

山东大卫 国际建筑设计有限公司
Shandong David International Architectural Design Co., Ltd.

地址：山东省济南市英雄山路166号
邮编：250002
电话：+86-531-82719857 / 82719858
传真：+86-531-82719879
网址：www.daweiarch.com
邮箱：dawei_sheji@163.com

Add: No. 166 Yingxiongshan Road, Jinan, Shandong
P.C.: 250002
Tel: +86-531-82719857 / 82719858
Fax: +86-531-82719879
Http: //www.daweiarch.com
E-mail: dawei_sheji@163.com

• 近期获奖作品

北京优山美地东韵生态住宅小区
获：2006年全国优秀工程设计金奖
2005年建设部城乡优秀勘察设计一等奖
首届全国民营优秀设计"华彩奖"银奖
2005年度山东省优秀工程勘察设计一等奖
2004年"中国人文珍品大院"称号
百年建筑优秀作品奖
全国人居奖铜奖

济南东环国际广场
获：2007年第二届全国民营优秀设计"华彩奖"银奖
2005年山东省优秀工程勘察设计一等奖

山东电力集团·鲁能中心
获：2003年山东省优秀工程勘察设计一等奖

北京优山美地东韵生态住宅小区A区
获：2008年全国优秀工程勘察设计行业二等奖
2009年第三届全国民营优秀设计"华彩奖"金奖
2007年山东省优秀工程勘察设计一等奖

山师附属中学幸福柳分校
获：2008年全国优秀工程勘察设计行业三等奖
2007年第二届全国民营优秀设计"华彩奖"铜奖
2007年山东省优秀工程勘察设计二等奖

泰安东尊·优山美地度假区
获：2008年全国优秀工程勘察设计行业三等奖
2007年山东省优秀工程勘察设计二等奖

鹤壁三和·淇林小镇高尚住宅居住区
获：2009年第六届中国人居典范建筑规划设计方案竞赛最佳设计方案金奖
2009年第三届百年建筑住宅类 综合大奖
2009年山东省优秀城市规划设计二等奖

临沂海纳置业曦园生态住宅小区
获：全国优秀民营设计企业优秀工程设计第四届"华彩奖"金奖
2010年全国人居经典建筑规划设计方案竞赛规划、建筑双金奖
山东省优秀工程勘察设计一等奖

临沂大学图书馆
获：全国优秀民营设计企业优秀工程设计第四届"华彩奖" 金奖
山东省优秀工程勘察设计一等奖
2010年山东省第二届城市设计精品工程二等奖

临沂四季春天国际公寓
获：全国优秀民营设计企业优秀工程设计第四届"华彩奖" 铜奖

• Award Works

Beijing Yosemite East Rhyme Ecological Residence
Gold Medal for National Excellent Engineering Design Award, 2006
First Prize of the Ministry of Construction Urban and Rural Excellent Engineering Survey and Design Award, 2005
Silver Medal of the First Huacai Award for Excellent Engineering Design of National Excellent Private Enterprise, 2007
First Prize of Shandong Province Excellent Engineering Survey and Design Award, 2005
Title of Chinese Cultural Treasures Compound, 2004
Excellent Works Award for Century Architecture
Bronze Medal for National Habitat Environment Award

Jinan Donghuan International Plaza
Silver Medal of the Second Huacai Award for Excellent Engineering Design of National Excellent Private Enterprise, 2007
First Prize of Shandong Province Excellent Engineering Survey and Design Award, 2005

Shandong Electric Power Group • Luneng Center
First Prize of Shandong Province Excellent Engineering Survey and Design Award, 2003

Beijing Yosemite East Rhyme Ecological Residence Zone A
Second Prize of National Excellent Engineering Survey and Design Award, 2008
Gold Medal of the Third Huacai Award for Excellent Engineering Design of National Excellent Private Enterprise, 2009
First Prize of Shandong Province Excellent Engineering Survey and Design Award, 2007

Xingfuliu Branch School of Attached School of Shandong Normal University
Third Prize of National Excellent Engineering Survey and Design Award, 2008
Bronze Medal of the Second Huacai Award for Excellent Engineering Design of National Excellent Private Enterprise, 2007
Second Prize of Shandong Province Excellent Engineering Survey and Design Award, 2007

Tai'an Dongzun Yosemite Resort
Third Prize of National Excellent Engineering Survey and Design Award, 2008
Second Prize of Shandong Province Excellent Engineering Survey and Design Award, 2007

Hebi Sanhe•Qilin Town Noble Residence
Gold Medal of the Sixth Classic Architecture and Planning and Design for Habitat Environment of China the Optimal Design Scheme Award, 2009
Residential Comprehensive Award for the Third Century Architecture Residence, 2009
Second Prize of Shandong Province Excellent Urban Planning Design Award, 2009

Linyi Haina Property Xi Garden Ecological Residence
Gold Medal of the Fourth Huacai Award for Excellent Engineering Design of National Excellent Private Enterprise
Gold Medal of Planning and Architecture Design in National Classic Architecture and Planning and Design for Habitat Environment Competition, 2010
First Prize of Shandong Province Excellent Engineering Survey and Design Award

Linyi University Library
Gold Medal of the Fourth Huacai Award for Excellent Engineering Design of National Excellent Private Enterprise
First Prize of Shandong Province Excellent Engineering Survey and Design Award
Second Prize of the Second High Quality Engineering of City Design Award in Shandong Province 2010

Linyi Season Spring International Apartment
Bronze Medal of the Fourth Huacai Award for Excellent Engineering Design of National Excellent Private Enterprise

建　成　项　目　选　登

临沂大学图书馆

建筑面积：68 000平方米，藏书400万册，7层
设计时间：2005–2006年
竣工时间：2009年
获奖情况：第二届山东省城市设计精品工程二等奖

设计说明

设计构思：如山体般自然生长，是建筑也是景观。

建筑与环境共生，图书馆既是一座建筑又是校园里的景观。

顺应校园高低起伏的地势，结合总体规划，将图书馆设计为形似山体的集中式建筑。建筑形体向中部的层层错落升起既解决了建筑功能的需要，又形成了丰富的外部形象，减轻了对环境压力，整座建筑坐落于台地之上，粗重的建筑实体从台地中升起，通透的玻璃又从实体中生长出来，重重叠叠显现出强烈的原生感与时代气息。

建筑各层向中部层层收进，形成多层次的屋顶平台花园，屋顶层层绿化，一直延伸至地面，与校园绿化融为一体，体现出与自然和谐共处的生态校园的设计理念。

建成项目选登

临沂海纳·曦园

建设地点：山东 临沂
设计时间：2007年
总用地规模：23.75公顷
总建筑面积：250 000平方米

临沂是历史文化悠久的文明古城，本项目的单体设计定位为现代新中式住宅，白墙、灰瓦、青砖、屋脊、山墙木构件及装饰构件等中式的建筑语言勾勒出传统中式建筑的神韵，完美的户型设计，充分考虑现代人的生活需求，整体造型简洁端庄，创造了适合现代中国文化审美的新中式居住建筑。

院落的设计，迎合了现代中国人的心理特点和需求，增加了邻里之间的和谐交往，单体设计更加注重细部及比例尺度，材料的运用也更加丰富，造型简洁又富有变化，符合现代中国人的生活需求和审美观，为低密度高档住宅项目的建造起到了一定的示范作用。

材料使用上，“曦园”内外墙体均采用煤矸石多孔砖墙，在建筑中大量运用的不锈钢栏杆、铝合金及木结构装饰构件和灰砖白墙形成了强烈的质感对比，做到了材料创新，符合现代人的审美观，赋予了中式建筑新的生命力和时代感。“曦园”的院墙采用玻璃及金属格栅和实体砖墙的有机搭配，创造出了虚中有实、朴实雅致、内外通透的独特空间。

在环保节能及对土地资源的合理利用上，“曦园”煤矸石多孔砖墙的采用，有利于节约资源和保护环境，60毫米厚挤塑聚苯乙烯外墙保温板及铝合金断热型材中空玻璃外门窗的使用有利于保温节能。

近　期　作　品　方　案

颜山国际社区规划设计

建设地点：山东 淄博
总用地面积：59.84公顷
总建筑面积：778 900平方米
容 积 率：1.3
设计时间：2010年至今，设计中

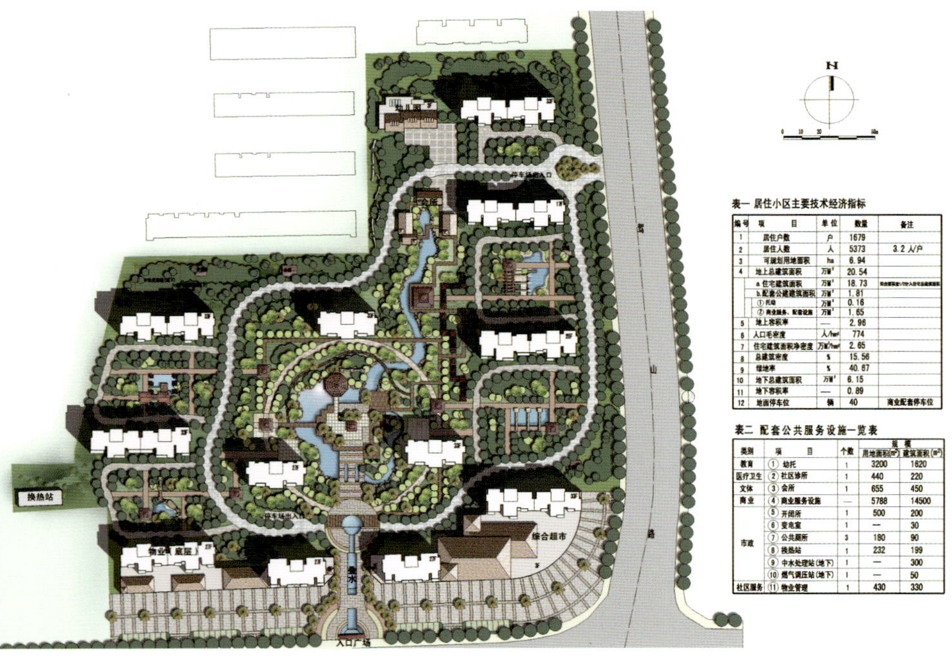

表一 居住小区主要技术经济指标

编号	项目	单位	数量	备注
1	居住户数	户	1679	
2	居住人数	人	5373	3.2 人/户
3	可规划用地面积	ha	6.94	
4	地上总建筑面积	万M²	20.54	
	a.住宅建筑面积	万M²	18.73	[illegible]
	b.配套公建建筑面积	万M²	1.81	
	①托幼	万M²	0.16	
	②商业服务、配套设施	万M²	1.65	
5	地上容积率	—	2.96	
6	人口毛密度	人/hm²	774	
7	住宅建筑面积净密度	万M²/hm²	2.65	
8	总建筑密度	%	15.56	
9	绿地率	%	40.67	
10	地下总建筑面积	万M²	6.15	
11	地下容积率	—	0.89	
12	地面停车位	辆	40	商业配套停车位

表二 配套公共服务设施一览表

类别	项目	个数	规模 用地面积(m²)	建筑面积(m²)
教育	①幼托	1	3200	1620
医疗卫生	②社区诊所	1	440	220
文体	③会所	1	655	450
商业	④商业服务设施	—	5788	14500
市政	⑤开闭所	1	500	200
	⑥变电室	1	—	30
	⑦公共厕所	3	180	90
	⑧换热站	1	232	199
	⑨中水处理站(地下)	1	—	300
	⑩燃气调压站(地下)	1	—	50
社区服务	⑪物业管理	1	430	330

邹城·钢山酒厂改造项目（投标中标）

建设地点：山东 济宁
总用地面积：6.94公顷
总建筑面积：205 400平方米
容 积 率：2.96
设计时间：2011年至今，设计中

高密豪迈城市花园（投标第一名）

建设地点：山东 潍坊
总用地面积：37.17公顷
总建筑面积：583 000平方米
设计时间：2010年

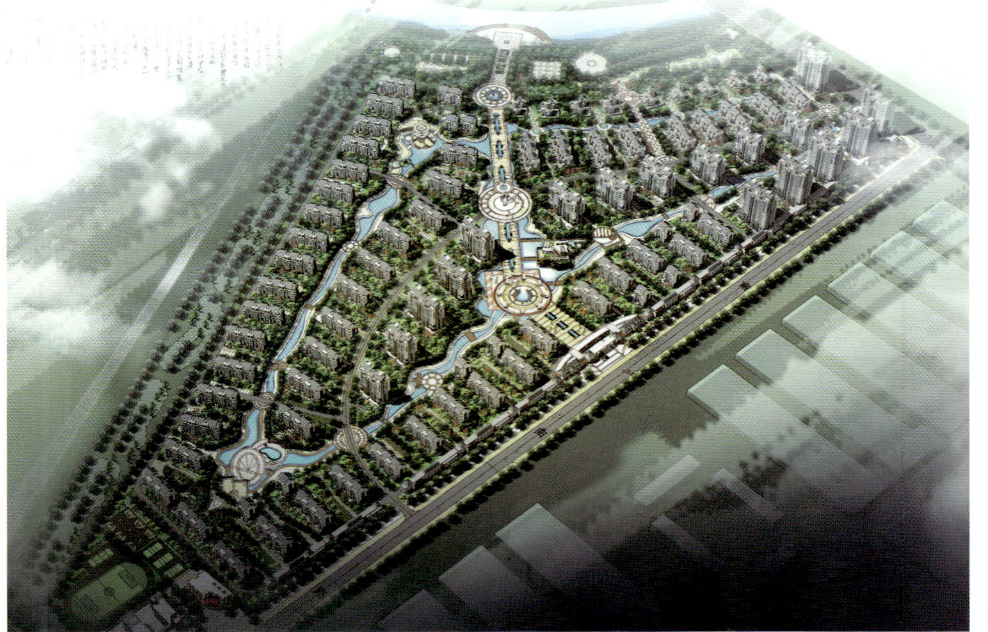

近 期 作 品 方 案

费县护城河沿岸城市设计

建设地点：山东 临沂
用地面积：67公顷
设计时间：2010年至今

1．莱钢金鼎.左岸水都

建设地点：山东 莱芜
用地面积：13.45公顷
建筑面积：103 200平方米
设计时间：2010年，设计中

2．乐陵翡翠绿洲小区

建设地点：山东 德州
用地面积：15.47公顷
建筑面积：266 000平方米
设计时间：2010年至今，设计中

3．龙奥金座商务大厦

建设地点：山东 济南
建筑面积：96 500平方米
设计时间：2010–2011年，在建

4．山东省城乡建设综合服务楼（节能示范楼）

建设地点：山东 济南
建筑面积：44 000平方米
设计时间：2010年至今，设计中

5．燕山新区7–1地块

建设地点：山东 济南
建筑面积：55 700平方米
设计时间：2010年至今，设计中

6．费县公共交通换乘枢纽

建设地点：山东 临沂
建筑面积：9 618平方米
设计时间：2010–2011年，在建

7．黄淮海大厦

建设地点：山东 济南
建筑面积：103 000平方米
设计时间：2010年至今

近 期 作 品 方 案

8．高密“高新产业园”中心片区

建设地点：山东 潍坊
建筑面积：245 000平方米
设计时间：2010年

9．泰安快乐驿站

建设地点：山东 泰安
用地面积：11.28公顷
建筑面积：58 900平方米
设计时间：2010年至今

10．翦云山酒店

建设地点：山东 泰安
建筑面积：11 861平方米
设计时间：2010年

11．金乡县人防指挥中心

建设地点：山东 济宁
建筑面积：28 852平方米
设计时间：2010年

12．日照正德大厦

建设地点：山东 日照
建筑面积：82 000平方米
设计时间：2010年

13．聊城东昌区医院综合楼

建设地点：山东 聊城
建筑面积：62 000平方米
设计时间：2011年

14．章丘长途汽车站

建设地点：山东 章丘
建筑面积：19 000平方米
设计时间：2010年

15．齐河盖世物流

建设地点：山东 德州
建筑面积：100 000平方米
设计时间：2011年至今

16．临沂公安局刑侦综合楼

建设地点：山东 临沂
建筑面积：60 000平方米
设计时间：2011年

17．金乡妇幼保健院综合楼

建设地点：山东 济宁
建筑面积：15 000平方米
设计时间：2010年

上海现代建筑设计（集团）有限公司
Shanghai Xian Dai Architectural Design (Group) Co.,Ltd.

地址：上海市石门二路258号
邮编：200041
电话：+86-21-52524567
传真：+86-21-62464000
网址：www.xd-ad.com.cn
邮箱：xiandai@xd-ad.com.cn

Add: No. 258 Shimen Er Road, Shanghai,
P.C.: 200041
Tel: +86-21-52524567
Fax: +86-21-62464000
Http: //www.xd-ad.com.cn
E-mail: xiandai@xd-ad.com.cn

上海现代建筑设计(集团)有限公司是一家以建筑设计为主的现代科技型企业，集团旗下拥有华东建筑设计研究院和上海建筑设计研究院等20余家专业公司和机构。公司在住房和城乡建设部历年全国勘察设计单位综合实力测评中跻身于前6名，在2005年中国工程设计企业60强排名中跻身于前3名，2001—2006年连续6年被上海市人民政府授予“上海市跨国经营先进企业”，同时连续6年被美国ENR“工程新闻记录”列入“国际工程设计公司200强”和“全球工程设计公司150强”。

集团现有职工3 000余人，其中包括中国工程院院士2名、国家设计大师5名。各专业技术人员占职工人数的83%，高级工程师占职工人数的24%，技术力量雄厚。

集团业务领域涵盖工程项目建设全过程的咨询服务，包括规划、建筑、结构、机电、建筑声学、市政设计、工程总承包、项目管理、工程监理、环境与装饰设计、工程地质勘察、项目可行性研究、科技咨询、投资开发等内容。2006年全集团营业净收入超过13亿元人民币。

至今集团累计完成了3万余项工程的设计与咨询任务，创作成果遍及全国29个省市及20余个国家和地区，其中800余项工程设计、科研项目和标准设计获国家、省（市）级优秀设计和科研进步奖。

集团已在北京、天津、重庆、武汉、青岛、深圳、西安、海南、厦门、大连、南宁、苏州、香港等地共设有25个分院(公司)或办事处，同时集团已与美国、加拿大、英国、法国、德国、澳大利亚、日本、新加坡、韩国、中国香港、中国台湾等国家和地区的设计机构合作设计的项目达400多个。

集团的专业设计人员以完美的创意和先进的建筑技术，将古典与现代、艺术与商业、舒适度与功能性、美感与科技完美地结合在一起，赋予建筑独特的风格和超凡的魅力，最大限度地满足了客户的需求。

Shanghai Xian Dai Architectural Design (Group) Co., Ltd. (herein after referred to as "The Group") is a modern scientific & technological enterprise majoring in architectural design. Over 20 professional companies including East China Architectural Design & Research Institute and Shanghai Institute of Architectural Design & Research are the members of The Group. The Group has been ranked Top 6 of "The Top 100 Nationwide Prospecting & Design Companies in China" in terms of comprehensive competition strength assessed by the State Ministry of Construction in recent years. The Group was ranked Top 3 of "The Top 60 Engineering Design Firms in China" in 2005. The Group was ranked by the People's Government of Shanghai as one of "The Top 20 Trans-national Business Enterprises in Shanghai" from the year 2001 to 2006. Furthermore, The Group has been ranked by US magazine *Engineering News Record* as one of "The Top 200 International Design Firms" and one of " The Top 150 Global Design Firms" from the year 2001 to 2006.

The Group, having a strong technical strength, has a staff of over 3,000 among which 2 are academicians of the Chinese Academy of Engineering, 5 are National-class Design Masters in Architecture, 83% are professionals of different disciplines, 24% are senior architects/engineers.

The service scope of the group covers the whole process of the construction of the project, including urban planning, architectural design, structural design, MEP design, architectural acoustical design, infrastructure design, general contractor, project management, engineering supervision, environmental & interior design, geological exploration, feasibility study, scientific & technological consultant, investment & development, etc. The gross revenue of the group reached the amount of RMB 1,329 million in 2006.

The Group has completed the design and consultation service for more than 30,000 projects which spread across 29 provinces and cities of China as well as more than 20 countries all over the world. Over 800 designs of projects, scientific & technological researches and standardized designs undertaken by The Group have been granted National or Provincial/Municipal Awards for Outstanding Design and Progressive Scientific & Technological Achievements.

The Group has 25 branches (subsidiary companies) or offices in Beijing, Tianjin, Chongqing, Wuhan, Qingdao, Shenzhen, Xi'an, Hainan, Xiamen, Dalian, Nanning, Suzhou, Hong Kong, etc. More than 400 projects have been completed in collaboration with the national and regional design institutions from USA, Canada, UK, France, Germany, Australia, Japan, Singapore, R.O. Korea, Hong Kong China, Taiwan China, etc.

Combining classic style with modern style, arts with commerce, comfort with functions and sense of beauty with science & technology by means of consummate skill and advanced architectural technique, The Group brings forth the unique style and charm to the architecture so as to fulfill the customers' needs to the utmost.

上海浦东国际机场Ⅱ期

建设性质：交通建筑
建设地点：上海
用地面积：1 099 000平方米
建筑面积：480 000平方米
设计时间：2004—2006年
主体建筑结构形式：钢结构屋面、钢筋混凝土框架
主要外装修材料：玻璃幕墙、清水混凝土、花岗石幕墙

浦东国际机场Ⅱ期工程建设总目标是建设一个枢纽型航空港，以满足2015年4 000万年旅客量（加上Ⅰ期2 000万年旅客量共计可达到6 000万年旅客量）。Ⅱ期工程建设分为两个阶段，第一阶段先建设T2航站楼、交通中心、总体道路及相关的配套设施，其中航站楼主楼部分可以处理年旅客量为4 000万人次，长廊部分为2 200万人次。Ⅱ期一阶段于2007年底完成。

Phase II of Shanghai Pudong International Airport
Construction Character: transportation building
Construction Location: Pudong International Airport
Site Area: 1,099,000 m^2
Building Area: 480,000 m^2
Design Time: 2004-2006
Structure Form of Main Body: steel roof, reinforced concrete frame
Main Outside Decoration Materials: glass curtain wall, exposed concrete, granite curtain wall

Phase II of Pudong International Airport aims at establishing "hub model" airport to accommodate 40 million passengers in 2015 (plus 20 million passengers of Phase I to reach a total passenger volume of 60 million). Phase II project is divided into two phases, during Phase I, T2 airport building, transportation center, general road and relative supporting facilities are planning to build firstly, and main body of airport building can handle 40 million person times per year, and part of gallery can handle 22 million person times per year. The second stage project in Phase I was completed at the end of 2007.

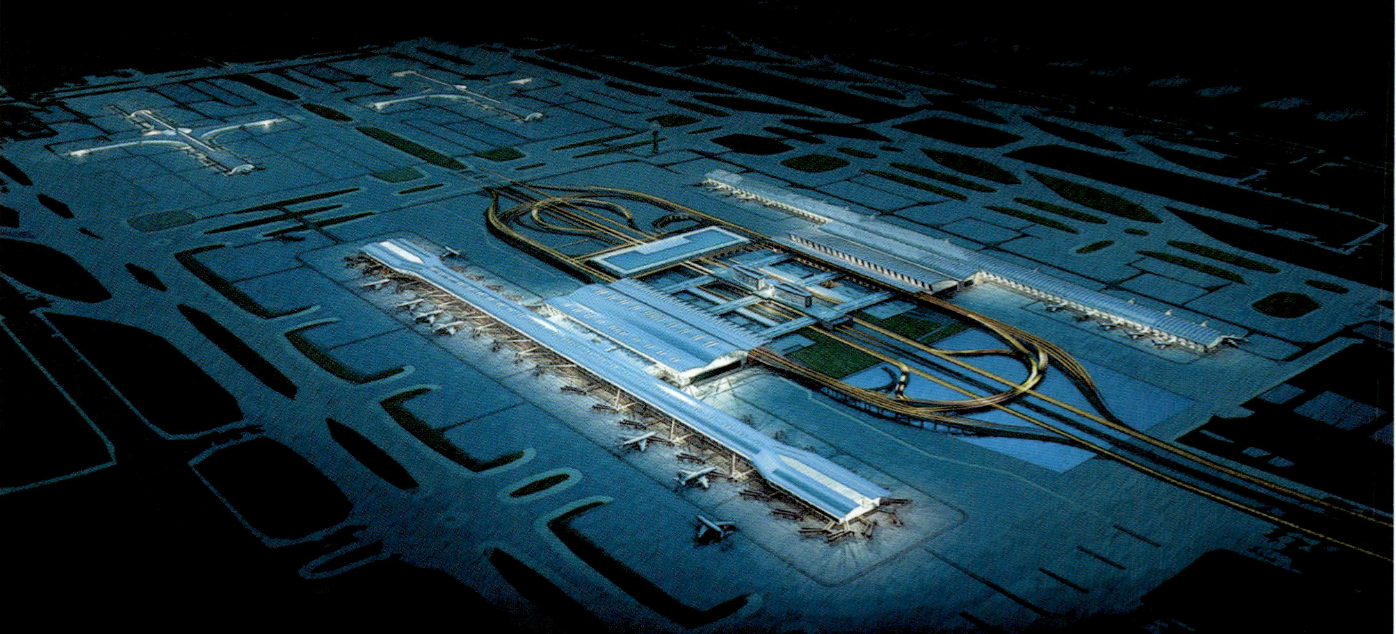

上海光源工程

上海光源工程总体布局和主体建筑的设计构思源于光源装置所特有的旋转与散射的轨迹，并结合自然界的鹦鹉螺曲线，在总体布局和主要建筑单体上产生空间上的旋转动线，使其产生强烈的动感，反映出光源项目所特有的美感。

Project of Shanghai Light Source

Overall layout of Shanghai Light Source and design concept of main building stem from the unique path of rotation and scattering, combined with the nautilus curve in natural world to create revolving line in upper space on overall layout of the main building and major single building, so as to have a strong sense of movement and reflect the unique beauty of the light project.

上海世博家园

项目总体规划充分考虑了世博会定向安置项目所面临的突出问题——经济投入与高品质的建设要求之间的矛盾。项目属于经济型住宅的开发项目，但无论在总体规划、景观构想，还是在单体设计方面都要求体现“超前性、先进性、整体性、示范性”的原则，创造出能够体现2010年世博会精神的居住区，这是本方案的特殊性和难点所在，也是公司在设计中努力寻求突破的主要着眼点。设计力求在功能布局上符合城区总体规划要求，并优化提升地域环境品质，强化资源共享及可持续发展。该项目的建成将为世博会动迁居民提供一个功能配套完善、生活和谐、具有新上海城市生活特色的大型居住区。

Shanghai World Expo Home

Overall planning of the project gives full consideration on the outstanding problems of resettlement projects — the conflict between economic input and high-quality building requirements. The project belongs to an economic-type residential development project. Both in the overall planning, landscape concept and in the single design, the principle of "foresight, advanced, comprehensive, exemplary " is required to embody to create residential district which can reflect the spirit of the 2010 World Expo. To try to be in accordance with the overall urban planning requirements, optimize the geographical environmental quality and strengthen the sharing of resources as well as sustainable development are the uniqueness and difficulties of the project, which are also the main focus of the company to seek a breakthrough. The project was built for the World Expo relocation residents to provide a new Shanghai's major urban residential district with completed functional system and peaceful living environment.

同济大学建筑设计研究院
Architectural Design & Research Institute of Tongji University

同济大学建筑设计研究院成立于1958年，是全国知名的集团化管理的特大型甲级设计单位。研究院持有国家建设部颁发的建筑、市政、桥梁、公路、岩土、地质、风景园林、环境污染防治、人防、文物保护等多项设计资质及国家计委颁发的工程咨询证书，是目前国内设计资质涵盖面最广的设计单位之一。

经过近50年的积累和进取，该院拥有了雄厚的设计实力、丰富的人力资源、先进的设计理念与技术。全院现有职工758人，一级注册建筑师98名，一级注册结构工程师114名。1986年以来，该院共有近200项设计作品获奖。

作为一所国际著名高校的设计单位，该院非常重视建筑教育，培养了一大批硕士生、博士生。此外，该院还具有丰富的对外交流合作的经验，曾成功地与来自美国、加拿大、德国、法国、西班牙等国的著名事务所合作，并互派员工交流学习。

按ISO 9001标准建立的质量保证体系通过中国SAC和美国RAB双重认证。自2001年起，该院与中国人民保险公司签订了每年累计赔偿一亿元人民币的工程设计险合同。

地址：上海市四平路1239号
邮编：200092
电话：+86–21–65987788
传真：+86–21–65985121
网址：www.tjadri.com
邮箱：info@tjadri.com

The Architectural Design and Research Institute of Tongji University, founded in 1958, is one of the leading design groups in China with the most comprehensive range of disciplines available in the country, allowing it to develop many types of design project. It holds design certificates issued by the State Ministry of Construction for China in architecture, municipal engineering, bridge engineering, highway engineering, geotechnical investigation, geology, landscaping, environmental engineering, civil air defence and cultural relic protection. It also holds an Engineering Consulting Certificate issued by the State Planning Committee.

The Design Institute has developed its strength in design, manpower and technology through its 50 years of experience. It employs 758 professionals including 98 National Class A registered architects and 114 State First Class registered structural engineers. It has won nearly 200 prizes for its design work since 1986.

Affiliated to a well known university, the Design Institute has a tradition in architectural education. It has cultivated a number of graduate students. In addition, the Design Institute has successfully collaborated with many well known design firms from the United States, Canada, Germany, France and Spain etc. Exchange of staff is also part of the collaboration.

The Design Institute has been approved to the ISO9001 Quality Assurance Standard by Shanghai Audit Centre and RAB, the American ISO assessment organization. In 2001, the Design Institute signed an Engineering Design Insurance Contract with PICC, one of the leading insurance companies in China. This gives total cover of 100 million RMB per annum.

Add: No. 1239 Siping Road, Shanghai
P.C.: 200092
Tel: +86–21–65987788
Fax: +86–21–65985121
Http: //www.tjadri.com
E-mail: info@tjadri.com

同济大学汽车学院

建设地点：上海
设计时间：2002–2003年
建筑面积：584 100平方米

项目概况

校园共分为六大片区：校前区（行政服务培训区）、公共教学区、二级学院区、研发实训区、生活区和体育运动区。各片区内建筑群体以院落式的方式组合，形成多院落的交往空间，使人与自然环境形成直接的对话。

考虑校园"统一规划，分期实施"的建设原则，校园规划不但要考虑一、二期工程完全实施后的整体性，还要保证校园仅在一期工程实施后的完整性。组团式建筑布局使各功能建筑分片建设，组团内部建筑可以分期建设，使校园规划形成可持续发展的、有机生长的设计结构。

Automotive Engineering College, Tongji University

Construction Location: Shanghai
Design Time: 2002–2003
Building Area: 584,100 m^2

Project Profile

The campus is divided into six areas including front area (administrative service training area), public teaching area, area of secondary institute, R & D practical training area and living area as well as sports area. Buildings in each area are located with courtyard type to form communication space of multiple courtyards, which makes people directly feel the natural environment.

Considering the principle of "integrated plan, phased implementation", the campus planning not only focuses on the integrity after the implementation of Phase I and II projects, but also needs to guarantee the completeness of campus after Phase I project. The group architectural layout ensures each functional building is built with areas and interior building is constructed in phases, becoming organically growing structure with the sustainable development.

秦皇岛体育场

建设地点：河北 秦皇岛
建筑面积：48 000 平方米

项目概况

体育场观众席内圈椭圆形与外圈椭圆形的中心线东西相距 15 米，这种设计可使视线较好的西看台容纳更多的观众。看台共设有 33 000 个坐席，其中有 2‰的伤残人士坐席，东西看台后侧还分别设有包厢。运动场内设有半径为 36.5 米的 400 米环形跑道标准田径场，跑道内设有 104 米 ×68 米的天然草皮足球场。

Qinhuangdao Stadium

Construction Location: Qinhuangdao, Hebei
Building Area: 48,000 m^2

Project Profile

The central line of inner ellipse spaces out 15 meters apart the outer ellipse from east to west, which can ensure the West Stand with better sight quality accommodate more audience. The stand sets up a total of 33,000 seats, of which 2‰ is for the disabled. Boxes are set up behind the East and West stands. The stadium is equipped with standard track and field with 36.5-meter radius and 400-meter circular track; meanwhile, inside the track there is a football field of 104 m × 68 m with natural sod.

中国银联上海信息处理中心

建设地点：上海
用地面积：20 984 平方米
建筑面积：15 387 平方米

项目概况

中国银联是中国货币与金融电子化的先驱，本次工作目标就是要为之提供一个与其品质相适应的具有典型现代主义特征的空间。

1．理性、可靠、坚固、现代的建筑造型，与中国银联的企业形象相吻合。

2．由于项目地处两条河道的交界处，自然环境优美，故采用园林式的建筑布局，为生产办公提供自然化的景观环境。在景观设计中运用了大量的水景，大面积平静的水面及设置于其上的休憩平台使工作者能拥有平和的心境，置身其间感受建筑与水景的相互辉映，使身体与心灵都融入到这种极富意境的体验中。

Shanghai Information Processing Center of China Union Pay

Construction Location: Shanghai
Site Area: 20,984 m^2
Building Area: 15,387 m^2

Project Profile

China Union Pay is the pioneer of Chinese currency and financial computerization, and its working goal is to furnish the qualified space with typical modernism characteristics.

1. With the rational, reliable, solid and modern architecture molding, which is matched the enterprise features of China Union Pay.

2. Because of the natural beauty brought by its location at the junction of two rivers, the project adopted garden-style building layout to provide naturalized landscape for the production and office work. In the design, a large amount of waterscape is used. A large area of calm water surface and the over platform are the supporting of the peace mind of staff. The building embraces the waterscape, and the peaceful environment matches the smart mind, which are all highly mood experience.

浙江省公安指挥中心

建设地点：浙江 杭州
建筑面积：32 000 平方米
设计时间：2000 年
竣工年份：2003 年

项目概况

项目的设计目的是为浙江省公安厅提供全新的建筑设施与功能布局。设计从城市设计入手，尊重总体规划和周围环境，充分考虑西湖风景区的自然景观，将已建公安厅大院内建筑与环境设施进行统筹设计、规划，满足在当前基本构架的基础上建立准确有效的工作模式，体现高效率、简洁的风格。

Zhejiang Police Control Center

Construction Location: Hangzhou, Zhejiang
Building Area: 32,000 m^2
Design Time: 2000
Completion Time: 2003

Project Profile

The project aimed at providing the brand new building facilities and functional layout for Department of Public Security Zhejiang Province at the beginning of the 21st century. Started from urban design and respected the master design and surrounding environment, the project gave full consideration to the visual landscape of the West Lake scenic area, and it made overall plan and design for the existed buildings and environmental facilities inside the compound to establish the accurate and effective working mode on the basis of the current basic framework, reflecting the high efficient and simple style.

上海创霖建筑规划设计有限公司
Shanghai Total Architectural Design & Urban Planning Co., Ltd.

上海创霖建筑规划设计有限公司（简称TDI），在全国设有新疆、烟台、厦门、宁夏、成都五家分公司，上海总部拥有两处自有产权的现代高级办公场所，持有国家颁发的建筑工程综合甲级资质、城乡规划乙级资质、园林景观设计乙级资质。公司现有专业技术人员320人，中高级职称及各专业注册设计人员占65%，公司拥有的一批学养深厚、经验丰富的专业骨干，是公司立足于行业、承诺于社会的基本力量。

作为建筑全过程设计的实践者，从前期策划、规划设计、建筑设计、景观设计、室内设计和施工监理等项目营建的各个环节提供控制设计和服务支持，利用广泛的技术协作网络，倾心为客户和社会奉献符合专业标准的建筑作品。

多年来，凭借对各类项目的全面透彻理解，依托创新的设计理念、上乘的设计质量、完善的管理体系、良好的服务口碑，承接了遍及全国20多个省份、自治区、直辖市的数百项设计工程，其中包括复杂、大型、超高层项目，并获得业内外广泛的认同和推崇，有多项设计项目在国内获奖。 在高科技、绿色建筑节能领域的研究与设计实践中，实施完成了多个国家绿色建筑、国家康居示范工程，获国家住房和城乡建设部高度评价。

Shanghai TDI Architectural Design & Urban Planning Co., Ltd. (TDI) has totally five branches in Xinjiang, Yantai, Xiamen, Ningxia and Chengdu. The headquarters in Shanghai owns two modern high-level office buildings. It has Class-A License of Architectural Engineering, Class-B License of Urban and Rural Planning, Class-B License of Landscape Design. TDI has 320 professional technicians, in which senior and medium engineers and all types of registered designers accounts for 65%. A group of educated experts with special experience are the basic strength of the company to compete in this industry and serve the society.

As a practitioner of whole-process architectural design, we provide design control and service support in pre-planning, planning and design, architectural design, landscape design, interior design, construction supervision and all other aspects of project construction. By using a wide network of technical cooperation, we bring customers and society the architectural works in accordance with professional standards.

Over these years, through comprehensive and thorough understanding of various projects, relying on innovative design, superior design quality, perfect management system and good service reputation, we have undertaken hundreds of complicated, large-scale and high-rise projects covering more than 20 provinces, autonomous regions and municipalities, and have been widely recognized and respected inside or outside the industry. A number of design projects have won national awards. In the research and design practice of high-tech, energy-saving green buildings, TDI has completed a number of national green buildings, national demonstration projects of healthy living, which are highly praised by the Ministry of Construction.

地址：上海市杨浦区国定东路275号汇创大厦8号楼3A03
邮编：200433
电话：+86-21-35315255 / 35315257
传真：+86-21-35315258-8000
邮箱：tdi@tdi-sh.com
网址：www.tdi-sh.com

Add: Room 3A03, Building 8, No.275, Guodingdong Rd, Shanghai
P.C.: 200433
Tel: +86-21-35315255 / 35315257
Fax: +86-21-35315258-8000
E-mail: tdi@tdi-sh.com
Http: //www.tdi-sh.com

东华大学

项目地点：上海
建筑面积：200 000平方米

江阴城市客厅三中心

项目地点：江苏 江阴
建筑面积：86 000平方米

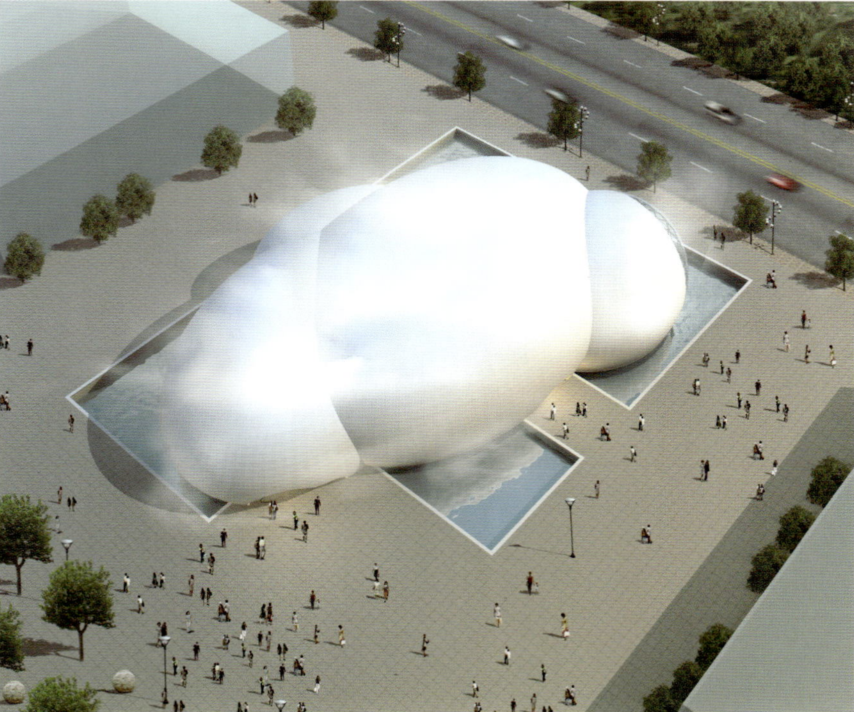

上海世博会世界气象组织馆

项目地点：上海
用地面积：2 200平方米
（荣获2010年世博会组委会优秀设计奖）

上海东海广场改、扩建

项目地点：上海
建筑面积：200 000平方米

澄星总部大厦

项目地点：江苏 江阴
建筑面积：239 356平方米

拉斯维加斯（南京）国际商务中心

项目地点：江苏 南京
建筑面积：130 880平方米

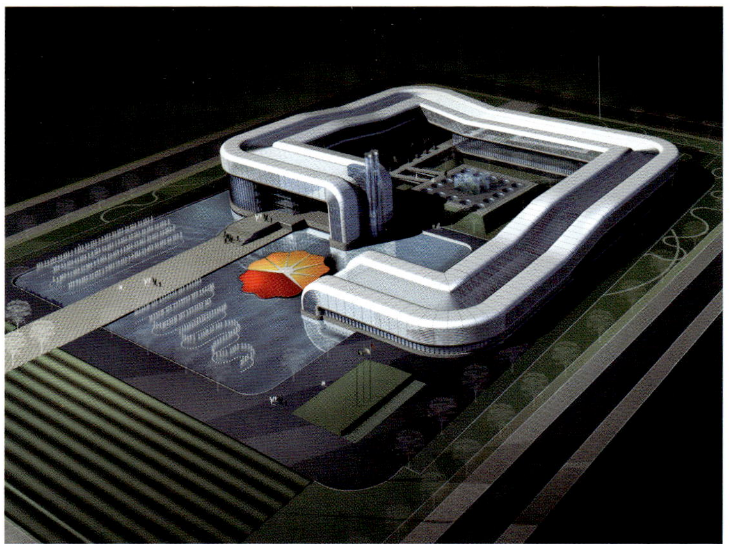

中石油北京科研基地

项目地点：北京
建筑面积：74 419平方米

新疆卷烟厂生产指挥中心大楼

项目地点：新疆 奎屯
建筑面积：42 597平方米

天禧嘉福酒店

项目地点：上海
建筑面积：150 000平方米

乔波滑雪中心

项目地点：北京

建筑面积：200 000平方米

松江影剧院

项目地点：上海

用地面积：29 000平方米

建筑面积：7 216平方米

上海彭越浦2号地块

项目地点：上海

用地面积：224 800平方米

1.5 设计理念

通过对项目的理解与分析，我们认为有三大问题与挑战

如何呼应上海世博会“人、城市与自然”和谐的主题？

如何体现上海规划公共绿地的导向？

如何连接被道路分割的六块用地？

针对这三个问题，设计提出了“绿叶”的理念

以叶片为本，由核心的脉络向周边延生，带来生机与活力

基地状似一片绿叶，受此启发，我们提出了“立体（3D）公园”的理念。

立体（3D）公园打破了传统的平面二维（2D）格局，在三维空间展开，注重垂直向度的变化，波浪起伏，宛若一片翻卷的绿叶，翩翩飘落在地块上，带来盎然绿意和全新感受。既丰富了空间形态，又提高了土地的综合利用率（部分绿地下空间开发），还可通过立交和环廊等将各个分散的地块联成一体。

立体（3D）公园与建筑物、滨水步道及室内商业街结合成有机共生的浑然整体，犹如绿叶通过脉络给各个组分提供营养。

纳帕尔湾

项目地点：江苏 昆山
用地面积：146 252平方米
建筑面积：73 101平方米
容 积 率：0.5
开 发 商：昆山建兴置业

新疆雪莲山地块

项目地点：新疆 乌鲁木齐
用地面积：239 253平方米
建筑面积：236 861平方米
容 积 率：0.99
开 发 商：新疆城建股份

盐城市大新河以东地块

项目地点：江苏 盐城
建筑面积：255 174平方米
开 发 商：盐城交通开发有限公司

桃园小区

项目地点：北京
用地面积：105 600平方米
建筑面积：276 899平方米
容 积 率：1.98
开 发 商：北京天坛房产

贝乌逸轩——国家康居示范工程

项目地点：新疆 昌吉
用地面积：169 000平方米
建筑面积：239 553平方米
容 积 率：1.4
开 发 商：华源集团

星威园

项目地点：江苏 泰州
建筑面积：750 000平方米
容 积 率：1.6
开 发 商：春兰集团

TIANHUA ARCHITECTURE | 天华建筑

让 | 设 | 计 | 创 | 造 | 价 | 值

天华建筑设计有限公司
Tianhua Achitecture Planning & Engineering Limited

www.thape.com

天华建筑设计有限公司创立于1997年，是中国第一批民营建筑设计公司之一，位居中国民营建筑设计公司前列。

公司具备建筑工程甲级、城市规划乙级和风景园林工程设计乙级专项资质，可为客户提供涵盖规划、建筑、室内、景观、工程和设计咨询、设计总包等全方位服务，以及从方案、扩初到施工图及施工配合等全过程一站式设计服务，亦可根据客户需求，提供规划、建筑设计方案和施工图等的阶段性服务。

天华总部位于上海，设北京、深圳、武汉3个全资子公司。已汇聚精英人才近1 500人；客户群由万科、中海、保利、绿地、金地、龙湖、招商、华润、九龙仓、瑞安、金融街等中国地产百强企业组成；项目分布于全国40余个省市，年完成建筑设计面积超过1 2 000 000平方米。

天华自成立伊始，便致力于对设计作品高品质和高完成度的追求，坚信“设计创造价值”，秉持“专业成就高度”，专注成为细分市场专家，并借此赢得在住宅设计领域的盛誉。

2010年始，天华全力进军商业地产。在保持住宅优势的前提下，天华建筑将战略眼光投向了城市综合体等新的设计细分市场，致力于在城市综合体、办公楼、酒店等综合开发领域再创辉煌。

成为综合地产开发设计领域的专家，将是我们未来长期奋斗的目标。为此，天华建筑汇聚了国内外多名一流人才，组建强大的公建团队。这些来自于国内外著名设计公司的优秀专家，均具备在多种公建设计领域的丰富经验和实践，分布于建筑、结构、机电等各个专业，可为客户提供全过程的专业设计服务。

完善的专业、技术配备为公司的发展提供了可靠的保障。公司自主研发的“基于项目管理系统的协同设计平台”运行成熟，为业界领先，为项目提供了强有力的技术保障；公司曾以在公共租赁房设计上的领先经验，参与制定上海公共租赁房设计的行业标准；公司亦不断增强研发团队实力，成立专业的BIM事业部；并加大力量，投入绿色技术及未来建筑技术方向的研究，为建筑设计行业的发展和公司愿景的实现而不懈努力。

14年的成就证明，专业与高品质，曾为我们创造辉煌，也必将使我们再攀高峰。

Established in 1997, Tianhua Achitecture Planning & Engineering Limited is one of the first group of private-owned architecture planning & engineering design companies in China and a front-runner among Chinese private-owned architecture planning & engineering design companies.

Tianhua has several special qualifications such as Construction Works Class A, Urban Planning Class B, and Landscape Architecture Class B, and provides our clients with all-sided services covering planning, buildings, interior, landscape design, engineering and design consultancy, and total package design as well as one-stop services for the whole process, from SD (Schematic Design), DD(Development Design), to CD (Construction Document) and CC (Construction Coordination). We also provide services at each stage of the works, including planning, construction architectural design scheme, and construction document, etc.

Tianhua's headquarters is located in Shanghai and has three wholly-owned subsidiaries in Beijing, Shenzhen, and Wuhan. We have recruited nearly 1,500 elites and our clients include those ranking top-100 Chinese real estate enterprises, such as Vanke, COLI, Poly, Greenland, Gemdale, Longfor, China Merchants Property Development, Crland, Wharf Holdings, Shui On Land, Jinrongjie, etc. Our projects reach every place in more than forty provinces and cities across China, and the area of our construction design per annum exceeds 12,000,000 m^2.

Since its inception, Tianhua has devoted itself in pursuit of high quality and high degree of completion. We strongly believe "Design Creates Value" and adhere to the principle of "Expertise Achieves Height", focusing on becoming an expert in niche markets and winning great reputations in the area of house design.

Since 2010, Tianhua has exerted its great efforts to commercial real estates. On the premises that our advantages in house design are kept, Tianhua puts its strategic vision on new niche design markets such as urban complex, committing ourselves to obtaining new successes in the comprehensive development area such as urban complex, office buildings, and hotels.

To become an expert in the area of development and design of comprehensive real estate will be a long-term objective that we will make arduous efforts to achieve. For this, Tianhua has gathered talented people at home and abroad, and organized a strong team for public construction. Those excellent experts, who come from famous domestic and foreign design companies, have sufficient experiences and practices in multiple areas of public construction design, and specialize in buildings, structure, electromechanical, etc., capable of providing our clients with whole process professional design services.

Complete professional and technical equipments provide reliable guarantee to Tianhua's development. The Coordination Design Platform Based on Project Management System independently developed by us is a proved and leading software providing technical support to our projects; Tianhua also continuously strengthens the power of our R&D development team, established a specific BIM Business Department and meanwhile we will conduct research works on green technology and future construction technology, making unremitting endeavor for the development of the construction design industry and the realization of our vision.

Our 14-years accomplishments prove that profession and high quality had created glories for us, and are bound to help us scale new height.

上海 SHANGHAI
上海市中山西路1800号兆丰环球大厦27楼
T: +86-21-64281588 F: +86-21-64281587

深圳 SHENZHEN
广东省深圳市福田区泰然车公庙工业区201栋3楼
T: +86-755-82571508 F: +86-755-82571500

北京 BEIJING
北京市西城区西直门外大街1号西环广场T1楼12层
T: +86-10-58301595 F: +86-10-58302926

武汉 WUHAN
湖北省武汉市江汉区新华路396号民生银行大厦
T: +86-27-85735158 F: +86-27-85735150

华润温州万象城

建设地点：浙江 温州

业　　主：华润置地（上海）有限公司

建筑面积：340 700平方米

用地面积：61 000平方米

设计时间：2011年

设计理念

项目位于温州市南三垟湿地旁，温瑞大道与南湖路交口处，包含区域级高端购物中心万象城，亲水时尚购物街和2栋100米SOHO塔楼。

设计源于基地周边特有的湿地环境：温瑞塘河的支流经基地注入三垟湿地，寓意着自然生态与代表着都市人生命活力的万象城相互交融。沿河两岸的亲水退台将为这个都市人生活与聚集的场所，以及丰富的空间与情趣。

设计旨在塑造一个拥有24小时活力的生活社区，集生活、工作、休闲、娱乐于一体，同时利用滨水退台、屋顶、广场等为城市提供一个聚会的客厅。

MIXC Wenzhou - The Urban Oasis

Construction Location: Wenzhou, Zhejiang

Owner: Crland (Shanghai)

Building Area: 340,700 m^2

Site Area: 61,000 m^2

Design Time: 2011

Design Concept

The project is located at the south of Wenzhou city, at the intersection of Wenrui Avenue and Nanhu Road, and facing the Sanyang Wetland at the east side. This high-end mixed-use development includes a regional shopping mall, which named as MIXC, a waterfront lifestyle center and two 100 m high SOHO tower.

The design is inspired by the unique context of wetland. A branch of Wenruitang River passes by the site and infuses to the Sanyang Wetland, which is a metaphor of fusion of beautiful natural resource and vibrant urban life. The waterfront terraces provide romantic gathering space for citizens.

We are trying to create a 24 hours vibrant community, which mixes living, working, recreation and entertainment. At the mean time, we also create an urban living room for people by integrating the terraces, roof garden, plaza with given natural landscape.

上海华润橡树湾

建设地点：上海
业　　主：华润置地（上海）有限公司
项目类别：多层、高层住宅
建筑面积：79 317.86平方米
用地面积：68 215.36平方米
设计时间：2007年
竣工时间：2009年

项目介绍

上海华润橡树湾位于上海市杨浦区新江湾城，是高起点规划、高标准建设的居住区。

项目设计要求既要符合城市空间的整体发展的要求，同时又要具有自身特色。

如今一期已经建成投入使用，为业主创造出富于浓厚的居住生活情趣的社区环境、便利的服务设施、多姿多彩的社区生活、丰富多样的功能和空间，景观优美的生态环境。同时，整个社区的整体氛围及建筑外观着力体现城市生活的风范，创造出了21世纪上海花园城区和生态居住区。

Shanghai Crland Oak Bay

Construction Location: Shanghai
Owner: Crland (Shanghai)
Type: multiple-level, high-rise residential house
Building Area: 79,317.86 m^2
Site Area: 68,215.36 m^2
Design Time: 2007
Completion Time: 2009

Project Introduction

Shanghai Crland Oak Bay is located in the New Jiangwan City, Yangpu District, Shanghai, which is a residential district having high starting level plan and high standard construction.

The project's design requires not only the compliance with the whole development of urban space, but having its own characteristics.

At present, the Stage I has been completed and was put in use, creating for the residents a community environment with deep fun and joy for living, convenient service facilities, colorful and varied community lives, various functions and space, beautiful landscape and ecological environment. At the same time, the whole community's atmosphere and architectural appearance strive to reflect the style of urban living, creating 21-century Shanghai garden urban area and ecological living area.

陆家嘴SB1–1地块金融中心

建设地点：上海
业　　主：上海陆家嘴（集团）有限公司
建筑面积：58 689平方米
用地面积：10 920平方米
设计时间：2011年

项目介绍

项目基地位于浦东新区主干道——世纪大道与浦东南路的交界处，介于陆家嘴金融中心区与竹园商贸区之间。作为3家金融机构总部的办公楼，设计以金融企业特点为出发点，目标营造一个整体、高效、高品质的企业形象和办公空间。

世纪大道作为浦东地区最为重要的城市景观道路，与现有路网形成45°斜角关系，打破了区域内原已形成的方格式路网，构成特有的城市肌理。体量概念根据陆家嘴地区城市肌理与景观主轴——世纪大道自然切割而成。主立面沿世纪大道，建筑体量契合城市肌理而形成3个相同的平行四边形的布局特点。大面积的景观广场位于地块东北部，提供入口进入感的同时，引入景观绿化。同时办公塔楼中嵌以采光中庭组织办公空间，通过节奏的变化强调办公楼的独立形象。立面灵感则来源于中国民间传统工艺"编织"，通过现代的演绎契合整体的形象。

Lujiazui Plot SB1-1 Financial Center

Construction Location: Shanghai

Owner: Lujiazui Finance & Trade Zone Development CO.,Ltd.

Building Area: 58,689 m^2

Site Area: 10,920 m^2

Design Time: 2011

Project Introduction

The base of the project is located at the road junction of the Century Avenue, the major road in Pudong New District, and Pudong South Road, between Lujiazui Financial Center Area and Zhuyuan Commercial and Trade Area. As the building for three financial institutions' headquarters, the project selected a design reflecting the characteristics of financial enterprises with an aim to establish integrated, efficient, and high-quality company image and office spaces. As the most urban landscape road in the Pudong area, the Century Avenue and current highway network shape a 45 degree oblique crossing, breaking the established square highway network in the area, and constituting a specific urban texture. The concept of dimension is based on the urban texture and landscape axis (Century Avenue) in the Lujiazui area under natural division. The façade is along the Century Avenue, and the building dimension complies with the urban texture, shaping an arrangement of three identical parallelograms. The large-area landscape plaza is located in the northeast of the plot, serving as an entrance and introducing landscape greening. At the same time, the office tower is filled in by daylight-flooded atrium to organize office space, emphasizing independent image of the buildings. The inspiration of the façade derives from "Weaving", a Chinese folk craft, trying to build a total image by modern deduction.

瑞安创智天地

建设地点：上海
业　　主：瑞安房地产发展有限公司
建筑面积：304 000万平方米
占地面积：96 000平方米
设计时间：2011年

项目介绍

本项目的规划目标：充分利用杨浦区人文、地理、历史的同时，以硅谷为借鉴，创造一种氛围和城市环境，带动杨浦大学城的建设，促进杨浦区高科技和教育事业的快速发展，增进文化交流、科技服务、休闲娱乐以及体育健身活动的开展。本地块位于创智坊的中心部位，是一个居住和工作相融合的新型社区。因此，在设计理念上基本遵循创智天地整体项目的规划理念，将创智天地打造成知识创新区，提出大学校区、科技园区、公共社区"三区融合、联动发展"的理念。并力图整合多种城市功能，延续城市文脉，努力营造符合创智理想、以知识生活为主题的大型综合社区。产品包括办公楼、住宅、酒店以及公寓式办公楼。

Shui On Chuangzhi Tiandi

Construction Location: Shanghai
Owner: Shui On Land Limited
Building Area: 304,000 m^2
Site Area: 96,000 m^2
Design Time: 2011

Project Introduction

The project's planning objective: to fully utilize Yangpu District's cultural, geological and historic advantages, and by refering to Silicon Valley, to create an atmosphere and urban environment so as to drive the construction of the Yangpu University City, facilitate rapid development of Yangpu District's high technology and education industry, and promote the carrying out of cultural communications, technical services, leisure and entertainment, and physical fitness. The plot is located at the central part of Changzhifang, and is a new-style community combining living and working. For this, the design concept basically complies with the planning concept of the whole Chuangzhi Tiandi Project, which aims to forge Chuangzhi Tiandi into a knowledge innovation area and raise a concept of "combination and joint development of university campus, technical park, and public community". The project also tries to integrate mutiple urban functions, continue the city's cultural heritage, and endeavor to build a large-scale comprehensive community, which complies with Chuangzhi's idea and makes knowledgeable living as its topic. The products include office buildings, residential houses, hotels, and apartment office buildings.

上海垣恒建筑设计咨询有限公司
Shanghai C@C Architectural Design Co.,Ltd.

地　址：上海市江场三路228号507室
电　话：+86-21-61077016
传　真：+86–21-61077018
邮　箱：shanghaiyuanheng@126.com

Add: Room 507, No. 228 of Jiangchang West Road, Shanghai
Tel: +86-21-61077016
Fax: +86-21-61077018
E-mail: shanghaiyuanheng@126.com

上海垣恒建筑设计咨询有限公司在成立之初就坚持走国际化的发展道路并采用先进的经营理念和完善的管理制度。公司与国际建筑市场接轨，实行人性化的管理模式。公司承接的项目以建筑、规划、景观、室内等的方案设计及施工图设计为主，并与国内多家甲级设计院及加拿大、德国的设计公司有良好的合作关系。

公司拥有专业的设计团队和扎实的技术力量，以有着20年国企总工程师工作背景的国家一级注册建筑师王一鸣为核心带头人，他拥有丰富的设计、投标、施工管理、工程技术等方面的经验，承接过许多大型公共建筑、住宅及工业建筑项目，并多次在上海及全国的工程竞标中一举中标，在业内享有良好的声誉。

Shanghai C@C Architectural Design Co.,Ltd. has been sticking to its international development principle, advanced business philosophy and perfect management system, linking itself to the international architectural market and implementing humanistic management mode. The company gives priority to architecture, planning, landscape, indoor scheme design and construction documents design, and has established sound cooperation ties with many domestic Class A design institutes and other countries such as Canada and Germany.

The company boasts a professional design team and huge technical strength. The company, led by Mr. Wang Yiming (a national Class A Registered Architect, with 20 years of experiences in state-owned companies as a chief engineer and with abundant experiences in design, tendering, construction management and engineering), has undertaken the construction of a great many of large-scale public buildings, residential buildings and industrial buildings and has won quite a number of bids for the projects in Shanghai and around China, and the company is broadly recognized in this industry.

上海市北工业园区商务办公综合配套项目

项目地点：上海
用地面积：22 600平方米
总建筑面积：86 150平方米
设计时间：2009年

设计构思

设计力求创造出一个通透别致，而又符合市北工业园区特点的，具有标志意义的综合办公建筑，建筑各个元素之间融为一体，但又不失去各自的独立性。

主体沉稳而又厚重的花岗岩与轻盈的玻璃材质形成鲜明的对比。主楼灵活的开窗形式，形成独特的表面肌理，细部设计与整体风格的完美搭配，使其顺应上海悠久的历史，捕捉到古典的幽雅之气，同时又体现出市北工业园区跨入新世纪的巨大动力和无限生机。

包头“东德·京师公馆”项目

项目地点：内蒙古 包头
用地面积：84 239平方米
总建筑面积：324 577平方米
设计时间：2011年

设计构思

在规划设计中我们采用了新城市主义理念，新城市主义提供了这样一个前景：邻里单元紧凑的、功能的混合有利于步行、尊重地域文化的独特性，有效和美丽的走廊地带能使自然环境和人造的社区环境融合成为一个可持续发展的整体；主张恢复都市地区中的现有城市中心和市镇，重新配置无序蔓延的郊区，使之成为具有真正社区和多样化的城区；保护自然环境，以及保护我们已有的传统遗产。

建筑外立面设计采用与新城市主义理念相一致的装饰艺术风格，考虑到包头当地并无此类风格的小区，采用装饰艺术风格在当地具有一定的独特性，能在包头市房地产市场脱颖而出，创造出独树一帜的高品质住宅区。

荆州水务·颐水澜天项目

项目地点：湖北 荆州
占地面积：50 101平方米
总建筑面积：104 991平方米
设计时间：2011年

设计构思

本项目有着得天独厚的自然环境与人文条件，基地北侧为荆州古城，东侧紧邻荆州古城护城河，文化底蕴深厚，景色优美，环境绝佳。在设计中我们注重地域文化，继承传统文脉，以院落式布局及山水园林的规划思想，努力打造"一城一园，一院一境"的高品质、高档次、传世文化住宅。

本项目的设计充分体现其独有的荆楚文化氛围，突出"文化–自然–生态–宜居"的特色，注重方案在整体结构上与城市肌理的有机融合，传承历史文脉，携取荆楚传统民居建筑的精髓，体现天人合一的园林生态观，以及建筑与自然融为一体的规划理念，运用现代的设计手法创作出以当代人生活观念为基础、以人类发展的眼光为主导、既具有时代感又尊重传统文化、重拾荆楚文化灵魂和神韵的生活空间。

奈曼旗新区核心地块修建性详细规划

设计构思

本项目位于内蒙古自治区通辽市奈曼旗大镇北侧，奈曼旗新区核心区域，舍力虎大街以南，诺恩吉雅大街以北，共有5个地块，从北到南依次为行政区、市政广场区、生态居住区和两块商业用地，总用地面积54.7公顷。

方案总体设计通过对基地比较全面的分析与理解，确定总体规划构思：以马头琴形中心景观轴为空间主轴，并以市政广场和商业广场为两个空间核心，使周边的奈曼旗产业发展研发综合楼、住宅和商业均能共享利用，并具有良好的景观，从而形成以核心景观轴为纽带、广场为中心、多点辐射的多组团布局。共同构成"一轴、两心、四区"，功能完善，结构合理的整体架构。

上海市北工业园区12号地块10号楼

项目地点：上海
总建筑面积：16 800平方米
设计时间：2007年
竣工时间：2009年

设计构思

上海市北工业园区12号地块位于整个工业园区的中心区域，是整个园区开发建设的核心地块。规划建设有7栋建筑，总建筑面积30万平方米。12号地块内10号楼总建筑面积16 800平方米，其以现代主义风格为主，运用铝板、玻璃、面砖等现代材料，在有机的立面线条分割下显得简洁、有序，充满时代感与文化气息，丰富的形体变化与完美的细部设计，透露出对现代主义的理解与把握，通过建筑将工业新区的理念完美地呈现出来，形成符合工业园区特点的且具有时代特色的和超前理念的现代主义高技派建筑群，使12号地块成为整个市北工业新区的文化标杆。

CDI上采国际设计集团前身为香港神采设计建筑装饰总公司，于1988年成立于香港。一直以建筑规划及室内设计为主业，并在相关领域不断开拓新业务，随着公司业务及规模的不断扩张，为了能更好地为客户服务，于2006年创立CDI上采国际设计集团，业务范围包括规划设计、建筑设计、高尔夫社区及度假村规划设计、室内设计、私人定制别墅、装饰工程、软装工程、工程管理、家居馆经营等。

CDI (Corporate Design International) Design Group was formerly incorporated in Hong Kong as Corporate Design Center since 1988. However, due to the ever expanding nature of our business and customers' specific demand, our services expanded into other areas of design aspects of the building and construction industries. In 2006, we have founded CDI Design Group that specializes in master planning, architectural design, golf course design & management, resort community planning and design, interior decorating, custom designed villas, interior decorating, home furnishings as well as project management etc.

地址：上海市闵行区虹梅路3125号302室
Add: Room 302, 3125 Hongmei Road, Minhang District, Shanghai
邮编 P.C.: 201103
电话 Tel : +86–21–6446 6696
传真 Fax: +86–21–6446 9866
邮箱 E-Mail: hemman.choi@cdidesigngroup.com
网址 Http: //www.cdidesigngroup.com

相关业务请联系市场部 唐佳芝
Miranda Tang: 139 1699 3006

GOLF COMMUNITY & RESORTS ARCHITECTURE 高尔夫社区及度假村

▲ QINGYUAN LION LAKE GOLF CLUB 清远狮子湖高尔夫会所

Total Site Area 总用地面积：10,005 m^2
GFA 总建筑面积：17,000 m^2
Plot Ratio 容积率：1.69
Design Time 设计时间：2009年
Completion Time 竣工时间：2010年

▼ GUANGZHOU NINE DRAGON LAKE GOLF CLUB 广州九龙湖高尔夫国王堡酒店

Total Project Area 总设计面积: 7,391 m^2
Design Time 设计时间: 2009年
Completion Time 竣工时间: 2009年

ARCHITECTURE 建筑规划

▼ NINGHAI SHANGJIN SPA RESORTS
宁海上金温泉度假村

Total Site Area 总用地面积：1,653,333 m^2
GFA 总建筑面积：307,311.13 m^2
Plot Ratio 容积率：0.18
Design Time 设计时间：2011年（设计中）

▲ NINGBO SHENGGUANG HOTEL(EXTERIOR RENOVATION)
宁波盛光大酒店（外观改造）

Total Site Area 总用地面积：30,000 m^2
GFA 总建筑面积：30,000 m^2
Plot Ratio 容积率：1.0
Design Time 设计时间：2008年
Completion Time 竣工时间：2008年

▼ DALIAN OCEAN WORLDVIEW MODEL HOUSE
大连红星海原墅

Total Site Area 总用地面积：48,683 m^2
GFA 总建筑面积：20,998 m^2
Plot Ratio 容积率：0.43
Design Time 设计时间：2008年
Completion Time 竣工时间：2010年

MODEL HOMES 样板房

◀ SHANGHAI SHESHAN GOLF VILLA
上海佘山高尔夫别墅

Total Project Area 总设计面积：1,005 m²
Design Time 设计时间：2002年
Completion Time 竣工时间：2003年

▼ NINGBO YOUNGOR SEA VIEW GARDEN
宁波雅戈尔海景花园

Total Project Area 总设计面积：2,300 m²
Design Time 设计时间：2006年
Completion Time 竣工时间：2006年

▼ SHANGHAI CHATEAU PINNACLE
上海华山夏都

Total Project Area 总设计面积：20,000 m²
Design Time 设计时间：2005年
Completion Time 竣工时间：2006年

► BEIJING LONGFOR SUMMER PALACE SPLENDOR
北京龙湖颐和原著

Total Project Area 总设计面积：794 m²
Design Time 设计时间：2009年
Completion Time 竣工时间：2009年

INTERIOR DESIGN & HOME FURNISHINGS 室内设计与家居摆设

◀ NANJING ZHONGSHAN GOLF VILLA
南京钟山高尔夫别墅

Total Project Area 总设计面积：1,000 m^2
Design Time 设计时间：2009年
Completion Time 竣工时间：2009年

▶ DALIAN NANSHAN 1910
大连南山1910

Total Project Area 总设计面积：800 m^2
Design Time 设计时间：2010年
Completion Time 竣工时间：2010年

◀ COFCO JUNDING HUAYUE CLUB
中粮君顶红酒会所

Total Project Area
总设计面积：4,000 m^2
Design Time
设计时间：2010年
Completion Time
竣工时间：2011年

Shanghai Tongwei Architectural Design Co., Ltd.

上海同为建筑设计有限公司

上海同为建筑设计有限公司起源于1999年成立的同济大学D＋M建筑设计工作室，2003年成立公司。公司从零开始，经过历年的风雨变迁、艰辛刻苦，始终坚持设计品质，不断研究世界先进设计思想，结合当代中国的社会需要、时代特征、人文品质，以埋头创作、不事张扬的精神，积极参与城市设计、城市规划、建筑设计、景观设计和室内设计的创作探索。

在业务发展过程中，秉承"兼容并蓄、大道同为"的企业文化理念，公司注重整合各类社会资源，形成了全方位的、综合性的设计服务能力。同时注重与境外设计师和设计团队的合作，先后引入加拿大、意大利、克罗地亚、日本、法国、美国、新加坡、菲律宾等境外设计师参与公司各类项目，为多个城市留下了值得称赞的作品，获得了一致好评。

作品多次获得各种奖项，先后获得了美国、俄罗斯、日本的设计奖项三项，国内设计奖项五项，各类项目投标中标。并与国内部分地区的开发商、政府以及NGO学术组织建立了长期战略合作伙伴关系。

公司机构包括TWUAD（上海）城市建筑设计艺术中心、TOPWAY城建开发投资顾问机构。2006年与浙江斯凯莱特装饰型材有限公司及厦门海石景观有限公司合作成立了上海尚石景观装饰设计工程有限公司。

目前公司业务侧重于大型房地产项目的前期策划和规划，政府土地利用及旧区改造更新的规划，大型城市景观工程及绿地规划，高档别墅区、高档住宅区、高档特色酒店、会所等建筑的室内外综合统筹规划设计及装修工程顾问，教育建筑及校园新区的规划建筑景观设计，高档特色办公楼宇的策划与建筑景观设计，创意产业园区的策划与规划建筑景观设计。

我们希望能继续拓展与海内外各类建设单位、设计机构的合作，共同为建设美好家园、创造人文生活空间尽心尽力。

大道同爲 TWUAD

兼容并蓄　大道同为

地址：上海市普陀区同普路1220号同普大厦8楼1座
电话：+86-21-6598 9200
传真：+86-21-6598 8766
网址：http://www.twuad.com/

From D+M Architectural Design Limited of Tongji University, Shanghai Tongwei Architectural Design Co., Ltd. was founded in 2003. Starting from scratch, after years of ups and downs, the company has always insisted on quality, constantly studied advanced design ideas in the world, and by combining designs with social needs, characteristics of the times and humanities quality, it actively participated in creation and exploration of urban design, urban planning, architectural design, landscape design and interior design under the unassuming spirit of hard-working.

In the process of development, with the cultural idea of "Inclusiveness & Cooperation", the company pays attention to integrating all kinds of social resources and forms an all-around comprehensive ability of design service. Meanwhile, it also focuses on cooperating with foreign designers and designing teams, and has invited designers from Canada, Italy, Croatia, Japan, France, USA, Singapore, Philippines etc to participate in all kinds of projects, which have left some amazing works for many cities and won a good reputation.

Designing works have won all kinds of awards, such as three awards from USA, Russia and Japan in sequence, five domestic awards and bid-winning of various projects. The company has established long-term strategic partnership with developers, governments and non-governmental academic organizations of some domestic areas.

The company includes TWUAD (Shanghai) Urban Architectural Design & Art Center, TOPWAY Urban Construction Development & Investment Consultants. In 2006, it cooperated with Zhejiang Skylight Decorative Section Co., Ltd. and Xiamen Ocean Rock Garden & Landscaping Co., Ltd. and founded Shanghai Shangshi Landscape Design Engineering Co., Ltd.

At present, the company focuses on pre-scheming and pre-planning of large-scale real estate projects, projects of government land use and reconstruction of old areas, large-scale urban landscape projects and vegetation planning, engineering consultant of indoor and outdoor comprehensive and integrated planning design and decoration of high-class villa areas, high-class residential communities, high-class special hotels and clubs, landscape planning of educational building and planned buildings of new campus, planning and architectural landscape design of high-class special office building, planning of creative industry park and landscape design of planned architectures.

We are hoping for further cooperation with all kinds of construction and design companies at home and abroad. By joining hands and sparing no efforts, we will build a beautiful place and create a humanity living space.

大连革镇堡基地位置

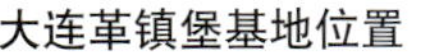

大连革镇堡基地位置

大连革镇堡基地位置

大连革镇堡现状肌理

大连革镇堡新城规划

主持规划师：李长君
主创规划师：杨文伟
项目团队：聂 蓉、王军峰、李树仁、
杨 磊、朱艺萍、国贞梅
设计时间：2010年
规划用地面积：220公顷
规划总建筑面积：3 050 000平方米
综合容积率：1.38
综合建筑密度：33%
综合容积率：45%

本规划采用“一心一轴一片二带三线三潭十组团”的规划布局结构，组合运用“居、庭、舍、院、墅、集、栈、庄、湖、园”的概念，结合山体水系，将住宅、学校、商业、酒店、办公、生活配套服务、绿化景观、体育休闲运动系统有机结合，构成“二龙戏珠，金龟抱蛋”的大格局。

大连革镇堡天际线分析图

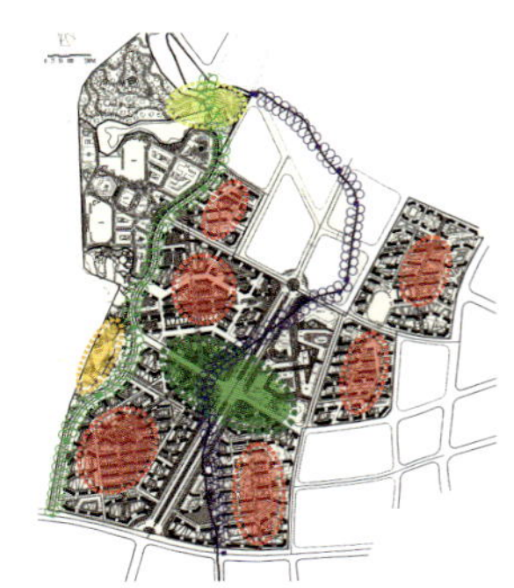
景观结构分析图

规划结构分析图

上海虹桥竹谷创意园

主持建筑师：李长君
主创建筑师：P. Svjetlana（意）、越 剑、
Seiki Mori（日）
项目团队：杨文伟、李梦谷、杨 磊、倪 琨、
吴帼英、李政华、聂 蓉、马燕萍、
尹必勤、国贞梅、李 冀、沈士钧、
李仕波、王 龙、张柯
委托业主：上海竹谷投资管理有限公司
设计时间：2007—2009年
用地面积：45 000平方米
总停车位：928个
容 积 率：2.2
建筑密度：35%

本项目意在打造虹桥商务区核心地带的时尚地标，为国际品牌设计师和创意机构提供环境品质优越的工作生活交流环境。

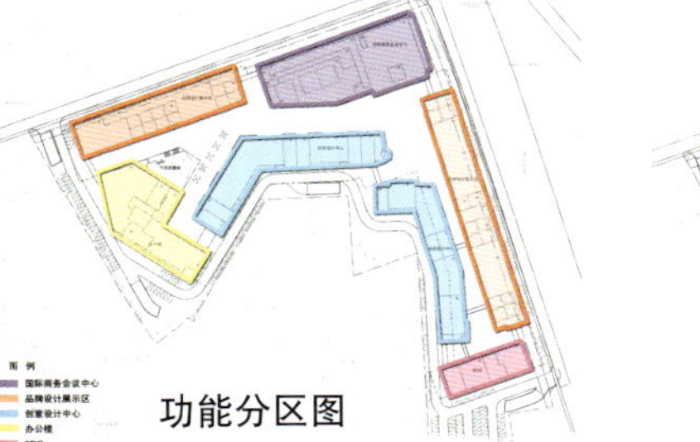
功能分区图

景观分析图

交通分析图

江苏江阴南闸如意滩公园

设计单位：上海同为建筑设计有限公司
甲方单位：南闸镇政府
主持设计师：李长君、聂　蓉
参与设计师：王军峰、马艳萍
项目面积：43 346.98平方米
项目地点：江苏 江阴
竣工时间：2011年

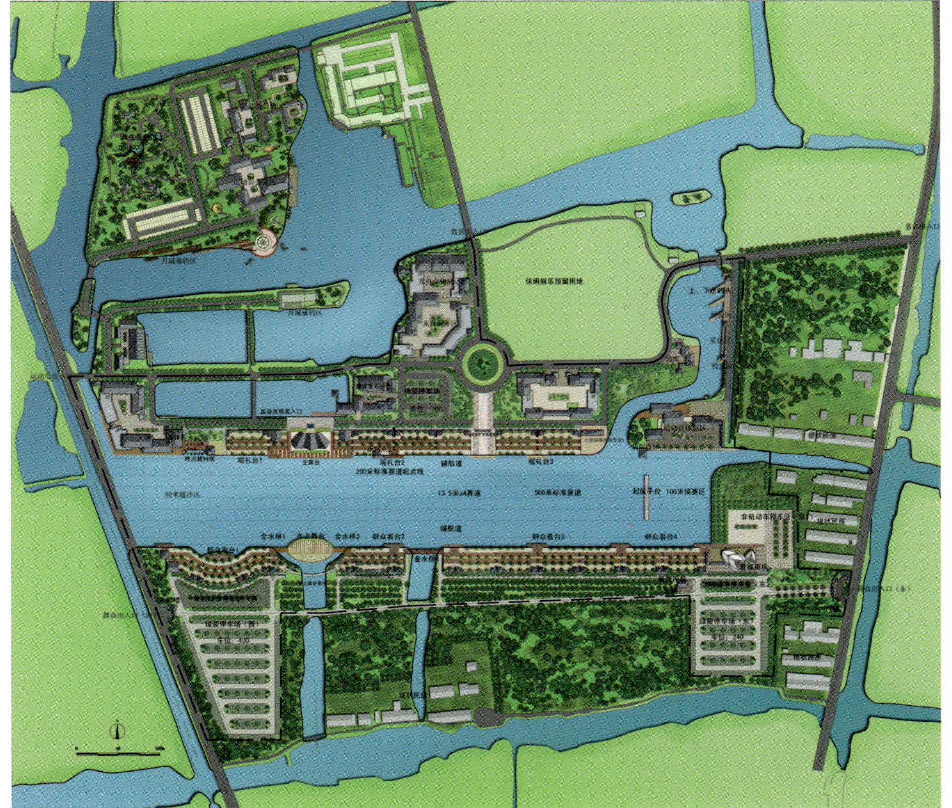

江苏省龙舟训练竞赛基地

“江苏省龙舟训练竞赛基地”是2010年经江苏省无锡市和江阴市体育部门批准的国家级重点群众性体育运动基础项目，位于月城镇双泾地区，周边交通便利、水网密布、水质优良，生态环境优美，并建设有林果园、农夫果园、农渔文化馆等生态旅游景点。

该赛道位于双泾生态园南部，总规划面积约为28万平方米。以龙骨泾河为基础，拓建长850米、宽100米的500米国际标准龙舟直道赛道，并配套建设有上下水码头、主席台、观礼台、观赛平台、运动员休息室、公共停车场、卫生间等硬件设施，可以承办大型国际级龙舟赛事。同时还可以作为皮划艇、游艇、赛艇等水上项目的训练基地。

本案是以景观规划为主的综合性项目，场地景观设计及植物种植（面积约9万平方米）以群众看台和绿荫停车场为主，比赛配套建（构）筑物（约6 000平方米）以主席台、水上舞台、裁判台、媒体发布中心为主。整体景观建筑设计主要体现江南风格，尊重该项目的地理现状，着重突出“水韵月城”的设计理念。

江阴森茂国际汽车文化城

项目地点：江苏 江阴
项目规模：139 092平方米
总 层 数：24层

该项目位于江阴市区东南部，背面紧靠城市快速路芙蓉大道并与江阴的总部经济功能组团相邻，东临东外环路，南靠澄杨汽车文化大道，西面为江阴车管所。地理位置优越，交通运输便捷。设计中坚持贯彻国家政策、法令的有关规定，积极协调业主需要、用地限制与国家政策法规之间的矛盾。综合采用新技术、新材料、新设备，以求建成一座高品位、高新技术、国际化、现代化的国际汽车文化城。

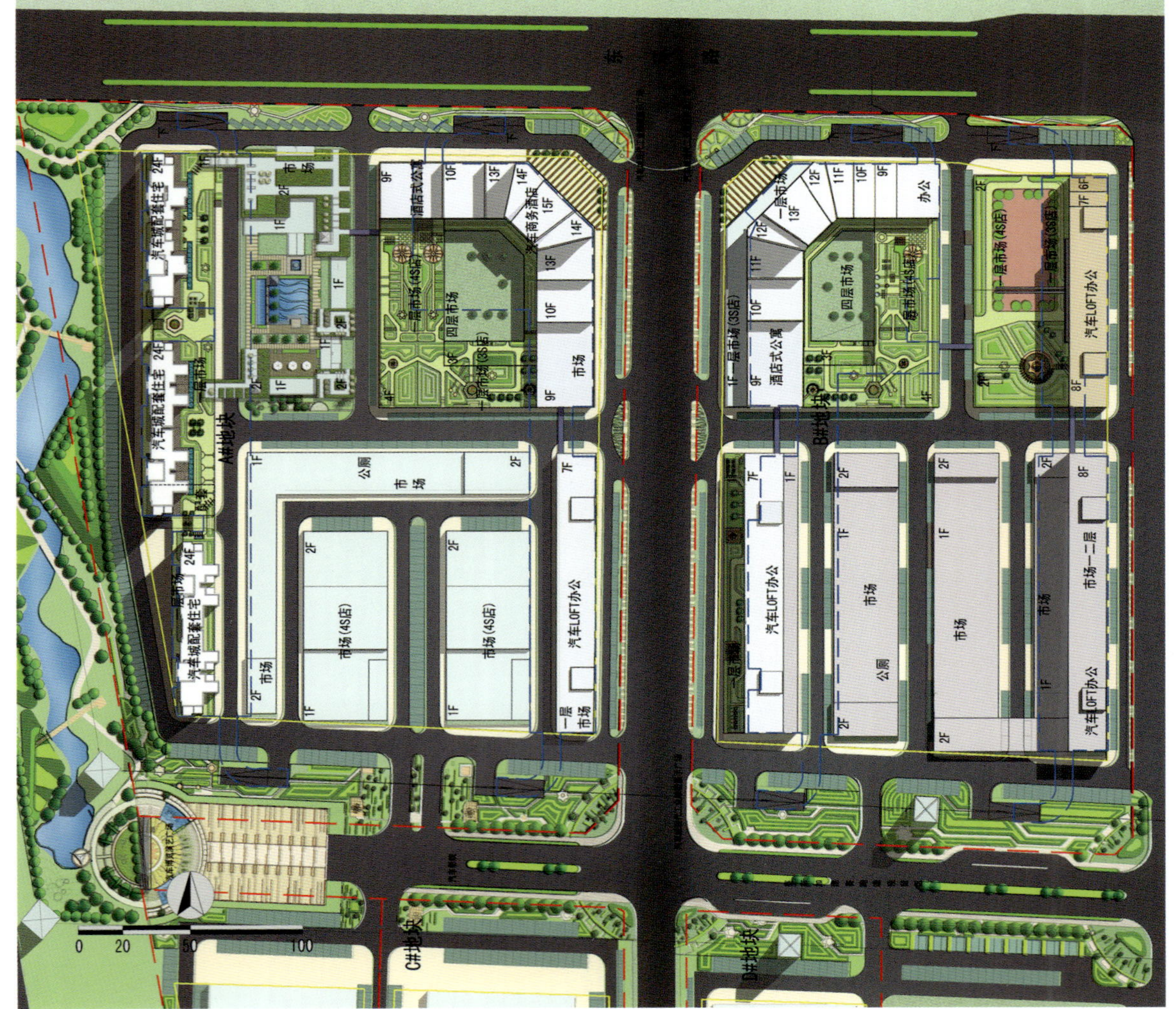

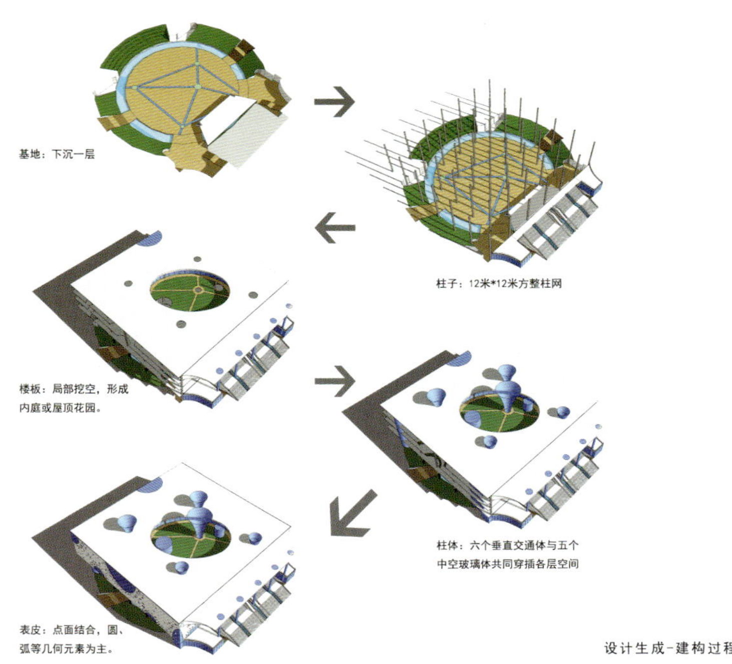

设计生成-建构过程

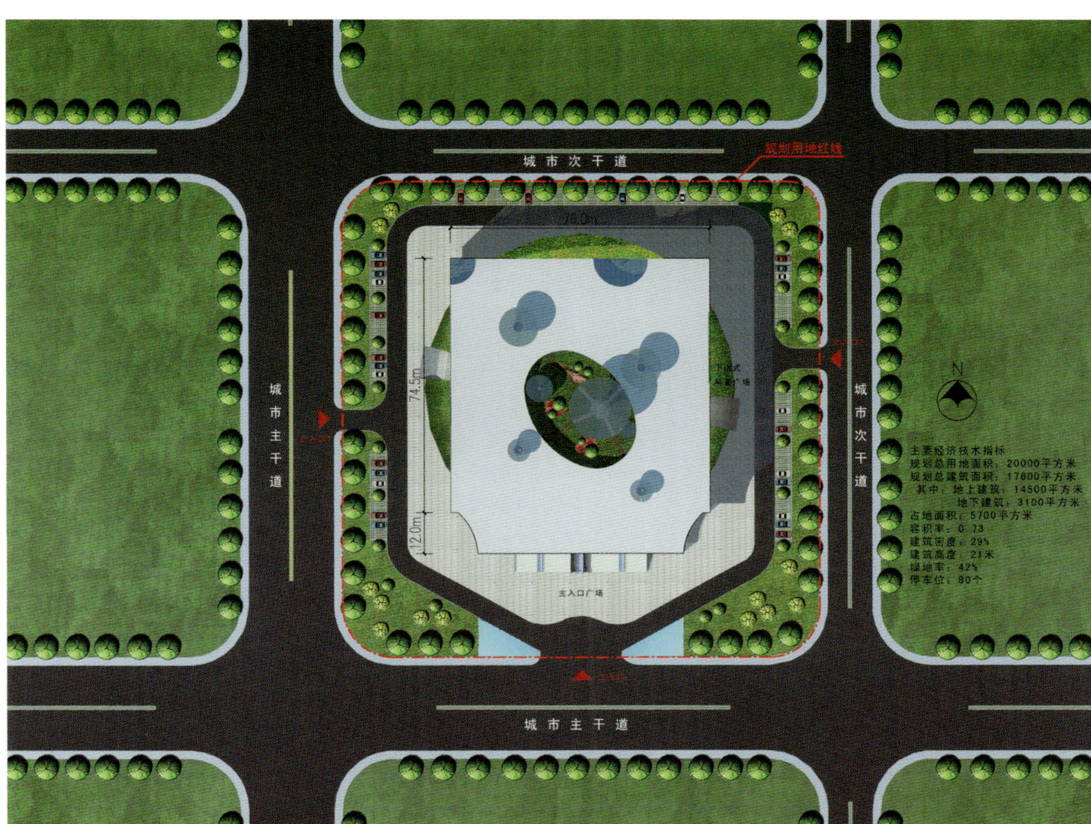

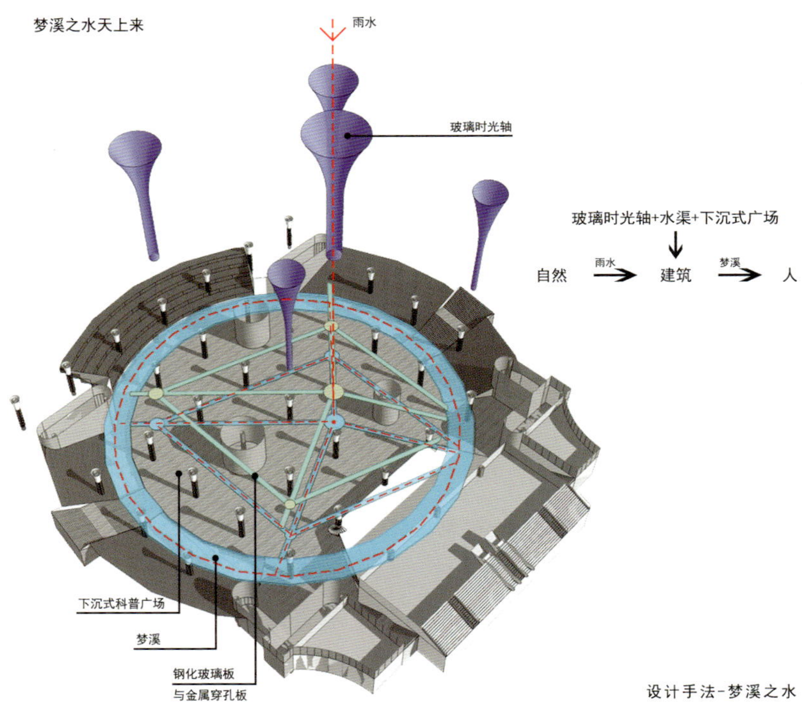

设计手法-梦溪之水

镇江科技馆

本案力求创造一个对话场景：即对古代的追忆与现代的感知以及未来的畅想紧密交织；人们在它们不断的对话中，感受着过去、现在和未来。若将下沉广场视为这个场景的舞台，而人是主角，那么串联过去、现在和未来的道具则必不可少——这就是“玻璃时光轴”。

建筑竖向空间设计为：地下一层、一层以及三、四层，层高均为5米，二层层高6米，建筑高度21米。6个筒形垂直交通体共同串联各层空间。另有5个竖向的喇叭形中空玻璃体穿插在各层空间，使各层内部空间丰富多变；并为各层空间带来流动的光照和通风。此外，局部空间形成上空的内庭，使得整个建筑形成极具流动性和多样性的展示空间。四层中部设有屋顶花园，为其后勤办公人员提供优质的空间感受。建筑表皮的设计采用现代流行语汇，点、面结合，塑造清新、活泼的立面形象。

Shanghai UAD Architectural Design Firm (General Partnership)

上海优爱建筑设计事务所（普通合伙）

地址：上海市普陀区陕西北路1438号1501室
电话：+86-21-62981500
传真：+86-21-62981536
邮箱：mail@uadgroup.com.cn
网址：www.uadgroup.com.cn

Add: Room 1501, No.1438 North Shaanxi Road, Putuo District, Shanghai
Tel.: +86-21-62981500
Fax: +86-21-62981536
E-mail: mail@uadgroup.com.cn
Http://www.uadgroup.com.cn

上海优爱建筑设计事务所（普通合伙）于2009年3月成立，2010年获得国家住房和城乡建设部颁发的建筑设计事务所甲级资质。

公司具有广泛的国际化背景，聚集众多优秀规划、建筑、景观、室内精装修设计专业人士，提供建筑设计、规划设计、室内设计、景观设计及项目管理的全面服务。在创作过程中，公司全面认知项目的土地价值、开发定位、文脉延续、艺术感知及心理体验，以开放、理性和积极的姿态，去寻求创作中的每一个机会和最佳答案。

每个项目设计组奉行〝以精心创精品〞的专业宗旨，把每一个作品都当做公司的代表作，在品牌建设上追求成功率。公司也把市场认同度作为衡量设计成败的重要尺度，处处着眼于项目的最佳效果，注重项目运行全过程的实操跟进，一切的努力在于把设计理念变成精彩现实，众多项目的业绩已证明优爱的设计为产品所带来的高效附加值和绝对竞争力。

公司坚持〝以优秀人才为基础，以周到、细致的服务为保证，以高质量的设计产品为生命〞的发展思路，通过人性化的设计理念，寻求富含人文价值的场所精神，创作具有时代精神的建筑艺术作品。

Founded in March 2009, Shanghai UAD Architectural Design Firm (general partnership) was qualified as Class-A Architectural Design Company by the Ministry of Construction PRC.

With extensive international backgrounds, gathering a great number of design professionals of planning, architecture, landscape, interior decoration, UAD GROUP provides comprehensive services of architectural design, planning and design, interior design, landscape design and project management. In the creative process, after knowing completely about value of the land, development orientation, culture extension, artistic perception and psychological experience of the project, the company will look for every opportunity and the best answers in creation with an open, rational and positive attitude.

Under the professional spirit of “hard work creates quality”, all project design teams take every project as the company’s representative work, pursuing the success rate of brand building. The company also takes market acceptance as an important yardstick to measure the success or failure of a design. Focusing on the best result of a project and follow-up of the entire practical-operation process of the project, the company spares no efforts to turning the design idea into a wonderful reality. Performances of many projects designed by UAD GROUP have been proved to be able to bring high added value and absolute competitiveness for the products.

Under the development spirit of “basis on talents, insurance by considerate and thoughtful services, and life for high-quality designs”, in pursuit of cultural value, the company has created artistic architectures with a modern spirit by humanized design ideas.

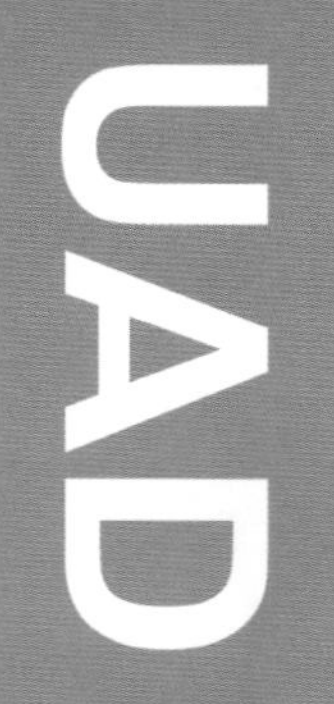

崇明县港西镇崇明酒店项目

业主单位：上海鹏昱置业有限公司
合作设计单位：泛太平洋设计集团（加拿大）

总经济技术指标：

规划用地面积：84 126.13平方米
总建筑面积：36 478.9平方米
特色精品酒店面积：22 390.7平方米
会所面积：6 756平方米
休闲商业面积：1 978.6平方米
独立酒店面积：5 353.6平方米
会议型别墅面积：488.7平方米
温泉型别墅面积：1 051.2平方米
派对居家型别墅面积：387平方米
收藏型别墅面积：1 002.7平方米
度假居家型别墅面积：2 424平方米
容 积 率：0.3
建筑密度：14.8%
绿 化 率：66.3%
机动车停车位：134个

港东公路

总平面图 1:1000

上饶老火车站地块综合改造项目

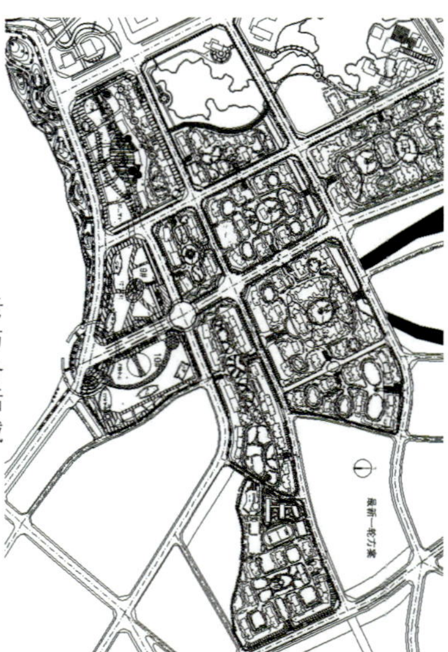

项目地点：江西 上饶
业主单位：绿地集团南昌事业部
设计时间：2010年
用地面积：461 610平方米
建筑面积：1 529 313平方米
建筑密度：28.35%
容 积 率：3.31

由“上乘富饶”而得名的上饶是一座历史源远流长的古城，它牵江浙，出沪宁，携八闽，达粤桂，素有“八省通衢”、“豫章第一门户”之荣称。随着国家城市高铁项目的进展，上饶老火车站已失去运输功能，原有站房面临城市功能的更新与改造，原有地块面临城市肌理的缝合与重组。老火车片区位于上饶市复合城市中心的核心地段。绿地集团通过提供居住和商业配套服务的支持，极大地提升了行政办公金融文体中心的核心功能，充分挖掘和激发出老城商业中心的活力潜质，并引领着三江片区文化娱乐的新风尚。通过综合改造将充分实现城市土地价值，弥补城市功能的缺失，完善城市基础设施，提高城市形象，实现上饶市跨越式发展的总体目标。

Integrated Reconstruction Project of Shangrao Old Railway Station

Project Location: Shangrao, Jiangxi
Owner: Nanchang Division of GreenLand Group
Design Time: 2010
Land Area: 461,610 m^2
Floor Area: 1,529,313 m^2
Building Density: 28.35%
Floor Area Ratio: 3.31

With the meaning of “a land of milk and honey”, Shangrao is an ancient city with a long history, which connects Jiangxi Province and Zhejiang Province, with highways to Shanghai and Ningbo, and reaches to Guangdong and Guangxi provinces crossing Fujian Province. Therefore, it has won the names of “City Connecting Eight Provinces” and “The No.1 Portal of Yuzhang County”. With the development of China’s high-speed railway project, Shangrao Old Railway Station has lost its function. Its building needs to be improved and reconstructed according to city functions, while its land lot needs to be recombined and divided according to city planning. The Old Railway Station is located in the complex downtown core area of Shangrao City. By providing supports of residential and commercial services, GreenLand Group has greatly improved the core function of Administrative Office Finance Culture Sports Center, fully explored and inspired the motivation and potentiality of old town commercial center, and has brought a new fashion of entertainments to Sanjiang New Area. Integrated reconstruction will fully realize land value of the city, make up the loss of city function, improve city infrastructures, raise city image, and achieve the overall target of leap-forward development of Shangrao City.

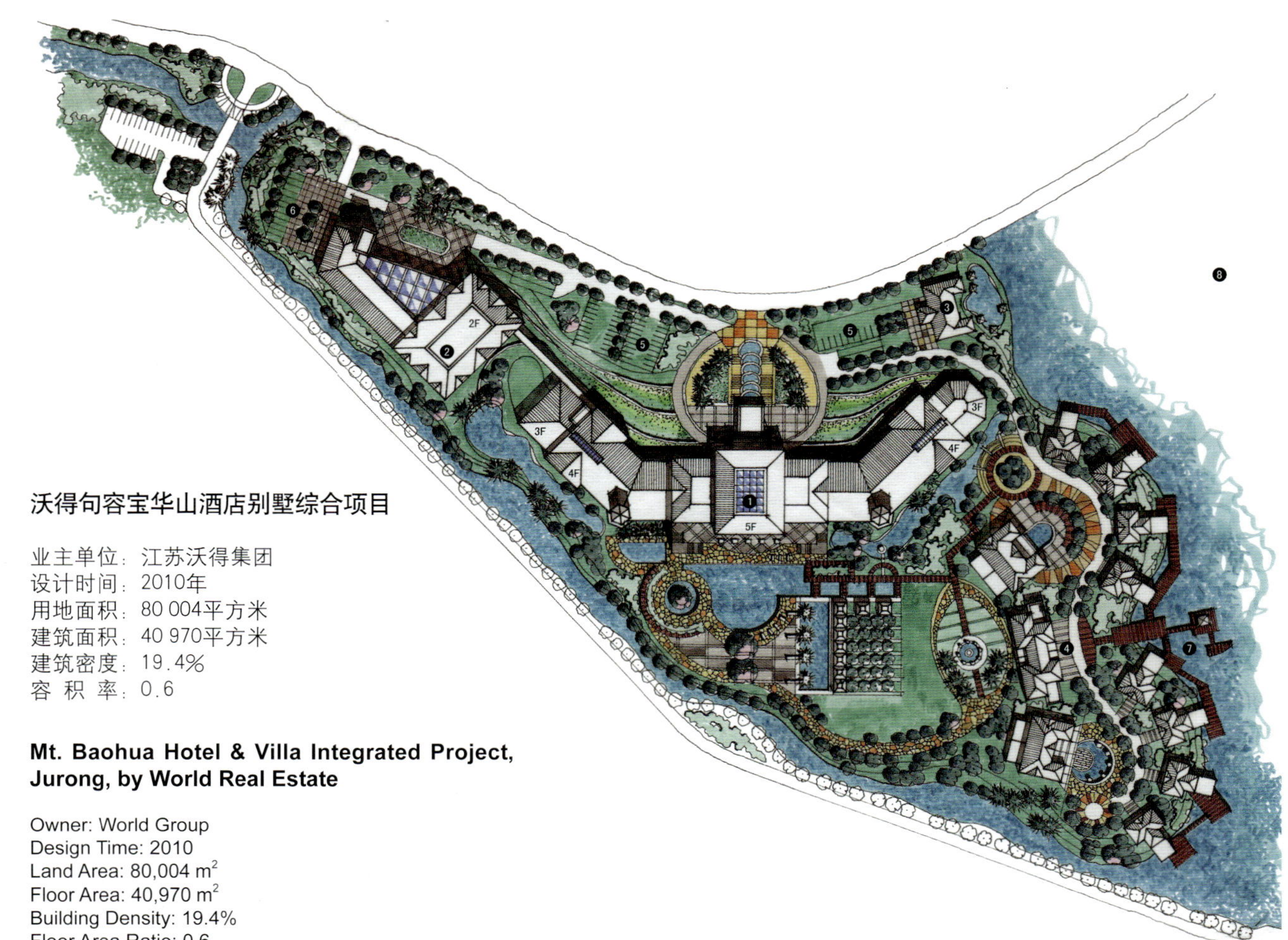

沃得句容宝华山酒店别墅综合项目

业主单位：江苏沃得集团
设计时间：2010年
用地面积：80 004平方米
建筑面积：40 970平方米
建筑密度：19.4%
容 积 率：0.6

Mt. Baohua Hotel & Villa Integrated Project, Jurong, by World Real Estate

Owner: World Group
Design Time: 2010
Land Area: 80,004 m^2
Floor Area: 40,970 m^2
Building Density: 19.4%
Floor Area Ratio: 0.6

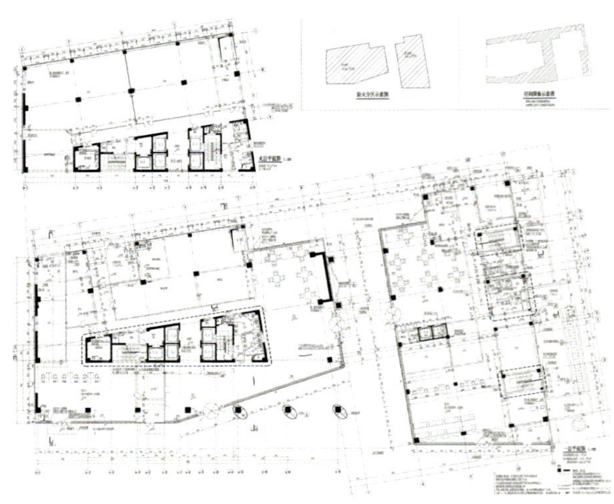

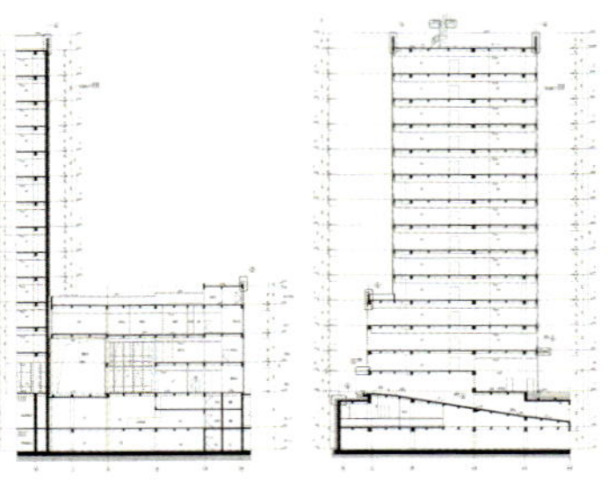

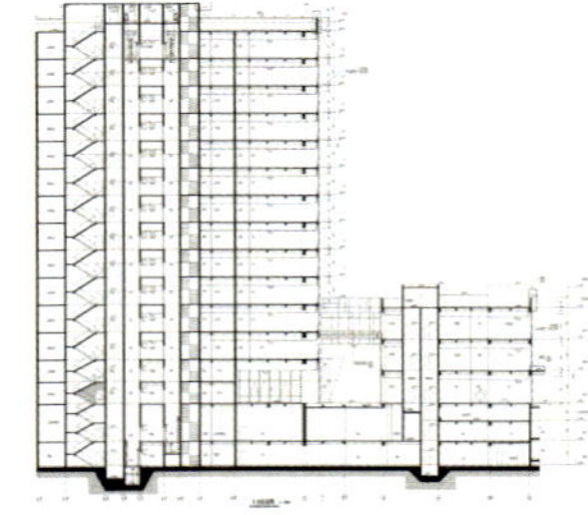

新华东国际大厦

项目地点：上海
业主单位：上海华东实业有限公司、上海国际集团
合作设计单位：上海泛巢建筑设计事务所
竣工时间：2010年
用地面积：5 557.4平方米
建筑面积：32 642平方米
建筑密度：44.2%
容 积 率：4.0

本项目位于上海浦东新区浦东南路1540号。东临浦东南路，西靠东格致中学操场，南北为城市现状居住社区。周边商业、商务等配套环境成熟完善，交通便利，是浦东新区的黄金地段。

方案合理地解决了新建办公与现有住宅、学校的冲突与矛盾，在建筑空间塑造、日照权益分配、交通组织安排上均有妥善设计。外立面采用工厂预制单元式幕墙体系，幕墙单元设计为三种尺寸规格，便于施工和安装。双层LOW-E中空玻璃、陶土板、金属板三种材料的组合形成错落有致的肌理效果，整体风格体现了现代高效的办公形象，简洁大方。建筑南侧、西侧和建筑底层减少玻璃的使用面积，有效地避免了办公建筑与南侧住宅之间的视线干扰，降低了西晒的不利因素。

New East China International Building

Project Location: Shanghai
Owner: Shanghai East China Industrial Co., Ltd., Shanghai International Group
Completion Time: 2010
Land Area: 5,557.4 m^2
Floor Area: 32,642 m^2
Building Density: 44.2%
Floor Area Ratio: 4.0

The project is located on No.1540 South Pudong Road of Pudong New Area, Shanghai, which faces South Pudong Road on the east, neighboring with the playground of East Gezhi Middle School on its west side, with residential communities on the south and north. With the mature and complete environment of commerce and business, as well as convenient transportation, it is the gold area of Pudong New Area.

The project has properly solved the conflicts and contradictions between newly-built office building and the existing residential communities, schools, with appropriate designs of architectural space, sunshine equity allocation and traffic organization. Facade of the building adopts factory prefabricated modular wall systems. Wall units are designed into three sizes to facilitate construction and installation. Three materials of Double LOW-E insulating glass, clay plates and metal plates are combined together to form a patchwork. The overall style reflects the image of modern and efficient office building with a simple and generous appearance. By reducing the use of glass on the southern, western and bottom parts of the building, we can effectively avoid sight interference between office building and southern residential communities, and reduce negative factors of western exposure.

Shanghai De-Sign Architectural Design Co., Ltd.

上海鼎实建筑设计有限公司

地址：上海市虹口区大连路1619号骏丰国际财富广场17F
电话（市场部）：+86-21-65150081
电话：+86-21-65156550 +86-21-65153550
传真：+86-21-65155560
邮箱：ds@de-sign.cn
网址：WWW.DE-SIGN.CN

Add: 1619 Dalian Road, Hongkou District, Shanghai
No. 17F Fortune Plaza Chun Feng International
Tel (Marketing): +86-21-65150081
Tel: +86-21-65156550 +86-21-65153550
Fax: +86-21-65155560
E-mail: ds@de-sign.cn
Http: //www.de-sign.cn

新加坡DS鼎实国际建筑设计集团是鼎实国际和谨阁两家以建筑设计为主的综合性设计公司，在新加坡、上海分别注册，目前在北京也设立了分公司。公司现有员工120余人，从事广泛的设计服务项目，包括全面的建筑、工程以及相关服务，如规划与建筑设计、景观设计、标志系统设计等。设计类型涵盖城市综合体、商业零售、度假酒店及居住建筑。DS鼎实国际在di 2009年全国民营企业排行榜中位列第33名，并被评为"2010年度最具潜力商业地产设计企业"。

DS鼎实国际的"绿地·无锡太湖大道商业中心"项目荣获2010年全国人居经典方案综合大奖；"西宁佳和世纪广场"项目荣获2010年全国人居经典方案规划金奖；"常熟琴湖国际购物中心"项目荣获2010年全国人居经典方案规划金奖。

现上海公司拥有中国建筑甲级资质，有专业从事商业地产的建筑设计师60余人，设计作品屡次获奖，已成为具有一定知名度和影响力的商业建筑设计、商业空间设计及商业后期运营管理的综合机构。为许多地产开发商、跨国公司、事业单位和政府部门完成了众多成功的作品。国际化的设计理念和丰富的设计管理思路使鼎实国际在城市综合体、商业建筑和商业空间领域为客户提供全程化服务，包括商业咨询、业态规划、建筑设计、商业空间塑造、商业景观、标志系统等。目前公司主要为万达集团、绿地集团、宝龙集团、北科建集团、金辉集团、恒开投资、青特集团、青岛中联置业、银鹭集团、上海悦合置业、上海苏河湾集团、徐州华厦集团、中国星宝控股集团、江苏佳和置业、江苏鸿裕投资、重庆鹏润置业等众多国内知名开发企业提供全程建筑与规划设计服务。DS鼎实国际致力于服务国内高端客户，在商业综合体领域，创造独具魅力的建筑空间。

鼎实人以自强不息的拼搏精神，遵循"强化科学管理，追求一流设计，力创鼎实国际品牌，确保质量第一"的质量目标，恪守"换位、沟通、创新、求精"的设计方针，与各界同仁精诚合作，奋力进取，共创美好未来。

DS Group is a Singapore-based architectural design-led practice, structured by DE-SIGN International and Attic China with approximately 120 employees. DS Group has offices in Singapore and Shanghai, and presently has branch office in Beijing. Today DS Group offers extensive design services, including the full range of architecture, project and related services, such as planning and architectural design, landscaping, and signage design. Its design patterns cover urban complexes, commercial & retail, resort hotels, and residential developments. DS Group was ranked No. 33 in the Top Private Enterprises of China in 2009, and earned "Award for the Most Potential Consultant at Commercial Real Estate Field". Its "Greenland Taihu Lake Blvd. Business Center, Wuxi" Project won the General Award of Typical Conception of China Human Habitat in 2010; its "Jiahe Century Plaza, Xining" Project won the Gold Prize in Planning of Typical Conception of China Human Habitat in 2010; and its "Qinhu Lake International Shopping Center, Changshu" Project won the Gold Prize in Planning of Typical Conception of China Human Habitat in 2010.

Now, its Shanghai firm which is licensed for PRC construction class-A, has over 60 architectural designers in commercial properties, with many design works winning numerous honors. And has become a well recognized and influential comprehensive organization of commercial building design, commercial space design and commercial after-sale operation & management. It has completed many successful works for many developers, MNCs, public services, and governmental agencies. Its international design philosophy and rich design management roadmap enable DE-SIGN International to provide customers with seamless services, including commercial consulting, industry planning, architectural design, commercial space building, commercial landscaping, signage system and others, in the fields of urban complex, commercial building and commercial space. Currently, the Company provides seamless architectural & planning design services to many famous domestic developers, including Wanda Group, Greenland Group, Powerlong Group, Beijing Science Park Development Group, Kamfei Group, Hengkai Investment, Qingte Group, Qingdao Zhonglian Property, Yinlu Group, Shanghai Yuehe Property, Shanghai Suzhouwan Group, Suzhou Kina Group, China SINB Holding Group, Jiangsu Jiahe Enterprises, Jiangsu Hongyu Investment, and Chongqing Pengrun Real Estate. DS Group is committed to serving domestic high-end customers, which creates a charming building space in the field of commercial complex.

DS people in the spirits of constant struggle follow the quality target "to strengthen scientific management, to pursue first-class design, to build DS brand, and to put quality in the first place", abide by the design guidelines of "tolerance, communication, innovation and perfection", and cooperate with all social communities, work hard to create a beautiful feature.

绿地集团无锡太湖大道东亭路商业综合体

设计单位：DS鼎实国际建筑设计有限公司（新加坡、上海、北京）
主创设计师：刁 睿、谢 璇、姬 涛、周 娜、吴旭辉
用地面积：76 800平方米
总建筑面积：366 260平方米
容 积 率：3.5
项目地点：江苏 无锡

项目说明

项目位于无锡市锡东新城，北侧紧邻太湖大道，东侧为东亭路，西侧为华夏南路，南侧为地块内规划道路。地块北侧为50米城市绿化带，其中30米为地块内可建设用地。地块内由一栋150米超高层写字楼、三栋100米LOFT办公、两栋100米公寓式办公、五层购物中心、多层商业街区，以及两层地下车库（含部分商业街、超市及设备用房）组成。

DS鼎实国际在实践中一直在探索具有前瞻性的商业空间，场所创造主题化是一个非常有效的手段。在鼎实的理念中，商业空间绝对不是单纯购物的场所，而是城市生活的舞台，它兼具着多重社会使命：创造新的公共空间、激发新的使用模式、提供全新的生活体验、赞扬纯粹美好的人性。

在绿地集团无锡太湖大道中央广场项目中，鼎实创造性地将基地内50米宽500米长的城市绿化带设计成一个令人耳目一新的绿地海岸线。项目的主题灵感来源于城市人心目中温馨的精神家园——海滨假日。“虽然你人在无锡，却可以体验到仿佛不在无锡的一天，体验到在绿地海岸度假的美好一天。”你能想象到的度假活动、运动、学习、看电影、婚礼、艺术展、阅读、遛狗、吃饭、聚会等都与商业活动顺畅地在海滨绿地上融为一体。购物即度假，度假即购物。商业开发脱离了粗浅的商场范畴，衍化成一个全新的城市人休闲娱乐度假的主题场所。

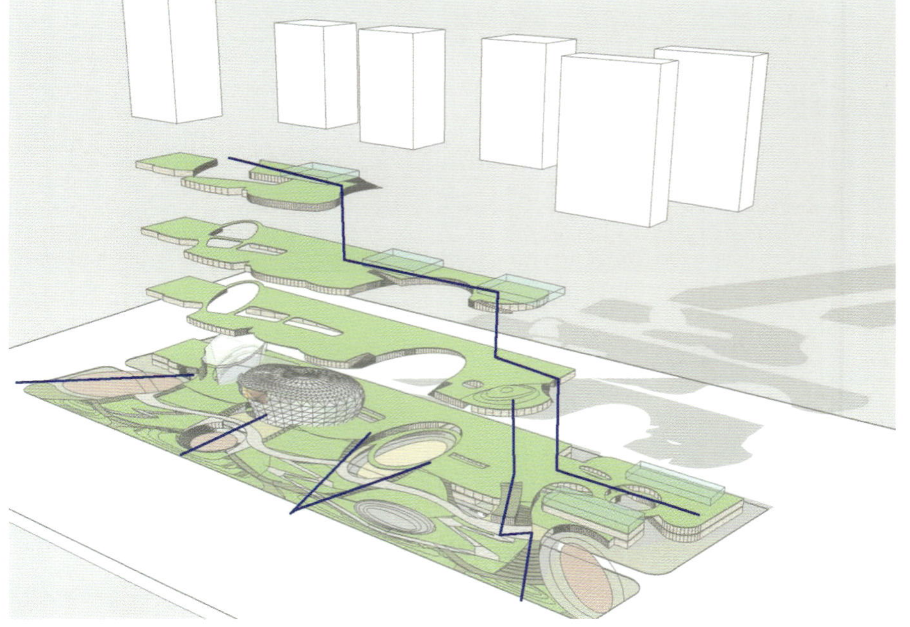

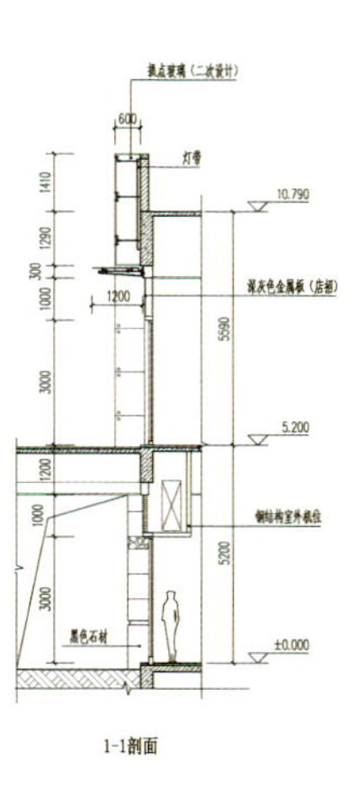

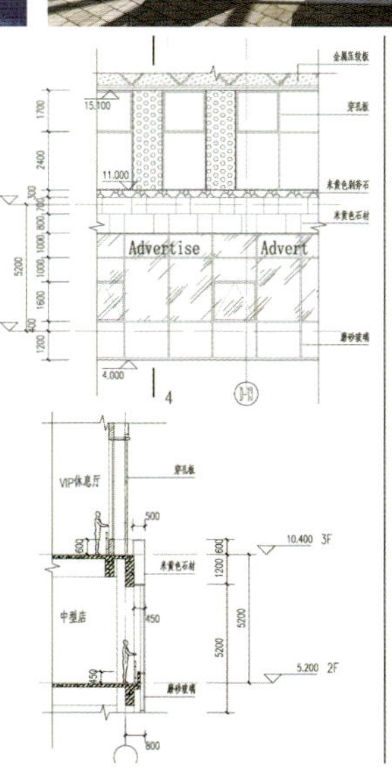

上海曹安商贸城悦合国际广场

设计单位：DS鼎实国际建筑设计有限公司（新加坡、上海、北京）+UDG
主创设计师：刁 睿、谢 璇、周 娜
用地面积：45 868平方米
建筑面积：215 900平方米
容 积 率：2.7
项目地点：上海

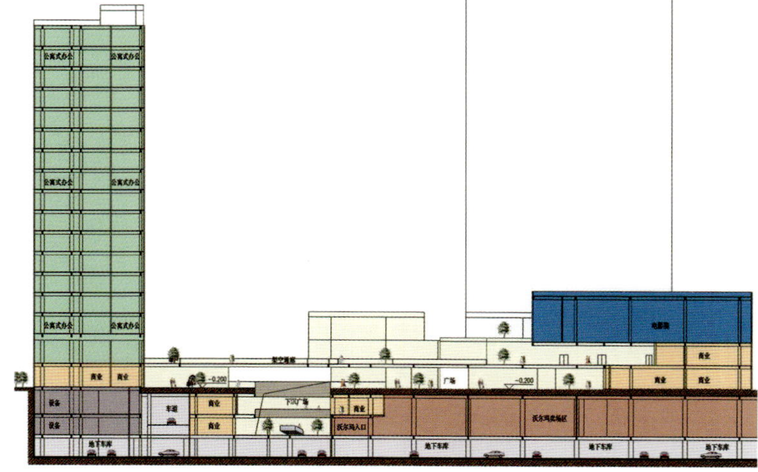

项目说明

项目位于上海市嘉定区真新社区曹安商贸城。DS鼎实国际首先在用地内由东至西设计了一条景观轴线，连接曹安商贸区内部轴线与城市公共绿地。其次，将景观轴线立体化，形成二层开放景观平台，有效地建构整个商业街区的互动立体人行系统。规划中接近45 000平方米的商业街区主要布局在一二层，局部三层。商业流线以景观轴为核心形成环线，保证商业价值的最大化。同时，在地下一层的沃尔玛会员店和三层的IMAX影院等几个大型主力店，形成强大的商业吸引力以保证商业的持续、良性运营。近8万平方米的办公楼设置成四栋高层，三栋为公寓式办公，以微曲的弧形形成一个连贯的沿河界面。2万平方米的甲级综合办公楼位于基地的西南角，形成造型上的标志节点。四栋主楼在设计上考虑充分呼应外环及沪宁高速交叉口的城市形象要求。

设计理念

城市——希望从城市的角度审视整个设计过程，通过与城市空间的对话，合理地规划项目的结构与空间，实现项目自身价值的同时也对城市的发展作出重要贡献，创造具有假日风情的居住社区和室外商业活动空间。
商业——希望最大限度地实现地面一层到三层的商业价值，同时与周边商业形成错位竞争，填补地区休闲类商业的空白，提升曹安商贸片区商业的档次和多样性。
生态——希望能与周边的绿地、公园相互渗透，融为一体，共同形成集绿化、景观、休闲为一体的综合性体验式购物公园。

青岛政建集团金地世纪城

设计单位：DS鼎实国际建筑设计有限公司
（新加坡、上海、北京）
主创设计师：刁 睿、吴旭辉、张春陆
用地面积：53 921平方米
建筑面积：20.15万平方米
容 积 率：2.49
项目地点：山东 青岛

项目说明

项目周边有世纪美居家具城、中韩小商品城等规模较大且较为成熟的专业性市场，因此，本项目在业态上引入五星级酒店、超市、公寓、影院、餐饮娱乐、儿童体验乐园等，希望能和现有的商业业态形成互补，成为城阳空港商贸区的中心。

由于基地内规划地铁线的原因，用地被分成南北两块，北地块面积较小，布置麦当劳等中小型主力店。其余大部分的功能均位于可建设用地较大的南地块。规划地铁线上部被设计成集散广场及绿地，并和城市主要道路有较好的连接，结合其余入口广场的设置，人流能顺畅及便捷地出入，激发出更大的商业价值。在规划中将"商业美学"的开发理念与城市发展规划相融合，全力打造商业核心区具有城市地标性、高附加值、多功能化的城市商业精品。同时，做足"环保、生态、科技"的文章，寓休闲于生活，寓娱乐于工作，从而推动"城阳空港商贸区"的快速发展，为项目周边及整个城阳区，乃至全青岛的商业腾飞和城市发展，创造巨大的影响力和拉动力。

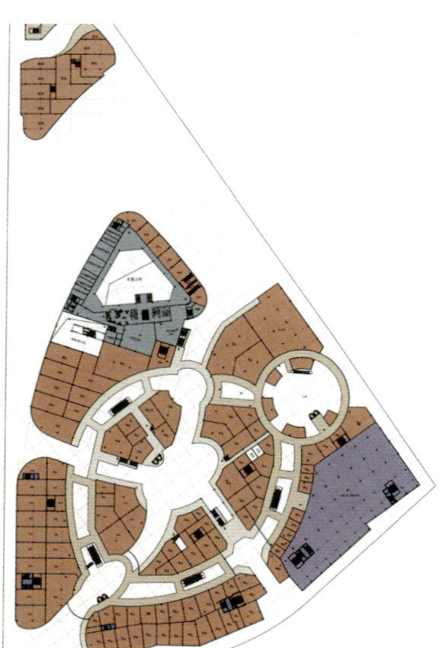

项目二层平面图

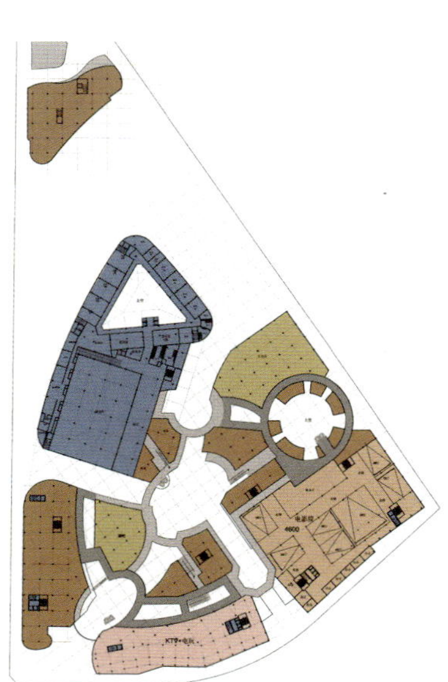

项目三层平面图

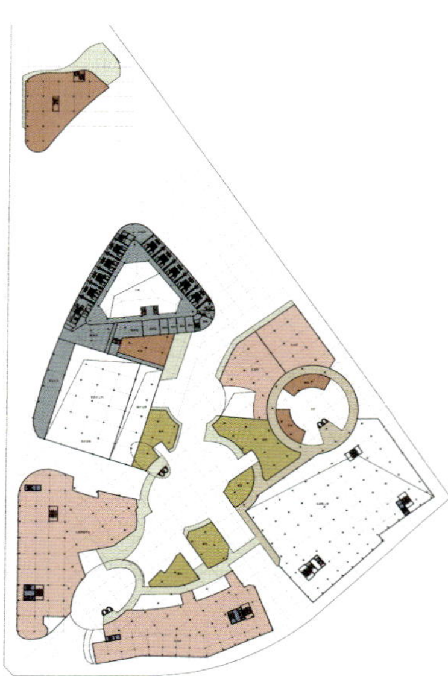

项目四层平面图

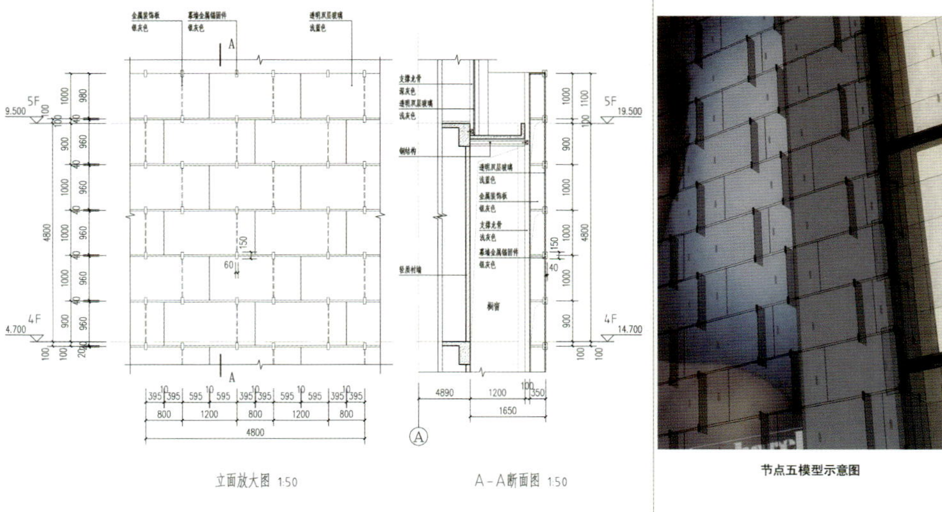

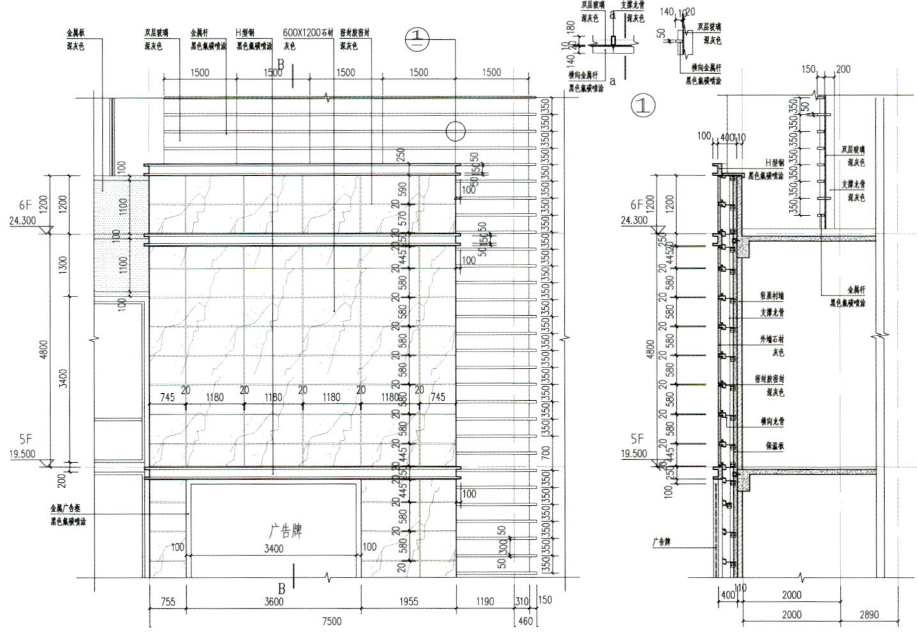

苏州佳和商厦

设计单位：DS鼎实国际建筑设计有限公司（新加坡、上海、北京）
主创设计师：刁 睿、吴旭辉、周 娜
建筑面积：32 464.03平方米
项目地点：江苏 苏州

项目说明

项目地处于苏州古城区中心地带金阊区石路步行街，区域位置优越。项目总建筑面积为32 464.03平方米，其中地上建筑面积为30 630.49平方米，地下建筑面积为1 833.54平方米。原建筑外立面主要为面砖和玻璃幕墙，材料已严重老旧，外立面造型已不能满足石路步行街的需求。本次立面改造重在突出公建化、标志化的设计主题，体现苏州新时代的建筑风貌，强调三段式美学构图，尊贵大气，稳重挺拔。立面材料主要以米灰色石材、深灰色玻璃、透明玻璃、深灰色金属板和金属百叶为主。设计手法上采用新古典的设计元素，裙房部分立面强调现代与古典的碰撞，采用新颖的装饰性玻璃幕墙并充分考虑广告位的设置，符合商业建筑的性格；塔楼部分通过竖向装饰构件突出建筑的挺拔感，强调虚实对比和材料质感色彩的搭配，亦能体现其高贵典雅的理性之美。

中关村长三角创新园商务功能区A-1地块

设计单位：DS鼎实国际建筑设计有限公司
(新加坡、上海、北京)
主创设计师：刁 睿、谢璇、朱炳臣
用地面积：45 981平方米
建筑面积：182 728平方米
容 积 率：2.7
项目地点：浙江 嘉兴

项目说明

项目位于中关村长三角创新园商务功能区中的A-1用地内，处于嘉兴市中心区与西部秀洲新区交接部位，占地面积约4.6公顷，项目由一栋超高层建筑和一个5层的购物中心组成。其中塔楼为31层，设计功能为5A级写字楼；商业分两大块功能，西边1-3层为集中百货、4-5层为影城，东边1-2层为精品商业，以中间的带形中庭相连，3层为餐饮与溜冰场及相关附属配套。地下室设为2层，地下1层主要为办公、商业的后台区、设备机房和超市、集中餐饮等功能。地下2层主要为集中地下停车。

购物中心围绕设置在3层的溜冰场组织商业流线，配以餐饮等功能，通过声光电的手法打造购物中心的主题空间，来吸引更明确的目标客群，带来更多的体验次数和更长的逗留时间，将多种不同的消费形式自然地融合在一起，使商业空间本身也成为商业定位及营销的重点。

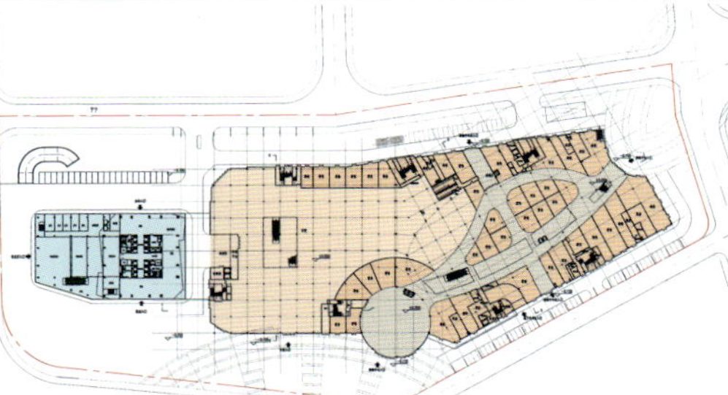

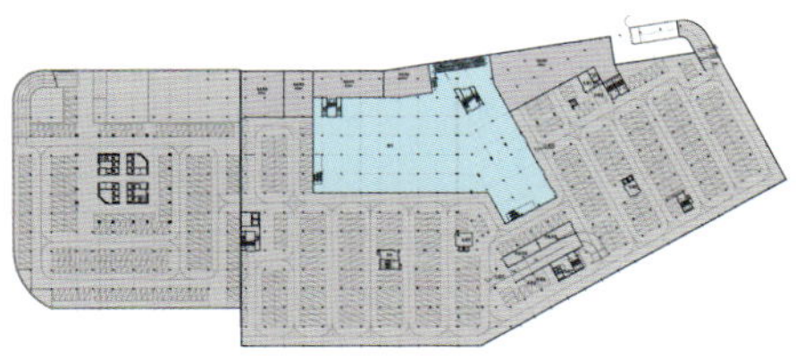

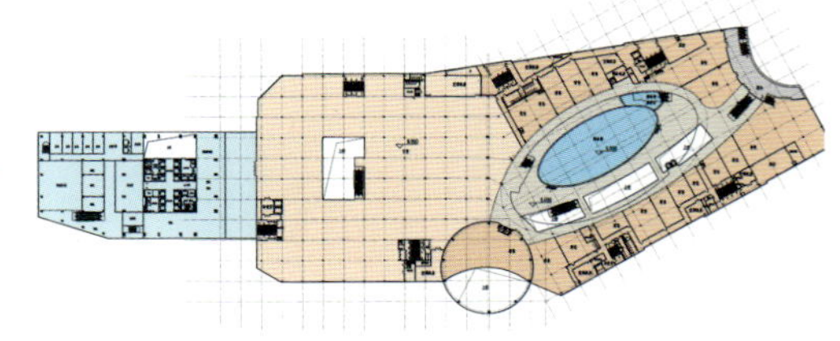

无锡洛社新城商业中心二期

设计单位：DS鼎实国际建筑设计有限公司（新加坡、上海、北京）
主创设计师：刁 睿、谢 璇、周 娜
用地面积：69 800平方米
总建筑面积：12 9947平方米
容 积 率：1.59
项目地点：江苏 无锡

项目说明

本街区位于无锡市惠山区洛社新城洛城大道以西，盛南路以北及以东，北临规划中的人工景观湖。地块地理位置及交通条件优越，场地地势平坦，地质条件良好。鉴于本案良好的地理位置，主要人流由东、南、西三个方向进入。

DS鼎实国际在整体空间布局上采用了点、线、面结合的设计手法，在D10地块设计了一个中心节点和一条商业街、景观带构成的主轴，在可自成体系的同时，又通过一条贯穿基地的景观道路和一条次级步行道路将本项目和D08地块有机联系在一起。在几条主轴将基地合理地划分之后，对于面状空间的处理则采用了较为灵活的街区模式。街道、内院的交叠出现共同构成了丰富的商业环境。

规划中以水为主题，引入了"水流"的设计理念，将商业人流比喻成水流、人流从洛城大道、新城西路的各个主要入口进入基地后，步行100米左右均能到达相应的中心广场，人流在广场汇集之后又能穿行于各个尺度宜人的街区商业中，空间序列富于节奏。与此同时，为了使商业街区中二、三层的商业价值最大化，设计中采用了多首层的有机设计模式，在面向湖面和广场的建筑设计中采用层层退台的处理手法，在延伸了景观层次的同时，也增进了消费人群的视线交流，将商业街区编织成了一个立体的交通网络。这样一个立体的商业街区，通过多层次的空间处理，力求形成热闹繁荣的商业氛围，最大限度地延长商业人流的逗留时间。此外，利用退台形成的步行交通体系较之于利用外挑的廊道作为交通的处理在商业的得房率上更具优势。

在整个商业街建筑形态上，采用了地面与空中两种相互平行又相互穿插的建筑手法来解决整个商业街的二至四层，空中商铺的吸引人流问题，在地面上，设置了很多景观外梯来吸引人流上二层或三层，又在空中商业区用外廊和过街廊形式把整个商业区的二层以上全部连通起来，使整个商业区形成了地面与空中相结合的立体式购物环境，更好地体现出了每一间商铺的商业价值。

开发管理\项目定位
规划设计\建筑设计
园林环艺\装饰装修
智能顾问\电力服务
照明设计\工程监理

上海市卢湾区复兴中路1号申能国际大厦2006-2008A室
Tel: +86-21-63919361 13976558740
Fax: +86-21-63919152
P.C.: 200021
E-mail: chenzhong@jia-design.cn
Http://www.jia-design.cn

福建省福州市东大路36号花开富贵A座29F
Tel: +86-591-87430666
Fax: +86-591-87430679
P.C.: 350001
E-mail: GB_JBSJ@163.com
Http://www.gbfj.com

▼ 长沙喜盈门城市广场

建筑类别：商业综合体
建筑面积：46万平方米
项目地点：湖南 长沙
投 资 商：湖南润领房产

▶ 台州黄岩北岸项目

建筑类别：商业综合体
建筑面积：15万平方米
项目地点：浙江 台州
投 资 商：台州市政府

开发管理\项目定位
规划设计\建筑设计
园林环艺\装饰装修
智能顾问\电力服务
照明设计\工程监理

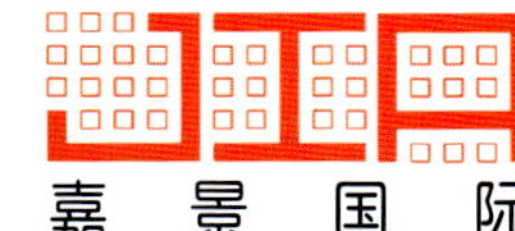

上海市卢湾区复兴中路1号申能国际大厦2006-2008A室
Tel: +86-21-63919361 13976558740
Fax: +86-21-63919152
P.C.: 200021
E-mail: chenzhong@jia-design.cn
Http://www.jia-design.cn

福建省福州市东大路36号花开富贵A座29F
Tel: +86-591-87430666
Fax: +86-591-87430679
P.C.: 350001
E-mail: GB_JBSJ@163.com
Http://www.gbfj.com

◀ 成都喜盈门城市广场

建筑类别：商业综合体
建筑面积：55万平方米
项目地点：四川 成都
投 资 商：上海喜盈门

▼ 南宁华南城1号广场

建筑类别：商业综合体
建筑面积：48万平方米
项目地点：广西 南宁
投 资 商：华南城

开发管理\项目定位
规划设计\建筑设计
园林环艺\装饰装修
智能顾问\电力服务
照明设计\工程监理

上海市卢湾区复兴中路1号申能国际大厦2006-2008A室
Tel: +86-21-63919361 13976558740
Fax: +86-21-63919152
P.C.: 200021
E-mail: chenzhong@jia-design.cn
Http://www.jia-design.cn

福建省福州市东大路36号花开富贵A座29F
Tel: +86-591-87430666
Fax: +86-591-87430679
P.C.: 350001
E-mail: GB_JBSJ@163.com
Http://www.gbfj.com

◀ 漳浦永嘉天地

建筑类别：商业中心、酒店式公寓
建筑面积：10万平方米
项目地点：福建 漳州
投 资 商：永嘉置业

▼ 福州大学城永嘉天地

建筑类别：城市商业综合体
建筑面积：18万平方米
项目地点：福建 福州
投 资 商：轩辉地产

◀ 重庆永辉生活广场

建筑类别：商业中心、总部大楼
建筑面积：10万平方米
项目地点：重庆
投 资 商：永辉集团

房地产开发管理及全专业技术服务集团

开发管理\项目定位
规划设计\建筑设计
园林环艺\装饰装修
智能顾问\电力服务
照明设计\工程监理

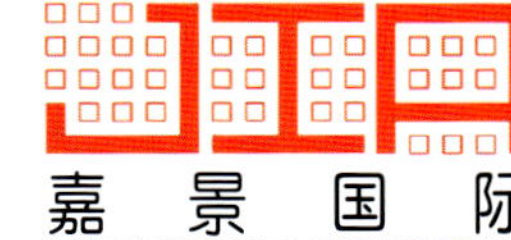

上海市卢湾区复兴中路1号申能国际大厦2006-2008A室
Tel: +86-21-63919361 13976558740
Fax: +86-21-63919152
P.C.: 200021
E-mail: chenzhong@jia-design.cn
Http://www.jia-design.cn

福建省福州市东大路36号花开富贵A座29F
Tel: +86-591-87430666
Fax: +86-591-87430679
P.C.: 350001
E-mail: GB_JBSJ@163.com
Http://www.gbfj.com

▼ 百联大厦

建筑类别：5A级写字楼
建筑面积：6.2万平方米
项目地点：福建 福州
投 资 商：百联集团

▶ 恒宇国际大厦

建筑类别：商业中心、写字楼
建筑面积：3.9万平方米
项目地点：福建 福州
投 资 商：福州永泰建设发展有限公司

开发管理\项目定位
规划设计\建筑设计
园林环艺\装饰装修
智能顾问\电力服务
照明设计\工程监理

上海市卢湾区复兴中路1号申能国际大厦2006-2008A室
Tel: +86-21-63919361 13976558740
Fax: +86-21-63919152
P.C.: 200021
E-mail: chenzhong@jia-design.cn
Http://www.jia-design.cn

福建省福州市东大路36号花开富贵A座29F
Tel: +86-591-87430666
Fax: +86-591-87430679
P.C.: 350001
E-mail: GB_JBSJ@163.com
Http://www.gbfj.com

▲ 福州三盛奥林匹克花园

建筑类别：住宅
建筑面积：146万平方米
项目地点：福建 福州
投 资 商：三盛房产

◀ 福州南屿康桥–丹提

建筑类别：住宅
建筑面积：44万平方米
项目地点：福建 福州
投 资 商：龙旺集团

开发管理\项目定位
规划设计\建筑设计
园林环艺\装饰装修
智能顾问\电力服务
照明设计\工程监理

上海市卢湾区复兴中路1号申能国际大厦2006-2008A室
Tel: +86-21-63919361 13976558740
Fax: +86-21-63919152
P.C.: 200021
E-mail: chenzhong@jia-design.cn
Http://www.jia-design.cn

福建省福州市东大路36号花开富贵A座29F
Tel: +86-591-87430666
Fax: +86-591-87430679
P.C.: 350001
E-mail: GB_JBSJ@163.com
Http://www.gbfj.com

◀ 石家庄辛集项目 ▼

建筑类别：住宅
建筑面积：33万平方米
项目地点：河北 石家庄
投 资 商：融晟房产

▲ 融侨福清项目

建筑类别：住宅
建筑面积：43万平方米
项目地点：福建 福州
投 资 商：融侨房产

开发管理\项目定位
规划设计\建筑设计
园林环艺\装饰装修
智能顾问\电力服务
照明设计\工程监理

上海市卢湾区复兴中路1号申能国际大厦2006-2008A室
Tel: +86-21-63919361 13976558740
Fax: +86-21-63919152
P.C.: 200021
E-mail: chenzhong@jia-design.cn
Http://www.jia-design.cn

福建省福州市东大路36号花开富贵A座29F
Tel: +86-591-87430666
Fax: +86-591-87430679
P.C.: 350001
E-mail: GB_JBSJ@163.com
Http://www.gbfj.com

▲ 龙华云锦 ▼

建筑类别：住宅
建筑面积：42万平方米
项目地点：江西 萍乡
投 资 商：乐升房产

◀ 华云山庄

建筑类别：山地别墅
建筑面积：55万平方米
项目地点：福建 福州
投资商：盛天房产

房地产开发管理及全专业技术服务集团

开发管理\项目定位
规划设计\建筑设计
园林环艺\装饰装修
智能顾问\电力服务
照明设计\工程监理

上海市卢湾区复兴中路1号申能国际大厦2006-2008A室
Tel: +86-21-63919361 13976558740
Fax: +86-21-63919152
P.C.: 200021
E-mail: chenzhong@jia-design.cn
Http://www.jia-design.cn

福建省福州市东大路36号花开富贵A座29F
Tel: +86-591-87430666
Fax: +86-591-87430679
P.C.: 350001
E-mail: GB_JBSJ@163.com
Http://www.gbfj.com

▼ 华润橡树湾

建筑类别：联排别墅、高层住宅　　项目地点：福建 福州
建筑面积：90万平方米　　开 发 商：华润置地

▲ 世纪金源 月亮岛

建筑类别：联排别墅、总部商住基地　　项目地点：福建 福州
建筑面积：70万平方米　　开 发 商：世纪金源集团

上海唯景建筑设计事务所
Shanghai First View Architecture Design Association

总经理：王黎华
地　址：上海市松江区九亭镇虬泾路126号(A)4楼
电　话：+86-21-67635800*8925
传　真：+86-21-67635800*8923

General Manager: Wang Lihua
Add: Fourth Floor, No.126(A) Qiujing Road, Jiuting Town, Songjiang District, Shanghai
Tel: +86-21-67635800*8925
Fax: +86-21-67635800*8923

上海唯景建筑设计事务所，系上海三盛宏业投资集团下属企业，成立于2002年，是一家拥有国家住房和城乡建设部颁发的具有甲级设计资质的建筑设计事务所，并已获得ISO9000质量管理体系认证证书。公司注重人文管理，形成了一套健全的激励机制，吸引了一批设计领域优秀人才的加盟，目前事务所共有建筑设计人员20余人，其中国家一级注册建筑师、高级工程师6人，主要设计人员都具有丰富的大中型建筑工程设计经验和较高的专业设计水平。

公司长期致力于大型居住区、酒店建筑、城市综合体、优秀历史保护建筑更新和改造等提供专业的设计服务，设计项目分布在全国各地。近年来，每年完成工程设计超过100万平方米的建筑面积，其中独创的“颐景园”品牌精品住宅小区，遍布上海、浙江、广东、安徽、山东、辽宁等地的十几个城市，得到了开发商、专家和业主的一致好评，且获得了多个国家级优秀设计奖项，在全国享有较高的知名度和美誉度，被业内誉为“第一园林地产”，并连续六年荣膺“中国房地产园林地产专业领先品牌”称号，代表作品有上海泗泾颐景园、佛山南海颐景园等。同时，公司也积极与众多国内外知名公司合作，积累了丰富的战略合作经验，为拓展公司在大型城市公共建筑、大型住宅小区、高端别墅产品等建筑设计业务方面提供了更高的运行平台，代表作品有上海衡山宾馆改造工程、上海大厦改造工程、上海东珠花园等。

公司非常关注新建筑技术和新建筑材料的应用，是“低碳联盟理事会”会员，根据低碳经济的目标，在工程系统中，积极采用新工艺与新技术，并得到了比较好的效果。同时，充分利用各种先进的专业软件，有效提高了质量控制和成本控制的效率。

公司成立至今，一直以“生成于自然，贡献于生态，亲和于人类，尊重于历史，经营于文化”为设计理念，以复兴建筑文化为使命，恪守“高品质、高品位、高品味”的设计原则，发扬“追求卓越”的企业精神，力争不断创造出令开发商满意、受业主欢迎的建筑作品。

主要设计作品

舟山中昌国际大厦、舟山海景时代广场、上海衡山宾馆综合楼改扩建工程、上海大厦改造工程、上海安信商业广场A块办公楼改造、苏州新宝厂房、上海泗泾颐景园、上海九亭颐景园、上海嘉定颐景园、上海美兰湖颐景园、上海东珠花园、佛山桂丹颐景园、沈阳三盛颐景园、沈阳三盛颐景园376地块、舟山海景颐园、杭州颐景园、富阳颐景山庄、南通七星花园、合肥三盛颐景园、舟山新晨颐景园、舟山檀东颐景山庄、山东淄博颐景园、杭州丁桥颐景园、舟山颐景名苑。

Shanghai First View Architecture Design Association was founded in 2002, which subordinates to Shanghai Sansheng Hongye Investment Group. The association is a Grade A architecture design association issued by the Ministry of Construction and has obtained the ISO9000 quality management system certification. The company has attracted a number of design talents for its consistent devotion to people-oriented management and formed appropriate incentive system. At present, we have more than twenty architectural designers, 6 of them have registered architect State and Senior Title, and all of the chief designers have high professional design standards and rich experience in large-scale engineering design.

Our company is devoted to providing professional design service to large-scale residence community, hotel building, urban complex, renovation & reconstruction of outstanding historic conservation architecture, and the projects distribute all over China. In recent years, our annual design projects have exceeded one million square meters. "Yi Jing Garden" boutique residential community with our original creation has spread more than ten cities including Shanghai, Zhejiang, Guangdong, Anhui, Shandong, Liaoning, etc. The brand has obtained high praise and reputation from all our developers, experts and clients, as well as many national excellence design awards. "Yi Jing Brand" has obtained high praise and reputation as "the First Landscape Garden Property" in the field of real estate industry and won the title of "The First-class Brand in China Real Estate Garden Landscape Architecture" in six succession years from 2005 to 2011. Our representative projects are Shanghai Sijing Yi Jing Garden, Foshan Nanhai Yi Jing Garden. What's more, the company cooperates actively with many famous companies at home and aboard and accumulates rich experience in strategic corporation,at the same time it builds a higher platform for the business of large-scale public building, residence community and advanced villa products etc. The representative projects are Shanghai East Pearl Garden, Shanghai Hengshan Hotel renovation project, Shanghai Tower renovation projects.

Our company is very concerned about applying the new architectural techniques and material to the due projects, and we are also the member of "Low-Carbon Union Council". According to our goal of low-carbon economy, we actively use new technology in the engineering system and have achieved relatively good results. At the same time, we make full use of kinds of advanced professional software, which effectively improved the efficiency of quality and cost control.

Since its foundation, our company takes the idea of "originating from the nature,dedicating to the ecology, harmonizing with the mankind, respecting to the history, managing the culture" as its design concept, takes renaissance architectural culture as its mission, adheres to the "High-Quality, High-Taste, High-Grade" design principle, takes the entrepreneurial spirit of "pursuing super excellence" as its purist and try our best to create better architecture design works to satisfy developers and clients.

Major Design Works

Zhoushan Zhongchang International Plaza; Zhoushan Seascape Times Square; Renovation and Expansion Project of Comprehensive Building of Hengshan Hotel, Shanghai; Renovation Project of Shanghai Tower; Renovation Project of A Block Office Building of Anxin Business Square,Shanghai; Suzhou Xinbao Workshop; Shanghai Sijing Yijing Garden, Shanghai Jiuting Yijing Garden; Shanghai Jiading Yijing Garden; Shanghai Meilan-Lake Yijing Garden; Shanghai East Pearl Garden, Foshan Guidan Yijing Garden; Shenyang Sansheng Yijing Garden; The 376th Block of Shenyang Sansheng Yijing Garden; Zhoushan Seascape Yijing Garden, Hangzhou Yijing Garden; Fuyang Yijing Villa; Nantong Seven Star Garden; Hefei Sansheng Yijing Garden; Zhoushan Xinchen Yijing Garden; Zhoushan Tandong Yijing Villa; Shandong Zibo Yijing Garden; Hangzhou Dingqiao Yijing Garden; Zhoushan Yijing Garden.

中昌国际大厦

建设地点：浙江 舟山
建筑面积：地上：52 610.77平方米
　　　　　地下：17 399.87平方米
用地面积：7 719平方米
设计时间：2006年
竣工时间：2011年
容 积 率：6.82

项目总建筑面积约7万平方米，楼高33层，建筑高度130米，包括5层裙楼，面积约15 524平方米。集楼宇自动化系统、办公自动化系统、消防自动化系统以及安防自动化系统于一体，是目前在建的舟山市建筑中高度最高、建筑单体体量较大、安全性高、舒适性好的高档综合性写字楼建筑，建成后将成为临城中央商务区首个5A级国际商务中心、也是临城新区的标志性建筑之一。

本项目建筑立面充分考虑城市景观要求、建筑物性质与功能、建筑物造型及特色等因素，考虑多个角度观察的视觉效果，以及该建筑与周边行政中心、财富大酒店、绿城五星级酒店等的关系，使之融合在其中。塔楼与裙房互相映衬，组合成完整的建筑造型，稳健而挺拔，成为舟山市具有显著特点的标志性建筑。

九亭颐景园

建设地点：上海
建筑面积：地上：370 629平方米　　地下：30 508平方米
用地面积：330 070平方米　　设计时间：2004年
竣工时间：2006年　　容 积 率：1.12

三盛九亭颐景园，位于上海松江区九亭镇，占地33.35万平方米，规划建筑面积约38万平方米，总体地形如一枚叶状书签。其地势由西北向东南延伸，南端为淀浦河，东端沿涞亭路展开，北端近邻九亭大街，西端近邻虬泾路。区内建筑走势北高南低，呈现负阴抱阳之势，充分体现中国传统造园布局原则。园区环境以颐景湖为中心，元宝形主干道“金带环腰”，配合蜿蜒水网，将园区划分成多个组团。各组团内建筑、溪湖与园林景点环绕映衬，营造出各具特点的组团意境。建筑以点式多层为主，兼有低层别墅和富有韵律感的小高层。各型建筑立面间色彩和谐，追求严谨法度的欧式建筑交替排序，彰显中西两种建筑风格乃至中西两种文明体系之意韵美的水乳交融。

该项目曾荣获“上海十大著名水景楼盘”、“全国新世纪人居经典住宅方案竞赛规划环境金奖”、“第二届全国优秀庭院环境社区”。

泗泾颐景园

建设地点：上海
建筑面积：119 183.49平方米
用地面积：224 670平方米
设计时间：2007年
竣工时间：2009年
容 积 率：0.53

上海泗泾颐景园别墅区位于上海市西郊松江区，总占地面积34.976公顷，此次规划的二期用地位于小区中、北部，占地面积22.467公顷。

规划设计充分地尊重自然、“以人为本”，从可持续发展的角度出发，运用中国古典园林的造园手法，将欧式别墅的阳刚之美和中国私家园林的阴柔之秀有机地结合起来，创造一个布局合理、功能齐备、设施完善、现代化的私家园林式的花园别墅区。整个规划布局及景观设计都围绕颐景湖展开，采用中国太极“八卦”的形式，水为阴，山、建筑为阳，小区主体湖面“颐景湖”和主体山体“颐景山”各为“八卦”阴阳两鱼的鱼眼，它们遥相呼应成对景，是整个小区景观设计的主体和组织者。

该项目曾荣获“2008联合国人居署迪拜国际最佳范例（中国）推动奖”。

中国建筑上海设计研究院有限公司 第二设计所

CHINA SHANGHAI ARCHITECTURAL DESIGN & RESEARCH INSTITUTE CO. ,LTD

中国建筑上海设计研究院（CSCEC）有限公司隶属于中国建筑工程总公司（CSCEC），是中国最大的建筑设计集团——中建设计集团核心成员。

中国建筑上海设计研究院有限公司是由中国建筑东北设计研究院、中国建筑西北设计研究院、中国建筑西南设计研究院的上海分院和中国建筑东南设计研究院于2005年底组建而成的，是一家秉承了中建几家大设计院50多年积淀的优良传统和技术实力，又充满勃勃生机的现代建筑设计企业；作为中建设计集团在上海及华东地区的核心设计企业，具有明确的“做大上海、面向全国、走向国际”的战略定位。

中建上海院二所是在原中国建筑西北设计研究院上海分院的基础上发展而成的。目前全所各类设计人员80多人，内部根据人员和专业分工划分为以下5个设计部：建筑创作部、建筑设计一部、建筑设计二部、结构设计部、机电设计部，并在西安、石家庄设有设计分部。

中建上海院二所秉承“创新、全面、专业、尽责”的设计理念，不断加强设计团队业务建设，大量汲取最新的设计成果，广泛建立与许多国内外知名设计事务所的合作交流，以高度责任感和高水平的专业服务，为客户创造最大的商业价值，为社会创造优美怡人的城市建筑。

“精心设计、优质服务，做有理想、有责任的设计团队”是中建上海院二所全体设计人员的追求与理想。

地址：上海市普陀区武宁路503号B楼3F
电话：+86-21-32020365
网址：Http://www.cscecshi.com
邮箱：zjsh002@163.com

1-2.广州2010年亚运会亚运村规划方案

建设地点：广东 广州
建筑面积：2.7万平方千米

该方案是2007年举行的面向国内外广州2010年亚运会亚运村规划方案竞赛中最终入围的3个优胜方案之一。

3.西安高新国际商务中心

建设地点：陕西 西安
建筑面积：14万平方米
建筑高度：160米
建筑层数：40层

4

5

6

7

8

9

11

12

13

15

4. 苏州恒业大酒店

建设地点：江苏 苏州
建筑面积：110 000平方米
建筑高度：160米
建筑层数：42层

5. 昆山温莎堡广场

建设地点：江苏 昆山
建筑面积：50 000平方米
建筑高度：99.95米

6. 延安延园中学

建设地点：陕西 延安
占地面积：15.47公顷
建筑面积：100 000平方米

7. 上海金明商务广场

建设地点：上海
占地面积：2.44公顷
建筑面积：70 000平方米

8. 上海陕西大厦

建设地点：上海
建筑面积：20 000平方米
建筑层数：19层

9. 南通国贸中心大厦

建设地点：江苏 南通
建筑面积：160 000平方米
建筑高度：220米
建筑层数：55层

10. 山西运城广电中心

建设地点：山西 运城
建筑面积：30 000平方米

11. 陕西日报新闻大厦及长安新区项目

建设地点：陕西 西安
建筑面积：250 000平方米

12. 石家庄“北国先天下”广场

建设地点：河北 石家庄
建筑面积：340 000平方米

13. 石家庄红星美凯龙家居广场

建设地点：河北 石家庄
建筑面积：340 000平方米

14. 西安融侨曲江观邸

建设地点：陕西 西安
建筑面积：450 000平方米

15. 昆明红星国际商业广场

建设地点：云南 昆明
建筑面积：860 000平方米

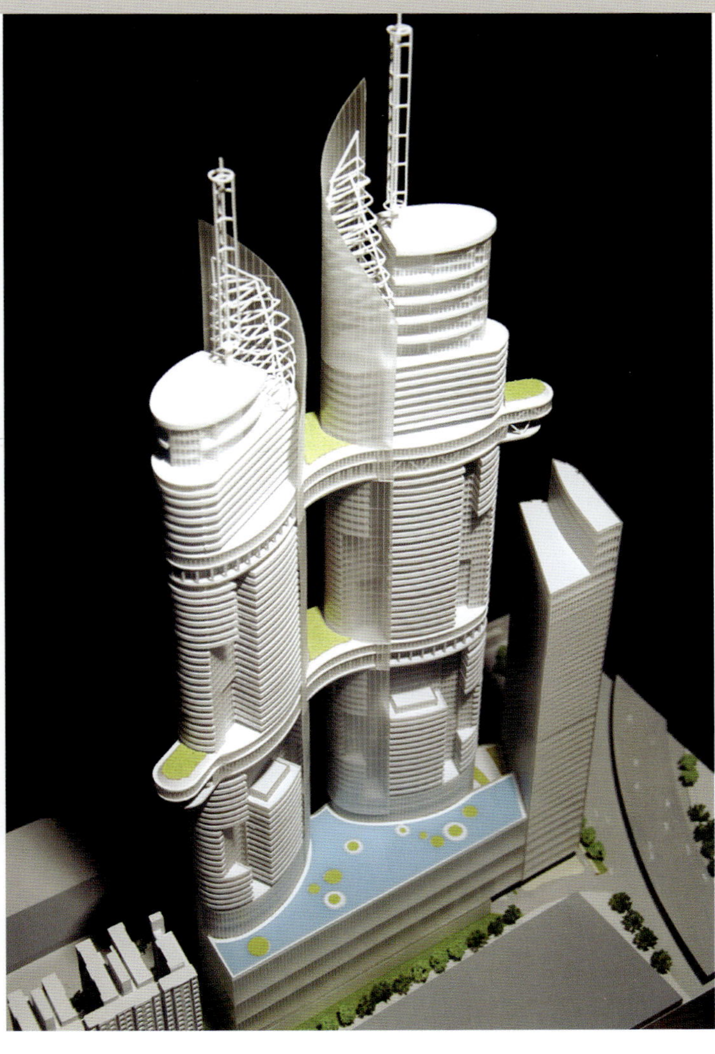

方 案 模型
规 划 模型
地 产 模型
桥 梁 模型

建延模型 JOYO Model Design

这是新时代的一股强大力量，一群富有朝气的战士，用满腔热忱的青春，描绘出建筑史上的浓浓希望；这是建筑业的一股中坚力量，一群英勇善战的骁将，用艰苦奋斗的决心谱写出模型设计的华丽乐章；这是模型界的一股劲爆力量，一群专业优异的技师，用精湛出色的技术，创造出行业领先的崭新风采；这是社会中的一股和谐力量，一群善于感恩的人士，用虔诚奉献的真心，造福于全社会的美好生活。

公司曾先后被美国CALLISON建筑事务所、美国GENSLER设计事务所、英国HALCROW集团、日本NIKKEN SEKKEI建筑设计事务所、中国香港P-T-GROUP设计事务所、新加坡AGA设计事务所、加拿大B+H设计事务所、澳大利亚JPW设计事务所、澳大利亚HASSELL设计集团及中国香港SHUI ON GROUP、中国无锡SPG LAND、美国UN+建筑设计事务所、中茵集团等择为长期合作伙伴。多年来被建筑设计界称为“建筑艺术的代言人”。

This is a powerful force in the new era. A group of vibrant soldiers describe the architectural history of the deep hope with the enthusiasm of youth. This is the backbone of the constru ction industry, a group of heroic general write a gorgeous model design movement with hard work determination. This is the world's best model of the power sector at the moment, a group of excellent professional technicians with excellent technical skills to create a new industry-leading style. This is the power of social harmony, a group of people who are good at thanksgiving, with pious devotion of the heart, benefit the whole society a better life.
The company establishes long-term partnership with numerous of renowned architectural design firms at home and abroad, such as Callison, Gensler, Halcrow, Nikken Sekkei, AGA, etc. It has been called "Spokesman of architectural art" by the architectural design circles for many years.

上海建延建筑模型设计有限公司
地址：上海市杨浦区军工路1300号（赛特工业园主楼5F）
电话：+86-21-65638880
传真：+86-21-65639990
网址：www.joyomo.com

Jiang Architects & Engineers

上海江欢成建筑设计有限公司

地址：宝矿洲际商务中心 / 上海裕通路100号 / 2602
邮编：200070
法定代表人：江欢成
电话：+86-21-32525400
传真：+86-21-32525418
邮箱：info@jiangs.com.cn
网址：www.jiangs.com.cn

Add: BM Intercontinental Business Centre / 100 Yutong Road, Suite 2602, Shanghai
P.C.: 200070
Legal Representative: Jiang Huancheng
Tel: +86-21-32525400
Fax: +86-21-32525418
Email: info@jiangs.com.cn
Http://www.jiangs.com.cn

上海江欢成建筑设计有限公司是以中国工程院江欢成院士命名并主持的具有建筑工程设计综合甲级资质的建筑设计公司。2000年，同名事务所于上海现代建筑设计集团内成立，2005年改制成股份制民营有限公司，它的前身是华东建筑设计研究院东方明珠设计组、金茂大厦顾问组、雅加达塔设计部和第五设计所。

公司目前有近60位员工，分为建筑、结构、机电设计三个工种；队伍精干，管理有序，技术力量雄厚，工种之间配合默契；项目以技术含量高的公建为主，尤其擅长于高层／超高层建筑。公司以设计创新、服务优质为立身之本。

曾任职英国RMJM设计总监的江春于2010年初加入，任职总经理／总建筑师。目前公司在他的带领下，全力打造国际化的质量、专业化的服务，追求艺术与技术的完美结合。

Jiang Architects & Engineers (JAE) is Chinese Grade A qualified architectural design firm. The firm is established by and named after professor Jiang Huancheng, a member of Chinese Academy of Engineering. It was set up in 2000 as a subsidiary of Shanghai Xian Dai Architectural Design Group. In 2005, it was privatized and became an independent company. Its core team is formed by the design team for Shanghai Oriental Pearl Tower, Jakarta Tower and the consulting team for Jin Mao Building.

JAE is a multi-discipline office that has architectural, structural, MEP design sectors. The company has around 60 staff. It is a well structured, technically strong design company. The company portfolio covers many technically demanding public buildings such as office, hotel, retail, civic building etc. High rise tower is one of its expertise. It is a design + service lead office.

In 2010, Jiang Chun, a former design principal of RMJM joined JAE as the managing director and the chief architect. Under his leadership, JAE is now being upgraded into a local design firm with international quality, professional service and is pursuing a perfect combination between art and technology.

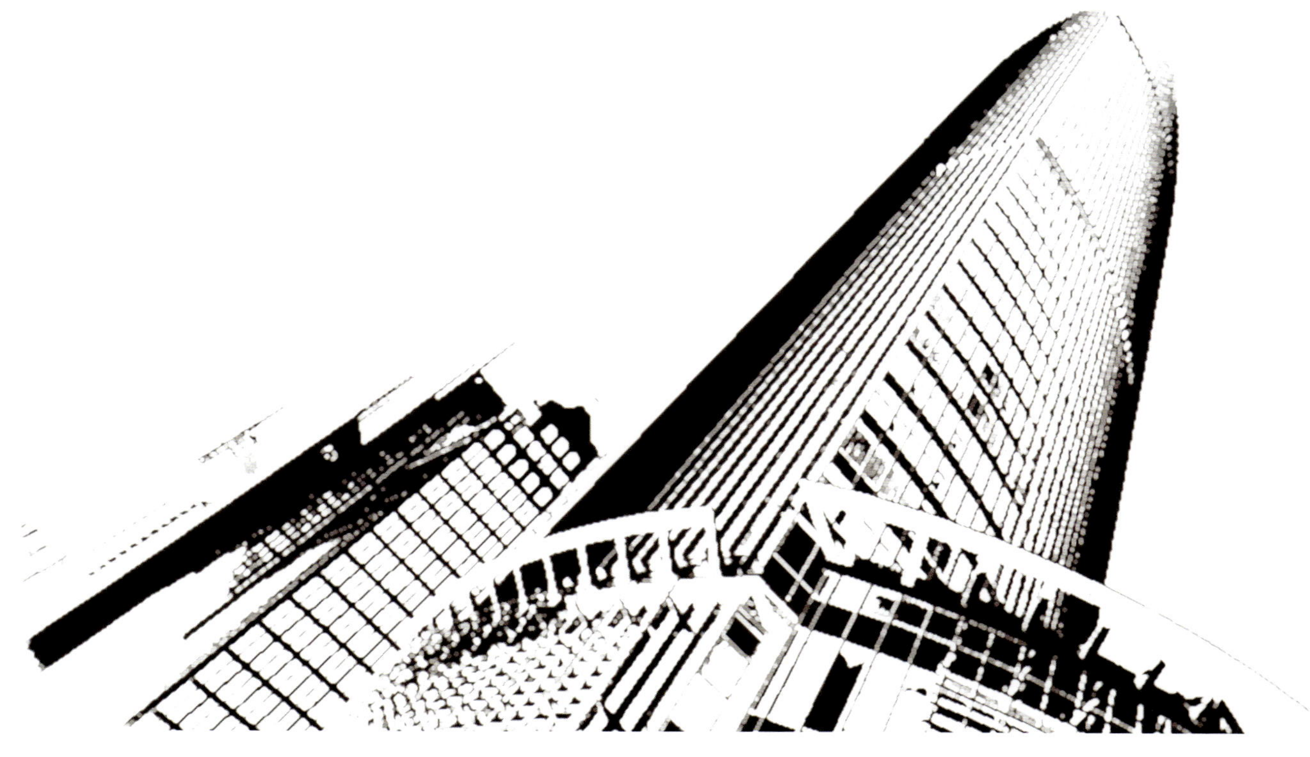

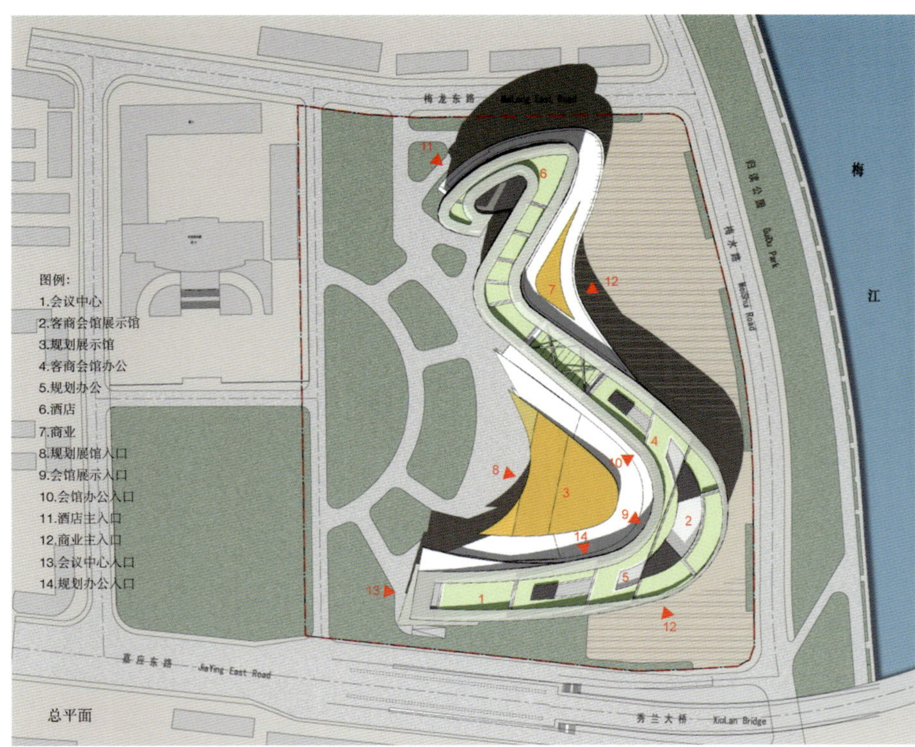

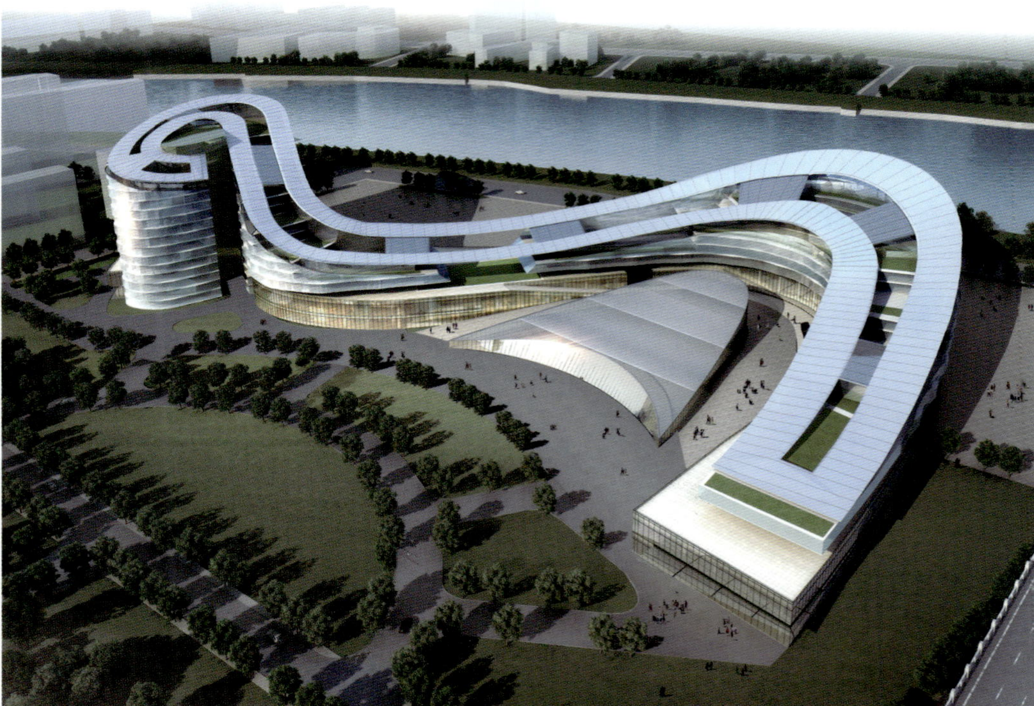

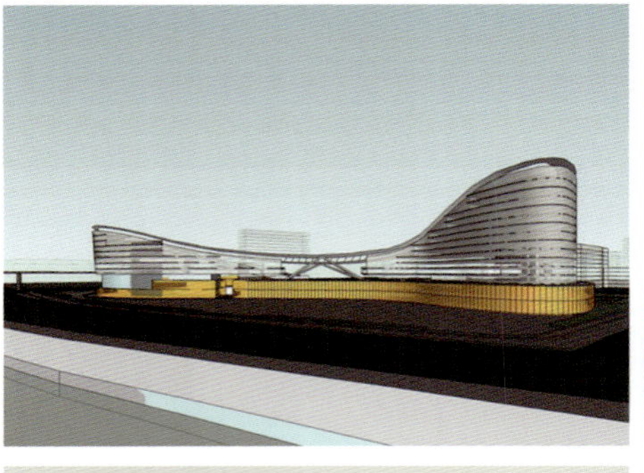

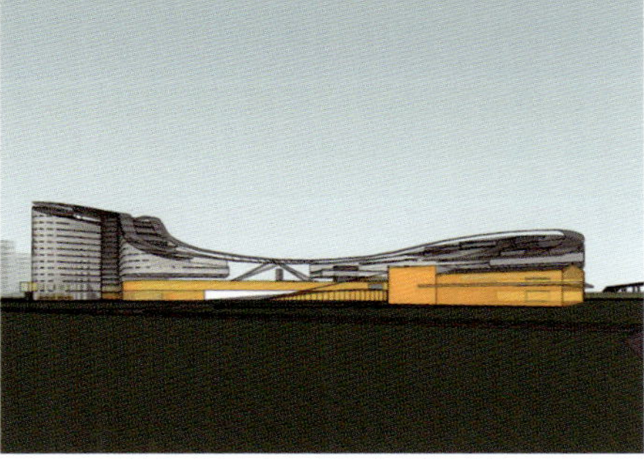

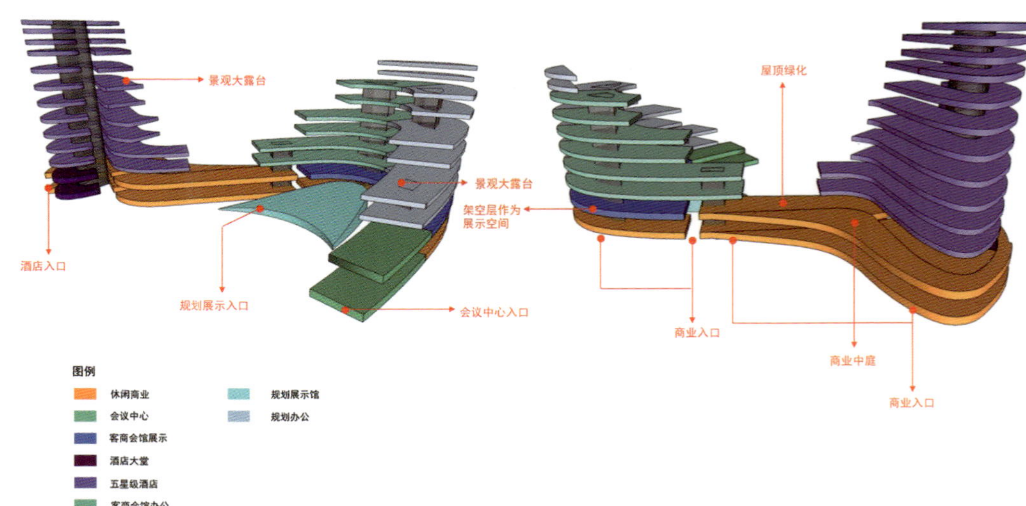

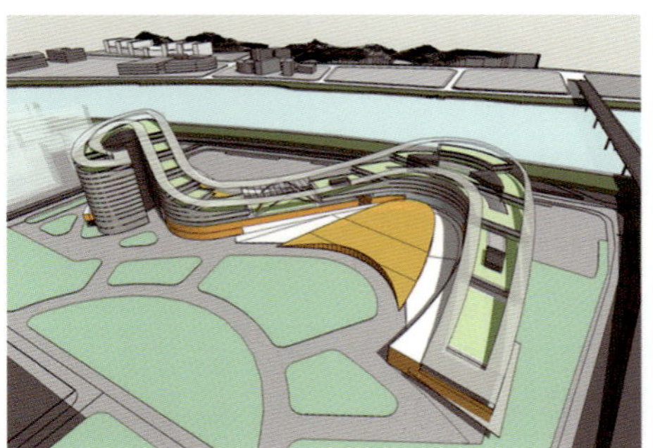

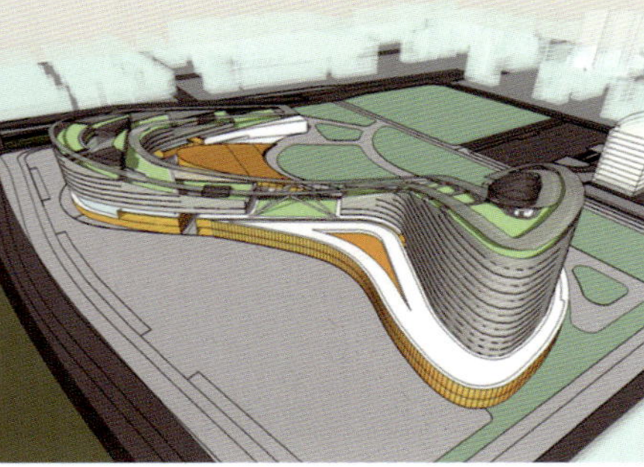

梅州世界客商会馆

建设地点：广东 梅州
占地面积：30 890平方米
建筑面积：11 760平方米
容 积 率：1.28
项目状态：2010年方案设计

项目说明

梅州世界客商会馆是以梅州市世界客商会馆的会展功能为主，兼有部分政府办公及出租办公的一栋独立的文化办公建筑。

此方案以自由曲线构成建筑主体，建筑各部分根据使用功能的不同围绕曲线进行高度、宽度的变化，最终形成一个连续、丰富、灵动的建筑形态。犹如一条飘逸的丝带飘在秀丽的梅江边，又如一座清丽的山峰与江水相映成趣。

思南公馆

建设地点：上海
占地面积：50 529平方米
建筑面积：56 592平方米
容 积 率：1.12
项目状态：2010年竣工

项目说明

思南公馆是一个主要由众多历史保护建筑组成的生活社区。

本项目在“保护该地区的城市肌理、空间布局、街巷尺度、绿化与优秀历史建筑等构成的历史风貌特征”的设计原则下：

1.保留原花园别墅49栋，共计30 282平方米。其中36栋为上海市优秀历史建筑，其余13栋为保留建筑及一般历史建筑；

2.拆除17 973平方米的危棚、简屋、违章建筑以及无法通过改造与整体环境相协调的建筑；

3.在拆除建筑旧址及原有空地上，依据社区原有建筑文脉新建8栋建筑。其中商业建筑5栋，住宅建筑3栋，总计约2.7万平方米，地下车库1座，约2.2万平方米。

项目在保护各栋近代优秀建筑的基础上，通过环境整治、功能置换及生活设施改善等方式保护和提升了这一地区的人文、历史、内涵与风貌，赋予旧建筑新的生命力，使其成为具有上海近代独特文化和历史特点的高品质的居住、商业休闲社区。

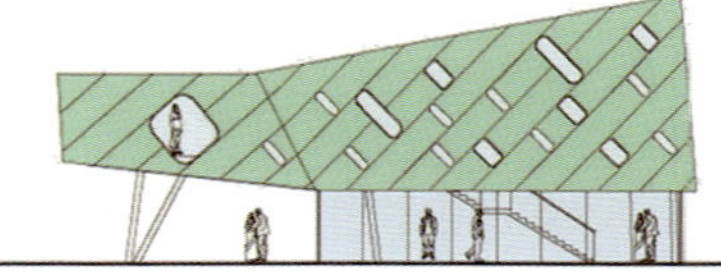

张家浜逸飞创意街

建设地点：上海
占地面积：22 596平方米
建筑面积：19 257平方米
容 积 率：0.80
项目状态：2006年建成

项目说明

地处花木的张家浜逸飞创意街定位于浦东新区第一条创意产业街，由艺术家陈逸飞先生生前总体策划，以文化创意为特征，彰显现代、时尚、创新、大视觉特色，形成具有影响力的、标志性的、休闲娱乐型的商业街区，建造出上海著名的休闲文化创意街。

该项目是国内最早、也是最长的新建于景观河岸边、集创意文化与休闲商业于一体的综合性街区，环境优美。同时也是国内领先、达到世界先进水平的知名预钝化铜板、钛锌板幕墙的应用案例。在建筑与环境景观的融合协调、商业休闲文化价值的充分挖掘、建筑技术对文化艺术内涵的诠释等方面完成得相当成功。

上海嘉兆嘉御庭

建设地点：上海
占地面积：2 224.51平方米
建筑面积：53 923.55平方米
容 积 率：3.65
项目状态：设计中

项目说明

租赁式公寓坐落于静谧的风貌区域，郁郁葱葱的树木将其环绕，闹中取静，精华其中。考虑到徐汇区的特质，其具有适度超前、示范、创新和实用的特点，结合高水平的设计，强化特色，塑造精品。同时，为尊重该风貌区，合理选用外立面用材，如大理石，以使建筑的外立面能和谐地融入周边具有历史风貌的街道景观，寻求推动社会创新和保护历史风貌的平衡。

酒店式公寓的暖灰色调的主体，完全融入了周围的环境，挺直简洁的线条巧妙地配合了错落的立面层次及不一样的玻璃颜色，勾勒出建筑的完美线条。粗细间隔的灰色石材框架庄重地分布于东西立面，恰似一本矗立在城市中的著作，充满着书卷气。裙房延续了塔楼的颜色对比，长短不一的杆件为街景增添了韵律和节奏，配合夜晚的灯光效果，好似条条流水，幅幅画卷，生动活泼而又庄重典雅。

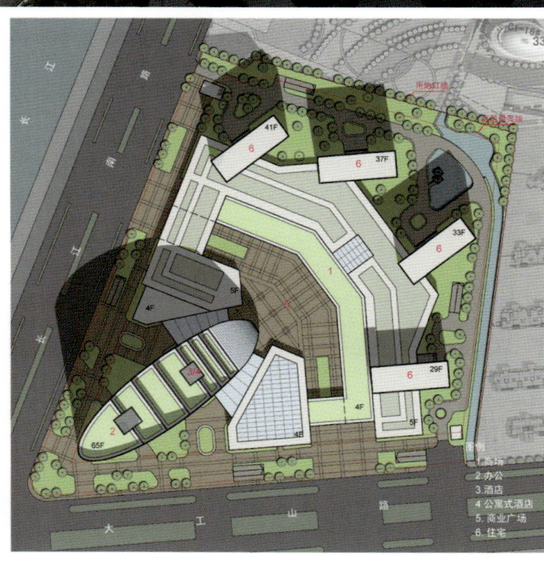

芜湖长江之歌C2地块城市综合体

建设地点：安徽 芜湖
占地面积：64 823平方米
建筑面积：536 500平方米
容 积 率：6.56
项目状态：2010年方案竞赛

项目说明

芜湖C2地块位于芜湖“长江之歌”整个项目的最南端，建筑的主要功能为酒店、酒店式公寓、办公、商业以及住宅。按照规划要求，该城市综合体应该成为该区域的标志性建筑物。因此在总体概念规划中，鉴于用地的限制且建筑功能比较复杂，分散成独立的一栋栋建筑物不利于突出标志性。于是，我们通过分析，将一些建筑功能，包括酒店、办公以及酒店式公寓整合到一起，形成双塔，并通过商业裙房将塔楼与住宅联系起来。结构上，双塔的平面由于受到建筑功能及景观的限制，核心筒偏置一侧，因此将双塔的结构相互联系起来，增强了结构的稳定性。同时，相连的双塔更加增强了建筑物的标志性。

天津市建筑设计院
TIANJIN ARCHITECTURE DESIGN INSTITUTE

Tianjin Architecture Design Institute

天津市建筑设计院

天津市建筑设计院（简称TADI）创立于1952年，历经50多年历史沧桑，现已发展成为技术实力雄厚、人才济济的天津地区最大的综合建筑设计院。

该院现有员工1 242人，其中全国工程设计大师3人，国务院批准享受政府特殊津贴专家12人，国家人事部批准有突出贡献的中青年专家2人，天津市中青年授衔专家4人，正高级建筑（工程）师53人，高级建筑（工程）师352人，中级建筑（工程）师323人，并有国家一级注册建筑师72人，国家二级注册建筑师32人，国家一级注册结构工程师71人，国家级注册监理工程师14人，造价工程师14人，注册城市规划师16人，注册岩土工程师4人，注册咨询（投资）工程师18人，注册电气工程师22人，注册公用设备工程师（给排水）25人，注册公用设备工程师（暖通空调）23人。

具有国家住房与城乡建设部颁发的甲级建筑工程设计资质；甲级城乡规划编制资质；甲级风景园林专项工程设计资质；房屋建筑工程监理甲级资质；工程招标代理机构甲级资质；工程造价咨询企业甲级资质；施工图设计文件一类审查机构资质；乙级人防工程设计资质；国家发改委颁发的工程咨询甲级资质；是国际建筑工程咨询协会（菲迪克）会员单位。

该院设有建筑、规划、结构、给排水、暖通空调、电气、智能化、经济技术分析、岩土工程、城镇规划、景观与环境设计和室内装修设计等专业。主要承接民用建筑设计、工业建筑设计、城市规划及居住区、住宅小区规划设计；承接民用与工业建筑项目的可行性研究、工程咨询、建设工程监理、工程项目总承包以及技术合作、技术成果转让等业务；承接境外工程的咨询、设计和监理、国外经济技术合作、智能建筑（系统工程）设计等业务。为更好地拓展业务，该院相继在海南、厦门、广州、上海、重庆等地成立分院。

工程技术人员积极进取，锐意创新，先后有10项设计获国家优秀设计奖，其中有三项获金质奖，近230项工程设计、科研和标准设计获得了国家建设部、市级优秀设计奖、优质工程奖及科研成果奖等奖项。其中：瑞景花园住宅小区工程、天津奥林匹克中心体育场工程荣获中国土木工程詹天佑奖。天津意式风貌保护区的保护和整修规划、天津海河两岸城市设计、天津津湾广场修建性详细规划均荣获全国优秀城市规划设计二等奖、天津市优秀城市规划设计一等奖。平津战役纪念馆，天津体育馆，天津第二南开中学，周恩来、邓颖超纪念馆、天津奥林匹克中心体育场、天津站荣获新中国成立60周年建筑创作、建筑设计大奖。近几年完成了天津奥林匹克中心体育场、天津医科大学总医院、中华剧院、泰达国际会展中心、泰达体育场、国际金融中心、天津市迎宾馆四号楼、哈尔滨市第十四中学、天津科技广场、天津港企业文化中心、天津空港国际汽车园、天津市东丽区华明示范小城镇、华夏未来少儿艺术中心——4D影院、天津电视台数字电视大厦、中国航天科工集团第三研究院六六八工程、天津城建管理职业技术学院、天津中新生态城服务中心、中船重工大厦、迁安市人民医院、天津市中心妇产科医院迁址新建工程、天津金融城津湾广场一期工程、天宾商务中心、小白楼音乐厅、天津站交通枢纽工程轨道换乘中心工程停车中心等一批大型重点公建项目。以其全新的设计理念和现代高新技术，展示了该院的技术实力和创新创优水平。

于1996年获得ISO9001国际质量体系认证，于2002年4月4日率先实现GB/T19001—2000—ISO9001：2000标准转换。2010年3月取得2008版IDSO9001标准认证证书。

质量方针：为顾客设计好每一平方米的建筑，提供优质的设计全过程服务，科技领先，锐意创新，实现质量管理体系的持续改进，达到顾客满意。

Founded in 1952, with over fifty years' history, TADI has grown to be the biggest integrated architectural design institute with strong skills and adequate talents.

TADI has 1,242 employees, including 3 engineering design experts, 12 experts with special subsidy of State Council, 2 middle-aged and young experts of special contribution approved by Ministry of Human Resources and Social Security 4 titled middle-aged and young experts of Tianjin, 53 senior architects (engineers), 352 deputy senior architects (engineers), 323 intermediate architects (engineers), 72 class-1 registered architects, 32 class-2 registered architects, 71 class-1 registered structural engineers, certified supervisory engineers, 14 cost engineers, 16 registered urban planners, 4 registered geotechnical engineers, 18 registered consulting (investment) engineers, 22 registered electrical engineers, 25 registered drainage engineers and 23 registered HVAC engineers.

TADI has Grade-A License for Architecture Engineering Design received from Ministry of Housing and Urban-Rural Development; Grade-A License for Urban-Rural Planning; Grade-A License for Design of Landscape Garden Special Project; Grade-A License for Housing Engineering Supervision; Grade-A Certificate for Engineer Bidding Agency; Grade-A License for Engineering Cost Consultants; Grade-1 License for Investigation Organization of Construction Drawing Design Documents; Grade-B License for Design of civil air defense; Grade-A License for Engineering Consultants received from National Development and Reform Commission; the member of China Association of International Engineering Consultants (FIDIC).

TADI has architecture, planning, structure, drainage, HVAC, electrical, intelligence, economic and technical analysis, geotechnical engineering, urban planning, landscape and environment design, interior design and other professions. It undertakes the design of civil building, industrial building design, urban planning, planning and design of residential area and residential community; provides feasibility research of civil and industrial construction project, engineering consulting, construction supervision, project contracting and technical cooperation, transfer of technological results and other services. It undertakes engineering consulting, design and supervision of foreign projects, international economic and technological cooperation, design of intelligent building (system engineering) and other works. In order to expand business in a better way, it establishes branches in Hainan, Xiamen, Guangzhou, Shanghai, Chongqing and other cities.

Engineering technicians of TADI are active, progressive, and innovative. 10 design works have won Excellent Design Award of China, three of which are Golden Awards. Nearly 230 engineering designs, scientific research and standard designs win excellent design awards, good-quality project award and achievements award of scientific research issued by Ministry of Housing and Urban-Rural Development. In which: Ruijing Garden Residential Community, Tianjin Olympic Center Stadium win Tien-Yow Jeme Civil Engineering Prize. Protection and Maintenance Planning of Italian Protection Area of Tianjin, Urban Planning of Two Banks of Tianjin Haihe River, Detailed Planning of Construction Characteristics of Tianjin Jinwan Plaza all win the Second Prize for Excellent Urban Planning of China, the First Prize for Excellent Urban Planning of Tianjin. Memorial of the Great Decisive War III, Tianjin Stadium, Tianjin No.2 Nankai Middle School, Memorials of Zhou Enlai and Deng Yingchao, Tianjin Olympic Center Stadium, Tianjin Railway Station win Big Prize for Architectural Creation and Architectural Design of the 60th Anniversary of China. In recent years, it has completed Tianjin Olympic Center Stadium, Tianjin Medical University General Hospital, China Theatre, Taida International Convention Center, Taida Stadium, International Financial Center, No.4 Building of Tianjin Yingbin Hotel, Harbin No.14 Middle School, Tianjin Technology Square, Enterprise Culture Center of Tianjin Harbor, Konggang International Automobile Park of Tianjin, Huaming Model Town of Dongli District, Tianjin, Huaxia Future Arts Center for Children – 4D Theatre, Digital TV Building of Tianjin TV, 668 Project of Research Institute 3 of China Airspace Science & Industry Corporation, Tianjin Urban Construction Management & Vocation Technology College, Tianjin Sino-Singapore Eco-city Service Center, Building of China Shipbuilding Industry Corporation, People's Hospital of Qian'an, New Construction of Tianjin Central Hospital for Gynaecology and Obstetrics, Jinwan Square Phase I of Tianjin Financial City, Tianbin Business Center, Little White House Music Hall, Parking Center of Track Transfer Central Project of Tianjin Railway Station Transportation Hub and a group of other large-scale important projects of public buildings. In virtue of brand-new design ideas and modern high technologies, TADI shows the level of skill, quality and innovation.

TADI received ISO9001 International Quality Management System Certification in 1996, and took a lead in realizing standard conversion of GB/T19001-2000-ISO9001:2000 in April 4, 2002. In Mar. 2010, it received IDSO9001:2008 Standard System Certification.

Quality policy of TADI: provide customer with good architectural design of every square meter and with high-quality one-stop design service; be advanced in technology, and innovative, and realize continuous improvement of quality management system to achieve customers' satisfaction.

地址：天津市河西区气象台路95号
邮编：300074
法定代表人：刘军
电话：+86-22-23543000
传真：+86-22-23345260
网址：www.tadi.net.cn
邮箱：tadi@tadi.net.cn

Add: No.95 Observatory Road, Hexi District, Tianjin
P.C.: 200074
Legal Representative: Liu Jun
Tel.: +86-22-23543000
Fax: +86-22-23345260
Http:// www.tadi.net.cn
E-mail: tadi@tadi.net.cn

天津健康产业园区体育基地项目

项目位于天津市静海县团泊健康产业园区，东至团泊大道，南至北华路，西至东海道，北至昆明湖路。总用地面积121.55公顷，可用地面积105.35公顷。东亚场馆用地主要位于基地的东侧，少数场地位于基地中北部。场馆包括自行车馆、射击馆、综合体育馆、曲棍球馆，除综合体育馆外，其余三馆均为东亚运动会比赛场馆；场地包括曲棍球场、棒球场、垒球场、射箭场。场馆总用地面积296 450.6平方米，可用地面积245 997.4平方米。

综合体育馆

建设面积：22 198平方米
占地面积：50 337.6平方米
容 积 率：0.44
设计时间：2010–2011年
竣工时间：未竣工

项目说明

本项目是天津2013年东亚运动会的辅助场馆之一，高约24米，主要功能为综合体育馆、练习馆以及体育博物馆。综合体育馆总坐席数约5 600座。建筑表皮由金属幕墙与穿孔金属板相间组成，纵向切割，穿孔金属板表皮凹进，在此处形成深深的阴影，体现了体育建筑的力量感。建筑造型给人以升腾向上的形态，寓意着该市体育事业的蓬勃发展。同时，利用金属幕墙表皮升起的部分形成入口空间，选用玻璃幕墙，通过材质的对比，更加突出了入口空间的导向性，并形成内部贯通的联系空间。此外，曲线形的网架结构与建筑功能在空间上完美结合，所形成的流线造型与自行车馆在形象上又取得了相得益彰的效果，使自行车馆、射击馆及综合体育馆在团泊大道上形成一道美丽和谐的街景。

自行车馆

建设面积：28 201.2平方米
占地面积：81 652.9平方米
容 积 率：0.34
设计时间：2010–2011年
竣工时间：未竣工

项目说明

本项目是天津2013年东亚运动会的主要场馆之一，高约41米，包括自行车馆、壁球馆，共约3 400个座位。建筑造型采用自行车头盔的意念，曲线流畅，造型舒展。椭圆形的建筑在环形草坡的烘托下更显轻盈，主体外观既为建筑的结构，主体所开的洞口缓和了建筑物的体量感，轻盈和速度贴切地反映了自行车馆的使用性质，人们通过坚固而优美的主体结构所看到的是容纳所有功能部分的比赛馆核心区，比赛场中透出的灯光使得它晶莹剔透、美轮美奂，都将给人留下深刻的印象。

射击馆

建设面积：38 000平方米
占地面积：61 692.4平方米
容 积 率：0.65
设计时间：2010–2011年
竣工时间：未竣工

项目说明

本项目是天津2013年东亚运动会的主要场馆之一，高约24米，包括预赛馆和决赛馆，共约2 600个座位。本方案以“速度”为主题，用两个不规则梭形筒体交错形成刚劲有力的独特体量。表现射击运动中子弹飞行的速度感。其独特的体量体型来表现建筑的形式美，强调建筑的体积感，使建筑美与运动美有机结合。

曲棍球馆

建设面积：4 555平方米
占地面积：52 335.9平方米
容 积 率：0.087
设计时间：2010–2011年
竣工时间：未竣工

项目说明

本项目是天津2013年东亚运动会的主要场馆之一，高约15.83米，包括曲棍球场地看台、一块曲棍球比赛场地、一块曲棍球练习场地、一块射箭练习场地及一片机动车停车场，看台规模为2 100个固定座位。本方案中，功能性空间与看台空间两个体量有机地交织在一起，建筑本身既有灵活的体量关系，又不失理性，在其外装修设计中，建筑主要材料为金属、玻璃，以及其他金属构件，体现了科技时代力量与美的结合，与运动的主题相吻合。

鸡西市第一中学

建设地点：黑龙江 鸡西
建筑面积：7.2万平方米
占地面积：15.3公顷
容 积 率：0.43
设计时间：2010年
竣工时间：未竣工

项目说明

1. 本工程场地是丘陵地形，规划设计采用多台地的处理，布置广场、运动场和建筑用地，形成多层次、不同形态的室外空间，达到节约建设成本，构建景观丰富的、逐步上升递进的校园空间。
2. 由图书馆、合班教室等公共教学用房建筑形成的半圆形校前广场进深70米，面阔260米，在满足学生、教师上下学的交通需要外，构建出景色优美、绿色环保的花园式校园模式。
3. 鸡西位于祖国的东北部，气候寒冷，结合建筑功能的组合，既达到联系紧密、流线短捷的目的，同时达到围合多个内庭院空间，利于学生课间户外的活动，减少寒风对学生的侵害，并有效实现建筑节能的目标。

天津美术馆

建设地点：天津
建筑面积：40 734平方米
占地面积：23 850.4平方米
容 积 率：1.0
设计时间：2009—2010年
竣工时间：未竣工

项目说明

1. 为美术馆设计开阔的室外雕塑展区，通过通透的玻璃幕墙与室内自然过渡，让建筑空间与环境融为一体。
2. 朝向湖面的玻璃肋玻璃幕墙，在四层为人们提供了俯瞰整个文化中心的视框，使人们能够享受到艺术与自然景观的充分融合。
3. 第五立面设计，将所有屋面设备隐藏，把立面的设计元素延续到屋顶，从各个角度表现统一简洁的建筑形象。
4. 综合公共区域位于首层和二层，设置有休息、培训、图书阅览等综合服务功能，为观众提供了充分的交流驻足的文化生活场所。
5. 二层主展厅室内空间净高达到8米，可以满足各种现代艺术展示的需要，三层、四层展厅形成环形，为灵活布展提供了必要条件。

丹东市新区医院——门急诊住院综合楼

建设地点：辽宁 丹东
建筑面积：204 140平方米，其中一期：91 140平方米
占地面积：201 600平方米
容 积 率：1.0
设计时间：2010年
竣工时间：未竣工

项目说明

1．体现现代化综合医院特色：住院楼一层有4个护理单元，最大限度地降低了住院楼层数，实现了医疗资源共享，方便患者救治。裙楼采用半集中式设计，运用水平动态组织人流、物流，在减少垂直运输压力的同时，方便患者救治。门诊模块化、采用医院街方式联系各功能布局，高效便捷。门诊设计采用大小庭院组合方式，既丰富了医疗空间，又改善了医院内部采光、通风，为患者提供舒适、宜人的救治环境。

2．体现“以人为本”的设计理念：采用“医患分开”、“单人诊室”等方式，充分保护患者隐私，体现空间设计的人性化，创造温馨、舒适的建筑环境。

3．体现环保、节能与经济可行性：在满足功能适用的前提下，选择最经济可行的空间组合、结构体系与设备运行方式；充分利用不可再生能源，采用太阳能节能技术。

消能减震支撑

屋顶绿化

太阳能光伏发电

天津市建筑设计院科技档案楼

建设地点：天津
建筑面积：4 585平方米
占地面积：1 263平方米
容 积 率：3.63
设计时间：2008年
竣工时间：2009年

项目说明

该项目是由天津市建筑设计院自主研发、设计、工程总承包及使用于一体的节能型综合办公楼。项目位于天津市建筑设计院东南角，总建筑面积4 585平方米。本项目在设计的过程中坚持绿色建筑设计理念，充分考虑建筑能源的有效利用，包括太阳能热水系统、太阳能光伏发电系统、地源热泵制冷制热系统等可再生能源利用；在建筑中创新性地使用了独立研发的消能减震支撑，用于改善建筑的抗震性能；建立建筑信息展示系统，用于监测建筑全生命周期的运行数据，作为设计人员的科研依据；同时空调系统末端采用毛细管网辐射系统，创造了舒适健康的办公环境。

本项目2009年获得国家绿色建筑二星级认证。

万丽天津宾馆内装修工程

建设地点：天津
设计时间：2009—2010年
竣工时间：2010年

项目说明

酒店的整体设计稳重、大气、清新、亮丽。本案注重细部刻画，通过细部表现高贵的品质，同时遵循简约、大气的原则。简洁的块面及材质灯光对比中包含了耐人寻味的细部处理，简单而不平淡乏味。在设计中遵循高雅、大方、柔和的原则，结合地域文化，体现人文、自然的理念，运用现代装饰手法、新型的施工技术，营造新颖别致的现代酒店。

万丽天津宾馆的设计主题是希望能呈现出不同的地域背景和文化特色，用大胆而当代的手法对传统元素进行现代的演绎。设计摒弃了常见的酒店大堂中奢华而又复杂的线条，采用大量的流畅线条和几何对称图案，极具视觉冲击力，营造出“空间里的空间”。最终完成的酒店内部设计，让客人虽身处繁华喧嚣都市的中心，却能独享一片优雅私密、现代舒适的灵动空间。

天津市文化中心商业体

建设地点：天津
建筑面积：332 353.5平方米
占地面积：7.51公顷
容 积 率：2.66
设计时间：2009—2011年
竣工时间：未竣工

项目说明

毗邻阳光乐园，在文化中心地块北侧，新天津乐园商业中心将为天津带来高品质的商业体验，商业中心提供了约35万平方米的建筑面积，功能包括零售、超大型自助商场、饮食、娱乐。建筑的尺度和周围的文化建筑呼应协调。建筑总高度在30米的限高内。一楼到五楼层高为5.3～6米，并设有两层地下层。对建筑体量，材料选择的考虑在于将歌剧院、阳光乐园和商业中心一起形成一个优雅和谐的整体。商业中心设计的构想是为游客和居民提供一个完整、安全的商业娱乐体验，而不仅仅只是购物。商业中心将为市民提供谐的环境氛围。

设计改变世界

Tianjin University Research Institute of Architectural Design & Urban Planning

天津大学建筑设计规划研究总院

天津大学建筑设计规划研究总院创建于1958年，是国家甲级建筑设计、甲级城市规划设计、甲级文物保护工程勘察设计、甲级工程咨询单位，天大设计总院依托于国家重点大学人才优势、学科优势和技术优势，凭借先进的技术装备和严谨的工作作风，近年来在全国范围内完成的各种类型、各种规模的建筑设计、城市规划设计中屡获各类奖项，受到项目委托单位及社会各界的好评。

天大设计总院现有一线设计人员400余人，其中45人为国家一级注册建筑师，35人为国家一级注册结构师，15人为国家注册城市规划师，拥有硕士学位的员工占全院总人数的40%以上。天大设计总院专业配套齐全，人员素质较高，单位在教育和科研建筑、旅游和疗养建筑、办公和商贸建筑以及大型医院建筑、体育建筑设计和城市规划、古建筑保护规划等方面都有不俗的成就。其中"天津美术学院美术馆"、"郑州大学工科园"、"天津市滨海新区响螺湾商务区城市设计"、"深圳欢乐谷"、"宝鸡市青铜器博物馆"、"天津蓟县地质博物馆"等项目在国内外影响深远。近几年来，天大设计总院在天津海河重大开发项目的国际竞赛中，曾力挫群雄，摘取一、二等奖的桂冠，同时天大设计总院在工业建筑设计、古建筑复原设计、设计监理、深基坑支护以及建设项目策划和可行性研究等方面也具有很强的竞争实力，业务范围已遍布全国各地，并已开始走出国门，迈进世界市场，取得了骄人的业绩。

"实事求是"是天津大学的校训，它也已经深深地融入本院的企业文化与服务理念之中。特色鲜明、风格独特的天津大学建筑设计规划研究总院，愿为中国的城市建设精心勾绘出最具地方特色、最具竞争力、最具时代精神的发展蓝图。

Established in 1958, Tianjin University Research Institute of Architectural Design & Urban Planning is ranked as national Grade A in architectural design, urban planning and design, survey and planning in heritage protection project, and engineering consulting unit. Relying on the advantage of national key university in talent, disciplines and technology, by virtue of advanced techniques, equipment and strict working style, our research institute has been successively awarded many awards in nationwide various types and scales of architectural design, urban planning and design in recent years, and has won praises from project commissioned units and community.

The institute now owns more than 400 front-line design staffs, of which 45 are registered architects at national Class-A, 35 are registered mechanical engineer at national Class-A, 15 are national registered urban planners, and the master's degree holders occupy more than 40% of the total number. Complete professional composition and high-quality personnel contribute a lot in achievements such as education and scientific research construction, tourism and convalescence construction, office and commercial construction, large hospital buildings, sports architecture design, urban planning as well as protection of ancient buildings, etc. wherein "Gallery of Tianjin Academy of Fine Arts", "Technology and Science Park of Zhengzhou University", "Urban Design for Tianjin Binhai New Area Xiangluo Bay Business District", "Shenzhen Happy Valley", "Chinese Bronzes Museum"," Tianjin Jixian Geological Museum" and other projects has won reputations at home and abroad, prevalent in the world. In the international competition of major projects on developing Haihe River in Tianjin in recent years, it has beat around and crowned the First and Second Prize respectively. It is also strongly competitive in respect of industrial building design, ancient architecture recovery design, design supervision, deep excavation support, construction project planning and feasibility studies, etc. Its business has spread all over the country as far as joining the world market, and has achieved remarkable results.

As the motto of Tianjin University, "Seeking truth from facts" has been deeply integrated into our corporate culture and service philosophy. With distinct features and unique style Academy of Architectural Design & City Planning Tianjin University is willing to provide city-building in China with the carefully drawn development blueprint which has the most local characteristics, the most competitiveness and most of the spirit of the times.

地址：天津市南开区卫津路192号
邮编：300073
电话：+86-22-27404753
传真：+86-22-27401845
邮箱：td_design@126.com
网址：www.aatu.com.cn

Add: Weijin Road, No.192, Nankai District, Tianjin
P.C.: 300073
Tel: +86-22-27404753
Fax: +86-22-27401845
E-mail: td_design@126.com
Http: //www.aaut.com.cn

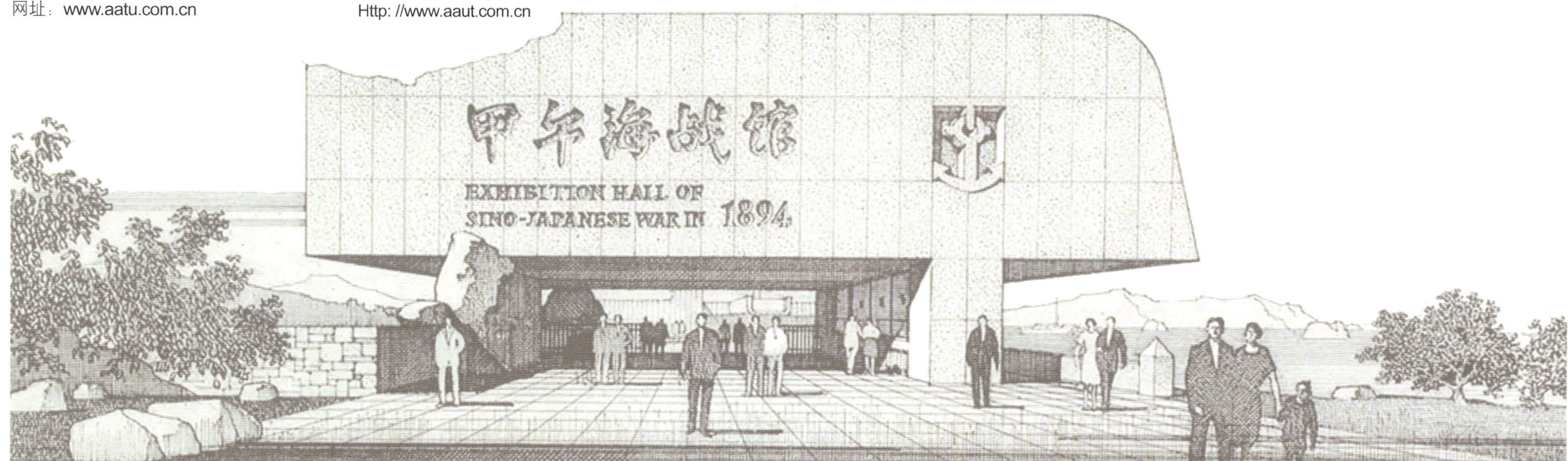

山东威海甲午海战馆

天津美术学院美术馆

宝鸡青铜器博物院

天津蓟县地质博物馆

南安烈士纪念碑

四川汶川映秀镇渔子溪村震后重建项目

中国民生银行总部

天津大学体育馆

天津滨海新区响螺湾商务区城市设计

天津西站城市副中心整体城市设计及西站站房建筑方案

Tianjin Tianzi Tuowei Architectural Design Co., Ltd.

天津天咨拓维建筑设计有限公司

天津天咨拓维建筑设计有限公司于1999年成立于天津，公司设计范围涵盖建筑设计和城市规划，包括酒店、办公、住宅、商业、学校、文化设施等内容，并重视发展城市景观设计、建筑设计与室内设计的衔接与渗透。在建筑设计日益专业化、开放化、市场化的背景下，公司于2004年进行了企业改革，并以现代企业制度的要求做到了公司的运作规范化、管理先进化、模式专业化。公司设计资质为建筑综合甲级资质。

公司拥有一支由60多人组成的优秀的设计团队，其中设计骨干多毕业于清华大学、天津大学等知名学府。通过流畅高效的团队合作，设计师将卓越的创作实力和丰富的工程经验融入公司的项目之中。设计团队的哲学是建立在理解客户需要、要求和目的的宗旨上的。每个精心为业主打造的项目，都体现了设计者对项目现实背景和文化环境的思索、对地域物理环境的尊重，以及对社会经济、艺术审美等方面的综合考虑。因为公司以此作为设计的出发点，所以公司的设计卓有成效。一流卓越的技术、在合理预算基础上建立的完美运作，以及与多方面长期愉快的合作关系是公司追求的设计目标。

成立十年间，公司通过对先进设计理念和建筑技术的整合，博采众长、精心研究，已完成各类重大工程项目百余项。建成项目多次获得天津市优秀建筑设计奖。

团队精神

方案创作强调以设计主创人为中心，设计组随时研讨，根据分工，同步协调展开设计工作，形成互动式的工作状态。工程设计强调以工程主持人为中心，主创人对整体空间处理及细部效果可根据工程设计进度同步把握。

创新意识

从机械的工作中跳出来，去体验活生生的、具有创造性的现实生活。我们主张多视角、多思维，不排斥急功近利的消费文化，推崇具有个性的原创文化。

服务意识

在建筑创作活动中，我们学会了换位思考，更多地去倾听、分析和满足业主的需求，建立诚信的合作关系。强调行为自律和职业道德。在创造精品与追求经济利益之间，强调道德与良好习惯的平衡。

Tianjin Tianzi Tuowei Architectural Design Co., Ltd. was established in 1999 in Tianjin. The company covers architecture design and urban planning, including hotel, office, residential, commercial, school, cultural facilities, etc., and values the contact and penetration between urban landscape design, architecture design interior design and interior design. As the design of architecture is becoming more and more specialized, open, commercial, our company takes some reformations in 2004 based on the requirements of modern enterprise system, to achieve standardization, advanced management, pattern of specialization. The company has the architecture design integrated Grade A qualification.

The company has an excellent design team consisting of 60 people, among whom many design backbones graduated from Tsinghua University, Tianjin University and other famous universities. Through the smooth and efficient team, the designers put the excellent creative strength and wealth of engineering experience into the company's projects. Design team's philosophy is based on understanding customer needs, requirements and purpose. Each well built project for the owners reflects the designer taking real thinking of the project background and cultural environment, respect for the physical environment of regions, as well as socio-economic, aesthetic and other aspects into account. The company takes this as a design starting point, so the design of the company is effective. First-class and excellent technique, based on a reasonable budget for a perfect operation, and long term happy relationship is the company's pursuit of design goals.

During the decade of its establishment, the company through the integration of advanced design concepts and construction techniques, absorbing, well-researched, has completed types of large and important projects over a hundred items. The projects which were completed have won Tianjin excellent architectural design awards many times.

The Team Spirit

For the schematic creation, the main designers are centered with coordination of the project designers.

For the project design, the project designer in charge is to control the space design and detail impression according to the design process.

Creation Consciousness

We expect to extricate ourselves from the mechanical working and experience lively and creative life.

We maintain the consumption culture with quick result and instant profit and assert multi-views, multi-thinking. And we appreciate the original creative culture with special characteristics.

Service Consciousness

During the architectural creation, we have understood how to listen, analyze and meet the requirements of the clients based on their conditions and to establish honest cooperation.

We emphasize incorrupt behavior and professional moral. Both of the moral and the sound habit should be balanced during creating the excellent works and concentrating on the economic benefit.

公司：天津天咨拓维建筑设计有限公司
地址：天津市河西区罗马花园A座1幢11楼
邮编：300204
电话：+86-22-23267966
传真：+86-22-23263966
邮箱：tztw2008@126.com
网址：www.tjtztw.com

Name: Tianjin Tianzi Tuowei Architectural Design Co., Ltd.
Add: Floor 11, Unit 1, Tower A, Roman Garden, Hexi District, Tianjin
P.C.: 300204
Tel : +86-22-23267966
Fax: +86-22-23263966
E-mail: tztw2008@126.com
Http: //www.tjtztw.com

巴彦淖尔市金川大道商业街

本项目用地位于巴彦淖尔市临河区中心地带,庆丰街、河套大街与金川大道交汇处。用地范围北至庆丰街，南至河套大街，东至金川大道。规划总用地面积133 567平方米。项目用地东侧毗邻金川河水系景观工程，是巴彦淖尔市政府的重点工程之一。工程由河道、园林小品、休闲广场、仿古建筑、各类亲水平台、水榭、长廊，以及大面积绿化等部分构成。

根据项目地形特点与邻近金川河水系的景观优势，我们意在通过对“金川大道商业步行街”的设计，再现历史上滨河商圈的繁荣与昌盛，全面打造出1 544米世界上最长的中式滨水商业步行街，一个集购物、休闲、娱乐与餐饮等多元化的商圈。项目建成后其规模足以载入吉尼斯世界纪录，必将成为提升临河综合实力的城市亮点，为临河的经济发展和城市建设起到带动作用。

商业街主体有四个主题区：时尚区、男人区、女人区、家居区；三条主街：天街、地街与内街组成。各条街的自动扶梯给予人们从容不迫的购物流动。序列的高潮聚焦在核心的大型生态广场。这里，像巨大的客流吸收器，将各个方向的消费人流吸纳进来，再将吸纳的人流通过各种交通方式向周边商业渗透、疏散，使内庭沿街店铺获得最优的商业效果。体验式阳光生态中庭将大体量、多功能区分布的商业业态整合为一体，并沿消费空间步道穿插绿化、水景、建筑、园艺小品等，形成休闲活动与购物互动，实现商业活动的室外延伸。广场、步行街、扶梯、两廊与丰富的业态、多元化的生活元素组合，形成完美内向型的城市街区。建筑风格上根据主题区特色分别运用了民族中式、传统中式和现代中式风格。

金川大道商业街民族中式风格

金川大道商业街传统中式风格

金川大道商业街现代中式风格

巴彦淖尔市国泰临狼路项目

此项目位于临狼路以南，安丰路以西；分为住宅和商业两部分，均为中式建筑风格；安北路两侧及成宜路以西为住宅部分。项目的总用地面积76.93公顷，总建筑面积115.18万平方米；其中住宅建筑面积101.25万平方米，公建建筑面积13.93万平方米。住宅区由多层住宅、高层住宅、洋房和临街商业底商等组成。成宜路东侧为商业区，总占地49万平方米，总建筑面积49.26万平方米，包括多层商业、快捷酒店、公寓等。

住宅部分以治丰街为界，北部为青砖徽派风格，治丰街南侧为红砖中式民居风格，均是传统中式建筑风格与现代建筑设计手法相结合，各具特色。商业部分也采用了较传统的中式建筑风格，比较到位地表达了传统木构建筑的特点。

秦皇岛耀华厂地块项目

秦皇岛耀华厂地块项目位于河北省秦皇岛市市区：东临友谊路，西临西港路，北靠和平大街。

明确的居住中心

规划中，为达到景观的均好性，结合四个地块的不同特点，布置了四个景观中心，同时用一条景观体系串联起来。为每一户住宅争取更多的景观，景观中心的每个元素都有一个合理的位置和体量，并由此建立一个可预见的、良好限定的生活空间。

贯通的道路网络

本案针对各个地块分别设置与城市道路的连接方式。各个地块结合商业分别设置车行出入口及人行出入口。由车行出入口环形散发，组织次一级支路，最终形成贯通的网络。由人行出入口经景观绿化分别抵达各个楼座，做到真正的人车分流，体现交通功能、交往功能和景观功能的丰富蕴含。

丰富的公共空间

本案为公共空间的设计与布置做了精心安排，高层、公寓、商业、幼儿园相结合，使空间错落有致。商业、公寓设置于住宅社区周边，与住宅社区互不连通，避免了对住宅的干扰。幼儿园则设置于社区内部，方便社区居民的使用，提高安全性和便捷性。通过视线通廊来界定各个功能的区域，用标志性的入口来连通。为了使住宅争取更多的阳光，住宅建筑均南北向布局，同时通过景观绿地和住宅建筑的互相渗透，使住宅与景观完美结合，提高居住品质。

深厚的文化内涵

本案在体现空间人文性的同时，注重建筑形式的塑造。装饰艺术风格和英伦风格的结合，体现了建筑的奢华及典雅，结合景观绿化塑造社区的整体氛围。

天津华厦建筑设计有限公司

TIANJIN HUAXIA ARCHITECTURAL DESIGN CO.,LTD.

天津华厦建筑设计有限公司始创于1992年，拥有建筑行业建筑工程设计甲级、城乡规划编制乙级、市政行业乙级、工程招标代理机构甲级、房屋建筑工程监理甲级以及房屋建筑施工图审查和工程项目管理等多项资质。业务范围涉及建筑设计、市政设计、规划设计、建筑咨询、施工图审查、招标代理、项目管理、工程监理等综合性服务，同时扩展到新型建材、混凝土、砂浆预拌、房地产开发等多个行业，现已形成以设计为主体、房地产为龙头、建筑业服务和建材开发为两翼的集团化发展态势。公司通过GB/T19001–2008质量体系认证。注册资金2 800万元，员工近500人，拥有中高级专业技术职称249人，其中一、二级注册建筑师36人，一、二级注册结构工程师42人，注册设备工程师6人，注册监理工程师28人，地方注册监理工程师75人，注册建造师38人，注册造价师16人，注册会计师8人。

公司本着以诚为本、锐意进取、举贤纳才的宗旨，旗下汇聚了一批资深专家和业界精英，其管理团队精诚干练、高效务实，设计团队创新求异、精益求精，开发团队熟习市场、勇于开拓，是一支充满活力、开放而多元的战斗集体。公司在学校、医院、宾馆、展厅、商厦、写字楼、住宅等民用建筑及各类厂区、大跨度钢结构工程等工业建筑的规划、设计、监理等业绩丰富、奖项云集，作品遍及华夏大地。

华厦人秉承精心设计、优质服务的宗旨，融贯东西方建筑之美学，以先进的设计理念、精湛的专业品质赢得广大业主的盛赞。公司以"爱岗敬业、忠诚守信、团结协作、以德兴业"的企业精神与各界友人精诚合作，共建美好家园。

Tianjin Huaxia Architecture Design Co., Ltd. was founded in 1992, has Grade A qualification of architecture engineering, Grade B qualification of urban and rural planning preparation and the municipal sector, Grade A qualification of project tendering agencies, housing construction works supervision, and housing construction drawings review and project management and many other qualifications. Involved in architecture design, municipal design, planning and design, construction consulting, construction drawing review, bidding, project management, project supervision and other integrated services, the company expands into new building materials, concrete, ready-mixed mortar, real estate development and other industries, forming the development trend with design as the main form, real estate as the pioneer, construction services and building material development as the two wings. The company has passed the GB/T19001-2008 quality system certification. Registered capital of 28 million, the total number of the staff is nearly 500, and 249 have senior professional titles, including 36 First and Second Grade Registered Architects, 42 First and Second Grade registered structural engineers, 6 registered equipment engineers, 28 registered supervision engineers, 75 registered supervision engineers, 38 registered construction engineers, 16 registered cost engineers, 8 certified public accountants.

Under the guidance of sincerity, enterprising spirit, staffing optimization, the company brings together a group of senior experts and the industry elites; its management team is sincere and capable, efficient and pragmatic; its design team is innovative, divergent and excellent; and the development team is familiar with market and full of courage, creating a dynamic, open and diverse collective battle team. The company with various awards has many achievements in the planning, design and supervision of schools, hospitals, hotels, exhibition halls, commercial buildings, office buildings, residential and other civil buildings and different types of plant, large-span steel structure engineering, and other industrial construction. The works are spread throughout the country.

Huaxia staffs are adhering to the well-designed, quality service purposes, combining eastern and western architecture aesthetic, and with the advanced design concept and excellent professional quality, has won the majority of the owners praises. The company with the spirit of enterprise for "dedication, loyalty and trustworthiness, solidarity and cooperation, ethics promote industry". cooperates with all friends to build a better home.

地址：天津市南开区华苑产业区榕苑路16号鑫茂科技园中心楼三层
电话：+86–22–58693336 / 58693341　+86–22–23672000
传真：+86–22–58598916
邮箱：tjhxjzsjgs@126.com
网址：www.tj-huaxia.cn

Add: Third Floor, Center Building, Xinmao Science and Technology Park, No. 16 Rongyuan Road, Huayuan Industry Area, Nankai District, Tianjin
Tel: +86–22–58693336 / 58693341　+86–22–23672000
Fax: +86–22–58598916
E-mail: tjhxjzsjgs@126.com
Http:// www.tj-huaxia.cn

华纳瑞都酒店
建设地点：天津
建设规模：68 000平方米

金岚山谷国际酒店（超五星级现代豪华酒店）
建设地点：福建 平潭
建设规模：42 000平方米（共226个客房，休闲、娱乐、餐饮及住宿）

华虹科技综合楼
建设地点：福建 三明
建设规模：35 764平方米

广西合浦隆鑫商业广场
建设地点：广西 北海
建设规模：105 000平方米

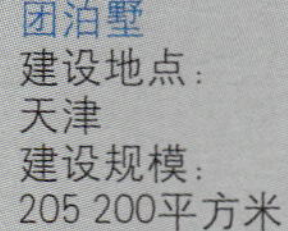

团泊墅
建设地点：
天津
建设规模：
205 200平方米

西部新城H地块项目
建设地点：天津
建设规模：地上108 000平方米　地下30 000平方米

天津奥建建筑设计有限公司

Tianjin ARGN Architectural Design Co.,Ltd.

天津奥建建筑设计有限公司于2007年由康庄总设计师创立于天津，是一家以创意为主的成长型公司，拥有一个由众多对建筑充满激情的年轻人组成的设计团队。

该团队一直坚持用一种不功利、完美主义的心态来做设计，对客户负责、对自己的理想负责；深刻理解和尊重建筑的多样性和地域性，同时也清楚地认识到从城市到所在地块的内在肌理和逻辑性，并在保证整体和谐的基础上进行个性化和创新性设计；坚持国际化的设计标准；保持清醒的头脑，不盲目崇拜，反对造作的模仿和脱离实际的晦涩理论。团队的设计追求最好的创意和表现，在建筑设计、城市规划设计、室内设计、景观设计方面，为客户提供最好的解决方案和专业化的设计服务。

Tianjin ARGN Architectural Design Co.,Ltd. which was founded by Chief Architect Louis Kang in Tianjin in 2007, is a growing corporation dependent on originality, and a design team made up of many young architects with passion for architectural design.

Our team has always insisted on designing with non-utilitarian, perfectionist attitude, responsible for customers and responsible for individual ideals. We have a deep understanding and respect for diversity and regionality of buildings. We also clearly recognize the internal texture and logic of the location from the city to block, and on the basis of ensuring the harmonious integrity undertake personalized and innovative designs. We adhere to international design standards, keep our head on our shoulders, do not blindly follow others, oppose to artificial imitation and unrealistic obscure theory. Our goal is the best creative design and performance in architectural design, urban planning and design, interior design, landscape design, providing customers with the best solutions and professional design services.

地址：天津市南开区华天道海泰信息广场F座南楼215室
邮编：300384
电话：+86–22–23708568
传真：+86–22–23708598
网址：www.argnnet.com

Add: Room 215, South Building of F Tower, Haitai Information Plaza, Huatian Road, Nankai District, Tianjin
P.C.: 300384
Tel: +86–22–23708568
Fax: +86–22–23708598
Http:// www.argnnet.com

1

1–2 天保二期住宅
地点：天津
规模：26万平方米
时间：2007年
（与美国CA公司合作）

1–2 Tianbao Residence Phrase 2
Location: Tianjin
Scale: 260,000 m^2
Time: 2007
(In cooperation with CA of America)

3–5 天津财富公馆
地点：天津
规模：24万平方米
时间：2010年

3–5 Tianjin Palais de Fortune
Location: Tianjin
Scale: 240,000 m^2
Time: 2010

2

3

4

5

6

7

8

9

10

6 天津森泰体育休闲广场

地点：天津
规模：2万平方米
时间：2011年

6 Tianjin Sentai Sports Recreational Square

Location: Tianjin
Scale: 20,000 m²
Time: 2011

7 迁西客运站

地点：河北 迁西
规模：4.7万平方米
面积：2.9万平方米
时间：2011年

7 Qianxi Terminal

Location: Qianxi, Hebei
Scale: 47,000 m²
Construction Area: 29,000 m²
Design Time: 2011

8–10 杨柳青・元宝岛

地点：天津
规模：48.5公顷
时间：2010年

8–10 Yangliuqing•Yuanbaodao Island

Location: Tianjin
Scale: 48.5 ha
Time: 2010

新疆民用建筑设计院有限公司

XINJIANG CIVIL BUILDING DESIGN INSTITUTE CO., LTD.

新疆民用建筑设计院有限公司成立于1985年6月。其前身为新疆机械电子工业设计院，隶属于原新疆机械电子工业厅。受行业滑坡的影响，1997年跌入低谷。1997年底，该院做出了转换经营机制、转型建筑设计的重大决策。2000年12月，在全疆同行业中第一家完成改企转制工作，成为行业深化改革的典范。自1998年以来，保持了连续高速健康发展的强劲势头。2001年顺利晋升建筑设计甲级资质，成为新疆第三家建筑设计甲级资质拥有者。目前拥有建筑设计甲级、工程咨询甲级、城市规划乙级和市政设计乙级资质。2000年进入"新疆勘察设计行业综合实力20强单位"。2010年经营收入是1997年的60倍。

由该院作为投资主体，新设立的西安创元建筑设计院有限公司，于2010年7月1日正式开业，已获得建筑设计乙级、城市规划乙级资质，并在中原地区有所作为。预计该院2011年度签订的设计合同将突破2 000万元。该院目前正在积极创造条件，力争晋升建筑甲级资质。

由该院作为投资主体，新设立的新疆创元华远房地产开发有限公司，已于2010年10月1日正式开业。斥资亿元开发的首批高层商住楼已经开工，市场情况良好。该院起步快、起点高，必将迅速实现规模经营。

在"以人为本，质量兴院"发展战略的指引下，该院致力于人才队伍建设，尽一切可能为人才的健康成长提供良好的平台，努力创建学习型组织，使一批又一批青年员工脱颖而出，成长为技术骨干，为企业的持续发展奠定了坚实的人才基础。

Xinjiang Civil Building Design Institute Co., Ltd. was founded in June 1985. It was preceded by Xinjiang Machinery Electronic Industry Design Institute, which was managed by former Machinery & Electronic Industry Department, it fell down in 1997 due to influence of industry downslide. In the end of 1997, the institute made a big decision of changing operational system and transforming architectural design. In Dec. 2000, it completed corporate transformation and system establishment as the first one in Xinjiang in this industry, and became the industry pioneer in deepening reform. Since 1998, it has kept a rapid speed of continuous healthy development. In 2001, it was promoted to be Class-A Architectural Institute, the third one in Xinjiang. At present, it is licensed for Architecture Class-A, Consulting Class-A, Urban Planning Class-B and Municipal Design Class-B. In 2000, it was selected as one of the "Top 20 Strong Enterprises of Xinjiang for Industry Integrated Strength of Investigation & Design". Its operating revenue in 2001 is sixty times that of in 1997.

Invested by this institute, the newly-founded Xi'an Chuangyuan Architectural Design Institute Co., Ltd. has won the Class-B Certificates of architectural design and urban planning. It was officially founded in July 1, 2010, and has made some achievements in central areas of China. It is estimated that total amount of money of design contract in 2011 will top RMB20 million. At present, it is actively preparing for the promotion of Class-A Certificate of Architecture in this winter and next spring. Invested by this institute, the newly-founded Xinjiang Chuangyuan Huayuan Real Estate Development Co., Ltd. formally went into operation in Oct. 1, 2010. With an investment of RMB100 million, the first batch of high-rise commercial & residential community is being constructed, the market situation of which is excellent. Starting fast and high, the company must realize scale operation soon.

Under the guidance of development strategy "Based on Human, Develop by Quality", this institute focuses on organizing talents, and spares no efforts in providing an excellent platform for talents to grow healthily, trying to create a learning organization. Young employees stand out one after another and grow to be technology backbone, which form the talent foundation for company's further development.

地址：新疆乌鲁木齐市新医路2号汇文大厦8楼(830054)
电话：+86-991-4338259, 4335977(Fax)
网址：www.mysjy.com.cn
邮箱：xjmyjzsjy@126.com

地址：陕西西安市朱雀大街中段58号陕西画报社七楼(710068)
电话：+86-29-88403772
邮箱：xacyjzsjy@126.com

喀什市华龙世贸中心

1	2	3
5	4	
6	7	

1 新疆兵团机关事务管理局居住区规划与建筑设计
2 恒大房产五家渠"金碧天下"居住小区规划与建筑设计
3 沙湾县第三期集资建房居住小区规划与建筑设计
4 河南许昌市长村居住区规划与建筑设计(西安创元作品)
5 河南许昌市城东区邓庄社区规划与建筑设计(西安创元作品)
6 西安阎良航空基地龙洋花园小区规划与建筑设计(西安创元作品)
7 河南许昌市三塔居住区规划与建筑设计(西安创元作品)

新疆印象
建设规划设计研究院（有限公司）

该院是集文化思考、城市导演、工程设计于一体的特色设计研究机构。

该院是中国环境艺术设计甲级，城市规划、旅游规划、建筑设计、室内设计、风景园林乙级资质的综合设计团队。

领导印象 高级建筑师、中国策划20年十大策划专家、中国环境艺术委员会专家委员会委员、全国首批注册高级环境艺术师、清华大学城市导演课题客座教授、新疆城市规划协会常务理事、新疆优秀中青年勘察设计工作者、董事长王元新；一级注册结构师、总工程师张晓民；享受国务院特殊津贴教授级高工、国家注册规划师、资深总规划师陆易农；中国科学院新疆地理与生态研究所研究员、著名旅游规划专家阎顺；一级注册建筑师、国家注册规划师副院长张世鹏；一级注册建筑师、总建筑师苗新育；高级建筑师、副院长时卫国、王冀东、王洋等多位资深骨干领导。

团队印象 该院设有建筑、规划、旅游、景观、装饰五个设计所，以及陆易农教授规划工作室、阎顺研究员旅游工作室、王洋创作事务所、徐乃钦方案室、影视动漫工作室、雕塑家工作室、新疆印象建设沙龙等。现拥有高、中级工程师，注册建筑师，注册结构师，注册城市规划师，注册环境艺术师，雕塑家及各类专业人才80余人。

作品印象 该院出版了新疆第一部规划专著《论城市的有机属性》（陆易农著），在区内外学术刊物发表论文80余篇；在全国第十二次建筑学术与文化讨论会上出版的《新疆印象创作思迹》（主编王元新、陆易农）一书作为大会学术交流作品专辑；受自治区建设厅、兵团等委托编辑了有关新农村建设规划及住宅图集三册；编撰制作了动漫电视片《伊宁市火车站片区城市设计》、《阿克苏紫金花苑》、《昌吉水木融城》等多部。

成果印象 近几年来，该院的作品有110件入选《中国建筑设计作品年鉴》、10件入选《中国创意优秀建筑设计作品集》；有30余项规划、建筑、景观设计成果获国家建设部、新疆维吾尔自治区、新疆生产建设兵团、乌鲁木齐市、昌吉市等各级优秀设计奖。其中："达坂城王洛宾音乐艺术城"获2008年中国策划20年经典策划金奖，"昌吉回民小吃街"和"莎车十二木卡姆故乡园"获2009年中国人居典范规划设计方案金奖，"沙湾大盘美食文化城""阜康市瑶池园""昌吉滨湖河"获2010年中国人居典范规划设计方案金奖。

文化印象 董事长王元新一直强调"创新与研究是设计院的第一生产力"，首次提出"故事建筑学"观点，认为建筑是一直讲下去的故事，要求新疆印象设计要高举"现代地域文化"大旗，唱响"新疆是个好地方"，势必让"新疆印象"文化传播开来，用优秀作品来阐释"新疆印象，不同创想"的境界。

Characteristics design institution with the cultural thinking, the city director, engineering design in one.

Integrated design team with Grade A qualification of China's environmental art and design, and Grade B qualification of urban planning, tourism planning, architectural design, interior design, landscape design team.

Leading Image Senior architect,China's top ten planning experts planning for 20 years, China Environment Arts Council Expert Committee, the national first batch of registered senior environmental artist, visiting professor at Tsinghua University, the city issues the director, executive director of the Xinjiang Urban Planning Association, Xinjiang outstanding young workers in the survey and design, chairman Wang Yuanxin; the First Class registered structural engineer, chief engineer Zhang Xiaomin; enjoy the professor-level senior State Council special allowance, the registered planner of China, senior chief planner Lu Yinong, Chinese Academy of Sciences Xinjiang Institute of Geography and ecology researcher, famous experts in tourism planning Yan Shun; the First Class registered architect of China, registered planner of China, vice president Zhang Shipeng; the First Class registered architect of China, chief architect Miao Xinyu; senior architect and vice president Shi Weiguo, Wang Jidong, Wang Yang and other the backbone of senior leadership.

Team Image The institute has architecture, planning, tourism, landscape, decoration five design studios, Professor Lu Yinong planning studio, Researcher Yan Shun Tourism Studio, Wang Yang Creative Firms, Xu Naiqin Program Studio, Film and Television Animation Studio, Sculptor Studio, Xinjiang Image salon construction and so on. At present, there are more than 80 kinds of professional talents of high, intermediate engineers, registered architect, registered structural engineer, registered urban planner, registered environmental artist, sculptor.

Works Image The institute published the first planning monograph *On the City's Organic Properties* (wrote by Lu Yinong) for Xinjiang, more than 80 papers published inside and outside the academic journals in the area. In the twelfth conference to discuss architecture academic and culture published *Creative Thinking Trace of Xinjiang Image* (Chief editor: Wang Yuanxin, Lu Yinong), the book as the academic works album for the conference. Commissioned by the Regional Department of Construction, the Corps and other agencies to Edit the new rural construction planning and housing atlas three volumes. Compilation of television animation production, *Yining City Railway Station Area Urban Design, Aksu Zijin Garden, Changji Mizuki Rong City* and any others.

Achievements Image In recent years, 110 works were selected in the *Annual Review of Chinese Architectural Design Works*, 10 were selected in *China's Creative Works Set of Outstanding Architectural Design*. More than 30 achievements in planning, architecture, landscape design won the Ministry of Construction, Xinjiang Autonomous Region, Xinjiang Production and Construction Corps, Urumqi, Changji City and other levels of the excellent design awards, of which Dabancheng Wang Luobin Music and Art City won the gold medal of Classic plan for 20 years China Planning Award in 2008, Changji Muslim Snack Street and Twelve Muqam Shache Home Park won the gold medal of China Habitat Model in Planning and Design Award in 2009, Shawan Food Market Cultural City, Fukang City Jade Pool Park, Changji Lake River won the gold medal of China Habitat Model in Planning and Design Award in 2010.

Culture Image Chairman Wang Yuanxin has been emphasizing that innovation and research is the Institute's first productive force", the first time proposed "story architecture" , the view that architecture is an everlasting story, asked Xinjiang image design to hold high the "modern local culture" banner and sing loudly "Xinjiang is a good place", be bound to make " Xinjiang image" cultured, with the outstanding works to illustrate the realm of "Xinjiang image, different Imagination".

新疆印象建设规划设计研究院（有限公司）
董事长：王元新
电　话：13325550916　　+86-991-7819196（办公室）
传　真：+86-991-7819189
网　址：Http://www.新疆印象.com (Http://www.xjyxgh.com)
地　址：新疆乌鲁木齐市高新区苏州东街568号金邦大厦15楼

Xinjiang Image Construction Planning and Design Research Institute Co., Ltd.
Chairman: Wang Yuanxin
Tel: 13325550916　　+86-991-7819196(Office)
Fax: +86-991-7819189
Http://www.xjyxgh.com
Add: Fifteenth Floor, Jinbang Building, No. 568 Suzhou East street, High-tech District, Urumqi, Xinjiang

千頃蒹葭十裡洲
溪居宜月更宜秋
鷗鳧棲水高僧舍
鸛鴒巢雲名士樓
蒼葡葉分飛鷺羽
荻蘆花散釣魚舟
黃橙紅柿紫菱角
不羨人間萬户侯

八盘水墨

建设地点：新疆 喀什
规划建设用地面积：123 649.362平方米
总建筑面积：296 776.9平方米
建筑密度：23.4%
容 积 率：2.1
绿 地 率：46%
主创人员：王　洋、赵新华
设计人员：戴金梅、罗　莎、冯艳丽、朱姜燕、
开 发 商：新疆鼎盛鑫业房地产开发有限责任公司

自然是上帝作品，其代表着和谐和完美，我们推崇并效法之。主轴线取"曲径"之意，形成随地势蜿蜒而上的栈道，弯曲的主轴线给营造中式空间带来机会。拾步阶而上，沿途移步换景，起、承、转、合、漏、借、对，适当运用，避免了西式轴线的单调直白，也为人们提供了进入各种活动场所的可能。休憩的小亭、弯曲的溪边小路和清澈的水池，让人流连其间而乐不知返。会所则巧借地形，层层叠落，别具一格。山水联排别墅自由灵活，依山就势，表达一种"在溪山优美处，宅院数间，比邻自然而居"的优美意境。

巴音布鲁克大营

建设地点：新疆　巴音布鲁克
建筑面积：19 829.09平方米
规划总用地：103 422.65平方米
容 积 率：0.19
绿 地 率：45.7%
主创人员：王元新、
设计人员：张世鹏、阎　顺、姬　畅、韦世武、冯　雪、吴建伟

巴音布鲁克大营为巴音布鲁克镇旅游景区的五星级大酒店，是一座集办公、餐饮、娱乐、休闲等功能为一体的智能化建筑。

规划分三个片区：宾馆区、篝火广场、蒙古包区。几个功能区通过环路有机地联系在一起。建筑结合周边规划及建筑形态，以蒙元建筑风格、造型及颜色上力求融入当地文化底蕴。建筑设计平面错落有致，竖向台地构图，体现草原文化，彰显地域特色。

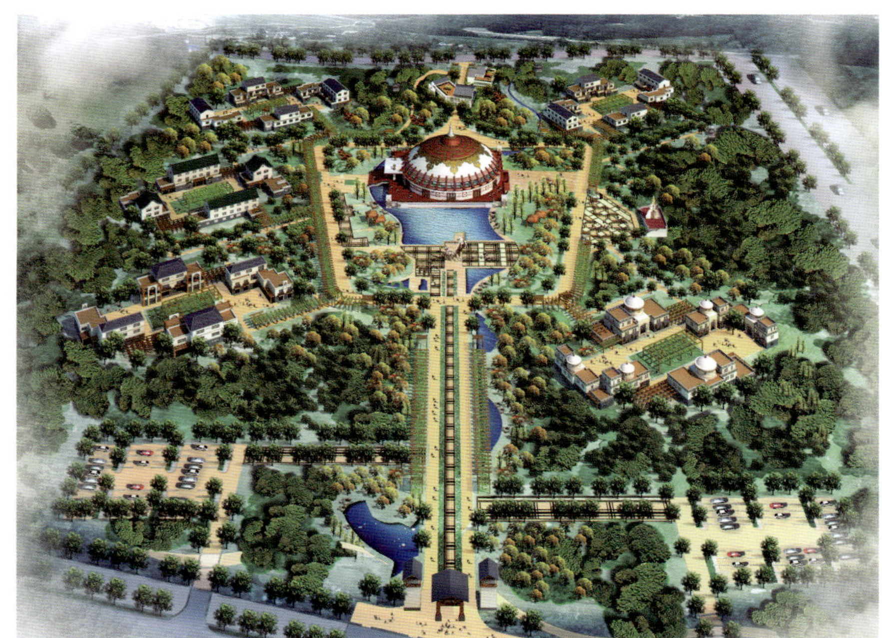

华夏第一包

建设地点：新疆 库尔勒
规划用地面积：19.66公顷
主创人员：王元新、薛　斌、张俊选
设计人员：姬　畅、张世鹏、韦世武、苗新育、
　　　　　徐乃钦、张　娟、陈　玲、袁　婷

梨园民街总平面道路骨架呈"马头琴"环状路网结构，外围四个民族的民街建筑色彩采用民间建筑形制，以灰、白、木色为主色调，与大自然融在一起。每组建筑均为退台形式，使游人可登上一、二层上人屋面平台，领略大自然香梨树丛之美景，享受巨大氧吧带来的健康与快乐。建筑纹样选择各族文化的结合，突显民族融合性。

蒙汉文化融合的"华夏第一包"是华夏第一州首府库尔勒市"梨园民街"的核心建筑，直径56米，建筑色彩采用宫廷金顶行制，以红、黄、白为主色调。建筑形态呈"天圆地方"设计元素，意在追溯中华汉文化——太阳：光芒四射；蒙古族崇尚车轮，承载历史。建筑纹样选择汉文化与蒙古文化的结合，突显巴州地域性与主体的融合性。

鄯善县鲁克沁镇街景风貌改造设计

建设地点：新疆 鄯善
项目规模：核心区长2千米
主创人员：徐乃钦、张雪坤
设计人员：王元新、薛冰峰、焦琪书、李 军

设计以"中国历史文化名镇——鲁克沁镇"柳中古城、双塔清真寺、郡王夏府三大历史文化遗存为核心，同时围绕中国历史文化名镇、中国民间艺术之乡、中国哈密瓜之乡的特点。通过汉唐与民族文化、造型对称与非对称、街道节点与建筑细部、饰面材料的粗与细、外部装饰与内部功能、设计的共性与个性、形态的简与繁七大关系，实现文化内涵、风貌改造、使用功能的三个结合，最终为揭开鲁克沁镇民居建筑的神秘历史面纱，将其历史建筑艺术和民族文化精神的悠久文脉优化地延续下去，并创造出深具价值的建筑艺术经典，体现鲁克沁汉唐历史文化与地域文化融合的风貌建筑特色。

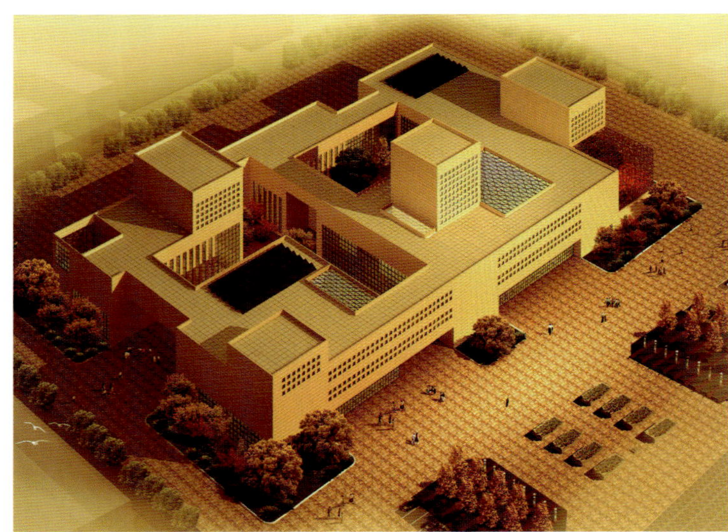

吐鲁番城市规划展览馆

建设地点：新疆 吐鲁番
行政服务中心的建筑面积：30 000平方米
规划展示馆的总建筑面积：20 000平方米
展示馆的面积：2 000平方米
主创人员：王元新、张世鹏、陆易农
设计人员：韦世武、贾婷婷、陈 玲、袁 婷

吐鲁番城市规划展览馆建设理念：低碳的示范区、绿色的示范区、宜居的示范区、和谐与节能的示范区。

设计上运用了当地最具代表性的交河故城、葡萄干凉房、坎儿井三大特殊形态为创作元素，建筑造型恢弘大气，具有鲜明的地域特色。

建筑成围合状组团式布局，且建筑高度不超过一棵树，在这世界上最炎热的地方，实现吐鲁番人向往的"园中有城、城中有园，一棵树下的城市"的愿景。

库车新区办公楼

建设地点：新疆 阿克苏
规划建设用地面积：101 167.6平方米
建筑占地面积：20 164.0平方米
总建筑面积：184 579平方米
绿地面积：31 665.4平方米
建筑密度：20.0%
容 积 率：1.62
绿 地 率：31.3%
停 车 数：600辆
主创人员：王 洋
设计人员：赵新华、罗 莎、吕松贤

园区以管委会大楼为核心，从南至北形成了一条主要景观轴线，南侧广场两侧形成次要景观节点，通过建筑间距的穿插，将整个园区的景观渗透于城市。

管委会办公大楼建筑主体由两个塔楼组合而成，形似库车著名的克孜尔尕烽火台，立面细节上将民族符号进行提炼再创作，体现出具有地域特色的现代建筑风格。

写字楼建筑造型上，点、线、面的结合，虚与实的对比形成了丰富的外部肌理，采用了玻璃与实墙的对比，体块上凸凹有致，形成了韵律，增强了光影感和立面效果。

住宅建筑以灰、白为主色调，适度点缀一抹暖色，清新淡雅，使居住者的情感回归宁静与自然，实现了花园式家园的居住理想。

Kunming Guanfang Architectural Design Co., Ltd.

昆明官房建筑设计有限公司

昆明官房建筑设计有限公司是一个具有甲级建筑设计资质和多项乙级设计资质的企业。经过15年的发展，现已成为云南省乃至全国著名设计企业。公司长期服务于房地产业的建筑设计。至今已完成近千万平方米的建筑设计，积累了房地产项目开发的前期策划、规划设计、方案设计的经验，善于和房地产商沟通并能对项目优化设计，以达到开发成本最低、效益最佳。尤其在住宅设计和酒店建筑创作方面具有丰富的经验和先进的理念。多个精品项目荣获国家建设部、科技部、云南省、市及亚洲优秀奖项达60余次，受到社会各界的赞誉。

设计公司审时度势，将发挥其专业特长走专业化、精细化的发展之路。主要以旅游度假酒店设计、商业地产设计为主，使其做精做专。

1996年设计的五星级“丽江官房大酒店”是官房建筑设计有限公司进军旅游酒店设计的起点。“丽江官房大酒店”作为云南省首家五星级酒店，曾无数次接待过国家领导人、外国元首及来自五湖四海的嘉宾。并成为丽江恢复重建的典范，推动了丽江的知名度和旅游业的发展。

2003年建成的“滇西明珠花园五星级度假酒店”，是一个典型的新中式建筑，是将纳西建筑风格和东巴文化内涵完美融合，具有地域文化的酒店。采用“管家式”的个性化服务，营造出休闲度假氛围，成功首创了云南“产权式酒店”的运作模式，官房设计公司所设计的旅游酒店也由此迈向新的高度。

此后，五星级的“红河官房大酒店”、“腾冲官房大酒店”、“曲靖官房大酒店”的相继建成，酒店设计使该公司走上专业化发展的道路。

官房建筑设计团队正努力为客户提供精细化、人性化、专业化的产品和有经济回报的酒店，力争在管理、服务、设计等各个方面与国际先进酒店及管理公司接轨，设计出具有国际先进水平的酒店品牌。

精品住宅设计是设计公司的另一大优势。

1999年，“金康园”被评为“国家小康住宅示范小区”，荣获国家建设部建筑质量最高奖——鲁班奖，总揽了建设部、科技部颁发的“规划设计优秀奖”、“工程质量优秀奖”、“科技进步优秀奖”、“室内装修优秀奖”、“环境设计优秀奖”、“物业管理优秀奖”6个单项大奖，成为全国第一个获得全部奖项的小康住宅示范小区。建筑面积23余万平方米，共有1 099套住房。

“金康园”是一个智能型的康居住宅示范小区。它的规划设计理念超前，“以人为本”是其最鲜明的特色，园内人车分流、动静分开，大面积的绿地花园，底层架空的开阔空间，为人们提供了休闲娱乐的场所。小区共对27项新技术，30项新材料、新产品进行了推广应用，从根本上提高了住宅的整体质量。整个小区布局科学合理，内有幼儿园、小学、儿童游乐场、游泳池、会所等配套设施，周边有中学、医院、康体娱乐中心、农贸市场、大型商场等场所。

“云康园”小区是设计公司的又一力作，小区占地19公顷，以双拼、联排、叠加别墅为主，总户数639户，绿地率52%。采用了大量绿色、环保、节能型建筑材料及多项新技术、新工艺，极大地提升了居住品质：新型、美观的外墙涂料，具有耐用、防火、防裂等多项功能；框架结构采用清水混凝土技术，杜绝了抹灰层开裂的质量隐患；采用倒置式屋面技术，具有较好的保温隔热效果，同时减缓了防水层的老化，大大延长了屋面防水层的使用寿命；新型太阳能辅助电加热系统，极大地提高了能源利用率，节约能源达40%。该公司自诞生之初就获得社会各界的高度赞誉，荣获“2004亚洲城市住宅及住区环境国际设计大奖”、“中国人居环境金牌试点项目”两项大奖。

2007年11月由云南省建设厅、世界华人建筑师协会主办，该公司和云南省规划院成功举办了的“2007年世界华人建筑师年会”和“小城镇保护与开发论坛”。论坛中由蒋鸿兴总建筑师发表的《新中式建筑的理念来保护开发小城镇》一文，受到业内和政府的认可。

2008年9月19日，由全国房地产商会、全国设计联盟主办，官房设计公司承办的“中国（昆明）旅游度假酒店设计与经营高端论坛”沙龙活动及昆明官房建筑设计有限公司成立十五周年的庆典活动在昆明成功举办。论坛上关于度假酒店在中国的发展与创新提出了很多崭新的理念与观点，其中包括将度假酒店与旅游地产业的开发紧密结合起来，促进我国度假酒店的发展。

设计质量方针

繁荣创作：建立良好的设计环境和创作氛围；激励创新，以人为本；设计新颖，特色鲜明，具有超前性和时代感。

争创名牌：公司设计作品以科技住宅、星级酒店为主，推行生态建筑、新中式建筑。做精、做专、做细，争取达到国内、国际先进水平，创出品牌。

精心设计：无论工程大小，确保设计项目功能适用、安全可靠、经济合理、造型美观、技术先进、可持续发展。

诚信服务：树立信守合同、热忱服务的理念，为业主提供完善的服务，积极配合建设方、施工方，及时处理有关事宜，促使工程项目顺利完成。

公司将致力于与全国的商业地产商建立长期友好的合作关系，并充分利用资源和发挥我们专长的优势，来服务好每一位合作伙伴，为业主创造更多的效益，为社会提供精品建筑。

地 址：云南省昆明市白云路258号官房广场21楼
Add: The 21st Floor, Guanfang Plaza, No.258 Baiyun Road, Kunming City, Yunnan Province
邮 编 P.C.: 650233
联系人 Contact: 叶先生，付先生 Mr. Ye, Mr. Fu
电 话 Tel: +86-871-5619888-62103 or 62115
传 真 Fax: +86-871-5613742
网 址 Http://www.kmgfsj. cn
邮 箱 E-mail: kmgfsi@126.com

Kunming Guanfang Architectural Design Co., Ltd. is an enterprise with Grade A architecture design qualification and a number of Grade B design qualifications. After fifteen years of development, it has become a well-known design firm in Yunnan Province and the country. The company offers service for real estate company in the architectural design for long term. It has completed nearly ten million square meters of building design, and accumulated experiences in real estate project development planning, planning and design, program design, good communication with real estate business and to optimize the design of the project to meet the minimum development cost, the most effective good. Especially in the residential and hotel construction creative design, has a wealth of experience and advanced concepts. Numbers of quality projects won more than 60 outstanding awards awarded by the Ministry of Construction, Ministry of Science, Yunnan Province, City, and Asia, received recognition by the society.

The design company takes use of situation and will take advantage of our professional expertise, refinement of the development. Mainly in the tourist resort hotel, commercial real estate design, so do the fine professionals.

In 1996, the design of five-star Lijiang Guanfang Grand Hotel is the starting point of Guanfang Architectural Design Co., Ltd. to enter tourist hotel design field. Lijiang Guanfang Grand Hotel as the first five-star hotel in Yunnan Province, has received numerous state leaders, foreign heads of state and guests from all over the world, and became a model for reconstruction of Lijiang, improving Lijiang visibility and promoting the development of tourism.

Built in 2003, Western Yunnan Pearl Garden Five-star Resort Hotel, a typical new Chinese architecture, is the combination of Naxi architectural style and Dongba culture, with a regional culture of the hotel. The company adopts a "housekeeping" type of personalized service, creating a leisure atmosphere, successfully pioneering the Yunnan "hotel property rights" mode of operation, making Guanfang Design Company in tourism hotel design move to a new height.

Since then, the five-star Red River Guanfang Hotel, Tengchong Guanfang Hotel and Qujing Guanfang Hotel have been built, the hotel design making the company embarked on professional development.

Guanfang Archirectural Design Team is working hard to provide customers with detailed, personalized, professional products and economic returns of the hotel, and strives to make management, service, design and other aspects keep up with the international advanced hotel management company standards, designed with international advanced the level of hotel brands.

Design quality of residential design is another major advantage of the company.

In 1999, the Golden Health Park was entitled "National well-off residential demonstration area", won the Ministry of Construction the highest construction quality award – Luban Award, and was awarded the "Planning and Design Excellence Award", "Project Quality Award of Excellence", "Award for Outstanding Scientific and Technological Progress", "Interior Decoration Excellence Award", "Environmental Design Excellence Award","Property Management Excellence Award" by the Ministry of Construction, Ministry of Science, becoming the first well-off residential demonstration area obtained all awards all over the country. Construction area is more than 23 million square meters, a total of 1,099 housing units.

Golden Health Park is an intelligent well-off residential demonstration area. The planning and design concept is advanced, "people-oriented" is its most distinctive feature. In the park, people and vehicles, static and dynamic are separated, large areas of green garden, the open bottom of overhead space, providing people with entertainment options. A total of 27 new community technology, 30 new materials, new products for the promotion and application, fundamentally improve the overall quality of housing. Scientific and rational layout of the entire district, there are kindergartens, primary schools, children's playground, swimming pool, clubhouse and other facilities, surrounded by schools, hospitals, sports and recreation center, farmers market, shopping malls and other places.

Cloud Health Park is another masterpiece of the design company, residential area of 19 ha, with larry, townhouses, villas superposition, the total number of 639 units, green ratio is 52%. With a large number of green, environmental protection, energy-saving building materials and a number of new technologies, new processes, the design greatly enhances the living quality: a new, beautiful exterior paint, with a durable, fire prevention, cracking prevention and many other features; framework used faced concrete technology, to prevent cracking of the plaster layer quality problems; inverted roof technology, has good thermal insulation, while slowing the aging of the waterproof layer, greatly extending the life of the roof waterproofing; new solar-assisted electric heating system, greatly improved energy efficiency, energy savings up to 40%. Since the beginning it has received high praise from the society, won the "2004 Asian Urban Residential and Residential Environments International Design Award", "China Habitat Environment Gold Pilot Project Award".

In November 2007, sponsored by the Yunnan Provincial Department of Construction and the World Association of Chinese Architects, undertook by Yunnan Province Planning Institute and our company, "2007 Annual World Chinese Architects Meeting" and "Protection and Development of Small Towns Forum" was successfully held. New Chinese Architecture Concepts to Protect the Development of Small Towns published by the chief architect Jiang Hongxing in the forum gained the recognition from the industry and government.

On September 19, 2008, sponsored by the National Real Estate Chamber of Commerce, National Design Federation undertook by Guanfang Design Company, China (Kunming) Tourist Resort Hotel Design and Management of High-end Forum Salon and the 15th Anniversary Celebrations of Kunming Guanfang Architectural Design Co., Ltd. was successfully held in Kunming. Forum on resort development and innovation in China emerged a lot of new ideas and perspectives, including the resort hotel and tourism industry development closely together to promote the development of China resort hotel.

Design Quality Principles

Prosperity of Creation: the establishment of good design and authoring environment; encouraging innovation, people-oriented; innovative design and distinctive features, advanced, and contemporary.

Create famous brand: the company works mainly in technology designed for residential and star hotel, the implementation of eco-building, the new Chinese architecture. To do the fine meticulous designs to reach domestic and international advanced level, to create the brand.

Elaborate Design: whether project size, ensure project function of safe, reliable, economical, attractive appearance, advanced technology, and sustainable development.

Integrity Services: abiding by the contract with the concept of dedicated service, providing adequate services for the owners to actively support the client side, the construction side, timely processing the matter, to promote the successful completion of the project.

We will work with the country's commercial real estate business to establish long-term friendly relations of cooperation and make full use of our resources and expertise to play the advantages to serve every partner, for the owners to create more benefits, to provide quality buildings for the society.

腾冲官房大酒店

总建筑面积：22 256.5平方米
酒店部分面积：16 560平方米
酒店综合楼：5 696.5平方米
容 积 率：0.46
备　　注：酒店共计33幢独立式别墅型酒店

设计理念

有“南方丝绸之路要冲”之称的腾冲，自古就有汉族、回族、傣族、阿昌族等23个民族孕育的悠久的民族文化和历史文化，此次设计除在秉承“延续历史，尊重文化”的整理构思下，还穿插体现了当地民族文化，如腾冲的热海洗浴文化和当地的旅游文化。面对着古老、开放的腾冲，全体官房设计同仁希望运用新的建筑语言、新的造型符号，呈现切合时代大背景且具有浓郁地方特色建筑群落，这就是此次设计的初旨。

热海“碧云杉”温泉酒店

总用地面积：32 400平方米
总建筑面积：6 421.7平方米
容　积　率：0.198

设计理念

云南腾冲热海风景区距离腾冲县城约15千米，车程不足半个小时。本项目位于热海景区用地东南面，与主入口处的大型停车场仅一水之隔，总用地面积为32 400平方米；先在停车场的侧面设计一座宽6.0米的桥梁，并通过盘山的道路蜿蜒而上至项目用地；项目位于半山腰处，海拔高度介于987.0～1025.0米，局部地势起伏较大；周边及内部植被茂盛，风景秀丽，具有独特的温湿气候。

本项目无论从总体布局到功能分区，还是尺度要求，均应充分体现“以人为本”的设计理念，满足不同空间及活动的功能要求，依据不同空间的不同使用功能和服务人群，设置尺度怡人的空间和环境小品设施，以期达到优美景观和完善的服务功能的高度结合。

金色抚仙湖国际会议酒店

酒店主楼总建筑面积：59 108.92平方米
酒店公寓A楼总建筑面积：11 791.40平方米
酒店公寓B楼总建筑面积：19 492.61平方米
酒店主楼及公寓总建筑面积：90 392.93平方米

设计理念

设计定位：主楼定位为豪华五星级国际会议度假型酒店。酒店公寓定为四星级酒店标准，共享主楼设施。

主题外观：一只展翅的雄鹰，俯瞰抚仙湖的美景。

主题定位：A.回归自然：将湖景引入每个房间，使客人在酒店的时光都可以享受到湖光山色。无论你在客房休息，还是用餐、饮茶、品酒、喝咖啡，甚至洗浴、健身、购物，都能时时刻刻享受到抚仙湖的美景。

B.豪华休闲：宽敞舒适的客房，通过在客房里突出干湿分离、半私密和通透开敞的卫生间设计，独具匠心地将浴池设在阳台上，并增添了整个空间的趣味性。

整体的酒店设计充分利用清澈的水面和有灵气的山脉等自然景观，营造了一个使人流连忘返的地方，并提升了抚仙湖度假风景区的接待能力，带动了该地区的经济发展。

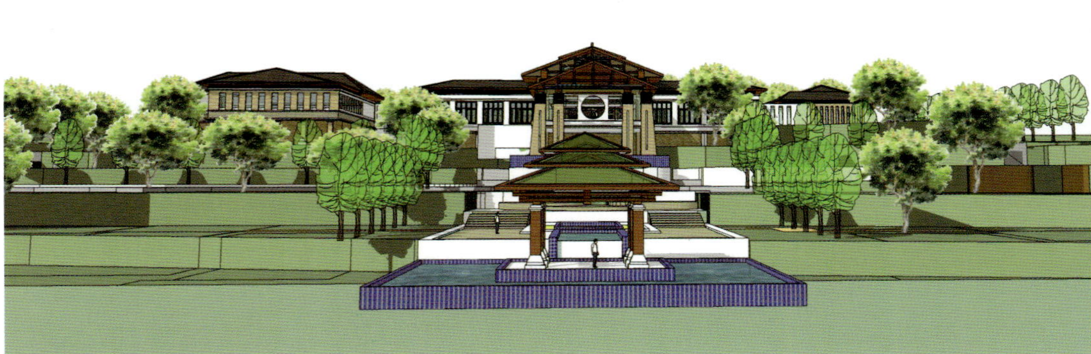

大山包春夏秋冬大酒店

总建筑面积：21 000平方米

设计理念

由昭通旅游投资公司投资1个亿兴建的大山包春夏秋冬大酒店，位于云南昭通的大山包风景区。大山包是乌蒙高原上的一颗璀璨明珠，它以自然、壮美而著称，它的山体峭壁落差纵横让人心惊肉跳，云海连天气象万千踩于足下……；鸡公山是岁月永恒的雕像屹立在此，鸟类珍宝黑颈鹤冬天来此栖息……；这里春夏秋冬景色各异，是摄影爱好者的天堂；这里大自然的奇观异景让人叹为观止、心胸视野开阔；这里还是一块尚未开放的生态风景区。酒店功能齐全，设计有全日制餐厅，（餐厅12个包房可同时容纳220人就餐）、100人多功能厅一个、15人贵宾室一间、300人会议室一间、30人会议室2间、酒吧、网球场、健身房、大众水疗馆和高尔夫练习场等。

安厦西双版纳—雨林·圣堤亚纳精品酒店

项目地点：云南 西双版纳
总建筑面积：22 445.70平方米
建筑占地面积：4 874.28平方米
构思主题：含蓄 神秘 脱俗

设计理念

1. 围绕酒店品牌创造温暖舒适的环境空间，突出私密浪漫的“雨林”度假氛围，体现商务休闲的空间特点，酒店采用独特的东南亚风格建筑形式。
2. 体现酒店的“身”、“心”、“灵”，做好硬件空间环境。
3. 注重节约投资和投资回报。
4. 采用耐用、低维护和高效的本土材料，使建筑集美观、实用、耐久和唯一性等特点于一体。

朝阳建筑

Yunnan Zhaoyang Architecture Design Institute
云南朝阳建筑设计院

地点：云南省昆明市西昌路688号长城证券大厦附三楼
电话：+86-871-5367288, +86-871-5367088
传真：+86-871-5367388

Add: Attached to the Third Floor, Great Wall Securities Building, No. 688 Xichang Road, Kunming,Yun nan
Tel: +86-871-5367288, +86-871-5367088
Fax: +86-871-5367388

云南朝阳建筑设计院由原云南省建设厅总工程师魏增阳创立于1992年，持有企业法人营业执照，具有独立法人资格，为云南省建设厅的下属设计单位。该院人员层次配备合理，老中青相结合，主要技术骨干为具有多年设计经验、超强开拓精神的中青年人员组成。该院多年来在魏增阳总工程师的领导下为云南省建筑事业作出了巨大的贡献，设计作品多次获奖。

设计理念是：
没有最好、只有更好，努力学习、不断创新。
立足实际、不放过每个细节，总结积累经验、服务好每一个用户！

Yunnan Zhaoyang Architecture Design Institute was founded in 1992 by Wei Zengyang, the former chief engineer of Yunnan Provincial Department of Construction, the institute holds the business license, is an independent legal entity, is a design unit subsidiary of Yunnan Provincial Department of Construction, with a reasonable level of staff, bring together the senior, the junior and the young, the main technical backbone are composed of the young staff with many years design experience, superior and pioneering spirit. Under the leadership of President Wei, they have made encouraging achievements for the construction industry of Yunnan province over the years, and the design works has won the awards many times.

Our design philosophy is:
There is no best, only better, study hard and innovate.
Based on reality, let off every detail, sum up the accumulation of experience, provide good service for each user.

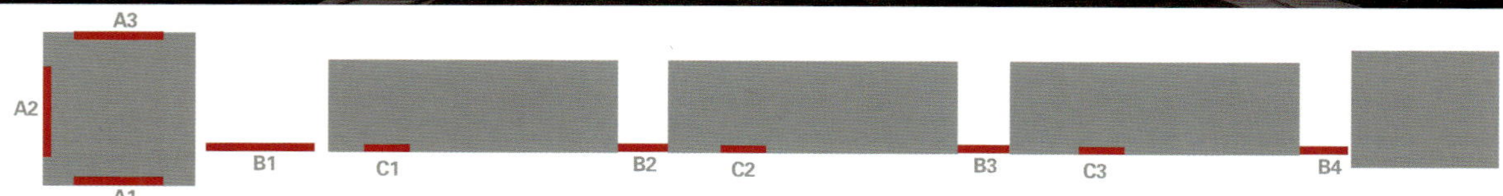

A1（东），A2（南），A3（西）：W30XH19.7米，屏底离地115.35米。 B1：W36XH17米；B2，B3：W23.5X17米；B4：W24XH17米；屏底离地20.05米。
C1，C2，C3：W14.4XH8米；屏底离地74.7米。
注：屏尺寸标注为外框尺寸。

北京盘古大观户外LED大型显示屏

规划范围

项目中显示系统设计为点间距30毫米规格；包含5种显示尺寸大小：写字楼A座的东面、西面、南面三面的A1、A2、A3；写字楼A座与公寓楼连接的B1；公寓楼顶的C1、C2、C3。

新河高速公路主线收费站、监控及通信分中心

建设地点：云南 河口
占地面积：13 513平方米
房屋建筑面积：4 722.8平方米
办公楼建筑面积：1 802.15平方米

维西傈僳族自治县人民医院整体搬迁建设项目

建设单位：维西傈僳族自治县人民医院
建设地点：维西傈僳族自治县城新区兴维大道中段
项目性质：国营非盈利性医院
项目类别：综合医院
设计单位：云南省朝阳建筑设计院
总建筑面积：16 153.04平方米
用地面积：33 241.13平方米

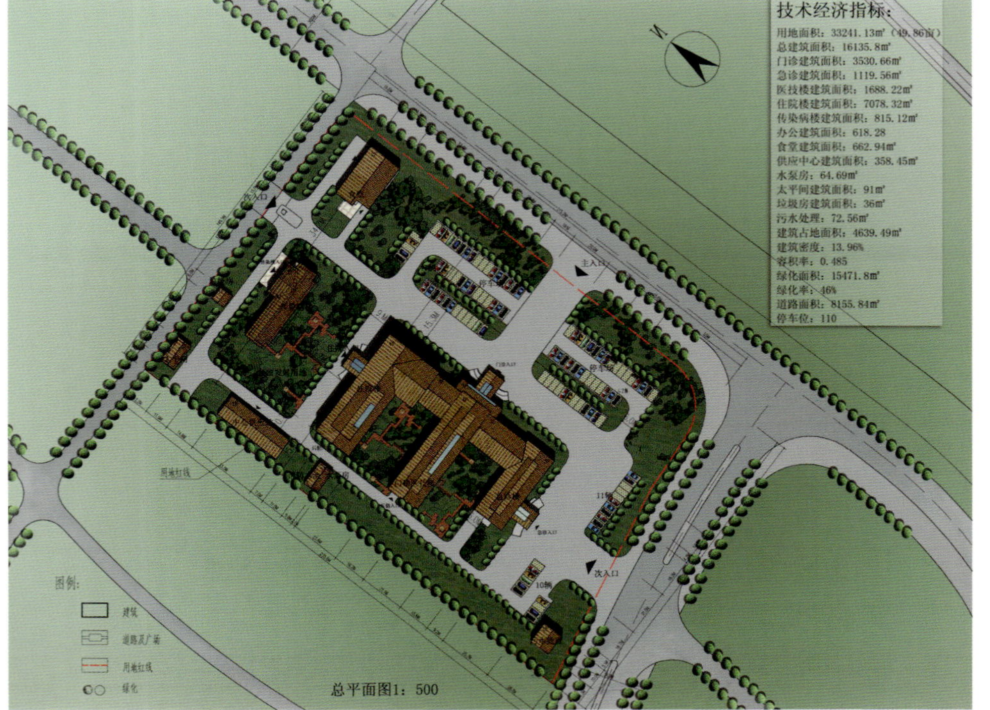

China United.Cheng Taining Architectural Design & Research Institute

中联·程泰宁建筑设计研究院

由中国工程院院士、中国建筑设计大师程泰宁主持的杭州中联·程泰宁建筑设计研究院是一个有着共同理想的集体。研究院本着"立足此时、立足此地、立足自己"的创作理念，努力创作有自己特色的建筑精品。该院的创作定位，是现代的、中国的、科学的；该院的服务态度，是认真的、负责的、及时的。该院愿意在与国内、外同行的竞争中验证该院的主张，树立该院的品质形象。

程泰宁建筑设计研究院原为中国联合工程公司的下属部门。中国联合工程公司是国内知名的大型科技型企业，在近年全国勘察设计百强企业中位居前列。为了适应勘察设计行业改革形势，优化资源配置，发挥大院与大师的综合优势，中国联合工程公司和程泰宁大师决定共同组建杭州中联·程泰宁建筑设计研究院有限公司。作为建设部特批的具有建筑综合甲级资质的设计单位，研究院主要承担城市及小区规划、工业与民用建筑设计，特别是在城市设计以及居住、公共建筑设计方面具有丰富的经验。根据客户多层次的需要，研究院可以在ISO9001质量管理体系认证的框架下提供项目策划、咨询、城市设计、总体规划、建筑及室内设计、景观及环境设计、结构设计、机电设计、照明设计和项目工程管理等多项、多种组合的优质服务。

以研究院成员为主要设计人的多项建筑作品，包括杭州黄龙饭店、加纳国家剧院、马里会议大厦、杭州铁路新客站、萧山绣衣坊商业街、杭州国际假日酒店等项目，曾多次获得国家、省、市颁发的奖项。研究院近期完成的优秀作品有浙江美术馆、宁波市高教园区图书信息中心、绍兴鲁迅纪念馆、上海市公安局办公指挥大楼、金都华府小区、金都城市芯宇住宅小区、浙江耀江大酒店、浙江宾馆商务别墅、常熟理工学院图书馆、银川市人民广场规划及建筑设计、杭州四季青服装交易中心、吴山博物馆、建川博物馆（俘虏馆）、湖州市南浔区行政中心、中国海盐博物馆、龙泉青瓷博物馆、宁夏大剧院等，也均以高质素的设计质量和服务水平得到了业主和业界的肯定和信任。

为了拓展服务空间，提高地域竞争优势，该院在上海设立了分院，并将根据项目的需要在全国重点城市陆续设立分公司，更好地服务客户。

Hangzhou China United. Cheng Taining Architectural Design and Research institute, held by the academician of China National Engineering Research Institute, renowned Chinese architect Cheng Taining, is a group with the same ideal. The institute takes the creative idea of "based on now and here and ourselves", tries its best to create the architecture with its own characteristics. Our creation orientation is modernized, Chinese and scientific; our service attitude is earnest, responsible and timely. We wish to certify our proposition in the competition with national and abroad craft brothers and build up our trait.

The original Cheng Taining Architectural Design & Research Institute is the subsection of China United Engineering Corporation. China United Engineering Corporation is a home famous large-sized technological corporation, which leaps into the front ranks of top 100 Corporations of China Investigation and Design in recent years. In order to adapt the innovative tendency of investigation and design industry, optimize the distribution of resources, exert the comprehensive advantages of great institute and great master, China United Engineering Corporation and Cheng Taining decide to build Hangzhou China United. Cheng Taining Architectural Design & Research Institute Co., Ltd. together.

As the design unit with integrated Grade A qualification in architecture authorized by MOHURD, the institute mainly takes on urban and community planning, industrial and civil architecture design, and it has especially rich experience in urban design and residence, and public architecture design aspect. According to the multiple-level demands of the clients, under the framework of ISO9001 Quality Control System certification, the institute can offer many items and many kinds of comprehensive and excellent services in project planning, consultation, urban design, overall planning, architecture and indoor design, landscape and environmental design, structural design, electromechanical design, illuminating design and project engineering management and so on.

Many architecture works designed mainly by the members in the institute, including projects like Hangzhou Huang Long Restaurant, Ghana National Theatre, Mali Conference Building, Hangzhou New Railway Station, Xiaoshan Xiu Yi Fang Business Street, Hangzhou International Holiday Hotel, have won rewards awarded by nation, province and city for many times. The excellent works finished by the institute in recent days are Zhejiang Art Gallery, the Library Center in Ningbo Higher Education Zone, Shaoxing Luxun Memorial, Office and Command Building for Shanghai Municipal Public Security Bureau, Jindu Huafu district, Jindu City Xinyu Residence Community, Zhejiang Yaojiang Hotel, Business Villa of Zhejiang Hotel, Library of Changshu Institute of Technology, Yinchuan People's Square planning and architectural design, Hangzhou four-season Dress Transaction Center, Wushan museum, Jianchuan museum (captive museum), Administration Center of Nanxun district, Huzhou city, China Sea Salt Museum, Longquan Celadon Museum, Ningxia Theater, etc. All have won the affirmation and trust of the owners and the circle for their high quality design and service level.

In order to extend the service space, advance zone competitive advantage, our institute has set up sub-institute in Shanghai and, according to the requirements of the project, it will continuously set up branch companies in national major cities for giving better service to the clients.

龙泉青瓷博物馆

浙江耀江大酒店

设计人员：程泰宁、徐东平、殷建栋、戴晓玲、石蔚天
项目地点：浙江 诸暨
委　　托：浙江耀江大酒店有限公司
用地面积：66 764平方米
建筑面积：62 325平方米
设计时间：2002年
竣工时间：2008年

项目说明

该项目位于浙江省诸暨市城东行政中心区块，总建筑面积为62 325平方米，共有客房424间，是集住宿、餐饮、会议、娱乐为一体的五星级酒店。

建筑位于城市干道转角处，裙房呈L形布置，转角处设入口大堂，围绕大堂布置了开放的绿化和休闲庭院，并连接三层裙房中的各类餐厅、会议、健身中心等公共设施。主体塔楼平面为方形，按五星级客房标准设置。

塔楼向上的动势与平面展开的裙房组合，强调了垂直与水平体量的对比，造型简洁舒展，立面处理精致，风格现代而独特。

银川国际会展中心

设计人员：程泰宁、程跃文、陈　玲、谢　曦、段继宗
项目地点：宁夏 银川
设计时间：2005年
竣工时间：2008年

项目说明

银川国际会展中心，位于银川城市核心区人民广场西侧，基地周边自成环路，人、货流线畅通，总建筑面积80 618平方米，国际标准展位1 742个。

建筑包括展厅、商务中心、办公以及辅助用房等。设计中将不同功能的用房分层分区，做到相对集中，独立进出又能相互联系。各展厅的主要出入口位于南侧。展厅贯通一层至二层，底层展厅为72米×72米，可以以3米为模块分割成多个展览单元。5个展厅连在一起可以扩展成长380米、宽72米的巨型空间，以满足特殊展览要求。二层展厅较小，可展出轻型物品。整个展厅可随季节、展品种类、展览规模的不同开放展厅空间，使用灵活。

会展中心以其现代而富有地域风格的造型、精致而富于雕塑感的细部处理，成为城市中一道亮丽的景观。

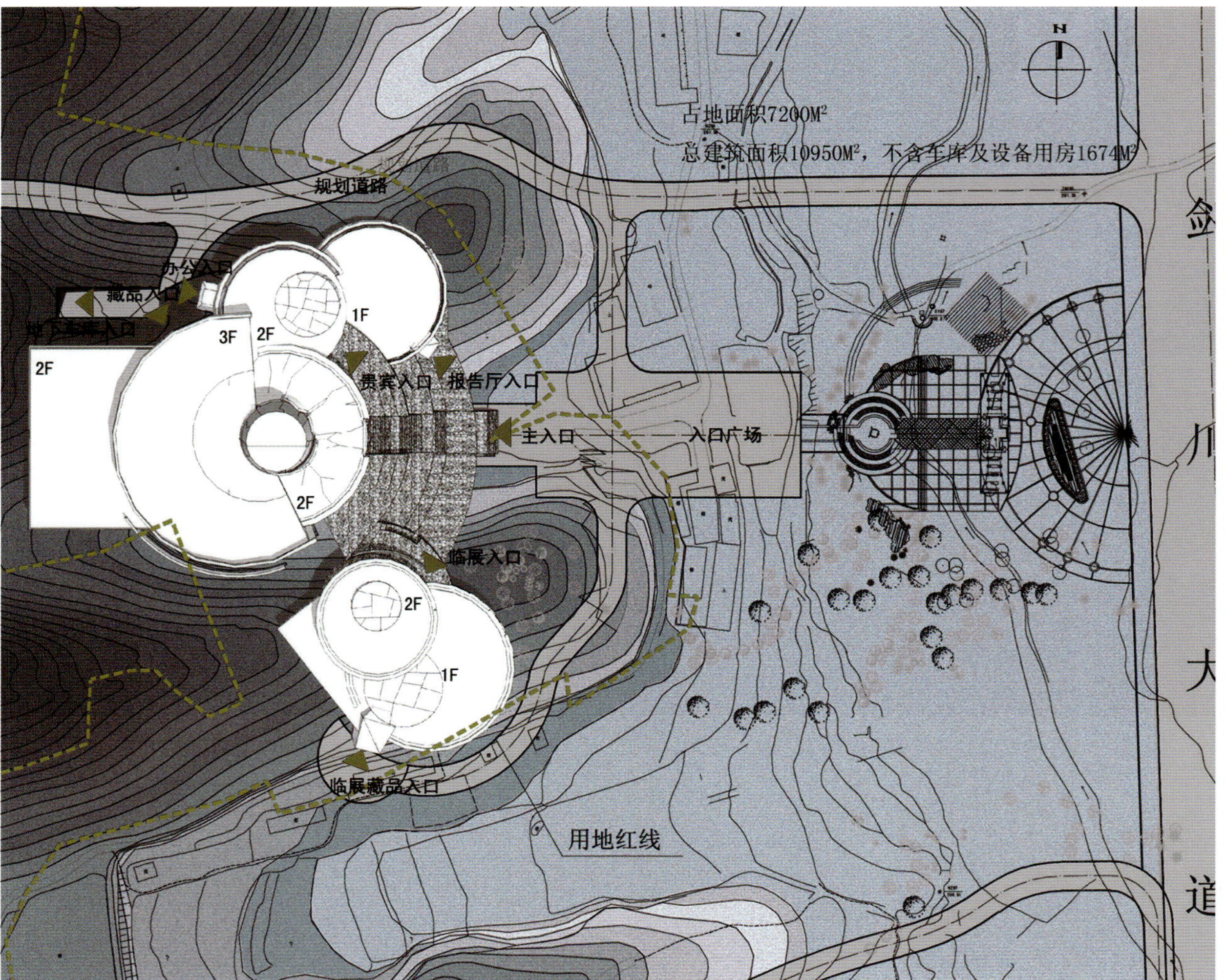

龙泉青瓷博物馆

设计人员：程泰宁、吴妮娜、刘鹏飞、李澍田、杨　涛
项目地点：浙江 龙泉
委　　托：龙泉市青瓷城建设（筹备）领导小组办公室
用地面积：6 000平方米
建筑面积：10 000平方米
设计时间：2007年
竣工时间：即将竣工

项目说明

龙泉青瓷博物馆位于龙泉市主城区以南的城市入口处。占地面积6 000平方米，总建筑面积10 000平方米。基地地貌由两个平缓的山脊以及中间的洼地构成。背面山体给建筑提供了优美的环境背景。

由于历史上青瓷生产高度发达，目前在龙泉四周的田野上随处可发掘到过去留下的窑址和青瓷碎片。处在这一特定的环境里，建筑以“瓷韵——在田野上流动”为创意，以一种非建筑的手法来表达这一博物馆的形象，如同考古发掘中层层叠叠的青瓷器物破土而出，自然地放置在田野之中。建筑造型是青瓷这一珍贵的文化遗产的抽象表达。换一个角度去欣赏，它又像绘画艺术家笔下的一组景物，展现出了一幅恬静、优美的画面。

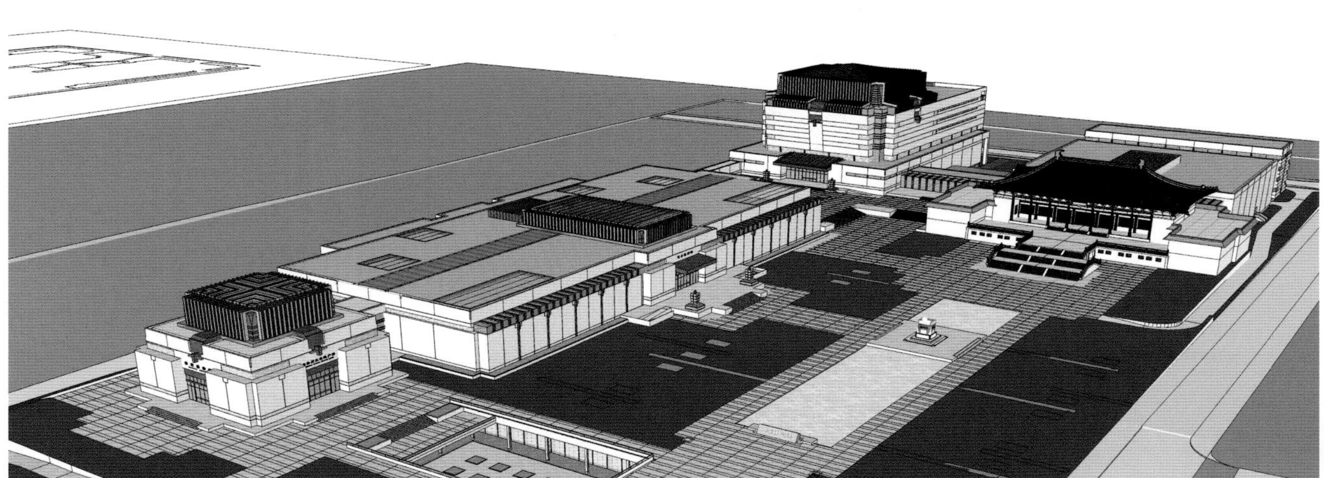

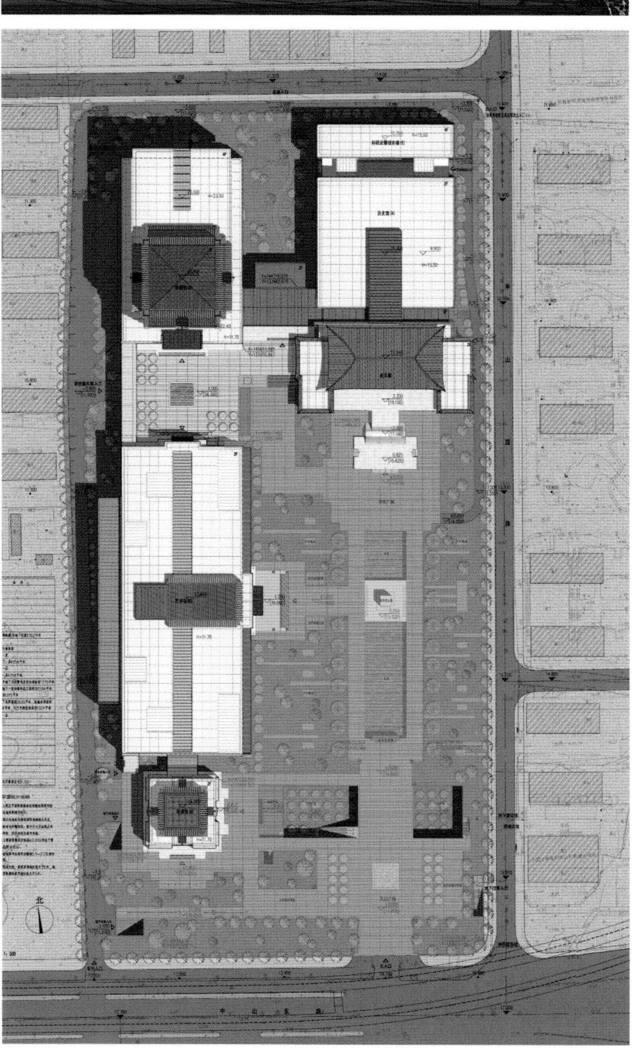

南京博物馆

设计人员：程泰宁、王幼芬、王大鹏、柴　敬、张鹏君、应　瑛、杨　涛
设计时间：2008年

项目说明

南京博物院位于中山门内西北侧，其前身系蔡元培等人于1933年倡建的国立中央博物院筹备处。因抗战爆发，当初规划的自然、人文、艺术三馆仅建成人文馆（历史馆），后在1999年新建了艺术馆。随着时代进步，原有展馆已无法适应现代博物馆展陈要求，因此南京博物院二期工程立项。

整个工程总用地面积67 273平方米，总建筑面积84 655平方米，其中地上建筑面积50 631平方米，地下建筑面积34 024平方米。二期工程要求对整个院域范围内所有建筑、设施、道路、环境进行整体规划设计。对历史馆仿辽式大殿按文物保护原则进行修缮，对文物库房拆除重建，艺术馆原则上不做大的改动，同时需要扩建特展馆、民国馆、非遗馆以及数字化博物馆，在此基础上还要整合文物库房、科研与武警综合楼以及停车设备机房等功能。

本方案针对南博的历史及现状问题逐一梳理并给出了应对策略：通过对整个博物院新老建筑的功能布局、交通流线、内外空间、建筑形式、休闲文化等五大部分的整合，使南京博物馆做到功能完备、设施先进、展藏研究与文化休闲一体化，从而与其作为重点博物院的地位相匹配。另外改建、新建建筑在力求自身整体化的基础上与老建筑在尺度、材质、色彩以及空间与形式上取得和谐统一，并且赋予南京博物院以强烈的历史文化特点和时代气息。

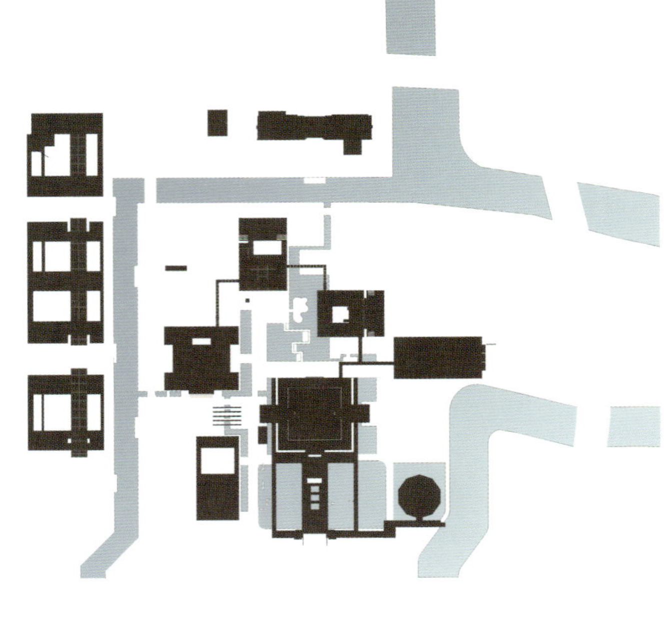

湖州市南浔区行政中心

设计人员：程泰宁、薄宏涛、于　晨
建设地点：浙江　湖州
建筑面积：23 094平方米
设计时间：2006—2008年
竣工时间：2010年

项目说明

行政中心主楼建筑面积为23 094.8平方米。

南浔古镇历史久远，文化底蕴深厚。如何使南浔江南水乡文化与行政建筑开放庄重的气质完美结合，是设计中的核心问题。

当下所谓的"新中式建筑"，基本仍然停留在手法主义的层面上。实际上，传统建筑空间更加注重一种与建筑功能、审美相联系的意境表达，而且，中国传统建筑基本上不存在单体建筑的概念，侧重的是群体建筑空间体系的建立。同时一个空间体系又包含在一个更大的体系之中，由单体到群体其实是一个"自相似性"的结构，建筑单体与群体的设计概念和寓意其实是同根同源的。

院落萦回，空间环绕，玻璃、钢构等一系列现代材料的引入又使建筑平添了一种生机盎然的时代感。中国园林的意境就在"似与不似之间"如画卷般展开……

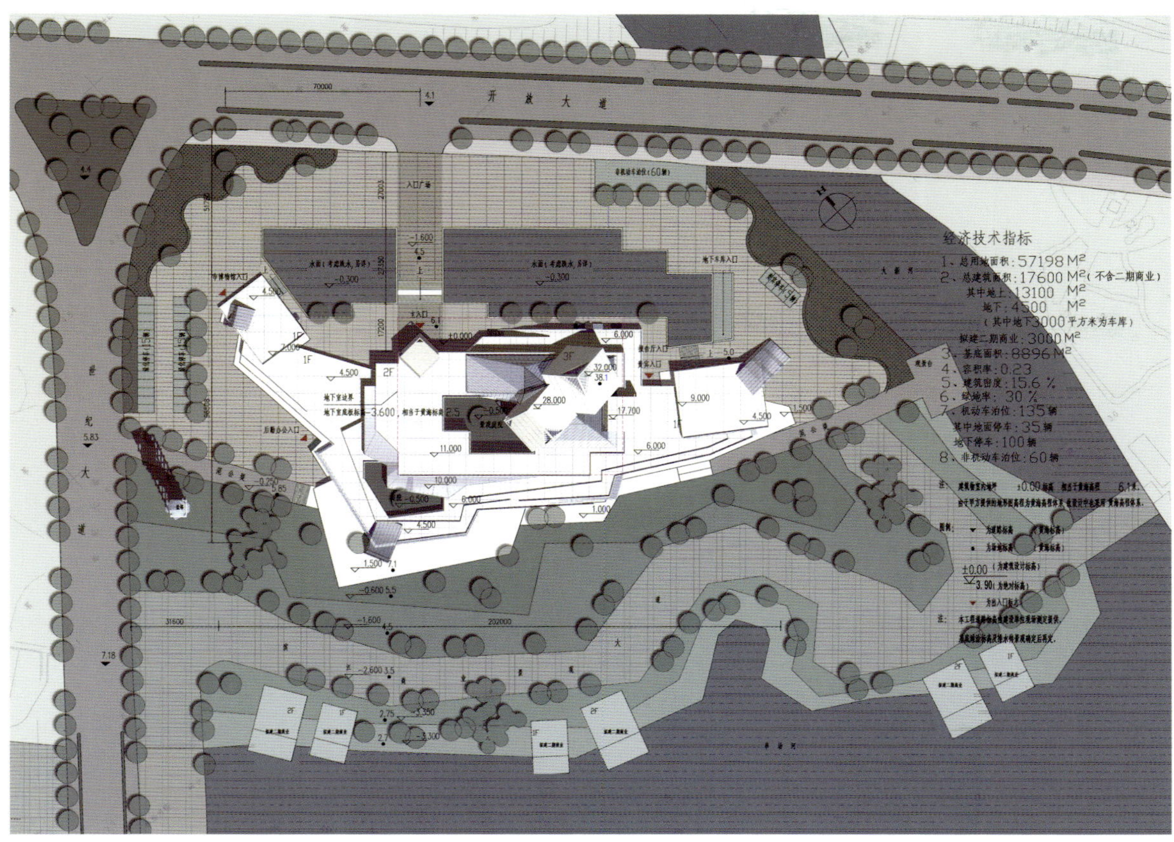

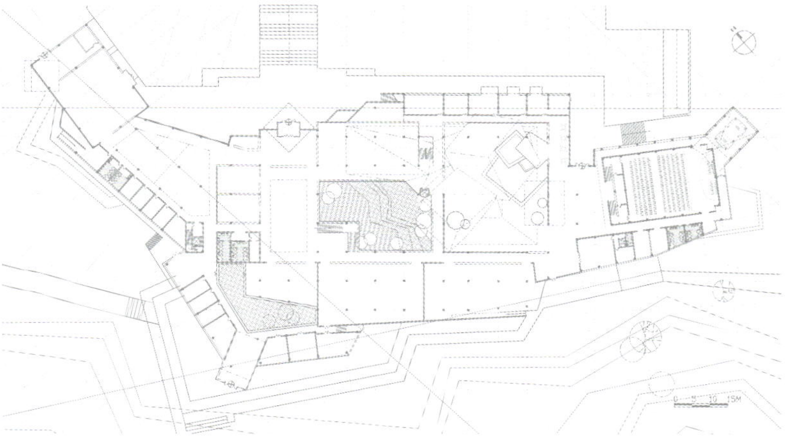
一层示意图

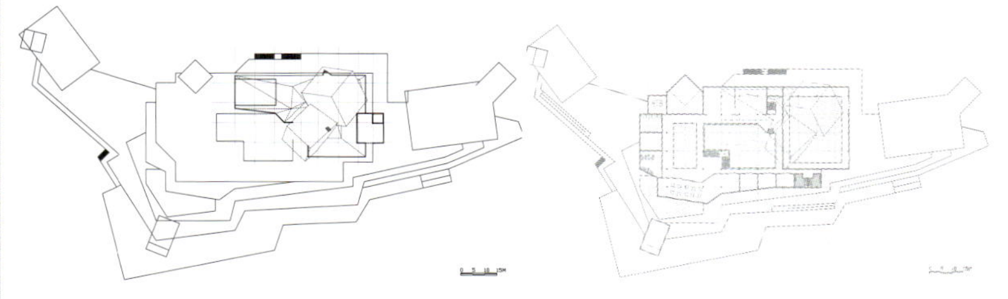
二、三层示意图

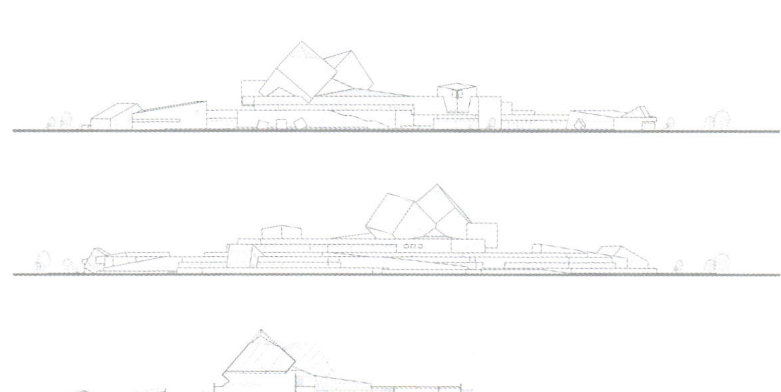
立面图

中国海盐博物馆

设计人员：程泰宁、程跃文、吴妮娜、杨　涛、李澍田、吴文竹
设计时间：2007年
竣工时间：2009年

项目说明

中国海盐博物馆位于江苏省盐城市，基地位于贯穿盐城市的重要河流“串场河”以东，总建筑面积17 800平方米。

建筑造型是对海盐结晶体的演绎，广阔的海边滩涂为海盐的生产提供了独特的环境，如何把这些元素融入建筑设计之中，是设计师首要研究的课题。

旋转的晶体与层层跌落的台基组合，就像一个个晶体自由地洒落在串场河沿岸的滩涂上，造型独特。

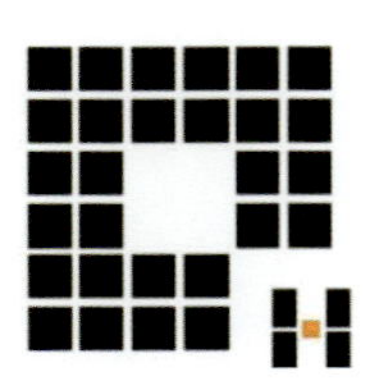

Hangzhou Huaqing Architecture & Engineering Design Co., Ltd.

杭州市华清建筑工程设计有限公司

杭州市华清建筑工程设计有限公司（原杭州市工业建筑设计院）成立于1984年，是一所具有建筑工程设计甲级资质（含钢结构、建筑装饰、建筑幕墙、建筑智能化、照明工程、消防工程）、建筑工程监理甲级资质、工程咨询乙级资质和压力管道专项设计资质的综合性设计单位。公司下辖全资杭州市工业危房鉴定站，控股杭州建申工程监理咨询有限公司。公司位于素有杭州“十里银湖墅”美誉的湖墅路上，交通方便，自有产权的办公大楼面积达1 800多平方米。

华清设计拥有建筑、结构、设备、电气、给排水、暖通、造价、技经等专业的工程技术人员队伍。现有中级职称以上工程技术人员50余名，其中各类注册执业人员40名。建院以来，公司先后承接完成各类工程设计1 500余项，项目遍布全国各地。

华清设计本着以人为本的理念，坚持“开拓、创新、务实、求精”的方针，遵循“为客户创造价值，以创意成就梦想”的宗旨，提供顾客满意的设计产品和服务，不断改进和完善体系业绩。公司广大技术人员积极学习和借鉴国内外同行的先进经验，推行和应用国内外最新的科学技术，不断提高自身业务素质与职业道德水平，积极参与市场竞争，力争以精湛的专业技术为业主提供高标准的服务，在许多重大工程设计投标中屡屡中标。公司所设计的项目因其精心设计、效率高、出图周期短、后续服务佳，得到了用户的充分信任和较高评价，使公司在社会上树立了良好的信誉。公司于2002年通过ISO9001国际质量管理体系认证，并于2006年被浙江省勘察设计行业协会授予“浙江省首批诚信设计单位”荣誉称号。

华清设计携手创基地产、浙商俱乐部共同打造地产产业链集成服务，以“整合产业资源，服务优质客户，为员工创造平台”为经营理念，以产业深度来确立行业高度，致力于成为地产产业结构链的“集成服务营运商”，为客户构架完美价值体系，开拓与众不同的服务领域。

设计理念

华于形 / 清于神 / 责于心 / 履于行
华者，美也；清者，至诚也。
春华秋实，至真至诚。
为客户创造价值，以创意成就梦想。
精雕细琢，坚持“开拓、创新、务实、求精”的方针，为顾客提供满意的设计产品和服务。

Hangzhou Huaqing Architecture & Engineering Design Co., Ltd. (former Hangzhou Industrial Architectural Design Institute), founded in 1984, is a comprehensive design organization with Grade A qualification in Architectural Engineering design (including steel structure, architectural decoration, building curtain walls, building intelligent, lighting engineering, fire protection engineering), Grade A qualification in Construction Engineering Project Supervision, Grade B qualification in chemical design and engineering consultation, and qualification in pressure piping engineered to order design. Hangzhou Industrial Dangerous Identification Station is wholly-owned under the jurisdiction of Huaqing Design, and the company also holds Hangzhou Jianshen Construction Project Management Consulting Co., Ltd. The company is located on South Hushu Road in Hangzhou which enjoyed a reputation of "Ten Miles Silver Hushu" and with convenient traffic system. The company office building is a self-owned proprietary with the building area of more than 1,800 square meters.

Huaqing has a clear design of buildings, structures, equipment, electrical, plumbing, HVAC, cost, and technology, supporting by a full range of engineering and other professional and technical personnel. Over the existing intermediate grade more than 50 engineers and technicians, among which 40 are certified professionals. Since the founding, the company has undertaken to complete more than 1,500 items of various types of engineering design, projects all over the country.

Huaqing designs in line with humanist, insist the policy of “development, innovation, practical, refinement”, and upholds the purpose of “creating value for the customer, building brand with quality”, provides the customer with satisfactory design of product and service, improves and perfects the performance unceasingly. The company's technical personnel study and profit from the domestic and foreign colleague's advanced experience positively, carry out and apply the domestic and foreign newest science and technology in order to improve their own professional quality and the occupational ethics level unceasingly, participate in the market competition positively, so as to vigorously provide the proprietors with high standard service with the exquisite specialized technology. At the same time Huaqing has won repeatedly in many important project bidding. The company designs obtained user's full trust and high appraises owing to its careful design, high efficiency, short chart cycle and good following service, which builds a good reputation for the company in the society. Huaqing Company has passed ISO9001 International quality management system certification in 2002, and in 2006, it was awarded the Certificate of First Integrity by Survey and Design Association of Zhejiang Province.

Hangzhou Huaqing Architecture & Engineering Design Co., Ltd. in cooperation with Change Real Estate Co., Ltd. and Zheshang Club have developed and created chain-integration services catering for the real estate industry. It is a comprehensive real estate service organization, the administrative concept is to "integrate industry resources, supply high-quality services for clients and create a suitable platform for the employees". Whilst focusing on becoming the "integration service providers" of the real estate industry, Huaqing Architecture & Engineering Design Co., Ltd. also strives to provide great value system for customers in addition to establishing unrivalled services in the field.

Design Concept

Delicate in image! Clear in spirit! Blame for heart! Completive in action!
Delicateness is beauty. Clearness is sincerity.
The literary attainment and moral cultivation. The truest and sincerest.
Creating value for the customer, originality make dreams reality.
To be Excelsior and insist the policy of “development, innovation, practical, refinement”, provide the customer with satisfactory design of product and service.

地址：浙江省杭州市拱墅区湖墅南路452号
邮编：310005
电话：+86-571-88056819
传真：+86-571-88056818
网址：www.hqsjy.com
邮箱：hq@hqsjy.com

Add: No. 452, South Hushu Road, Hangzhou, Zhejiang
P.C.: 310005
Fax: +86-571-88056819
Tel: +86-571-88056818
Http://www.hqsjy.com
E-mail: hq@hqsjy.com

永康总部中心

设计时间：2010年
项目状态：施工中
建设地点：浙江 永康
建筑面积：52 808平方米
占地面积：13 843平方米

Yongkang Headquarters

Design Time: 2010
Project Status: Constructing
Location: Yongkang, Zhejiang
Construction Area: 52,808 m^2
Occupied Area: 13,843 m^2

项目概况

永康城市建设目标定位："融千年古城之典雅、滨江城市之秀美、五金名城之恢弘于一体的，基础设施完备、文化气息浓郁、市场经济繁荣的现代化中等城市，国内外知名的五金之都"。永康的崛起，离不开几千年的五金文化底蕴。当前，永康市按照"提升工业化，加快城市化，推进现代化"的战略要求，做出了建设"五金之都"的决策，而建设"五金之都"，非一时之计，其任重而道远。因而，永康市将在中国科技五金城与永康经济开发区相邻的66.7公顷土地上建设总部中心、会展中心、通关中心，在永康经济开发区建设物流与科技创新中心。

杭州钱江经济开发区管委会大楼

设计时间：2010年
项目状态：完成设计
建设地点：浙江 杭州
建筑面积：58 900平方米
占地面积：13 405平方米

Management Committee Building of Hangzhou Qianjiang Economic Development Zone

Design Time: 2010
Project Status: Design Completed
Location: Hangzhou, Zhejiang
Construction Area: 58,900 m^2
Occupied Area: 13,405 m^2

项目概况

该项目包含办公用地和广场用地。办公大楼用地的矩形地块内，南北短、东西长，在主体高层建筑设计构思上考虑居中设置一栋点式高层，两边对称设置三层裙房，中部是上抬的景观广场。点式布局能把对北侧地块的日照、通风影响降到最小，同时对称的布局和点式的挺拔向上能使建筑在该地域起到统领一方的作用，代表了地域特征和开发区形象，更能突出建筑的本身性质和作用，以及开发区的引导和管理者地位。裙房以朝南开的内弧线平面形式突出广场和建筑体量开放、透明的特有内涵，传达开发区海纳百川的包容和强烈的凝聚力。

中部广场抬高2米左右，在空间层次上丰富了地域特性，创造城市阳台的概念。向南隔规划路眺望广场用地，以抬高的视觉感受给人以更高层次、更丰富的空间体验。广场在南侧沿路以阶梯绿化、叠水、人行台阶等形式由地面自然衔接到抬高广场，使人在一个向上行进的过程中体验整个建筑群体的恢弘有度，具有强烈冲击力的气势。

广场地块设计时以环保、节能、利用新科技为原则，以保留生态环境为首要条件，对现有环境和植被进行改造以符合该广场的特有性质，充分利用原有水系的自然景观，在广场用地范围内设置广场、露天剧场、健身场地、景观小品等景观元素，与办公大楼的抬高广场形成呼应和延续，空间上自然衔接，体现出南北向的中轴关系，打造出具有集会、健身、园林休闲等多功能于一体的综合性广场。

杭州运河新城C–R22–03地块30班中学、C–R22–02地块36班小学

The Project of Middle School in NO.C-R22-03 Land and Primary School in NO.C-R22-02 Land in Hangzhou Cannel Newtown

设计时间：2010年
项目状态：完成设计
建设地点：浙江 杭州
建筑面积：17 100平方米（中学）
　　　　　15 420平方米（小学）
占地面积：28 500平方米（中学）
　　　　　25 700平方米（小学）

Design Time: 2010
Project Status: Design Completed
Location: Hangzhou, Zhejiang
Construction Area: 17,100 m^2 (Middle school)
Construction Area: 15,420 m^2 (Primary school)
Occupied Area: 28,500 m^2 (Middle school)
Occupied Area: 25,700 m^2 (Primary school)

项目概况

本项目位于杭州市康桥镇谢村。“群居讲学遁迹著书之所”应该“无市井之喧，有泉石之胜”，由此得知学校不仅仅只是一个教书传授知识的场所，更多的是伴随学生成长接受新事物、新知识，相互之间交流情感、体会认知，共同学习、游戏，共同提高成长的场所。为此，我们提出了一个“有交流中心的学校”的规划设计理念，让学校成为学生身心健康成长的中心园地，创造符合学生独特心理个性的人性化空间。

遵循“有交流中心的学校”这一设计理念，在学校校园的具体设计中，分析提炼出以下设计构思要点：一是理性的功能组织与感性的环境体验相结合。将学校正常运作模式的合理性与师生校园生活体验的情趣性结合起来，并将合理的功能结构、空间逻辑与感性的建筑形态、环境气氛有机融汇。二是校园环境、建筑空间的体验与教育教学文化相结合。对中小学生在德、智、体和情感教育并重的文化的教育，在考虑校园建筑和景观设计中更多地注入人文内涵和情趣体验式空间和场所。提出中小学共享空间这一人性化的设计理念。设计中注重考虑孩子交流、玩耍的建筑景观空间和休息游戏景观平台。三是设计创造现代校园气息。贴近时代气息，力图创造一个多样化体验所交融的校园环境氛围，从儿童自身的生理和心理需求成长的角度出发，合理设计现代学校发展模式下所需要的校园环境景观空间，无论是对建筑空间的把握还是建筑立面风格基调的设计，均让整个学校充满现代文明校园气息。

杭州下沙“天下游戏”项目

Project of Game of the World in Xiasha, Hangzhou

设计时间：2010年
项目状态：完成设计
建设地点：浙江 杭州
建筑面积：63 267平方米
占地面积：9 900平方米

Design Time: 2010
Project Status: Design completed
Location: Hangzhou, Zhejiang
Construction Area: 63,267 m^2
Occupied Area: 9,900 m^2

项目概况

本项目建筑造型从一款众所周知、老少皆宜的游戏“贪吃蛇”抽象出来，提出“动漫电子光带”的理念，通过一个连续的发光带，将公寓、办公、商业、娱乐等各自独立的功能体加以整合。发光带可以表现出众多的质感、电子形态及颜色，乃至于时下流行游戏的电子宣传广告，在凸显建筑本身的地标性的同时，将各功能区块加以区分和联系，使整个建筑浑然一体。

台州市商业银行

Taizhou Commercial Bank

设计时间：2010年
项目状态：完成设计
建设地点：浙江 台州
建筑面积：44 837平方米
占地面积：9 007平方米

Design Time: 2010
Project Status: Design Completed
Location: Taizhou, Zhejiang
Construction Area: 44,837 m^2
Occupied Area: 9,007 m^2

项目概况

台州市中央商务区位于台州市中心区，是推动台州城市经济发展的重要砝码。本项目地处中心商务区东北角入口广场，紧邻城市景观斜轴，与市民广场相连，北侧与东侧是市府大道、中心大道两条城市干道。本项目作为中心商务区启动区块的重要项目对台州中央商务区的开发建设起着不可估量的典范作用。因此，我们在尊重整体规划及城市设计的前提下，力求打造成台州标志性建筑，成为中央商务区的门户象征。

杭州碧桂园住宅工程 Hangzhou Biguiyuan housing Project

设计时间：2010年
项目状态：完成设计
建设地点：浙江 杭州
建筑面积：267 347平方米
占地面积：65 711平方米

Design Time: 2010
Project Status: Design completed
Location: Hangzhou, Zhejiang
Construction Area: 267,347 m^2
Occupied Area: 65,711 m^2

项目概况

本案位于杭州下沙经济技术开发区东北部，总用地面积为6.57公顷，地块所处位置景观优越，北望18洞高尔夫球场，交通通达性强，到杭州中心区仅25千米。地铁1号线从杭州市中心东西走向至下沙大学城区内。本案充分研究高端收入群体的居住行为与居住习惯，利用现有区位优势，以高层、高档住宅为主，适当布置少量底层，合理组织各种人流、车流，形成简捷、畅通的交通体系；同时注重景观设计，处理好公共开放空间、建筑和环境之间的关系，以提高环境质量，为城市建设增添亮点。

扬州九龙花园 Yangzhou Jiulong Garden

设计时间：2010年
项目状态：施工中
建设地点：江苏 扬州
建筑面积：5 144.4平方米
占地面积：25 073.7平方米

Design Time: 2010
Project Status: Constructing
Location: Yangzhou, Jiangsu
Construction Area: 5,144.4 m^2
Occupied Area: 25,073.7 m^2

下沙高教园区东区景冉小区

Jing Ran Area of Eastern District of Xiasha Higher Education Campus

设计时间：2010年
项目状态：施工中
建设地点：浙江 杭州
建筑面积：198 600平方米
占地面积：67 000平方米

Design Tim: 2010
Project Status: Constructing
Location: Hangzhou, Zhejiang
Construction Area: 198,600 m^2
Occupied Area: 67,000 m^2

项目概况

充满现代感的立面设计凸现出简洁明朗的精英品位。中央景观设计贯穿社区整体建筑，环境营造以物衬景、以景带面，彰显了“把建筑景观融入生活”的设计理念。设计沿袭了构图的完整性，维护了强烈的对称感，形成了既具有私密性又具有开阔视野的建筑体。居住区内完全人车分流，并很好地贯彻了节能、绿色设计的理念。

黄山绿园半岛

设计时间：2009年
项目状态：完成设计
建设地点：安徽 黄山
建筑面积：44 750平方米
占地面积：100 292.5平方米

Huangshan Green Garden Peninsula

Design Time: 2009
Project Status: Design Completed
Location: Huangshan, An'hui
Construction Area: 44,750 m^2
Occupied Area: 100,292.5 m^2

项目概况

黄山绿园半岛项目位于太平湖风景区，黄河山北麓、九华山东南麓，地理位置得天独厚，是镶嵌在“两山一湖”旅游黄金线上的一颗璀璨明珠。环湖四周青山环绕，湖水澄碧幽静，犹如翡翠般的世界。而太平湖这一条长达80千米的山水画廊，以一泓碧绿和无雕无饰的自然风光沁人心脾，与黄山如情侣般，被人们称为“黄山情侣”。本方案的基本出发点是尊重自然环境，调整生态结构，建设性地维护自然生态状况，尊重山体及坡地，用建筑适应现有地形，尽可能避免大规模的山体开挖。

杭州钱江经济开发区工业项目

设计时间：2010年
项目状态：施工中
建设地点：浙江 杭州
建筑面积：184 100平方米
占地面积：100 868平方米

Industrial Project of Hangzhou Qianjiang Economic Development Zone

Design Time: 2010
Project Status: Constructing
Location: Hangzhou, Zhejiang
Construction Area: 184,100 m^2
Occupied Area:100,868 m^2

项目概况

以50米×50米为模数网格将地块均匀划分，再运用经典十字形构图，两条轴线将地块均分为四份，南北方向和东西方向连贯成一条收放有序的空间序列。加上对称式的建筑布局，让整个园区更显得庄重大气，并且空间变化丰富。

为加强整个园区的整体感，采用相似空间的布置手法，引入“block”概念——四栋不同面积的标准厂房围合一个近乎方的中庭，通过连廊组成一个组群（block），通过模数化的移位，形成中心区大小不等的广场和绿地空间，成为整个园区的“绿心”，各标准厂房单元组群以及配套服务建筑围绕中心广场布局。

既自成体系又相互渗透，各组团之间可自由穿行，景观相互渗透，共同组成高品质的工业园区；结合中心景观布置食堂、超市、管理办公等配套设施。

“Block”车行流线外环布置，实现地块内的人车分流，内部形成环境优美、绿意盎然的园区环境。沿河边点状布置9栋小型厂房，并将南北轴线打通，将河道景观延伸至园区内部，使景观得以渗透和最大化。

HSS Architectural Design Corporation Ltd.

嘉和设计咨询有限公司

嘉和设计咨询有限公司（HSS）作为甲级设计公司中的精锐团队，HSS向客户提供深化而值得信任的专业服务。“深度专业化”和“五心信任”是HSS的两大特色。

特色1——“深度专业化”：表现在HSS有能力精准地满足客户变化的需求，这种信心源自于拥有专业优势的团队以及具有专业优势的工具。HSS采用和国际接轨的先进模式运营，打造出在相关项目领域内具有优势的专业团队，同时，每一个项目都由合伙人主持或总监，以合伙人本人的职业水准和公司标准对项目客户双重负责。HSS特别强调精品意识，为了确保顺利执行，在公司内部运营共享着一个宝贵的不断发展变化的专业工具库系统，包括设计标准、流程、模式、经验总结等。

特色2——“五心信任”：表现在HSS为客户提供的服务附加值，即：设计主持放心；项目经验信心；口碑品牌安心；沟通服务省心；契约精神舒心。

HSS以“精品、人和”为价值主张，以“用专业的方法做成功的项目”为发展理念，多年来已积累了六大项目设计领域的成功经验：房地产住宅、商业市场和商业综合体、园区及建筑风貌规划、酒店、公共建筑、景观设计。以建筑为龙头，HSS的专业业务范围包括规划景观、结构、机电，从方案到施工图。公司还投资参股了国内最早提出“房地产全程价值整合”理念的机构——动力地产，架构起以市场需求为导向的模式。近年来，通过创新及和国内外的合作，HSS的业务发展迅猛，业务遍及全国，赢得了广泛的赞誉和众多的客户。获得的荣誉奖项包括“中国最具影响力品牌设计机构”、“建设部优秀设计单位”、“当代中华建设名家”、“中国城市建设与管理60年百名人物”；设计的项目获得“联合国人居署迪拜国际最佳范例（中国）推动奖”、“全国人居经典建筑规划双金奖”、“中国最佳山水别墅大奖”、“中国最佳创新户型楼盘”、“华彩奖”等；HSS 的项目作品还被选入《中国建筑设计作品年鉴——建国六十周年优秀设计成果专集》。

HSS是浙江省首家甲级建筑设计事务所，作为走专业化道路最早的成功实践者，获得了“中国最具创新力建筑设计事务所”的称号。“做到更好”是HSS无止境的追求。

As an elite team among Grade A design companies, HSS provides deep and trusted professional services to customs. "Deep specialization" and "five-heart trust" are the two characteristics of HSS.

Characteristic 1—“deep specialization”. It reflects that HSS has ability to satisfy customs' changing needs precisely , which comes from professional team and tools in HSS. HSS uses international advanced operation mode to create professional teams in relevant projects. At the same time, every project contains chair or director architects who are responsible for projects and customs with company and professional standards. HSS emphasizes "excellence awareness". In order to ensure the implementation, a valuable and evolving professional tool database system is run and shared within the company which includes design standards, processes, models, and experience.

Characteristic 2—“five-heart trust”. It reflects that HSS provides the value-added services to customs. Namely: reassuring for chairing design; confident in project experience; secure in virtue of reputation brand; worry-free due to communication based service; comfortable serving according to contracts.

HSS takes “excellence, teamwork” as its value proposition and “a professional approach creates successful projects” as development concept. HSS has accumulated successful experience in the field of six projects for many years: estate residential, commercial market and commercial complex, park and architectural planning, hotels, public building, landscape design. Taking architecture as emphasis, HSS professional services include planning landscape, structure, electric and machine, from scheme to working drawing. HSS also makes the investment in POWER+ who is the first agency that put forward the concept "value integrated in estate" in china, constructing the market demand-oriented model.

In recent years, through innovation and the cooperation at home and abroad, HSS business has boomed all over the country and won a wide reputation and many successful customs. The awards have included the “the Most Influential Brand Design Agency in China”, “Good Design Construction Unit”, “Master of Contemporary Chinese Building”, “China Urban Construction and Management of 60 people 60 years”. “Designs of the project have won "UN-HABITAT Dubai International Best Practices (China) to Promote the Award”, “Outstanding Contemporary Chinese Urbanization Project Blueprint”, “National Habitat Classic Architectural Planning Double Gold”, “Best Landscape Award Villa”, “Best Innovation Unit properties for sale”, HSS designs of project have also been selected into the *Yearbook of China Architecture Design Works - the Sixtieth Anniversary of the Founding of the Outcome of Outstanding Album Designs*.

HSS is the first Grade A architectural design office in Zhejiang province. As the first successful practitioner engaged in the professionalization, HSS won the title “china's most innovative architectural design firm”. "Do the best" is the endless pursuit of HSS.

香溢集团德清会所别墅

设计时间：2009年
设计规模：2 830平方米
设计深度：方案、初步、施工图设计
项目地点：浙江 德清

项目说明

本项目以高品位休闲度假基地的定位为出发点，满足特定客户需求，为其量身设计。方案在建筑布局与地势利用、平面功能、建筑造型等方面都进行了多种尝试，最终确立了本项目的三大设计特色：以景观资源为导向的建筑布局设计；以精致高档餐饮、度假为核心的功能设计；融合休闲、优雅、尊贵气质的造型设计。

白鹿温泉别墅

设计时间：2011年
设计规模：302 610平方米
设计深度：方案、初步设计
项目地点：河北 石家庄

项目说明

本项目为平山县白鹿温泉度假村内的延伸项目，定位高端，在充分整合白鹿温泉度假村现有配套资源以及本项目地形优势的基础上，力争打造一个以温泉主题为特色的集休闲、健身、养生、商务功能为一体的度假会所型旅游产品——温泉度假合院。

本项目规划设计特点：
——尊重地形，依山就势；
——以景观资源为导向；
——组团式布局。

本项目建筑设计特点：
——立体全景观温泉主题花园设计；
——融合景观的平面功能设计；
——汉唐主题风格的造型设计。

九州1品

设计时间：2010年
设计规模：120 000平方米
设计深度：方案、初步、施工图设计
项目地点：浙江 安吉

项目说明

项目坐落在安吉灵峰山脚，浒溪水畔，以111 322平方米的总占地面积、120 000平方米的建筑面积，延续安吉珍贵山水文脉。以2—4层为主的低密度精品建筑、特色庭院景观，与地块周边茂密清幽的树林相结合。运用现代建筑理念，以一种低调、沉静而不失华贵的散发着优雅气息的规划设计，打造安吉最具价值的精品豪宅，让人得到精神与物质的双重享受。

湖南省烟草公司邵阳市公司综合用房

设计时间：2009年
设计规模：83 000平方米
设计深度：方案、初步、施工图设计
项目地点：湖南 邵阳

项目说明

本方案为邵阳烟草局量身定做，解决三大设计难点，提供了一个高起点、高品位、高效率的建筑方案。方案设计时充分结合了山丘地形，为办公楼和住宅营造"山环水绕、坐北朝南"的地势境界，同时注重内部环境和外部环境的紧密联系，将阳光、空气、水和绿化与各种空间进行一体化设计，营造私密型、半私密型等不同庭院空间，创造了造型错落有致、立面不奢华张扬，但大气、亲切、有品位的建筑物和优美的环境。

金华五百滩地块概念性规划

设计时间：2010年
设计规模：17公顷
设计深度：概念性规划
项目地点：浙江 金华

项目说明

本项目位于金华市区核心地块——五百滩，在"心中之心，星流星岛"设计理念的指引下，在其中心部位规划一个"心"形岛——"星岛新天地"，围绕此，由西向南再延伸向东，依次布置四组副星，形成一股流动状的星流，"国际名品之星"、"国内名品之星"、"创意工作室之星"、"五星酒店之星"，且炫动的灯光设计，引出无可复制的星光璀璨的金华不夜城，流线型的建筑形态，引出滨水休闲、商务的最佳处所，无序搭配的新旧建筑，更是引出了金华不朽的过去、繁荣的现在、蓬勃的未来。

爱山广场二期综合体

设计时间：2011年
设计规模：35 000平方米
设计深度：方案设计
项目地点：浙江 湖州

项目说明

本项目位于湖州市中心，南街与红旗路交叉口的东北角。本项目所在地是湖州市唯一的市级商圈，业态最全、档次最高、人气最旺。传承本地块现有建筑特色，本方案力求突破创新，以满足各类需求为基础，实现大面积的开放空间、清晰的项目定位、合理的业态搭配，最终打造湖州地标级建筑群。

禾泽都林设计机构

HESOMS DESIGN INSTITUTE

诠释理想城市，打造商业地产设计第一品牌，是禾泽都林人始终的追求。

禾泽都林关注城市，研究文化。秉承“东方文化现代化”的观点，顺应“大规划、大建筑、大景观”的设计理念，把工程学、美学、哲学三者有机结合，探索当今及未来城市发展道路，构建城市和谐之美。

禾泽都林关注结果，尊重客户。帮助客户实现理想，为客户提供超值服务、为社会提供传世精品是禾泽都林人的奋斗目标。每完成一个项目，即成为一方地标，引领一个城市的发展，用凝固的作品赢得社会的赞誉及各界的好评。

禾泽都林关注市场，追求质量。以中国长三角为中心，向其他城市不断拓展，业绩现已遍布北京、天津、上海、杭州、武汉、成都、合肥、昆明、扬州等30余座城市。

公司的作品大气睿智、浑然天成；我们的设计精雕细刻、精益求精。近年来禾泽都林先后获得了全国人居经典综合大奖、IACE2006国际人居建筑金奖、第二届中国国际建筑艺术双年展建筑设计一等奖等15个国家及国际大奖。2009年禾泽都林被中国房地产协会评选为中国商业地产最佳设计机构。

禾泽都林始终以高标准、严要求为员工打造平台，为客户和社会提供服务，创造价值！

Concerned about city, Hesoms researches culture. Adhering to “modernization of oriental culture”, together with our design principle of “great plan, great building, great landscape”, Hesomes combines the engineering, aesthetics and philosophy vibrantly, explores the city development path from present days to future, and designs the harmonious and beautiful city.

Concerned about results, Hesoms respect customers. Hesoms’ goal is to help customers to realize their dreams, to provide value-added services, and to create new city legend. Every finished project with our design becomes the landmark at that area and leads the development trend. We win the praise and appreciation with our full devotion, efforts and expertise.

Concerned about market, Hesoms pursues quality. Hesoms provides efficient design art-work and services through our standardized system from process to module.

Our design art-works are majestic, extremely fine and naturally built with the wisdom of driving excellence. Focusing the Yangtze River Delta, Hesomes’ business is continuously expanding to more than 30 cities which locate in different provinces such as Beijing, Tianjin, Shanghai, Hangzhou, Wuhan, Chengdu, Hefei, Kunming, Yangzhou and etc. Hesomes had won 15 national and international prizes including the National Classic Comprehensive Habitat Award, IACE2006 International Habitat Architecture Award, Architectural Design Prize in the Second China International Architectural Art Biennial Exhibition, along with other awards. In 2009, Hesoms was selected by China Real Estate Association as the best design agency of commercial real estate in China. In 2010, Hesoms was rated as the best architectural design firm in China’s real estate industry.
Abiding by the highest standards and strict requirements, Hesoms has always been striving to build solid career development platform for our employees, providing services and creating values for our customers and society!

禾泽都林国际・上海
地址：上海市沙泾路10号上海1933创意园1号楼418-420
邮编：200080
总机：+86-21-5505 9505
传真：+86-21-5505 9507
邮箱：sh@hesoms.com

禾泽都林・杭州
地址：浙江省杭州市文二路391号西湖国际科技大厦B2座8层
邮编：310012
总机：+86-571-8721 9177
传真：+86-571-8721 5877
邮箱：hesoms@hesoms.com

Hesoms International Design Institute-Shanghai
Add: Room 418-420, Building 1, Shanghai 1933 Creative Park, No. 10 Shajing Road, Shanghai
P.C.: 200080
Tel: +86-21-5505 9505
Fax: +86-21-5505 9507
E-mail: sh@hesoms.com

Hesoms International Design Institute-Hangzhou
Add: 8F of Building B2, West Lake Plaza of International Science & Technology, No. 391 Wen'er Road, Hangzhou City, Zhejiang
P.C.: 310012
Tel: +86-571-8721 9177
Fax: +86-571-8721 5877
E-mail: hesoms@hesoms.com

项目地点：江苏 泰兴
项目规模：667 177平方米
项目进程：施工中

Project Location: Taixing, Jiangsu
Project Area: 667,177 m^2
Project Process: Under Construction

泰兴 · 新天地
Taixing New World

商业街区透视图一

商业街区透视图二

商业街区鸟瞰图

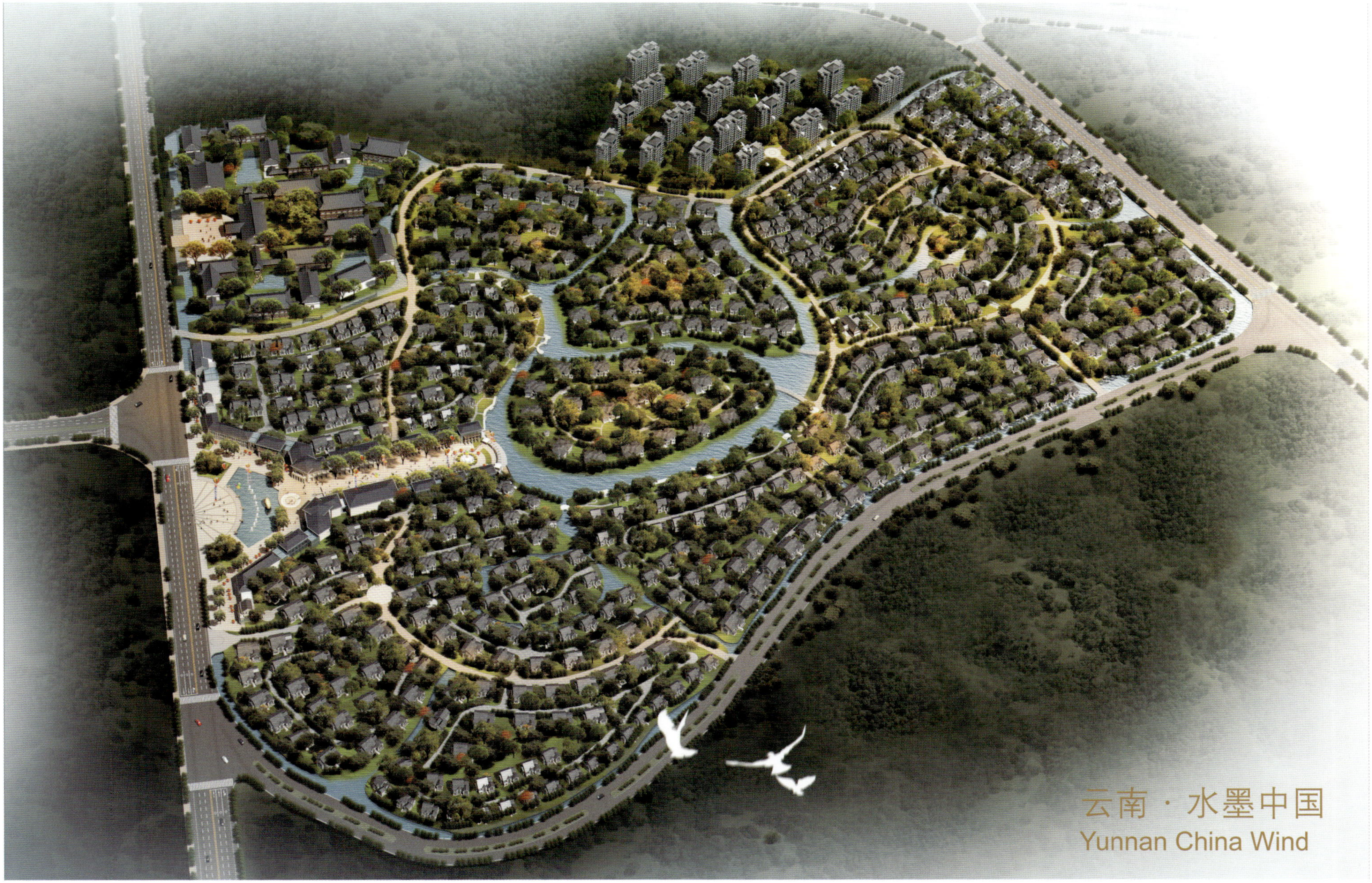
云南 · 水墨中国
Yunnan China Wind

VERSACE
FENDI

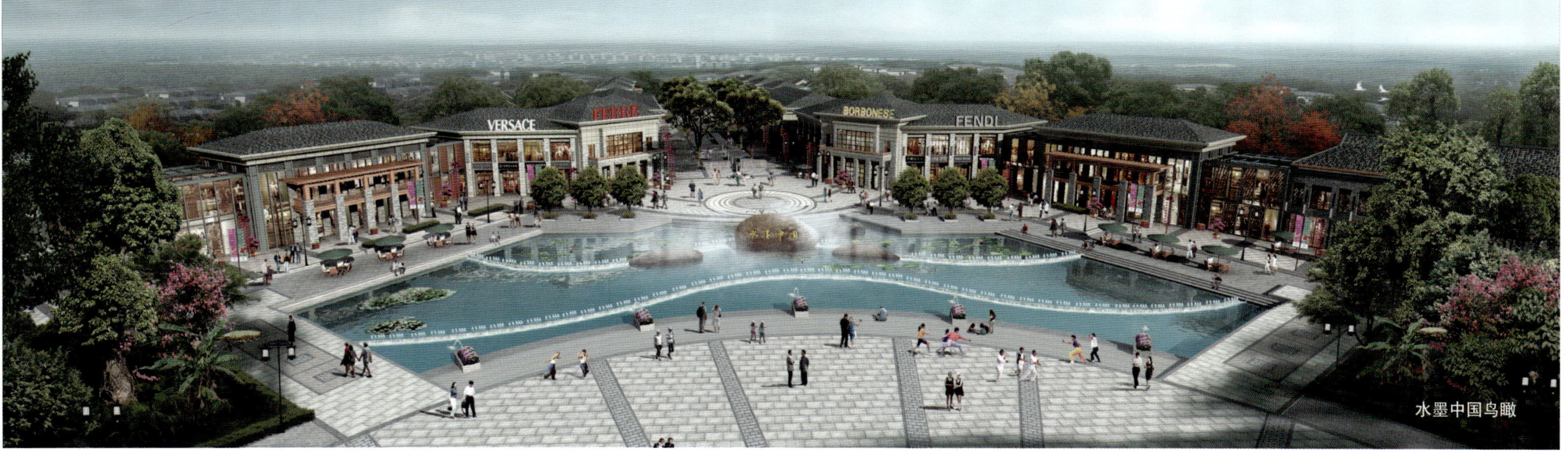

水墨中国鸟瞰

水墨中国景观

水墨中国酒店夜景

项目地点：云南 腾冲
项目规模：678 250平方米
项目进度：施工阶段

Project Location: Tengchong, Yunnan
Project Area: 678,250 m^2
Project Process: Under Construction

运用新城市主义的设计理念，将其建设成为融酒店、旅游、居住、商业、休闲于一体的开放式顶级城市片段。

成为最能体现腾冲自然、文化、旅游、可持续活力的代表，进而成为云南乃至全国的名片。

The designer follows new urbanism design philosophy, integrates the hotel, tourism, residence, commerce and leisure altogether as the most open and top class city segment. Let it become the representative of nature, culture, tourism and sustainable energy of Tengchong, furtherly become the landmark of the Yunnan Province and even the whole country.

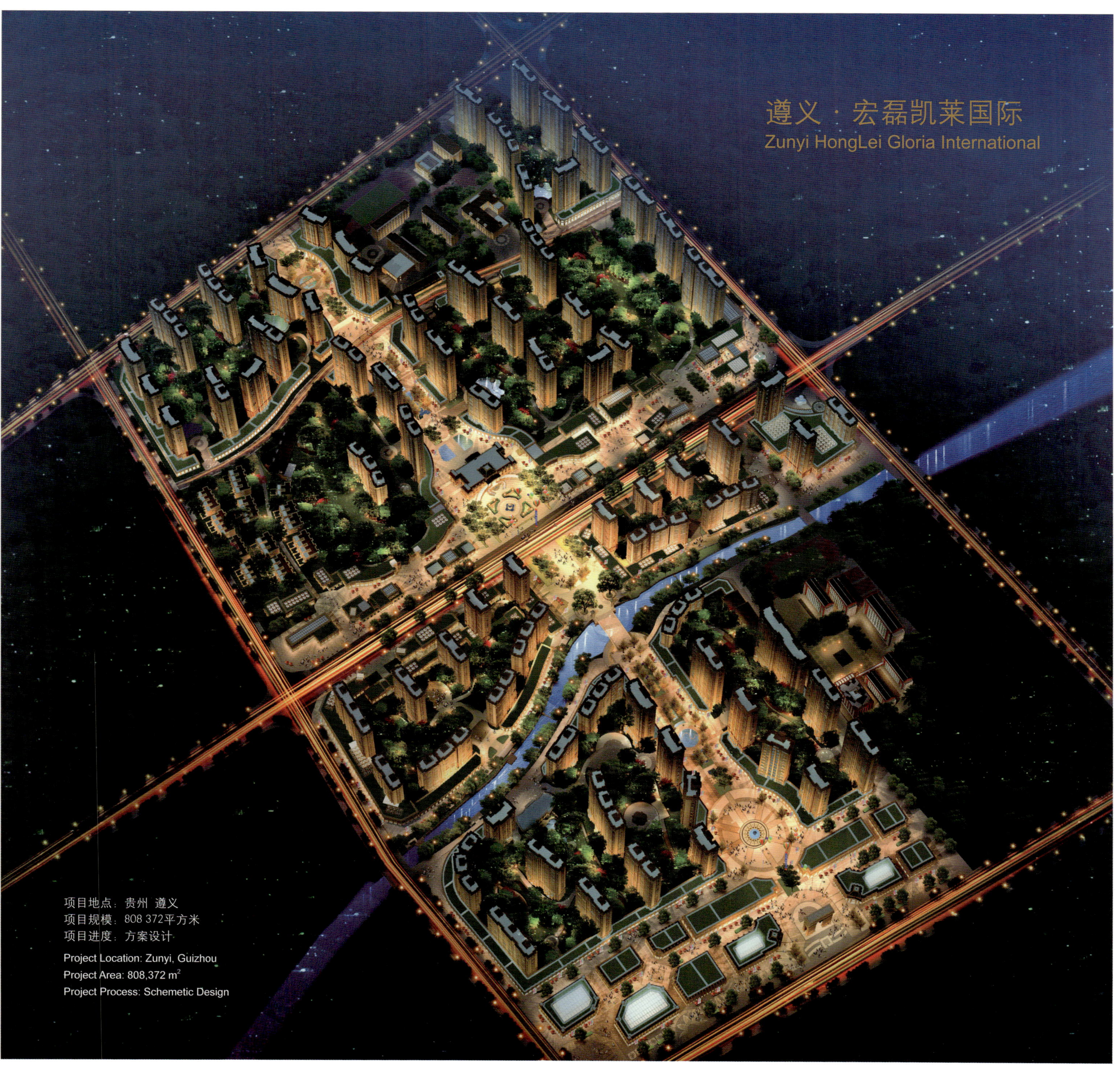

项目地点：贵州 遵义
项目规模：808 372平方米
项目进度：方案设计

Project Location: Zunyi, Guizhou
Project Area: 808,372 m^2
Project Process: Schemetic Design

遵义大道商业街透视

共青二路透视

CENTALAND Design Co.,Ltd.

CENTALAND 森拓设计机构

森拓设计机构(CENTALAND)近年来致力于内地高端建筑产品设计。森拓除在加拿大温哥华及中国香港设立分支机构外，还分别在中国上海、重庆、杭州三地独立注册成立了分公司。

森拓设计擅长将优秀的建筑理念、技术及设计手法融为一体，积极并能正确地理解业主及市场的要求、地域环境条件及文化背景，同时关注项目的投资成本与经济因素，力争将业主对建筑产品的物质追求与艺术价值完美结合。森拓设计在多种建筑类型上均有实践，其中特别擅长于高端别墅、大型综合居住区、高尔夫及旅游休闲度假区、酒店、复合型商业建筑等。

机构的代表作品

· 休闲度假类别墅

上海城开 · 万源城御溪、上海保利 · 十二橡树、北京保利 · 垄上、重庆常青藤人文别墅、重庆龙湖 · 香樟林、重庆保利 · 国际高尔夫花园、重庆金科 · 太阳海岸、重庆中渝 · 梧桐郡、成都龙湖 · 长桥郡、广州保利 · 林语山庄别墅居住区、杭州绿城 · 桃花源、杭州新湖 · 香格里拉、南京瑞基 · 山河水、南京鸿信 · 云深处、南京华隆 · 紫金七号、包头保利 · 南海高尔夫庄园、安吉恒励 · 龙王溪谷等；

· 大型综合居住区

重庆庆鸿 · 郁金香国际公寓、重庆武夷滨江、重庆天骄 · 美茵河谷、成都朗基 · 欧城、成都朗基 · 望今缘、成都远洋 · 朗郡、成都龙湖 · 弗莱明戈、昆明俊发 · 滇池华府、昆明俊发 · 滨江俊园、杭州万科 · 魅力之城、郑州建业 · 森林半岛、石家庄路劲 · 蓝郡、常州莱蒙水榭花都国际城、许昌建业 · 帕拉迪奥等；

· 商业办公综合体

重庆帝景摩尔、重庆龙湖 · 蓝湖郡商业街、重庆旭阳 · 朗晴广场、重庆金科 · 太阳海岸时代广场、成都朗基 · 和芯科技楼、山东蓬莱酒庄、漯河建业 · 福朋酒店、常州莱蒙国际城假日酒店、绍兴县行政中心等。

如您希望了解更多的关于森拓设计机构(CENTALAND)的资料，请上网浏览我们的网页，网址为www.centaland.com

上海公司
公司地址：上海市虹口区吴淞路297号甲5楼
邮政编码：200080
联系电话：+86–21–63831225 (26) (27)
传　　真：+86–21–63831622
邮　　箱：centaland-sh@163.com

重庆公司
公司地址：重庆市渝中区双钢路3号科协大厦7楼
邮政编码：400013
联系电话：+86–23–89065518 (28) (38)
传　　真：+86–23–89065516 / +86–23–63659826
邮　　箱：centaland-cq@163.com

杭州公司
公司地址：浙江省杭州市西湖区天目山路176号11号楼二层
邮政编码：310012
联系电话：+86–571–88270595 (96) (97) (98)
传　　真：+86–571–88270593
邮　　箱：centaland-hz@163.com

Centaland Design Consultants CO., LTD (CENTALAND) is dedicated to high-end architectural design in mainland in recent years. In addition to Vancouver, Canada and Hong Kong, China, it set up three independent branches respectively in Shanghai, Chongqing and Hangzhou.

Centaland Design is good at integrating excellent architectural design concepts, technologies and design techniques, actively and correctly understands the owners and market, regional environmental conditions and cultural background, at the same time concerns about the project's investment cost and economic factors, and strives to combine perfectly the owners' pursuit of building material products with the artistic value. Centaland Design has practised in a variety of building types, particularly high-end villas, large-scale integrated residential, golf and tourism and leisure categories, hotels, commercial complex buildings and so on.

Representative Works

• Recreational Villas

Shanghai Urban Development·Wanyuan City Yuxi, Shanghai Poly·Twelve Oaks, Beijing Poly·Longshang, Chongqing Ivy Humanities Villa, Chongqing Longfor·Fragrant Forest, Chongqing Poly·International Golf Garden, Chongqing Jinke·Sun Coast Villa, Chongqing Zhongyu·Phoenix County, Chengdu Longfor·Bridge County, Guangzhou Poly·Linyu Mountain Villa, Hangzhou Greentown·Tao Hua Yuan, Hangzhou Xinhu ·Shangri-la, Nanjing Ruiji·Romanvision, Nanjin Hongxin·Invisible Class, Nanjing Hualong·Zijin Qihao, Baotou Poly·South-sea Golf Manor, Anji Handnice·King Valley, etc.

• Large-scale Integrated Residential

Chongqing Qing Hong·Tulip International Apartments,Chongqing Wuyi·Riverside, Chongqing Tianjiao·Meiyin River Valley, Chengdu Langji·Europe City, Chengdu Langji·Wangjinyuan, Chengdu Ocean·LangJun, Chengdu Longfor·FlamencoSpain, Kunming Junfa·Dianchi Huafu, Kunming Junfa·Binjiang Junyuan, Hangzhou Vanke·Glamorous City, Zhengzhou Jianye·Forest Peninsula, Shijiazhuang RoadKing·Blue County, Changzhou Le Leman City, Xuchang Jianye·Palladio,etc.

• Commercial Office Complex

Chongqing Dijing Mell, Chongqing Longfor·Blue Lake County Commercial Street, Chongqing Xuyang·Langqing Place, Chongqing Jinke·Sun Coast Times Square,Chengdu Langji·Science And Technology Page, Shandong Penglai Winery, Luohe Jianye·Fupeng Hotel, Changzhou Le Leman City Holiday Inn, Shaoxing Country Adminstration Center, etc.

If you want to know more about CENTALAND design institution.Please visit our internet websits at www.centaland.com

Shanghai Branch
Add: Fifth Floor, No. 297 Wusong Road, Hongkou District, Shanghai
P.C.: 200080
Tel: +86–21–63831225 (26) (27)
Fax: +86–21–63831622
E-mail: centaland-sh@163.com

Chongqing Branch
Add: Seventh Floor, Kexie Building, No. 3 Shuanggang Road, Yuzhong District, Chongqing
P.C.: 400013
Tel: +86–23–89065518 (28) (38)
Fax: +86–23–89065516 / +86–23–63659826
E-mail: centaland-cq@163.com

Hangzhou Branch
Add: Second Floor, No. 11 Building, No. 176 Tianmushan Road, West Lake District, Hangzhou, Zhejiang
P.C.: 310012
Tel: +86–571–88270595 (96) (97) (98)
Fax: +86–571–88270593
E-mail: centaland-hz@163.com

CENTALAND
森拓设计机构

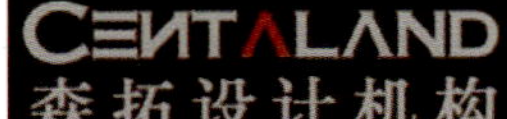

规划

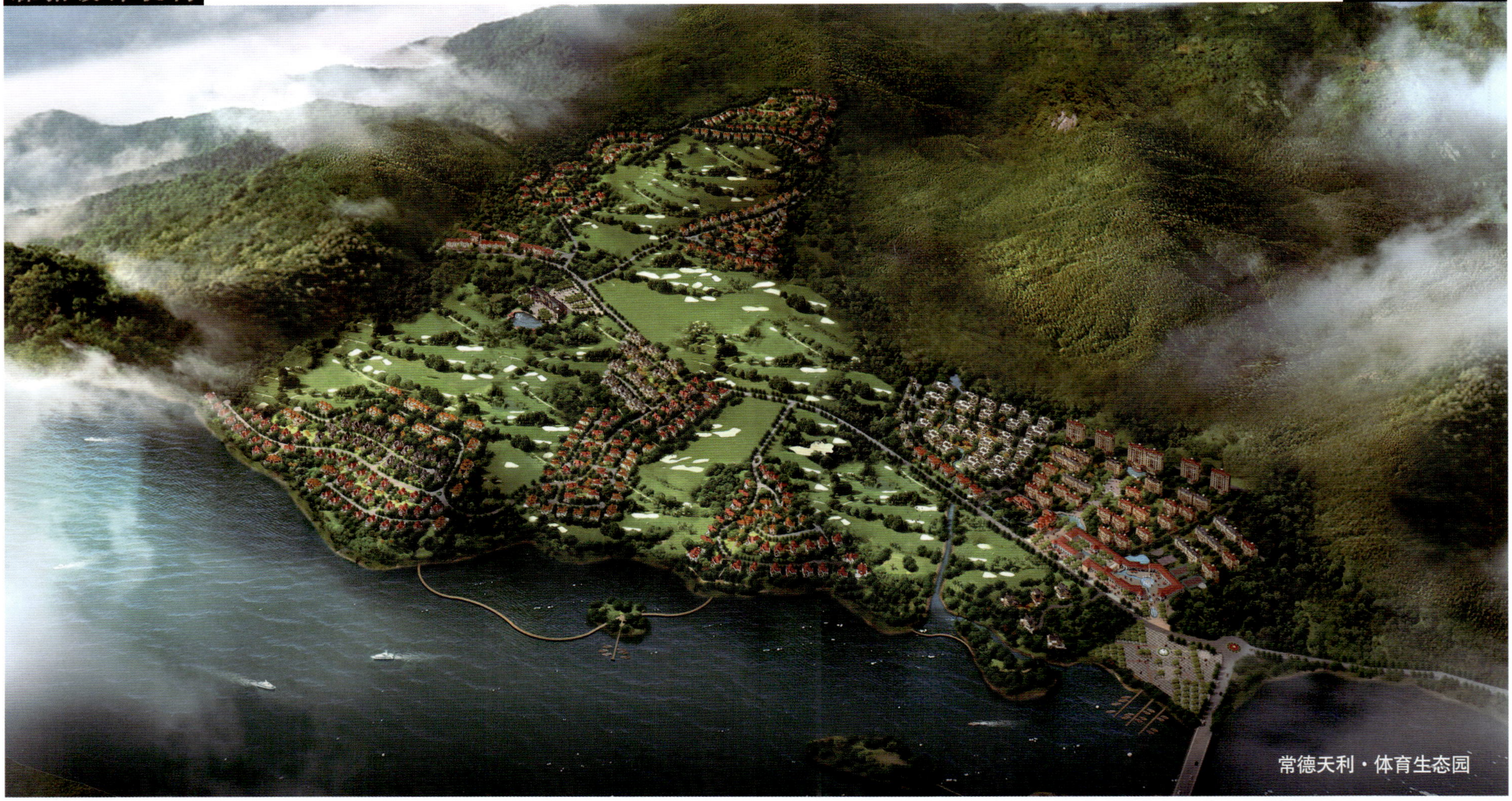

常德天利·体育生态园

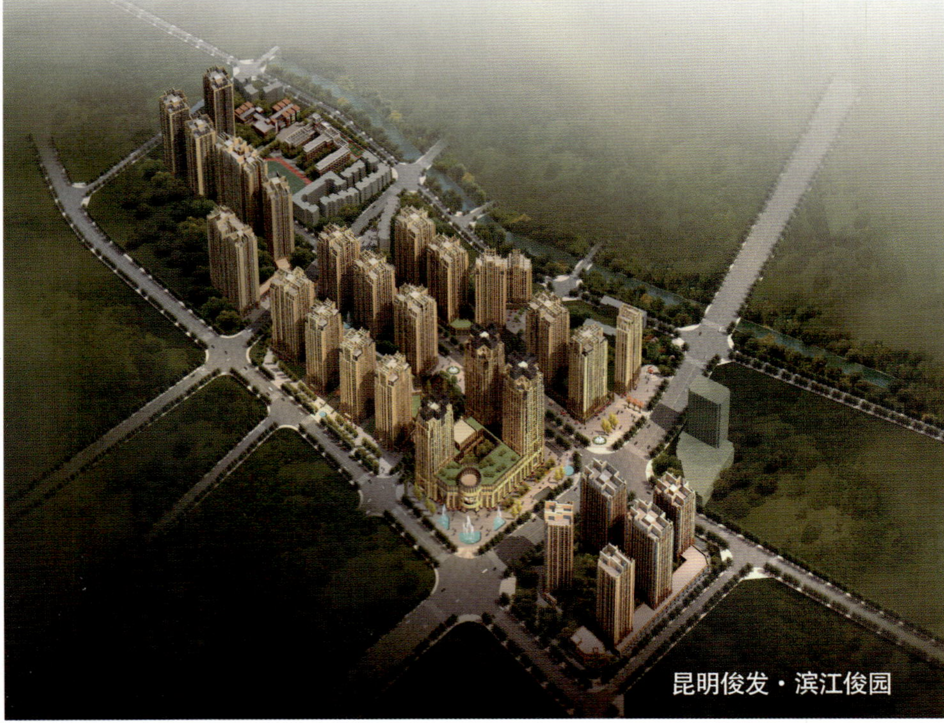

昆明俊发·滨江俊园

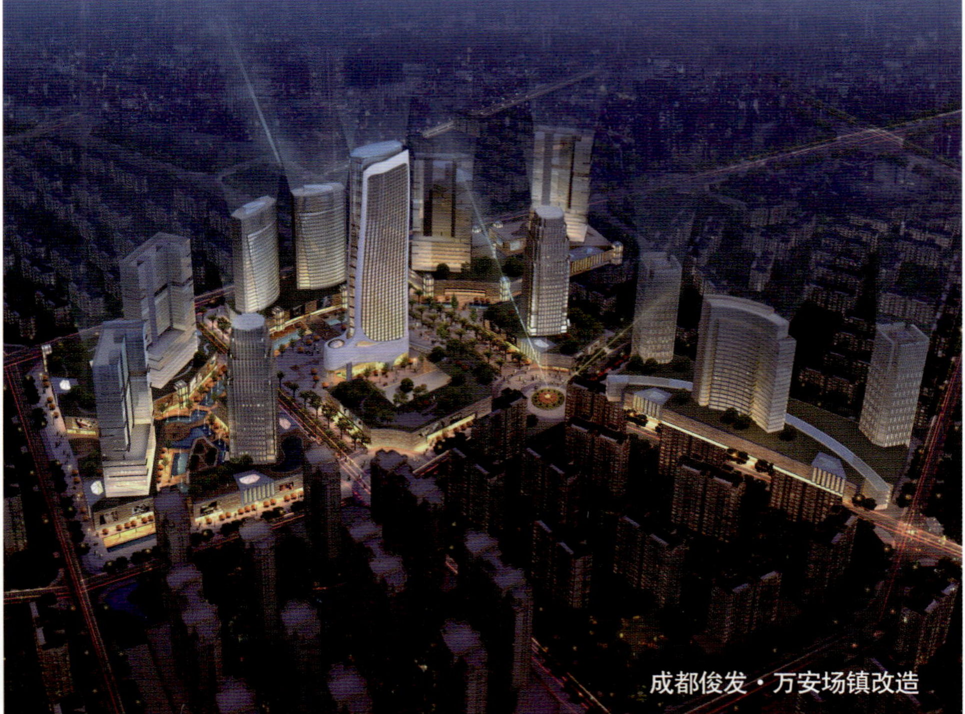

成都俊发·万安场镇改造

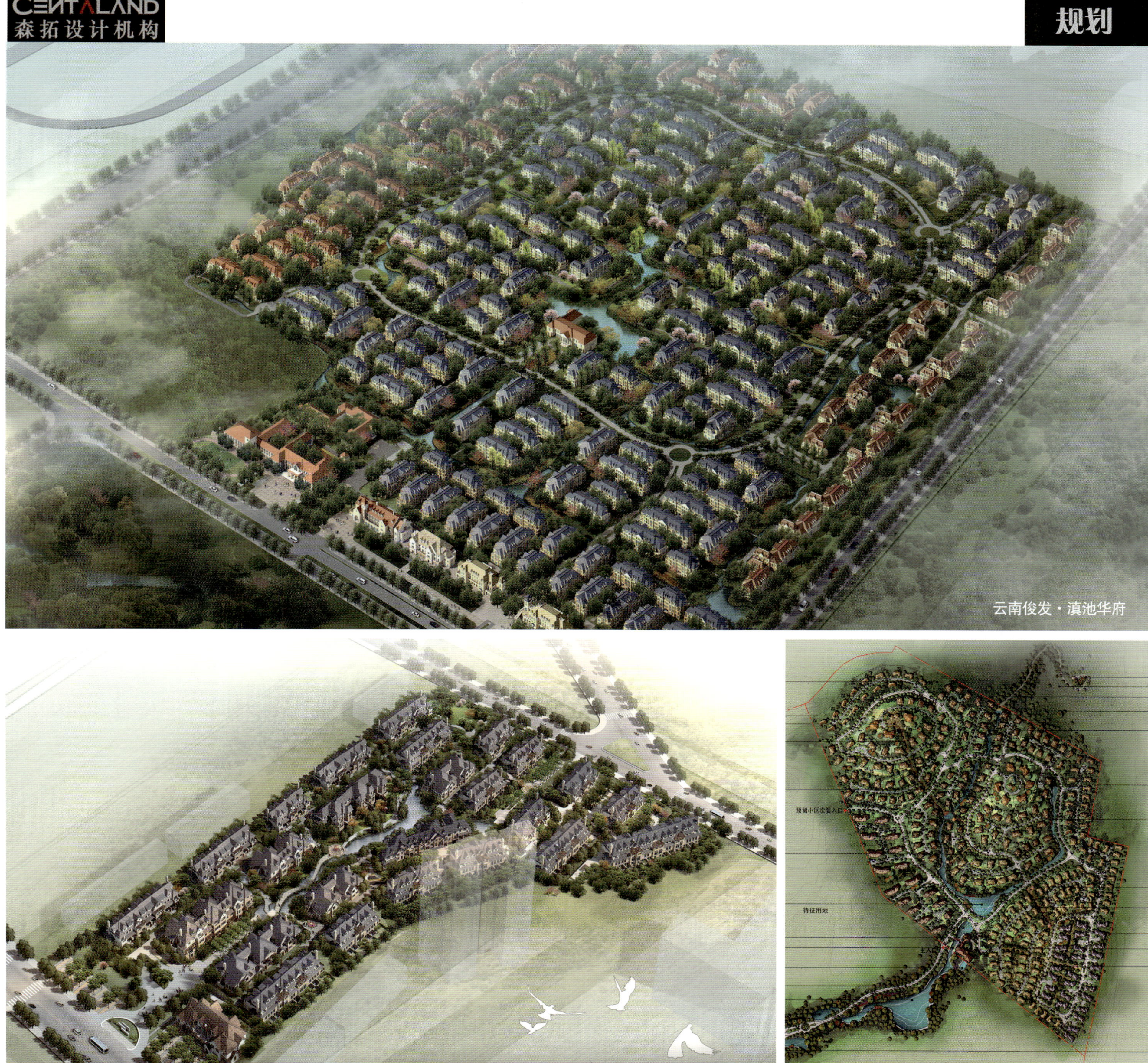

云南俊发・滇池华府

郑州建业・壹号城邦

南京华隆・紫金七号

CENTALAND
森拓设计机构

别墅

南京鸿信·云深处

南京鸿信·云深处

南京鸿信·云深处

成都龙湖·长桥郡三期

南京华隆·紫金七号

南京华隆·紫金七号

联排别墅

重庆金科·太阳海岸

重庆金科·太阳海岸

重庆金科·太阳海岸

浙江仙居别墅区

浙江仙居别墅区

CENTALAND
森拓设计机构

云南俊发・滇池华府

云南俊发・滇池华府

云南俊发・滇池华府

云南俊发・滇池华府

云南俊发・滇池华府

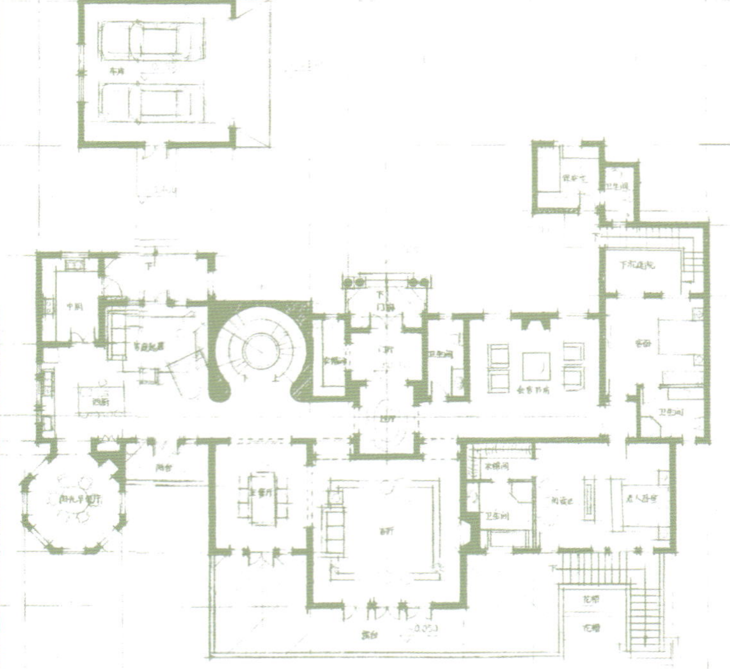

CENTALAND 森拓设计机构

花园洋房

重庆中渝・梧桐郡

重庆中渝・梧桐郡

成都龙湖・弗莱明戈

成都龙湖・弗莱明戈

CENTALAND
森拓设计机构

高层

成都朗基·望今缘

石家庄顺驰·蓝郡

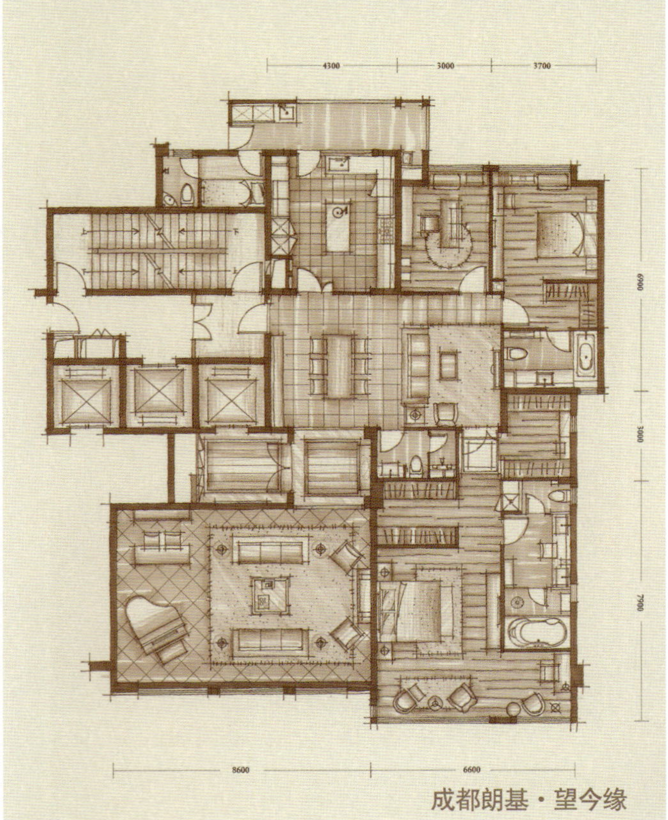
成都朗基·望今缘

成都远洋·朗郡

CENTALAND
森拓设计机构

昆明俊发・滨江俊园

昆明俊发・滨江俊园

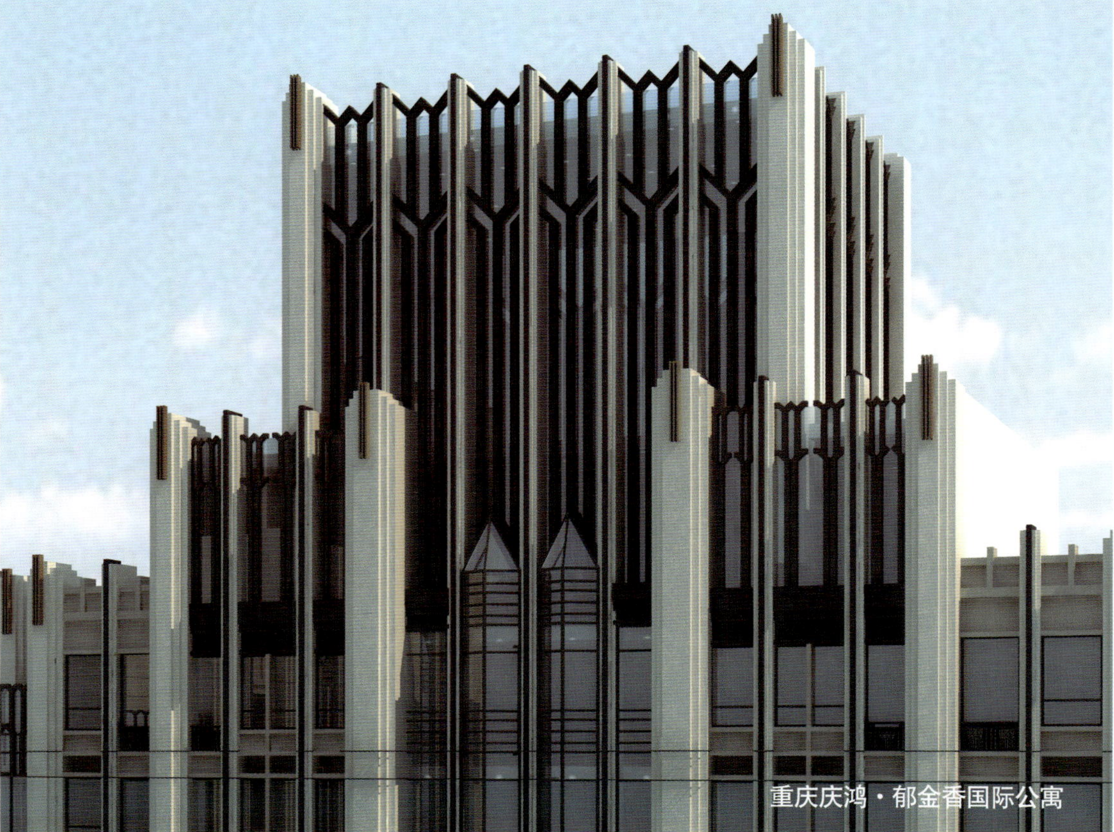
重庆庆鸿・郁金香国际公寓

重庆庆鸿・郁金香国际公寓

CENTALAND
森拓设计机构

公建·商业

重庆中渝·梧桐郡

南京华隆·紫金七号会所背面

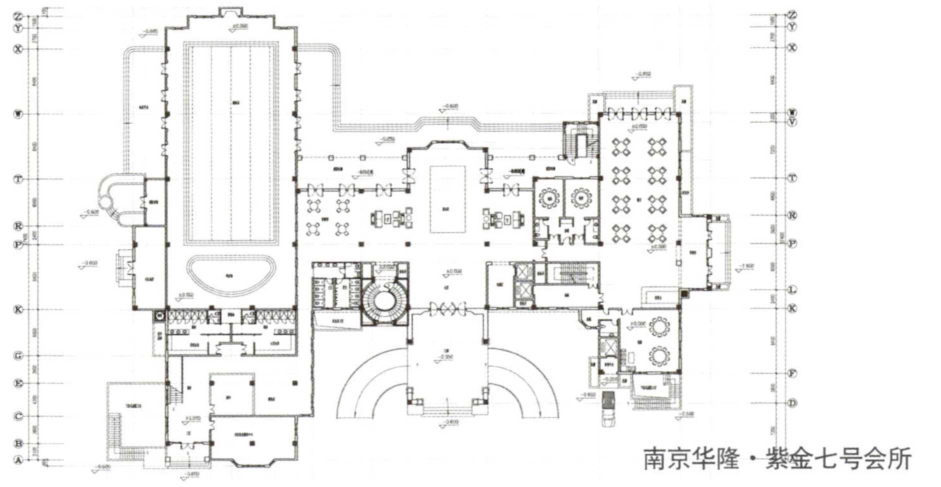
南京华隆·紫金七号会所

南京华隆·紫金七号会所正面

CENTALAND 森拓设计机构

公建·商业

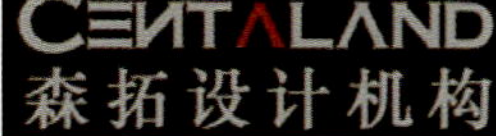

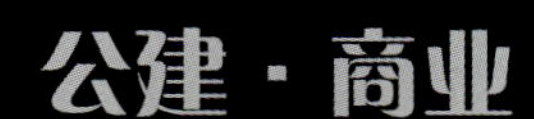

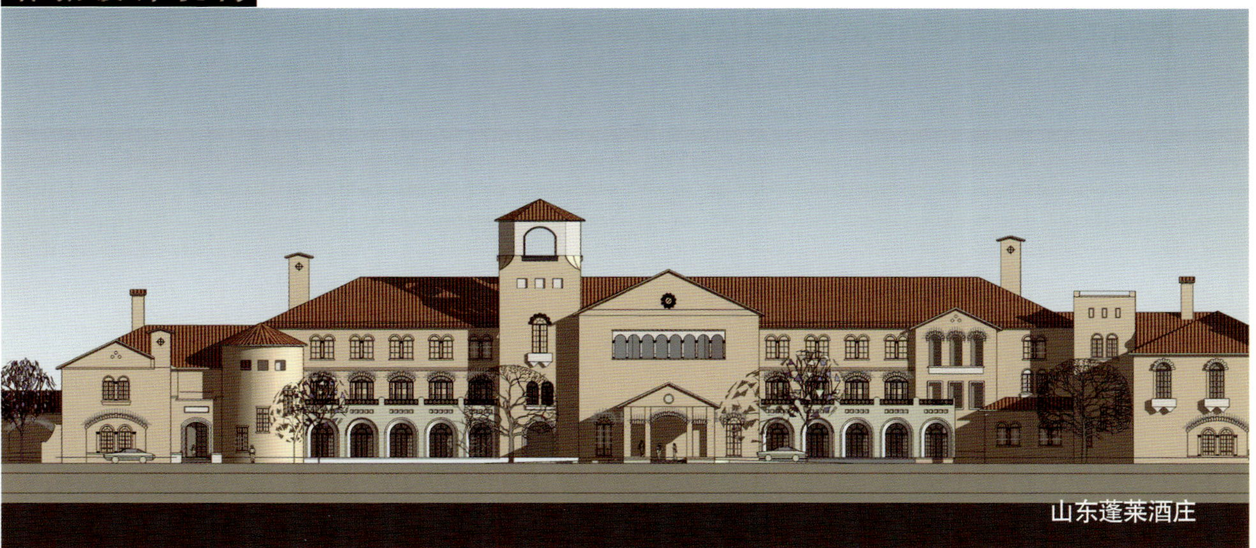

山东蓬莱酒庄

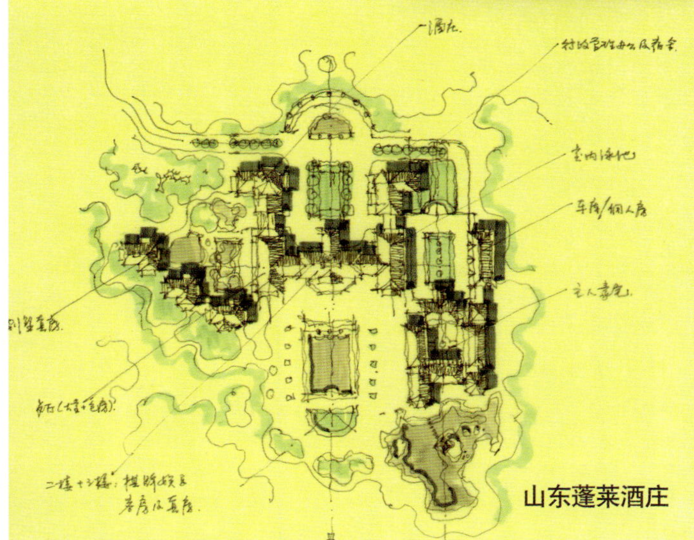

山东蓬莱酒庄

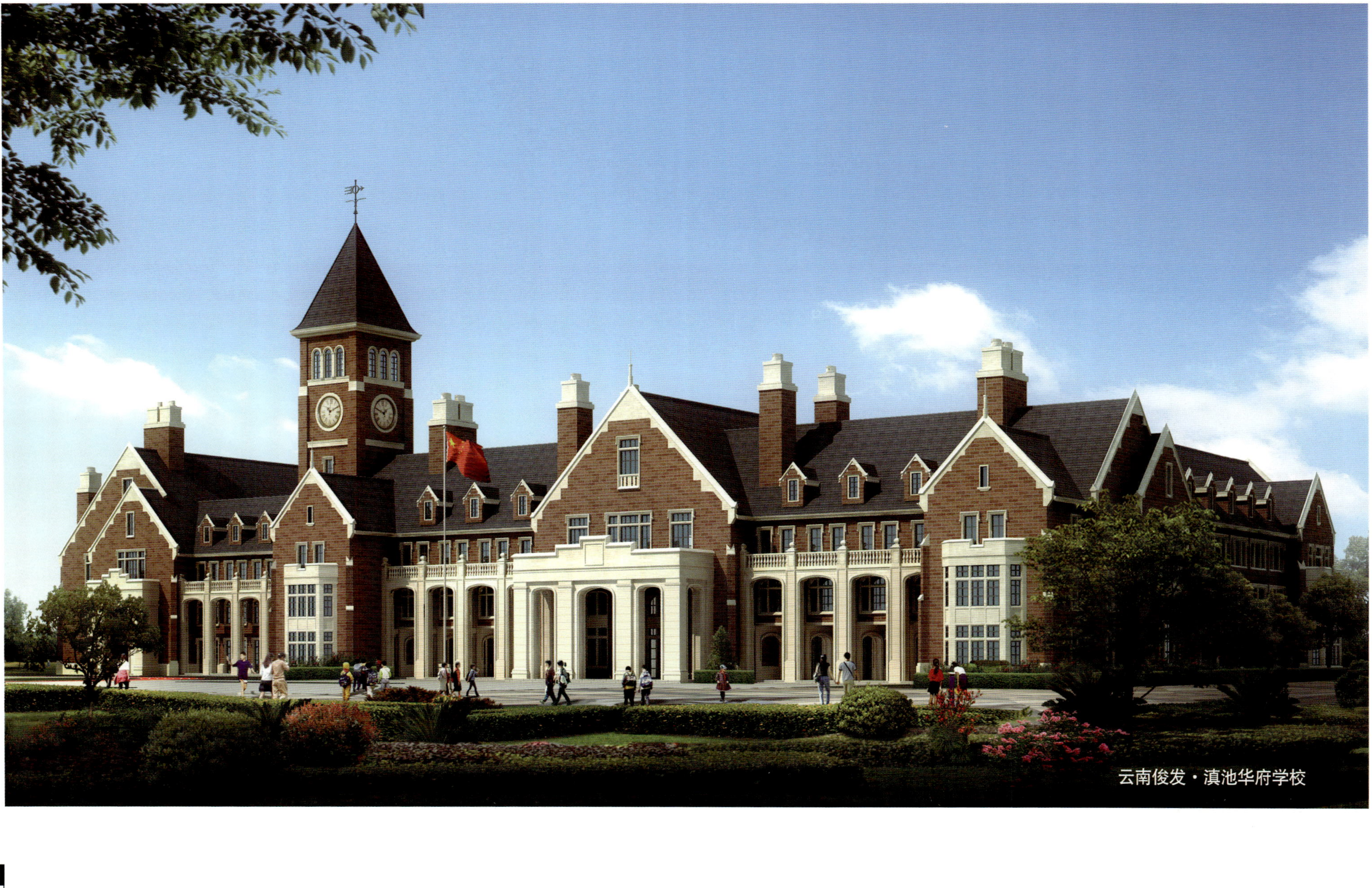

云南俊发·滇池华府学校

CENTALAND
森拓设计机构

公建·商业

云南俊发·滇池华府商业街

重庆金科·太阳海岸时代广场

浙江天人境界建筑设计事务所有限公司

ZHEJIANG TIANREN JINGJIE ARCHITECTS & ASSOCIATES

地　址：浙江省杭州市西湖大道150号
金泰商务大厦12楼
邮政编码：310009
联系电话：+86-571-8792 3037
电话总机：+86-571-8782 0945
传　真：+86-571-8781 1641
电子邮箱：tianren9@vip.sina.con

■ 杭州西湖高尔夫别墅

■ 杭州苏黎世小镇别墅

■ 浙江丽水・白云花苑坡地建筑

■ 杭州地中海别墅

■ 安徽黄山・趣园

■ 杭州青山湖郡原・美树

景观别墅

浙江天人境界建筑设计事务所有限公司

ZHEJIANG TIANREN JINGJIE ARCHITECTS & ASSOCIATES

■ 杭州·水岸青云住宅小区

■ 江苏常熟恒泰国际花园

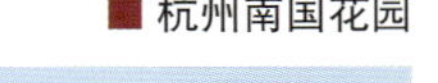

■ 杭州南国花园

■ 温州金色尚品住宅小区

■ 浙江安吉新嘉园住宅小区

■ 杭州滨江观澜公寓

浙江天人境界建筑设计事务所有限公司
ZHEJIANG TIANREN JINGJIE ARCHITECTS & ASSOCIATES

地　　址：浙江省杭州市西湖大道150号
金泰商务大厦12楼
邮政编码：310009
联系电话：+86–571–8792 3037
电话总机：+86–571–8782 0945
传　　真：+86–571–8781 1641
电子邮箱：tianren9@vip.sina.con

■ 温州中学民族部

■ 浙江商业职业技术学院二期实训楼

■ 杭州职业技术学院二期公共教学楼

■ 温州实验中学

■ 宁波鄞州高级中学

文化教育

浙江天人境界建筑设计事务所有限公司

ZHEJIANG TIANREN JINGJIE ARCHITECTS & ASSOCIATES

■ 温州华盟大厦

■ 上海东方国际文化贸易中心

■ 杭州通达大厦

■ 杭州钱江新城城投大厦

■ 上海歌林春天写字楼

■ 温州尚品国际大厦

浙江天人境界建筑设计事务所有限公司

ZHEJIANG TIANREN JINGJIE ARCHITECTS & ASSOCIATES

地　　址：浙江省杭州市西湖大道150号
金泰商务大厦12楼
邮政编码：310009
联系电话：+86-571-8792 3037
电话总机：+86-571-8782 0945
传　　真：+86-571-8781 1641
电子邮箱：tianren9@vip.sina.con

■ 浙江宋城集团云和湖度假酒店

■ 浙江安吉丰华白金五星级酒店

■ 上海空中管制中心青浦基地宾馆

■ 江西婺源国际大酒店

■ 浙江安吉半岛酒店

■ 江苏常熟虞山锦江五星级酒店

■ 杭州千岛湖715所基地宾馆

浙江天人境界建筑设计事务所有限公司

ZHEJIANG TIANREN JINGJIE ARCHITECTS & ASSOCIATES

■ 浙江安吉丰华花园酒店综合体规划

■ 安徽黄山金龙山湖名舍规划

■ 浙江舟山松山岛旅游规划

■ 江苏句容湖总体规划

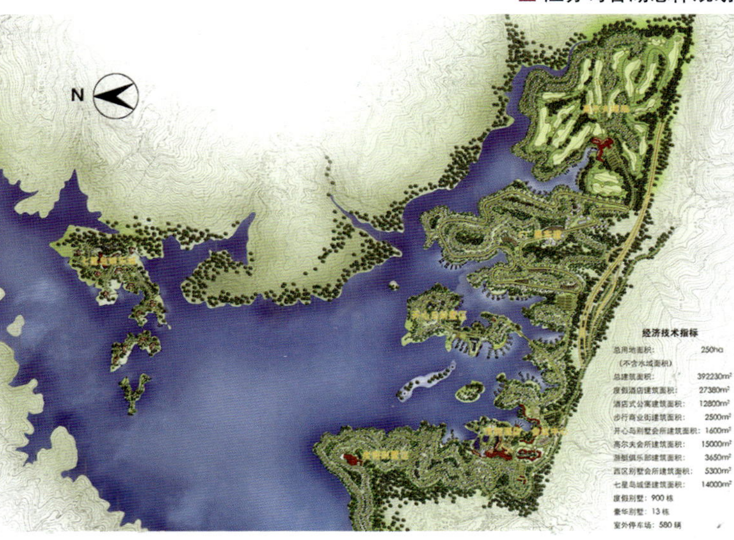

■ 浙江宋城集团云和湖度假区规划

■ 江苏常熟湖畔现代城规划